ɔn
ng
University ake

David T. Hein

The Motorola MC68000 Microprocessor Family
Assembly Language, Interface Design, and System Design

Second Edition

Prentice Hall
Englewood Cliffs, New Jersey   Columbus, Ohio

Library of Congress Cataloging-in-Publication Data

Harman, Thomas L., 1942–
    The Motorola MC68000 microprocessor family : assembly language,
interface design, and system design / Thomas L. Harman, David T.
Hein.—2nd ed.
        p.   cm.
    Includes bibliographical references and index.
    ISBN 0–13–158742–0
    1. Motorola 68000 series microprocessors.   I. Hein, David T.
II. Title.
QA76.8.M6895H37   1996
004.165—dc20   95-9116
                                                                       CIP

Cover photo: © Phil Matt
Editor: Charles E. Stewart, Jr.
Production Editor: Linda Hillis Bayma
Production Coordination: Betsy Keefer
Cover Designer: Julia Zonneveld Van Hook
Production Buyer: Deidra M. Schwartz

This book was set in Times Roman by Bi-Comp, Inc. and was printed and bound by Quebecor Printing/
Book Press. The cover was printed by Phoenix Color Corp.

Printed in the United States of America

10 9 8 7 6 5 4 3 2 1

ISBN 0-13-158742-0

Prentice-Hall International (UK) Limited,London
Prentice-Hall of Australia Pty. Limited, Sydney
Prentice-Hall Canada Inc., Toronto
Prentice-Hall Hispanoamericana, S.A., Mexico
Prentice-Hall of India Private Limited, New Delhi
Prentice-Hall of Japan, Inc., Tokyo
Pearson Education Asia Pte. Ltd., Singapore
Editora Prentice-Hall do Brasil, Ltda., Rio de Janeiro

Preface

The introduction of the Motorola MC68000 family of microprocessors ushered in a new generation of processors. This family first included the MC68000 and the MC68010. Both were 16-bit microprocessors considered powerful enough to function as the central processing units (CPU) of sophisticated computer systems. Since the introduction of the basic processors in 1980, the family has grown to include not only advanced microprocessors but also microcontrollers and other integrated processors that use basically the same CPU. All of these devices constitute what is now called the *68000 family* of devices. The second edition of this textbook was written to introduce the reader to the characteristics of this Motorola family of processors. For these processors, the book covers assembly language, programming, interface design, and system design.

One important purpose of this book is to introduce the student or the practicing computer professional to all of the significant aspects of design using the MC68000 and the other processors in the 68000 family. In addition, the book can serve as a reference in which topics are organized according to function and importance for the design of programs, interfaces, or systems.

This book is organized into three parts as indicated in the chapter descriptions below. Chapters 1 through 4 present the 68000 family to the reader and also cover other introductory material. Chapters 5 through 9 cover assembly language programming techniques for the processors. Chapters 10 through 13 discuss system design and development for 68000-based computer systems. Appendices I through IV summarize pertinent material about the MC68000, MC68008, and MC68010. Appendix V presents a comparison of the 68000 family members. Finally, selected answers to problems within the chapters are included at the end of the book.

A detailed summary of the major topics covered in this book plus a list of the acronyms, a list of programs by topic, and a list of additional materials that may be used in conjunction with this book are given on the following pages.

CHAPTER DESCRIPTIONS

Chapters 1 through 4 introduce the reader to the 68000 family of microcontrollers and microprocessors, fundamentals of machine arithmetic, and the MC68000 central processing unit.

- Chapter 1 presents a number of applications of the Motorola MC68000 microprocessor. The hardware and software support for the MC68000 family of integrated circuit devices is also described. Motorola's single-board computer, the Integrated Development System, is introduced. Finally, the chapter presents a brief introduction to product design and development.
- Chapter 2 discusses the organization of typical microcomputer and microcontroller-based systems. The function of the major system components (CPU, memory, input/output) is described and the importance of the CPU word length and addressing range is presented. The chapter concludes with three views of the system as seen by the system designer, the assembly language programmer, and the interface designer.
- Chapter 3 explains the internal representation of numbers and characters as used in MC68000 systems. Binary, decimal, and floating-point notations are treated along with details of arithmetic operations. The ASCII code for alphanumeric characters is also presented.
- Chapter 4 is devoted to a discussion of the characteristics of the MC68000 processor. The MC68000 is first introduced as a Complex Instruction Set Computer (CISC) chip before its characteristics as a programmable processor are presented. The organization of memory in a typical MC68000-based system is also covered.

Chapters 5 through 9 are devoted to programming techniques using the MC68000. The assembly language for the processor is used to explain the many capabilities of the MC68000.

- Chapter 5 introduces the MC68000 assembly language. Software development, assembly language features, and the various addressing modes of the MC68000 are covered.
- Chapter 6 presents three important categories of instructions for the MC68000: data transfer, program control, and subroutines.
- Chapter 7 contains explanations and program examples concerning the arithmetic capability of the MC68000. Binary arithmetic, decimal arithmetic, and conversions between ASCII, binary, and BCD are covered.
- Chapter 8 introduces logical operations, shift and rotate instructions, and bit-manipulation instructions.
- Chapter 9 completes the study of basic programming techniques. Methods of creating position-independent code are covered. Program examples are given for manipulation of data structures, including arrays, lists, strings, and queues. More advanced subroutine techniques are presented.

Chapters 10 through 13 are devoted to aspects of the MC68000 that determine the operation of the computer system. Chapters 10 and 11 cover material of interest to the system designer and the programmer who create supervisor programs. Chapter 12 covers I/O programming. Chapter 13 treats the details of hardware design with 68000 family processors.

- Chapter 10 considers the processor's various states and modes of operation. The assembly language instructions to control the processor and examples of initialization procedures are presented.
- Chapter 11 covers exception processing. The exceptions include interrupts, traps, and various error conditions recognized by the CPU during program execution.
- Chapter 12 presents the interfacing requirements and I/O programming for MC68000 systems.
- Chapter 13 first presents instruction execution times for the MC68000. The timing and hardware characteristics of the CPU bus are also covered.

Appendices I through V contain summary material. Appendix I presents the ASCII character codes and a table of powers of 2 and 16. Appendix II, Appendix III, and Appendix IV contain a summary of the MC68000 family characteristics, the assembly language instruction set, and the machine language instruction set for the MC68000, MC68008, and MC68010. Appendix V presents a comparison of M68000 family members.

SUMMARY OF MAJOR TOPICS

General Programming

Data Types	Chapter 3
Machine Language	Section 4.5
Memory Addressing	Section 4.6
Assembly Language	Chapter 5
Summary	Appendices

MC68000 Processor

Addressing Modes	Sections 4.4, 5.3, and 9.2
Instruction Set	
Introduction	Section 4.3
Major Instructions	Chapters 6, 7, and 8
Register Set	Sections 4.2 and 10.2
Summary	Appendices

Programming Techniques

Arithmetic and Logical	
Data Types	Chapter 3
Condition Codes	Sections 6.3 and 7.1
Arithmetic Operations	Chapter 7
Logical and Bit	Chapter 8

List of Acronyms

ACIA Asynchronous Communications Interface Adapter
ALU Arithmetic and Logic Unit
ASCII American Standard Code for Information Interchange
ASIC Application Specific Integrated Circuit
BCD Binary-Coded Decimal
CAD Computer-Aided Design
CAM Computer-Aided Manufacturing
CISC Complex Instruction Set Computer
CPU Central Processing Unit
CRT Cathode Ray Tube
DIP Dual-in-Line Package
DMA Direct Memory Access
I/O Input and Output
MCU Microcontroller Unit
MMU Memory Management Unit
MOS Metal Oxide Semiconductor
PIA Peripheral Interface Adapter
PLA Programmed Logic Array
PTM Programmable Timer Module
RAM Random Access Memory (Read/Write)
RISC Reduced (Reusable) Instruction Set Computer
ROM Read Only Memory
VLSI Very Large Scale Integration

LIST OF PROGRAMS BY TOPIC

General

Arithmetic

Lists and Tables

System Operation: I/O, Interrupts, and Traps

ADDITIONAL MATERIAL

The processor manuals provide a more complete treatment of certain processor characteristics than what we have covered in this book. The *User's Manual* for the MC68000 and other processors can be obtained from one of the many Motorola Sales offices located in major cities around the world. Because development systems vary considerably, the reader should refer to the reference manuals for the specific system being used to develop programs. These manuals include the *Assembly Language Reference Manual* for the assembler being used, as well as the manuals that describe the use of the system, i.e., how to edit, assemble, and execute programs. Similarly, the hardware manuals for the system and the manufacturer's data sheets for specific components should be used because these documents cover details peculiar to these items.

Motorola provides excellent support for students and instructors through their University Support program. Also, a "Freeware" telephone line for modem access provides callers with free information and programs for Motorola products. There are also a number of newsgroups on the Internet devoted to 68000 family devices. For more information, contact a Motorola representative.

An instructor's manual is available from Prentice Hall. We include answers to problems and programs as solutions to questions in the textbook as well as suggested laboratory exercises and other teaching hints.

ACKNOWLEDGMENTS

First, we would like to thank the designers and other people at Motorola for their support in many ways over the years. It has been a pleasure working with the courteous and professional staff at Motorola.

A number of people have made special contributions to this book. The material was developed over several semesters with many helpful suggestions for improvement from the students in the Microcomputer Design Class at the University of Houston, Clear Lake.

We would like to thank the following reviewers for their assistance and helpful suggestions in the preparation of this edition: R. G. Deshmukh, Florida Institute of Technology; Shahram Latifi, University of Nevada–Las Vegas; and Pratapa Reddy, Rochester Institute of Technology. Of course, the staff at Prentice Hall did a fine job of putting the book into production. Our sincere appreciation goes to all the people who were involved in one way or another in the project.

Finally, we thank our families and friends for their patience and support during the long process of producing a textbook. We apologize to any other people who helped in the endeavor but were not cited here. Please send any comments or criticisms to the authors in care of Prentice-Hall, Inc., Englewood Cliffs, New Jersey, 07632, or contact us at the University.

T. L. Harman (harman@cl.uh.edu)
D. T. Hein

Brief Contents

Contents

Introduction to the Motorola Microprocessors and Controllers

A decade of advances in integrated circuit technology was crowned at Motorola by the announcement in 1980 of the commercial availability of the MC68000 microprocessor. This microprocessor was identified as a 16-bit microprocessor intended to function as the central processing unit (CPU) of a computer. Since that time, the MC68000 has become the foundation for the *68000 family*. This family of microprocessors and microcontrollers has grown to contain dozens of members, from newer versions of the 16-bit MC68000, to the powerful 32-bit MC68060.

As microprocessors evolved, each new generation represented a higher speed of operation as well as increased capability. The emphasis for the latest generation is high speed of execution for instructions and support of advanced operating systems and applications software. To form a complete system, the microprocessor CPU requires a number of other integrated circuit chips for memory and to perform Input/Output (I/O) operations. Well-known examples of computers based on the 68000 family include the Apple Macintosh personal computers.

The original family has also added a number of special-purpose processors called *Embedded Controllers* (EC) and another family of computer chips called microcontrollers or integrated processors. Processors such as the MC68EC000, designed as embedded controllers, are usually simpler versions of the parent processor used in low-cost products. Members of the 68300 family of *microcontrollers* contain a core processor based on the 68000 family and a number of modules such as timer units and on-chip memory. These microcontrollers are also used in a variety of products such as laser printers and automobile engine controllers. Thus, the purpose of the embedded controllers and microcontrollers is to control the operation of a system or product for a particular application. The user normally does not view the product as a "computer" system.

From the point of view of many product designers and programmers, compatibility is one of the most important features of the 68000 family members. For example, binary versions of application programs written for the MC68000 will also execute on the more advanced members of the family. Once the instructions of the basic MC68000 are well understood, it is generally an easy task to program other processors in the family. The advanced processors will have additional instructions and features but the programming model is based on the MC68000.

The development of programs (software) for these microcomputers has also proceeded at a rapid pace. A number of operating systems and related programs are available for 68000 family computers, including software supplied by Motorola as well as that created by independent suppliers. The combination of software and the extensive line of hardware components offered by Motorola and other manufacturers form a family of products that support the development of 68000-based systems.

This introductory chapter first introduces the Motorola family of microprocessors and microcontrollers and supporting chips. This chapter also discusses the software and hardware support available for the 68000 family. Finally, the chapter presents an introduction to product design and development using microprocessors.

1.1 THE FAMILY NOTATION AND APPLICATIONS

When discussing a processor family, careful notation is necessary to distinguish between specific processors and the entire family of microprocessors, embedded controllers, microcontrollers, and peripheral chips. It is generally Motorola's convention to refer to the complete range of products as the 68000 family or occasionally in the literature as the M68000 family. This includes the individual members such as the MC68000, MC68EC000, M68020, and the MC68EC020. Thus, the prefix MC before the numerical designation generally indicates one particular member of the family. However, several versions of each processor are available, perhaps with different physical packages or speed of operation. The variations are also said to form a family of processors. For example, the MC68000 and its variations such as the MC68EC000 could be referred to as the MC68000 family. In any case, the context should indicate whether an individual processor or a group of processors is being discussed.

Table 1.1 lists some of the applications for members of the 68000 family processors. The MC68000 or MC68EC000 is typically used in relatively low-cost products that need the computational capability of a 16-bit microprocessor. The advanced processors are often used as the CPU in personal computers and engineering workstations. The 68000 family microcontrollers are found in a wide range of products.

Another approach to product design employs a 68000 family CPU and custom logic on a single chip. For general-purpose functions, the standard CPU offers the advantage of easy programmability and a great deal of development support. For special applications, designers can create circuits called *Applications Specific Integrated*

Table 1.1 Typical Applications of the 68000 Family

MC68000 (16-bit CPU)	Instruments, programmable controllers
MC68020, . . . , MC68060 (32-bit CPUs)	Personal computers, workstations and other sophisticated products
MC68302 (MC68000 CPU)	Communications processor
MC68332 (Microcontroller)	Automobile engine control, robotics, medical instrumentation
FlexCore (Combined CPU and logic)	Laser printers, memory controllers

Circuits (ASIC). The circuit implementation for an ASIC is often termed custom logic because the circuit design is created for a specific purpose such as memory control. Once the ASIC is produced, its function cannot be changed.

By combining a standard CPU and the custom logic on a single chip, the best features of CPUs and ASICs can be utilized. The result is a custom processor for applications such as laser printers. This design approach can result in a lower-cost product with fewer chips. Motorola terms the program to create the combined chips as FlexCore. The result is a single chip with custom logic, memory, peripheral modules, and a 68000 family CPU. References at the end of the chapter discuss these applications and others in greater detail.

1.2 THE FAMILY CONCEPT

A specific processor is chosen as the CPU in a product for a variety of economic and technical reasons. The specific capabilities of the processor were without doubt of primary importance in the choice, particularly when the finished product exhibits enhanced performance when compared to similar predecessors employing a less powerful processor.

Another vital factor in the choice for most manufacturers of products is the *support* given to the processor line. This support comes from the manufacturer of the processor and a number of other sources. The support consists of hardware, software, development systems, and other items. Such support is provided to enable a product manufacturer to design, build, test, and produce the product in the most economical manner. Also, as technological advances allow improvement of performance and lower cost of the product, the manufacturer must modify the product in various ways to remain competitive.

The family concept, as applied to microcomputer systems, assures that the processor line is adequately supported and is improved with time. One who is familiar with the basic processor, for example the MC68000, has little trouble learning the characteristics of newer processors and various items that constitute other members of the family. This justifies the effort to study the programming and hardware features of the MC68000, even if a different member of the family is to be used in an application.

Table 1.2 summarizes many of the support criteria discussed in this section. The processor line of integrated circuits contains processors and controllers, circuits to facilitate design of interfaces, and special devices to improve the performance of a computer system. The software support for the 68000 family consists of operating systems, development programs, and other programs collectively termed system software.

Programs are also available for specific applications such as computer-aided design. These programs are usually provided by companies that specialize in software support for a particular processor.

For those programmers developing application software, a number of development systems are available to facilitate software production. In addition to allowing the creation and testing of software, a few development systems aid the integration

Table 1.2 Support for Various Families of Processors and Controllers

Type	Support
Processors and support circuits	
8-, 16-, and 32-bit microprocessors and microcontrollers	Basic processor and enhanced versions available in various packages and speeds of operation
Embedded controllers	Lower cost processors for product design
Microcontrollers	Single chip with CPU, memory and other peripheral functions for product design
Peripheral interface circuits	Circuits for interfacing CPU to a wide range of peripheral devices
Special devices	Devices for floating-point mathematics, network control, and other applications
Software	
Operating systems	Various operating systems for program development, real-time applications, time sharing, or special purposes
Development software	Editors, assemblers, compilers, debugging programs
Applications software	Special-purpose programs for accounting, engineering analysis, etc.
Documentation	Manuals, application notes, data sheets
Product development	
Development systems	Complete systems for software development and hardware/software integration
Single-board computer modules	Processing units, memory modules, and other complete hardware subsystems

of the application software with the prototype hardware of a system developed by the user. This capability is vital if the final product will be a complete system, as was the case with the examples presented in Section 1.1.

1.2.1 The Motorola Families

A processing unit of a microcomputer consists of the CPU and a number of auxiliary integrated circuits (chips) for interfacing, but does not include the memory chips. The chips are connected together on a circuit board. Overall capability and the speed of operation of the system is determined by the characteristics of the CPU when I/O transfers of data are not required during execution of a program. For convenience, the processors are classified as 8-, 16-, and 32-bit processors. The category for a processor is generally based on the addressing range and data size of the chip. However, there are some exceptions to the classification scheme, as discussed later.

When I/O transfers are required, the flexibility of the system in communicating with various peripheral devices depends on the type of interfacing chips provided with the system. To allow the greatest range of applications, Motorola and other suppliers produce a number of different CPUs and interfacing chips. As the family of chips evolves, enhanced processors and additional interfacing chips will become available.

The integrated processors or microcontrollers are single chips containing both a CPU and various modules for memory, I/O, and other functions. The type of CPU specifies the family.

The 8-bit Family. Motorola's family of 8-bit processors was introduced in 1974 with the MC6800 microprocessor. This CPU has eight signal lines for data transfer, which classified it as an 8-bit device. Four major families of microcontrollers have evolved from the MC6800, which are the M6801, M6804, M6805, and M68HC11 families. In these families, the basic MC6800 CPU is augmented with various modules on a single chip to satisfy different requirements in product design. The reader is reminded that the prefix "MC" designates a particular chip within a family of processors. Thus, as shown in Figure 1.1, the M68HC11 family contains the MC68HC11A8, the MC68HC11E2, and the MC68HC11E9. Other members of this family are produced by Motorola but are not shown in the figure.

Different versions of the basic microcontroller are offered with Random Access Memory (RAM) or Read-Only Memory (ROM) on the chip. Several versions have Analog-to-Digital Converter (ADC) circuitry. Most microcontroller chips have timer and counter circuitry in a single integrated-circuit package. The timers may generate interrupts to the CPU, measure external events, or generate output waveforms via the signal lines of the microcontroller.

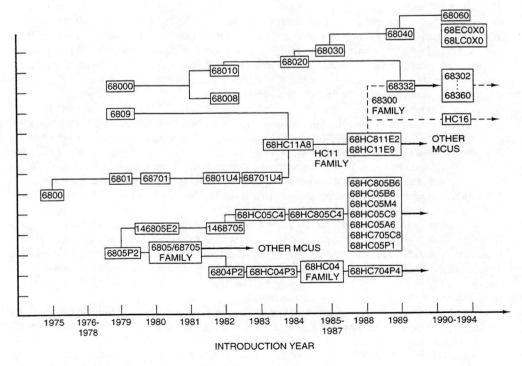

Figure 1.1 Motorola processors. (Courtesy of Motorola, Inc.)

The 68HC16 Microcontroller Family. The various microcontrollers in the 68HC16 family contain a CPU designated the CPU16 that extends the capability of the 8-bit 6800. These controllers have a 16-bit data bus and a number of special modules such as timers and communications interfaces. For example, the MC68HC16Z1 contains the CPU16, on-chip memory, ADCs, a general-purpose timer, and a system integration module for interfacing to external units such as memory modules.

Thus, the various 8- and 16-bit microcontrollers in Figure 1.1 have different designations according to their CPU type, features and even the semiconductor fabrication technique used to manufacture them. References listed in the Further Reading section at the end of this chapter describe each chip in detail. The Motorola *Semiconductor Master Selection Guide* is the best source of information about the complete line of chips available from Motorola.

The 16-bit and 32-bit Families. The 8-bit MC6800 CPU forms the basis for many processors and microcontrollers. This processor and its derivatives have been used in countless products, including appliances, cameras, and automobiles. For product designers who need a more powerful CPU, the MC68000 and its enhanced versions are available. The more powerful processors derived from the MC68000 will be discussed shortly.

Example 1.1 _____

The MC68332 microcontroller, as the first member of the M68300 family of microcontrollers, evolved as shown in Figure 1.1. Its features include capabilities of the 32-bit MC68020 CPU and the 8-bit MC68HC11 microcontroller. The M68HC11 family had previously evolved from the M6801 and M6809, 8-bit processor families. The evolution shown in Figure 1.1 was possible because of the rapid advance in integrated-circuit technology since the MC6800 microprocessor was introduced. The increasing capability of the 8-bit microcontrollers can be explained by describing the evolution of the M6801 and M6805 families.

The MC6801 microcontroller and its derivatives include RAM, timers, parallel I/O, and serial communications capability on the chip. The MC6801 also has 2,048 bytes of factory-programmable ROM. The MC68701 has the same features but replaces the ROM with Erasable Programmable ROM (EPROM) so that the product designer can modify programs during the product development cycle. When the program is correct, the program can be sent to Motorola to be incorporated in the ROM of a less-expensive MC6801. This chip would be included in the product when the product is produced in volume. The MC6801U4 has 4,096 bytes of ROM but is otherwise practically equivalent to the MC6801.

The parts in the M68HC11 family include most of the features of the M6801 family with the addition of analog inputs to the on-chip ADC as well as enhanced timing and communications functions. The fabrication technology is termed the High-speed Complimentary Metal Oxide Silicon process (HCMOS). This process yields lower power consumption compared to that possible in the earlier M6801 microcontrollers. The HCMOS process is indicated by the "HC" in the middle of the part numbers for the devices.

The M6805 family offers a variety of on-chip memory selections and I/O functions. These processors and those in the M6804 family are intended for products produced in large volume for which low-cost components are mandatory. The chips do not have many of the powerful capabilities of the MC68HC11 or MC68332 microcontrollers. In fact, the vertical axis in Figure 1.1 can be regarded as an arbitrary scale indicating the relative performance of the parts specified in the figure. Additional information concerning Motorola's processors and microcontrollers can be found in several of the references in the Further Reading section of this Chapter.

The 16-bit Family of Microprocessors.   Processor manufacturers respond to increasing needs of product manufacturers and to technological advances by enhancing the design of a microprocessor such as the MC68000 and producing new versions. Later versions, such as the MC68010, are distinguished from the basic MC68000 by their numerical designation. These processors are normally compatible with the MC68000 in many ways but offer different features.

Also, various MC68000 processors are available with different physical packaging, different operating temperature ranges, and with various speeds of operation. Otherwise, their instruction set and their electrical characteristics are identical. Other processors in the family may show more significant differences, as is the case with several advanced versions of the MC68000. The evolution of the family by year of introduction of each processor is shown in Figure 1.1. Table 1.3 describes the characteristics of the example versions.

Variations of the MC68000.   The range of operational speed available in the MC68000 family is evident in the Motorola processors designated as MC68000L8, MC68000L10,

Table 1.3   68000 Family of Processors and Microcontrollers

Processor	Characteristics or Use
16-bit family	
MC68000L8, MC68000L10, MC68000L12	Different speeds of operation indicated by the suffix
MC68008, MC68010, MC68EC000, MC68HC000	Versions of the processor for special purposes
32-bit family	
MC68020	32-bit data transfer and addressing
MC68030	On-chip memory management unit
MC68040	On-chip floating point unit
MC68060	Latest generation of the 68000 family CPU
MC68EC020, . . . , MC68EC040	Embedded controller version of the 32-bit CPUs
MC68882	Floating-point coprocessor
68300 family of microcontrollers and integrated processors	
MC68302 (MC68000 CPU)	Integrated communications processor
MC68322	MC68EC000 CPU and processor for graphics
MC68331, MC68332, MC68333, . . . (CPU 32)	Microcontrollers with 32-bit CPU
FlexCore	
(Choice of CPU)	Special-purpose processors with CPU and custom logic

and MC68000L12. The last numeric designation indicates the number of fundamental hardware operations per second in millions of clock cycles. The speed of the processor clock is typically measured in megahertz (MHz) or millions of cycles per second. The L12 represents 12.5 million cycles per second, or 12.5 MHz. Each assembly language instruction requires at least four clock cycles.

When comparing processors, the L12 device will execute the same program 1.25 (12.5/10) times as fast as the L10 and 1.56 (12.5/8) times as fast the L8. In a given product, the replacement of the CPU by a faster (or slower) processor will change the performance proportionally if the speed of operation is determined by the processor alone.

The MC68EC000 is a low-cost version of the MC68000. Versions are available with clock speeds of 10, 12.5 and 16.67 MHz. The MC68HC000 is an MC68000 requiring 1/10 of the power to operate. Such processors frequently are found as the CPU in portable instruments.

Motorola has also introduced the MC68008, which is an 8-bit data bus version of the MC68000. This processor retains most of the characteristics of the original MC68000 but is designed for applications employing 8-bit data transfers. This reduction in the number of data signal lines reduces the cost of a system or product and simplifies the interface to certain peripheral units. An important advantage over earlier 8-bit processors in such an application is that the MC68008 executes programs written for the MC68000.

The 32-bit Family: Beyond the MC68000. The Motorola MC68020, MC68030, MC68040 and MC68060 processors represent enhanced versions of the MC68000. The MC68020 has a 32-bit data bus which classifies it as a 32-bit microprocessor, whereas the MC68000 has a 16-bit data bus. In terms of programming, the enhanced processors are very similar to the MC68000. The floating-point coprocessor in Table 1.3 is used with the MC68020 or MC68030 and other family numbers to increase the capability of a system used for scientific or engineering applications.

From the programmer's viewpoint, the MC68030 and its MC68882 coprocessor are compatible in many ways with the MC68020 and its coprocessors. The MC68030 has a memory management unit on the chip to speed up address translations required in virtual-memory systems. The MC68882 coprocessor is designed for parallel operation. For example, concurrent operation of the CPU and the MC68882 mathematics coprocessor is possible with the MC68030.

The MC68040 includes the features of the MC68030 and adds a floating-point arithmetic unit on the chip. Thus, CPU and memory management and floating-point capability are combined on a single chip. The MC68060 is a high-performance version of the MC68040. This CPU is referred to as a *superscalar* processor because it can decode and execute several instructions at the same time. It also combines some features of both Complex Instruction Set Computer (CISC) and Reduced or Reusable Instruction Set Computer (RISC) architectures discussed in Chapter 4.

There are also lower-cost versions of several of the advanced processors designated MC68EC020, MC68EC030, and MC68EC040. These processors may have fewer features than the primary processor. For example, the MC68EC040 has only an integer arithmetic unit without floating-point capability.

FlexCore processors are available with various 68000 family CPUs. These chips are intended for high-volume products such as laser printers. The chips are created by a collaboration between Motorola designers and the product designer.

1.2.2 The M68300 Family and the MC68332 Microcontroller

The MC68332 was the first member of the M68300 family of microcontrollers from Motorola. A much simplified block diagram of the MC68332 microcontroller is shown in Figure 1.2. Integrated on a single chip are a 32-bit CPU and various other modules for parallel and serial I/O timing, motor control, and system expansion. The intermodule bus connects these components internally.

The CPU 32 is a modified MC68020 processor with an instruction set and other features compatible with Motorola's popular 32-bit CPU. This modular approach to microcontroller design allows Motorola to produce various versions of the microcontroller with new or modified modules without changing the basic characteristics of the device.

Another advantage deriving from the modular construction of a microcontroller lies in the capability of the individual modules that can be integrated onto the chip. The MC68332 modules are designed to perform specific functions with a minimum of CPU intervention. Thus, once a module is initialized by the CPU, the CPU program is only interrupted when a module completes a given task, rather than at the occurrence of minor events during the activity. For example, the Time Processor Unit (TPU) of the MC68332 has a motor controller function that is capable of accelerating, driving and halting a stepper motor without CPU intervention once the module is programmed to perform these operations.

When the modules on the chip cannot meet the requirements of an application, additional memory or interfacing chips can be added to expand the capabilities of the

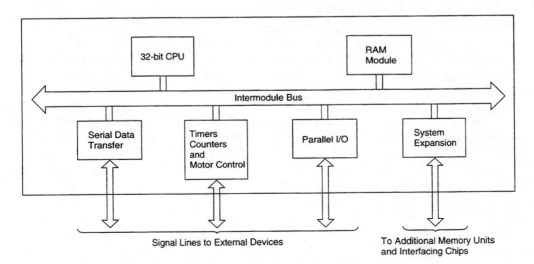

Figure 1.2   Simplified diagram showing the functions of the MC68332.

system. The chip also has a low-power standby mode of operation useful when a battery-backup power supply is needed in a product.

M68300 Microcontroller Family Members. Table 1.4 lists several members of the M68300 microcontroller family. Each family member contains the CPU32 and a System Integration Module (SIM). The MC68331 contains four modules including a General-Purpose Timer (GPT) and a Queued Serial Module (QSM) for serial I/O. The primary member of the family is the MC68332. This microcontroller has on-chip RAM and a powerful TPU in addition to the CPU32, QSM, and the SIM. The MC68F333 has all of the modules of the MC68332 and in addition an Electrically Erasable Programmable Read-Only Memory (EEPROM) as well as an eight-channel ADC. An array of these processors will be used to control the braking system of the Boeing 777 airplane.

Other members of the 68300 family using the CPU32 include the MC68334, MC68340, and the MC68360. The MC68340 is targeted for products that require high-speed data movement such as compact disk players. Thus, a two-channel Direct Memory Access (DMA) module is included in the MC68340.

Motorola defines their microcontrollers with the CPU16 or the CPU32 as *modular microcontrollers.* Each member of the HC16 or M68300 family contains the CPU, a system integration module, and a number of special-purpose modules. The modules include programmable timers, serial communication interfaces, ADCs, and various memory modules. The purpose is to allow the product designer to choose the appropriate microcontroller for an application.

In summary, the 68000 family of processors provides a choice for a system designer in that the designer can select the CPU or microcontroller that best fits the requirements of an application. Refer to Motorola literature listed in the references at the end of this chapter for more information about these microcontrollers.

1.2.3 M68300 Family Integrated Processors

Several members of the M68300 family employ the MC68000 CPU with other modules to create processors primarily for communications applications. These chips, such as the MC68302, MC68306, and MC68356, are called *integrated processors* to distinguish them from the microcontrollers just discussed. The distinction is based on the intended use because the microcontroller's strength is implementing timing functions and motor control.

The MC68302 was designed not for control applications but for incorporation in data concentrators and modems and similar communications products. The MC68356

Table 1.4   Several M68300 Family Microcontrollers

Microcontroller	Features
MC68331	CPU32, GPT, QSM, SIM
MC68332	CPU32, QSM, RAM, SIM, TPU
MC68F333	CPU32, QSM, RAM, SIM, TPU, EEPROM, ADC
MC68340	CPU32, DMA, serial I/O, timers, SIM

Table 1.5 Interfacing Devices for the 68000 Family

Application	Examples of Interfacing Chips
Parallel and serial input/output	General-purpose interface devices (MC68901, MC68230, MC68681)
DMA Controller	Direct memory access (MC68450)
Communications and Local Area Networks	Controller and conversion devices (MC68605, MC68824, MC68184)
System control	Interrupt control and bus arbitration (MC68153, MC68452)

Note: Each interfacing device listed is manufactured as a single-chip integrated circuit. They are called interfacing "chips" using the jargon of the industry.

is intended for communications and Digital Signal Processing (DSP) for voice, audio, and video signals. Thus, the trend in many products is to use integrated processors that employ a standard CPU, such as the MC68000, for computations combined with special-purpose modules on a single chip.

1.2.4 Interface Chips for the MC68000 Family

Motorola and other manufacturers also produce a variety of interfacing chips that simplify the interfacing task of computer system development. The purpose of these chips is to minimize the need for specialized hardware design.

A sampling of the numerous types of interfacing chips for the 68000 family is listed in Table 1.5. For example, the MC68901 is referred to by Motorola as a multi-function peripheral and serves as the interface between the MC68000 CPU and a wide range of peripheral units. This chip is programmable to provide a flexible I/O capability as well as timing functions under the control of the CPU. The MC68230 has parallel interfaces and timers. The MC68681 provides two serial I/O interfaces.

The other interfacing devices listed in Table 1.5 are also programmable, with each chip responding to coded instructions according to its design and purpose.

1.3 SOFTWARE SUPPORT FOR THE MC68000 FAMILY

After the hardware of a computer system is constructed, programs must be developed to control the overall operation of the system. The various programs that are executed by the microcomputer are referred to generically as *software* to distinguish them from the physical equipment (hardware) of the system. For our purposes, it is convenient to discuss three categories or "levels" of software, as shown in Table 1.6.

Table 1.6 Software Levels

Level	Examples
Applications software	Programs tailored to solve a specific problem
Development software	Editor, assembler, or compiler to create applications programs
Operating system	Program to control CPU, input/output, and disk storage (files)
Hardware level	CPU, interfacing chips, memory, and peripheral devices

At the level closest to the hardware, the operating system manages the hardware resources of the system. The operating system is frequently called the executive or supervisor program. At the next level lies the development software used by the programmer to create and debug applications programs. The applications software is written to tailor the computer system to solve a specific task. In the 68000 family, an extensive list of software products is available from Motorola and independent software suppliers.

A product design and development cycle includes both hardware and software development. The software production cycle for the development of programs that require hardware/software integration is shown in simplified form in Figure 1.3. Program coding, debugging, and testing are typically performed by a program designer using a software development system. If the programs execute correctly on the development system, they are then transferred to the memory of the microprocessor-based product for further testing with the actual hardware of the product. This constitutes the hardware/software integration portion of the development cycle. The software discussed in this section includes programs to aid software development as well as operating systems and applications programs that become part of the final product. A complete product development cycle is described in Section 1.5.

Figure 1.3 Software development and hardware/software integration.

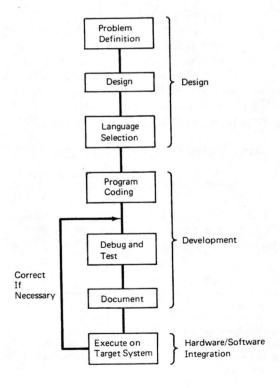

1.3.1 Development Software

Under the control of the operating system, various programs to aid software development can be executed on a general-purpose computer system. Software for development, as listed in Table 1.7, includes a text editor, language translators, and various utility programs. The *text editor* is used to create and edit the program under development before it is translated into machine language statements by the appropriate language translators. An *assembler* and many high-level language *compilers* can be purchased for the 68000 family CPUs to allow a selection of computer languages.

Several special-purpose programs, often called utility programs, supplement the development software for the 68000 family. A *debugger* program is a useful tool for isolating programming errors (bugs) by allowing execution of a small portion of a program and testing the effects. The *linkage editor* serves to combine program modules that have been assembled (or compiled) separately so that a complete program ready for execution is created. The executable binary instructions for the CPU are called the *machine language code*. Later chapters present further details about many of these programs that aid software development.

During program development, the operating system is used to control program execution, as well as for I/O operations such as printing the program text or the results after execution. The file management capability allows programs under development to be stored and loaded into memory for execution using the disk storage unit of the computer. These features of the operating system are required by the development software and may also be used by the applications programs if needed. In the latter case, the applications programs are intended to be executed with a particular operating system.

Cross-Software for Development. In the previous discussion, it was assumed that the applications programs were developed and targeted for execution on the same 68000-based system. An alternative approach allows applications programs to be created on

Table 1.7 Development Software for the 68000 Family

Software	Use
Operating systems	Control program execution and file management
Programs to create software	
Text editors	Create and edit source programs
Assemblers and debuggers	Translate and debug assembly language programs
Compilers	Compile programs in Ada, C, Fortran, etc.
Linkage editors	Link separate program modules and create machine language code
Cross-software	
Cross-assemblers, cross-compilers	Create 68000 family machine language programs on another computer
Simulators	Simulate execution of 68000 family programs

another computer and translated into executable (machine language) code for the 68000 family processor. This technique, called cross-development, is possible for assembly language programs using a cross-assembler and for high-level languages such as C with a cross-compiler, as listed in Table 1.7.

When using cross-development software, the programmer creates the program in the chosen language for a product with a 68000 family CPU, even though the compiler or assembler executes on another computer called the *host* computer. Thus, a cross-assembler when executed on the host computer processes assembly-language statements and converts those into either executable code or an equivalent machine-readable form that can be executed on a 68000-based system.

The advantage of the cross-development approach is that a programmer without a 68000-based development system can use the program development facilities of a host computer. These facilities include a text editor for preparing the program, a disk storage unit to store completed programs, and a printer unit to print the program for correction or documentation.

Simulators. To execute the cross-assembled or cross-compiled program, the machine language instructions can be processed by either a simulator program or a 68000-based computer system. With a simulator, the operation of the 68000 family processor is simulated on the cross-development computer. This approach is useful when the system hardware is not available but the programs must be tested. In effect, the simulator provides a software model of the 68000 processor. However, hardware-dependent aspects such as timing must be tested on the target 68000-based computer. This is accomplished by transferring the program from the development system to the memory of the target system.

Operating Systems and Applications Programs for the Product. Product designers may employ operating systems in two different contexts. The first, which was previously discussed, employs an operating system during the software development phase of a project. The other use of an operating system is to direct the overall operation of a product and the execution of applications programs in particular. In this context, the operating system becomes part of the product. Its purpose is to respond to external events such as interrupts, perform I/O, and schedule the sequence of execution for programs that perform a specific function. These applications programs are typically organized into independent modules called *tasks*. An operating system used to control the operation of a product in this manner is usually referred to as a *real-time* operating system or a real-time kernel.

Real-time Operating Systems. Many companies provide real-time operating systems specifically designed to control a 68000 family CPU. In most cases, a product designer will purchase an operating system that is available commercially rather than develop a new one because of the difficulty involved in creating a new operating system. The applications programs or tasks are usually developed by a programmer or programming team as part of the product development effort because the operations performed by the tasks are so specific to a particular product.

1.3.2 Development Systems

The 68000 family elements described in the preceding subsections can be combined with appropriate peripheral devices to create a computer system that is suitable for a software development system. A typical system includes an operator's terminal, disk units, and a printer. Some development systems can be used by multiple users.

Once applications programs are designed, they can be created, debugged, and tested on the development system. If the completed programs are to be executed on the development system itself, the testing assures that the entire system meets the requirements of the application. For example, with the proper application software, the development computer might serve as a business system with the capability to do inventory control, job scheduling, and other required tasks.

Hardware/Software Integration. The software production cycle for the development of programs that require hardware/software integration was shown previously in simplified form in Figure 1.3. If the programs execute correctly on the development system as far as they can be tested on it, they are then transferred to the target computer for further testing with the actual hardware of the final product. Errors are corrected until the product performs as required.

When hardware design and development is required for an application, the development effort must include integration of the application software with the hardware of the target computer. The target computer is usually a different computer than the one chosen for the application development system. Integration consists of extensive testing of the software components that control the hardware created for the target system, as well as testing of any hardware developed for the product.

In typical applications, the product may contain special-purpose interfaces that are not available as part of the Motorola family of interfacing chips. In this case, custom software and hardware must be developed and tested. When using the family devices, an operating system for the target computer can be chosen that includes programs to control the interfaces so that a minimum amount of development and testing is needed. If no operating system is to be included with the target system, development of interfacing programs for the family chips must be created or purchased and integrated into the target system. This is frequently the case when the target is a product such as a laboratory instrument that performs computer processing.

Figure 1.4 shows a development station that includes the connection between a single-board computer (target computer) and the development system. For hardware/software integration, program execution on the target system is controlled with the development system via appropriate commands entered by the user. These commands allow debugging of the entire target system, much as the debug program discussed previously allows debugging of programs executed on the development system itself. For example, single instructions or a small portion of the program can be executed on the target system. Then, information for debugging purposes can be displayed for the user.

More details of product development and testing will be presented in Chapter 13. The reader interested in the complete range of development support for the 68000 family should contact a Motorola representative to determine the latest products that

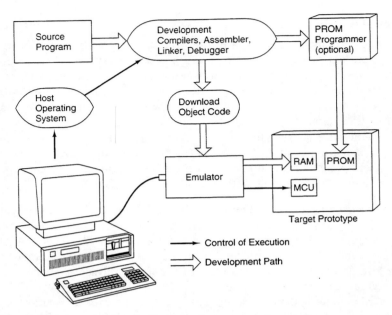

Figure 1.4   Development system for hardware/software integration.

are available. Also, several references in the Further Reading section of this chapter list companies that produce support products for the 68000 family.

1.4   BOARD-LEVEL COMPUTER MODULES

The chips discussed in Section 1.2, together with other necessary circuits, form the basic elements necessary for the hardware implementation of a computer system. The CPU, interfacing chips, and memory chips are electrically connected on one or more circuit boards to create the hardware of the processing system and memory storage.

A manufacturer of a product or a computer system with adequate engineering and manufacturing facilities can design, test, and manufacture the complete boards if necessary, purchasing only the required chips from a supplier such as Motorola. But to simplify the initial design and creation of a prototype of a product, Motorola and other manufacturers produce hardware modules that contain a 68000 family CPU, memory, and interfacing chips connected together on a single circuit board. These modules are often called *single-board computers* because they contain all of the chips and circuitry needed to form a complete system when peripheral devices and a power supply are connected to the board. If the single-board computers are used in product development for hardware/software integration, they are described as *evaluation modules*. The Motorola Integrated Development Platform is used as an example in this section.

Another approach used in some cases by a product manufacturer is to purchase board-level modules from Motorola or other sources. The manufacturer then only has to assemble the product using the modules that suit the application. The modules are connected together electrically using a standard bus to transfer data, status, and commands from module to module. The VME bus is an example of a standard bus system.

The Motorola IDP Single-Board Computer.   The Motorola Integrated Development Platform (IDP) is shown in Figure 1.5. With the addition of a power supply and peripheral units, this module becomes a low-cost evaluation system designed for development and testing. The main module is constructed with a circuit board, called the motherboard, that contains monitor programs in ROM, memory (RAM) for program storage, and several connectors for external devices.

A second module containing the CPU is plugged into the IDP motherboard. Such modules are available for a number of microprocessors in the 68000 family. The user communicates with the module via a CRT terminal attached by cable to the IDP. In most cases, the operator's station will be a host computer such as the popular PC. This addition allows the IDP to be used as a development system as previously described.

The IDP may be used to evaluate the characteristics of the 68000 family CPU during program development and execution. These programs may be created using a monitor program provided in ROM chips on the module. Because no assembler or compiler is provided, this approach requires that programs be written (coded) in machine language. Alternatively, programs can be created on another system which generates machine-readable code for the CPU (e.g., with a cross-assembler or cross-compiler). The executable code can then be loaded into the memory of the IDP for execution.

Both of these approaches to software development are described elsewhere in the book. For present purposes, it should be noted that the IDP represents the necessary hardware to provide the CPU, memory, and interfacing capability of a computer system all on a single-printed circuit board.

Standard Bus Systems.   Figure 1.6 shows a computer system assembled from VME bus modules. Such a system will typically contain a single-board computer as the processing unit, together with modules containing memory and perhaps other modules to control peripheral devices. Details of how the chips on each board are connected and how the modules themselves are connected together are presented in Chapter 2, where bus structures for computer systems are discussed. The VME bus itself is discussed in more detail in the author's book covering the MC68020 and MC68030 microprocessors, which is listed in the Further Reading section of this chapter.

1.5   INTRODUCTION TO PRODUCT DESIGN AND DEVELOPMENT

Although this textbook is primarily concerned with software and hardware design techniques for products based on the 68000 family CPUs, a brief discussion of a complete product creation cycle is presented in this section. An understanding of the entire cycle influences a designer when making choices about the implementation

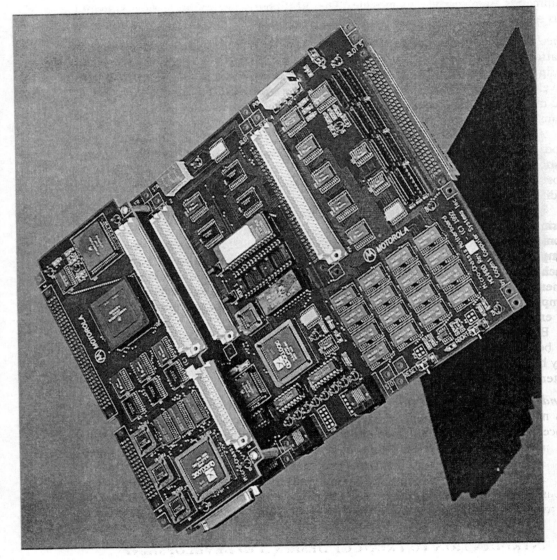

Figure 1.5 MC68000 IDP module. (Courtesy of Motorola, Inc.)

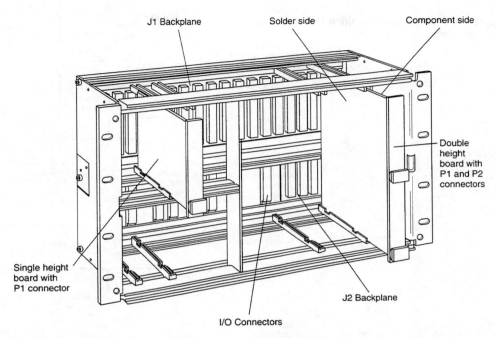

Figure 1.6 Modular construction of a computer system showing VME bus example. (Courtesy of Motorola, Inc.)

concerning both software and hardware. Figure 1.7 shows the documents and the stages in a typical product cycle.

At each stage, a designer must keep in mind the intended application, which is important to the usér, as well as the technical and economic factors involved in the production of the product. The first step is to define the *general specifications* for the product. These are typically presented in a document or report that describes the operation and capabilities of the product but does not usually define how the required functions are implemented. The emphasis is on the requirements of the application or market that the product is to serve rather than any considerations of hardware or software.

The specifications of an automobile engine controller, for example, would include not only the functions to be performed, but also any size, weight, or power-consumption constraints imposed by the operating environment. In many products, such constraints may determine the final hardware implementation as much as the operational requirements that the product must meet. For example, with a microcontroller-based product, the required speed of response as control actions are performed is always defined as part of the general specifications.

Depending on the complexity of the product, one or more additional design documents may be written to further specify the *detailed requirements* and the *functional specifications* of the product. It is at this stage that the product designer defines the various functions to be performed in terms of a hardware or a software solution.

Figure 1.7   Product design, development and testing cycle.

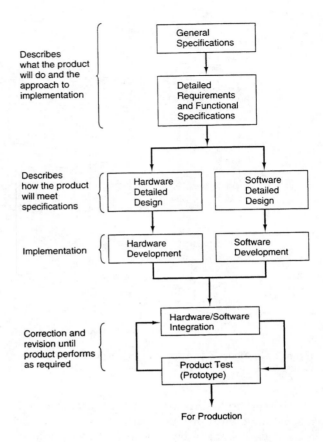

This is a critical stage, when the designer must have a thorough knowledge of the characteristics of the various hardware components that are available as well as an understanding of the software solutions that are possible.

For example, the detailed requirements might specify how often a value being monitored must be sampled. The functional specifications might require a hardware subsystem to acquire and store the values if the required time between samples is very short, perhaps a microsecond or less. The routine for data acquisition in this case might consist of only a few instructions to initialize the subsystem. On the other hand, if the time between samples is relatively long, a design that relies upon simple hardware but more sophisticated programs could be appropriate. The ability to make correct *hardware/software tradeoffs* is one skill possessed by a successful product designer.

Once the functional specifications for a product are written, the design process divides into two paths. A detailed hardware design is created that leads to a prototype unit to be tested as independent of the software as possible. The hardware design includes the selection of the microcontroller, memory, peripheral chips and the other circuitry needed to complete the prototype of the product. Detailed electrical circuit diagrams indicate how the components connect together to meet the requirements of the specifications. Engineering drawings showing the physical layout of the product are also included in the hardware design document.

The second path leads to the creation of the software modules for the product. These modules are typically combined and tested independent of the hardware under development. The purpose of the separate development and testing of hardware and software is to avoid errors the cause of which may be difficult to determine if untested hardware is controlled by undebugged software. The appropriate test procedures and expected results of independent tests should be specified before the development phase of hardware or software has begun.

The final phase of the development and testing cycle for a product consists of executing the specific applications programs for the product using the prototype hardware. In this integration phase, errors caused by the interaction of the software and hardware are discovered and corrected. As with the independent tests of software and hardware, the procedure and expected results for the integrated tests should be well defined before testing begins. An appropriate development system to aid hardware and software integration is essential to allow the correction of errors that arise during this phase.

Selection of the Processor.   The product designer chooses a CPU that can meet the programming and timing requirements of the product. As far as possible, the processor should include all of the capability that permits it to serve to control the product. Furthermore, whether a microprocessor or a microcontroller is chosen, the product will require additional support chips such as special-purpose peripheral chips to expand the resources of the system.

This textbook concentrates on software and electrical circuit design using the 68000 family of microprocessors. To produce an effective design using these processors, the reader should understand the devices in complete detail. For that purpose, every important characteristic of the MC68000 is presented in the text because it is the basic microprocessor in the 68000 family. After the processor is described, several chapters apply that knowledge of the CPU to the design and development of products using the MC68000 and the other family members.

REVIEW QUESTIONS AND PROBLEMS

1.1 List some of the target applications for each of the processor types
   (a) microprocessors
   (b) embedded controllers
   (c) microcontrollers

1.2 Microcontrollers consist of a microprocessor integrated with several modules providing such functions as analog-to-digital conversion, timing, etc. Give several advantages of this integrated approach versus a design using separate components (consider the target applications of microcontrollers). Are there any disadvantages to this approach?

1.3 An ASIC provides even greater integration than a microcontroller. What additional features and functions do these chips provide over a standard microcontroller? What might limit the application of ASIC devices (consider economic as well as design issues)?

1.4 What meaning does the prefix "MC" convey?

1.5 What is the significance of the processor family concept to a company that develops and manufactures products containing microprocessors and microcontrollers? Discuss both the business implications and the design implications.

1.6 What are the four major families of Motorola 8-bit microcontrollers?

1.7 What major function does the 68HC11 family add to those functions provided by the 6801 family?

1.8 The 68HC16 is a popular microcontroller family. From what microprocessor family was it derived—the MC6800 or MC68000?

1.9 List two members of the 68000 family which are
   (a) microprocessors
   (b) microcontrollers

1.10 What characteristics and features do the members of the 68000 family have in common?

1.11 List the significant changes from the previous generation for each processor
   (a) MC68020
   (b) MC68030
   (c) MC68040
   (d) MC68060

1.12 What is the difference between the Motorola MC68000L8, MC68000L10, and MC68000L12?

1.13 The Motorola M6804 microcontroller is used for which type of product
   (a) workstations
   (b) automobile engine control
   (c) high-volume products

1.14 List the advantages and disadvantages of product development using a cross-assembler and simulator.

1.15 Describe the critical skills possessed by a successful product designer.

1.16 Assuming that each instruction takes an average of 5 clock cycles, how many instructions can the MC68000L10 execute per second?

1.17 Considering the criteria discussed in this chapter, which of the items are of greatest importance to the designer when selecting a processor for the following products:
   (a) a low-cost consumer product
   (b) an automobile engine controller
   (c) a portable oscilloscope
   (d) an engineering workstation

1.18 Describe the components of a single-board computer. Which components need to be added to create a complete computer system?

1.19 Describe the hardware and software layers of a general-purpose computer system. List each component and define its purpose.

1.20 Why do computers such as the Apple Macintosh use peripheral expansion slots to add peripheral controller boards rather than having the function implemented on the main computer board in the product?

Design Problems and Further Exploration

(Some of these problems may require reference to the manufacturer's literature or other references.)

Processor specifications. By consulting the Motorola specification for two members of the 68000 family, compare the characteristics of the processors. Include specifications for clock speed, addressing capability, data bus size, power consumption, and other relevant information.

Microprocessor-controlled devices. Study some device, instrument, or appliance that uses a microprocessor to control its operation. Write a description of the function of the microprocessor in the product.

Design of an instrument. Modern measuring instruments such as oscilloscopes and spectrum analyzers incorporate microprocessors or microcontrollers to provide accuracy and flexibility in measurements. Features such as automatic calibration and range selection are possible with a microcontroller-based instrument. Front panel selections by the operator can be recognized and processed by the CPU through its interface to a front-panel keyboard or a touch-sensitive display. The CPU also controls the display of numerical values or other information for the operator of the instrument.

Although special-purpose circuitry is typically employed in an instrument to actually perform a measurement, the CPU serves to sequence the operations required. Another use of the microprocessor within an instrument may be to transfer data from the instrument's memory to another computer system. The values are transferred for permanent storage or detailed analysis by programs executed on the computer.

Select a product to design and write a simplified version of the detailed requirements for the product. Then, design the system in block diagram form, including the hardware chips or circuit boards and software modules. List the peripheral devices that might be needed.

FURTHER READING

Various products and support for the Motorola families of microprocessors and microcontrollers are announced monthly in many electronics and computer journals such as *Byte, Computer Design, IEEE Computer,* and *IEEE Micro.* The IEEE publications are available from the Institute of Electrical and Electronics Engineers, Inc. These publications cover the latest developments in products and components of interest to designers who use microcontrollers. Specific recommendations for further reading follow in this section.

The entire range of 68000 family of products from Motorola are described in a number of publications from that company. For a particular chip, a data sheet or User's manual is the best source of technical information. New product announcements appear in Motorola's *Semiconductor Data Update,* listed below. This and other periodic publications such as *The 68K Connection* are available by subscription from Motorola.

The Motorola *Semiconductor Master Selection Guide* describes Motorola's line of semiconductor products, including microprocessors and microcontrollers. The Motorola publication *The 68K Source* is a catalog of development, hardware, and software products for the 68000 family, including operating systems, development systems, and applications programs from independent suppliers as well as from Motorola. The *Microprocessor Development Tools Directory* contains similar information for the families of microcontrollers.

There are news categories on the Internet that deal with the 68000 family. Motorola also maintains a bulletin board service accessible by modem that contains product information and programs for the microprocessors and microcontrollers. Motorola representatives in your local area will be happy to explain how information from Motorola can be obtained.

Information from Motorola

Semiconductor Data Update (Phoenix: Motorola, Inc.)

The 68K Connection. (Phoenix: Motorola, Inc.)

Semiconductor Master Selection Guide. (Phoenix: Motorola, Inc.)

The 68K Source (BR729/D) (Phoenix: Motorola, Inc.)

Microcontroller Development Tools Directory (MCUDEVTLDIR/D).

Applications

There is an extremely large number of products that utilize 68000 family processors, including the winning Indy 500 car in 1993. The references present a short list of products mentioned in Chapter 1.

Many of Allen-Bradley's programmable controllers contain a 68000 family CPU. The *Alsys* article describes the use of MC68300 family microcontrollers in the braking system of the Boeing 777. *The Apple Catalog* describes the Macintosh line of computers that use 68000 family processors.

Programmable Controller product literature. (Milwaukee: Allen-Bradley.)

"Smart Safety." *Alsys World Dialog,* 7, no. 1 (Spring 1993): 6–7.

The Apple Catalog. (Cupertino: Apple Computer, Inc.)

68000 Family Processors

One of the author's other textbooks covers the 32-bit MC68020 and MC68030 microprocessors offered by Motorola. The book treats assembly language programming, interface design, and system design for the 68000 family microprocessors and related chips. The VME bus structure is also presented in detail.

Harman, Thomas L. *The Motorola MC68020 and MC68030 Microprocessors.* (Englewood Cliffs, N.J.: Prentice Hall, 1989).

Microcontrollers

The author's textbook about the MC68332 covers all of the aspects of that microcontroller in detail. Lipovski's textbook treats the Motorola MC68HC11 in detail. Many of the features of the MC68HC11 are employed in the MC68332. Peatman covers product design using two popular 8-bit microcontrollers. He describes the Motorola 68HC11 and the Intel 8096 in a very readable manner.

Harman, Thomas L. *The Motorola MC68332 Microcontroller.* (Englewood Cliffs, N.J.: Prentice Hall, 1991).

Lipovski, G.J. *Single- and Multiple-Chip Microcomputer Interfacing.* (Englewood Cliffs, N.J.: Prentice Hall, 1988).

Peatman, John B. *Design with Microcontrollers.* (New York; McGraw-Hill, 1988).

FlexCore

The FlexCore system is described in the article by Child.

Child, Jeff. "System-on-a-chip strategy threatens old-line ASICs vendors." *Computer Design* 33, no. 7 (June 1994): 46–50.

CHAPTER 2

Microcomputer and Microprocessor Characteristics

Microcomputer systems operate on sequences of binary digits (bits) that represent either coded machine instructions to be executed by the CPU or a data value to be transferred, transformed, or otherwise manipulated. The ease and speed with which the CPU and other elements of the system treat the binary sequences determine the basic characteristics of the computer system. Is it fast? Can it handle large numbers of data values as well as large programs? Is it cost-effective for a particular application? These questions and similar ones can be answered at least in part by examining the structure of the system and the capability of the CPU.

The microcomputer system can be characterized in terms of its CPU, memory elements, and I/O circuits, all of which are connected electrically by a system bus. When the microcomputer is considered functionally without regard to the precise physical structure, the system may be described in terms of its *organization*. This organizational view focuses on the major elements of the system, such as the CPU and memory, and their interconnections. The system designer or assembly language programmer is concerned with the system at this level.

In contrast to the organizational view, more detailed descriptions of the hardware are needed by the interface designer, who must know precise electrical and mechanical details about the system components. This chapter concentrates on the organization of typical microcomputer systems. Chapters 12 and 13 present the hardware description.

Two important characteristics of the CPU, which are useful in determining its capabilities, are its *data bus size* and its *addressing range*. The data bus size refers to the number of bits in a data item that may be transferred at one time (in parallel) between the CPU and other elements of the system. This measure of the CPU capability is also called the processor's *word length* in some contexts. The addressing range defines the number of memory locations addressable by the CPU. The range is determined by the number of signal lines on the address bus. The Motorola MC68000, for example, has a 16-bit data bus and an addressing range of over 16 million memory locations.

This chapter defines the most common microcomputer organizations and describes the major elements of a microcomputer system. The differences between typical 8-bit, 16-bit, and 32-bit processors are then illustrated by describing their data bus sizes and addressing ranges. These descriptions pertain to systems used to perform basic computer functions such as instruction execution and I/O operations. The information in these two first two sections is presented to acquaint the reader with a microprocessor in its role as the CPU. Emphasis is on the interaction between the various components of the computer.

The final section in this chapter presents three views of the microcomputer and introduces the characteristics of the MC68000 and similar microprocessors. These views are those of the *system designer,* the *assembly-language programmer,* and the *interface designer.* The MC68000 was developed to satisfy three needs: as a CPU in a computer system, as a programmable processor, and as a hardware chip.

2.1 MICROCOMPUTER ORGANIZATION AND BUS STRUCTURE

The elements of a simple microcomputer are shown in Figure 2.1. This block diagram shows the microprocessor (CPU), a memory unit, and I/O circuitry. The CPU communicates with its memory and interface circuits via parallel signal lines which, taken together, constitute the internal bus. Using parallel signal lines, each signal line may be used simultaneously with and independently of the others to transfer electrical signals representing information. Thus, 16 data signal lines would allow the simultaneous transfer of 16 bits. In contrast, a serial data line would require 16 transfers of one bit at a time to accomplish the same thing.

In Figure 2.1, the internal bus is shown separated functionally into *address* signal lines, *data* signal lines, and *control* signal lines. For simplicity when discussing only

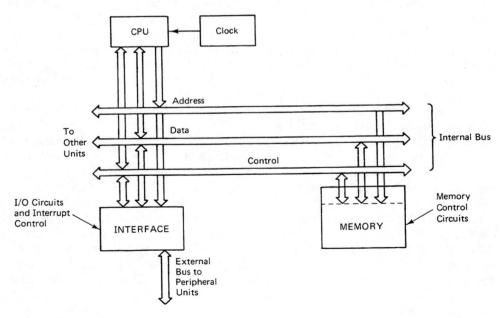

Figure 2.1 Simplified microcomputer organization.

one part of the internal bus, these sets of signal lines can be described as the address bus, data bus, and control bus, respectively. The internal bus serves to connect the CPU electrically to circuits for the purpose of transferring data into and out of the microcomputer, as well as to connect a memory unit holding instructions and data for the processor. When several different bus structures are present in a complete system, the internal bus is often called the CPU *bus* or the *local bus*.

The system may be constructed in a number of ways using components of the microprocessor family described in Chapter 1. At the most elementary level, individual integrated circuit chips for the CPU, memory elements, and I/O circuits can be combined by a hardware designer using knowledge of the electrical characteristics of each element. A higher-level approach, representing board-level design, is implemented by combining a single-board computer with various other circuit boards that contain memory subsystems and I/O subsystems. In such products, the external bus could be a standard such as the VME bus. This is referred to as the *system bus* to distinguish it from the internal CPU bus present on a single board with a CPU.

Figure 2.2 illustrates a possible structure for a board-level system such as a VME bus system. The single board (monoboard) modules and CPU modules contain an internal bus and other chips to interface to the system bus. The modules plug in to connectors in a chassis containing the system bus structure and a power supply. Additional modules, such as an arithmetic unit for scientific applications or I/O modules for special purposes, can be added to enlarge the system. The organization is defined by the system designer, who specifies the memory and I/O capability necessary to meet the needs of a given application.

The completed hardware system can serve as the basis for a variety of products, from the controller of an automatic welding machine to a general-purpose computer.

Figure 2.2   Board-level implementation of a microcomputer.

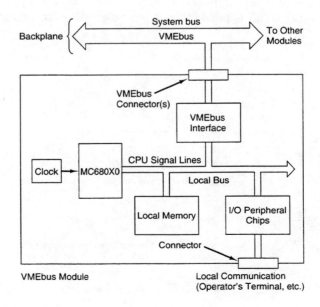

Software is developed to meet specific requirements. A portion of the software in machine language may be stored in a local memory, as indicated in Figure 2.2. The CPU can also access data or programs stored in memory located in another module on the bus.

In a board-level system, interfaces to special peripheral units may be required. Because the manufacturer of the other modules may not supply the unique interface needed, a custom interface may be developed by an interface designer. The integration of the hardware and the software routines for these special interfaces is a vital part of the system development.

Microcontrollers. Figure 2.3 shows the internal organization of a typical microcontroller chip. As in any computer system, the CPU controls and coordinates all activities of the microcontroller. It executes machine language instructions fetched from memory and performs all the arithmetic, logical operations, or other operations required by the instructions. The CPU communicates with the other modules via the intermodule bus. This bus serves the same purpose as the internal bus in a microprocessor-based system.

If the on-chip memory module cannot contain all of the instructions and associated data, the CPU can access external memory units using the external CPU bus. This bus can also be used to connect one or more peripheral units to the microcontroller. In many applications, interfacing circuitry is required to expand the system in this way just as it is required for the microprocessor-based system in Figure 2.1. Since microcontrollers are used primarily as embedded controllers in products, they are usually not employed in standard bus systems such as the VME bus.

An advantage of the microcontroller over the microprocessor in some applications, however, is derived from the inclusion of the various modules on the microcontroller chip. These modules can control and interact with external units without CPU intervention. This is normally not the case with microprocessor-based systems.

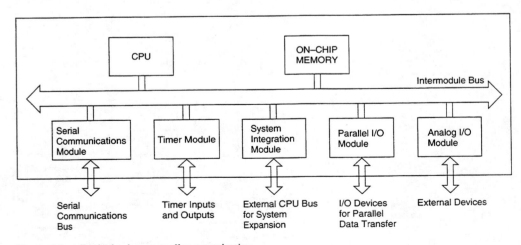

Figure 2.3 Typical microcontroller organization.

For the microcomputer of Figure 2.1, interrupts are directed to the CPU when a peripheral unit needs attention. With a microcontroller, a request by a peripheral unit to transfer data might be serviced by one of its modules without disturbing the instruction execution of the CPU. The ability to respond rapidly to external events without interrupting CPU program execution is the hallmark of modern microcontrollers such as the MC68332.

Assuming that the microcontroller and its modules are capable of meeting the functional and performance specifications of a product, the use of a microcontroller rather than a microprocessor in the product can lead to lower cost, lower power consumption, smaller overall product size and higher reliability. These benefits arise because a microcontroller-based product generally requires fewer additional chips, and hence fewer circuit connections than one based on a single-chip CPU. Design using a microprocessor is more appropriate when the microcontroller is not fast enough for an application or if many of the modules on the microcontroller are not needed in the product.

2.1.1 The Central Processing Unit and Clock

In a single CPU system, the CPU controls and coordinates all activities in the microcomputer. The primary purpose of the CPU is to execute machine language instructions fetched from the memory unit on the data signal lines as binary sequences. The CPU can *read* data values (operands) from memory or *write* values into memory by sending the appropriate electrical signals (commands) via the internal bus. The CPU can also initiate an I/O operation to transfer data to or from peripheral devices. The external bus shown in Figure 2.1 connects the system to one or more peripheral units and allows the CPU to perform such I/O operations.

The occurrence of events is precisely coordinated by the system clock. The clock "ticks" to indicate the passage of a time interval that is much shorter than one millionth of a second. In each interval, the CPU or one of the other components of the system performs a precise function, such as presenting a data value on the data signal lines. The rate at which the clock runs (number of ticks per second) determines the fundamental speed of operation of the system. Doubling the clock rate should double the operational speed of the system unless other factors, such as memory response time, limit the speed. The maximum clock rate for the system is determined by the CPU. Versions of the MC68000 are available with different maximum clock rates, as discussed in Chapter 1. A more precise definition of the system clock for MC68000 systems is given in Chapter 13.

2.1.2 The Memory Unit

In the single-memory system shown in Figure 2.1, the memory unit holds both instructions and data to be used by the CPU. Each memory cell in the memory unit contains 1 bit and the memory cells are organized as shown in Figure 2.4. Each memory location containing m bits (binary digits) is referenced by a positive number, its address, that indicates the position of that location in memory. The CPU can reference an individual memory location by placing the address on the address bus signal lines and can control

Figure 2.4   Memory organization for an *m*-bit memory.

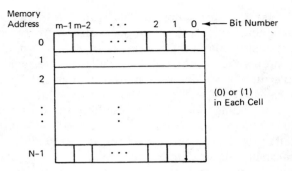

the operation of the memory with selected control signal lines. For addressing purposes, the memory of a modern microcomputer is divided into 8-bit locations that each hold a *byte* (binary term) of information. For programming purposes, the memory locations are numbered sequentially from 0 to the maximum number of bytes in memory.

Because the MC68000 has 16 signal lines for data transfer, the size of memory addressed can be specified in terms of the number of 16-bit words it contains as well as the number of bytes. The MC68000 memory word, for example, is 16 bits or 2 bytes, as shown in Figure 2.5. However, the MC68000 is said to be a *byte-addressable* CPU because it can address either one of the 2 bytes (bits 0-7 or bits 8-15) of a memory word. The CPU can also address *word* operands, consisting of 2 bytes, or 32-bit operands called *longwords*.

One feature of 68000-family CPUs is their ability to operate on byte, word, or longword operands according to program instructions. The programmer specifies the length of the operand as byte, word, or longword in any CPU instruction that references an operand. One advantage of specifying the operand length in the instructions is that the actual physical arrangement of memory is normally not of concern to a programmer.

Figure 2.5   Organization of memory in an MC68000 system.

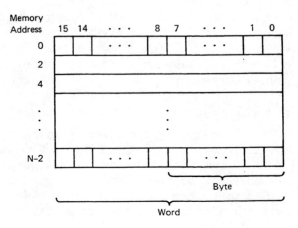

In the MC68000, a word or longword address must be an even address. Chapter 4 explains the exact details of memory addressing for the MC68000.

The contents of any memory location can be obtained from memory on the data signal lines (CPU read) or the value on the data signal lines can be stored into the addressed location (CPU write). Actual operation of the memory unit is directed by memory control circuits not shown in Figures 2.4 and 2.5. The memory length, given as N locations in Figure 2.4, is typically a power of 2, such as 1,024 or 4,096. In general, the number of addressable locations is $2^k = N$, where k is an integer. The MC68000 uses 24 bits to represent an address. Thus, the MC68000 can address 2^{24} different memory locations, each containing a byte of information. Its word-addressing capability is therefore 2^{23}, 16-bit words. Memory length is often given as a multiple of 1,024 locations, which is termed 1K of memory.

The number of bits that can be transferred in parallel is determined by the number of data signal lines connecting the memory and the CPU. A typical 8-bit microcomputer has eight data lines and the memory is organized in bytes. The MC68000 has a 16-bit-wide data path that allows either a byte transfer on eight lines or a 2-byte transfer (16 bits) on the 16 signal lines. A longword transfer would require two 16-bit transfers.

Figure 2.6 shows a possible organization of the programs in memory for a typical product employing a microprocessor. The operating system and the applications programs are assumed to be stored in a ROM for this example. The read/write memory holds values that are likely to be changed during the course of operation of the product.

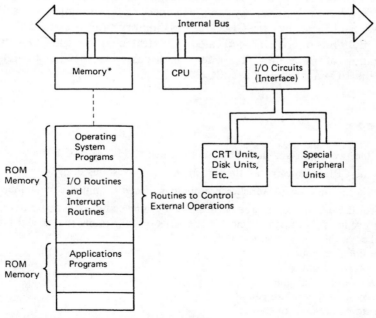

* Memory Areas not designated ROM are Read/Write areas.

Figure 2.6   Hardware and software organization for a microcomputer-based product.

An alternative approach used in most general-purpose systems is to store the bulk of the programs on an external disk unit and load the memory with appropriate programs as needed. The operating system controls the use of the disk in this case.

Notice that the operating system and its routines to handle I/O and interrupts are separated from the applications programs and their read/write memory area. If an application program requires data transfer to external units, the transfer is typically controlled by the operating system to ensure orderly operation of the system.

2.1.3 Input/Output

The I/O circuitry shown in Figure 2.6 is controlled by the CPU to allow transfers of data between the internal bus and an external bus connecting peripheral devices to the system. Such interfacing circuitry is designed to meet the needs of the peripheral device. Also, the operation of the external bus is generally independent of the CPU operation in terms of speed and data path width. A CRT terminal, for example, normally requires data bits to be transferred in a serial manner at a rate that is very slow compared to the rate at which data values can be transferred in parallel on the internal system bus. The major function of the I/O circuitry in this case is to resolve this mismatch in speed and format. The interrupt control lines (not shown) from the interface circuitry notify the processor when the device is ready for data transfer.

Peripheral devices of the microcomputer in the figure are addressed by the CPU in the same manner as the memory is addressed because the memory unit and I/O unit are effectively in parallel. Systems designed in this manner are said to have *memory-mapped* I/O because the CPU accesses memory and I/O circuits based on these addresses only. Thus, there is no dedicated I/O interface or special I/O instructions in the 68000 family of processors. This scheme is discussed further in Chapter 12, which deals with the I/O capability of the MC68000.

EXERCISES

2.1.1 Name the two important characteristics by which CPUs are classified.

2.1.2 Name the CPU characteristics that are important to systems designers.

2.1.3 What are the three major functional groupings of microcomputer bus signals?

2.1.4 Consider the differences between board-level design versus single-board design. For each of the products listed, give some requirements or constraints that would justify one or the other of the two design methods:
   (a) Desktop computer
   (b) Laptop computer
   (c) Laboratory instrument controller
   (d) Pressure/temperature gauge
   (e) Automobile control system

2.1.5 If the CPU clock is running at 2^{24} clock cycles per second (16.67 MHz) and the CPU requires 4 clock cycles to write a word to memory, how long would it take to clear the entire address space of the MC68000, ignoring all other considerations? (The term "clear a location" means to write zeros to the location.)

2.1.6 Can the 68000 read a word beginning at address 1001?

2.1.7 If an operating system is available, data transfer with external units by an application program is typically controlled through the operating system. What are the advantages of this approach? What are the disadvantages, if any?

2.1.8 Compare board-level design with chip-level design of a microprocessor-based product. Include technical and financial considerations. Referring to manufacturer's literature to answer the question is encouraged.

2.1.9 The MC68000L8 (8 MHz) can execute an instruction to clear a register in 0.75 microseconds. How fast can the same instruction be executed by the following CPUs:
  (a) MC68000L10 (10 MHz)
  (b) MC68000L12 (12.5 MHz)
  (c) MC68HC000 (20 MHz)

2.2 DATA BUS SIZE, OPERAND LENGTH AND ADDRESSING RANGE

Processors such as the MC68000 are classified as 16-bit microprocessors because they have 16 data signal lines for data transfers. For currently available processors in this class, the number of addressing lines is between 16 and 32. The MC68000 can address over 16 million byte-length memory locations. Processors classified as 32-bit devices generally have 32 data signal lines as well as 32 signal lines for addressing. The capabilities and characteristics of microprocessors based on their data bus size and addressing range are discussed in this section.

2.2.1 Data Bus Size

Although other definitions can be applied, the *m-bit processor* can be defined as a processor whose data paths external to the processor are parallel *m*-bit paths. In the microprocessor arena, data bus sizes of 4, 8, 16, and 32 bits serve to classify the majority of processors. The data bus size determines the number of bytes per second that can be transferred by the CPU. A 16-bit processor can obviously transfer 16-bit quantities at least twice as fast as an 8-bit unit using two transfer cycles, all other things being equal.

However, the data bus size does not determine the length (in bits) of the instruction and operands for these instructions that can be processed by the CPU. For example, some processors with only eight data signal lines can operate internally on 8-bit, 16-bit, or 32-bit quantities. The MC68008 can read or write 32-bit operands in memory, but the complete data transfer requires four 8-bit transfers. The penalty for multiple-word operations is a loss of operating speed compared to operations using only 8-bit quantities.

Thus, the MC68008 described in Chapter 1 is more difficult to classify in this scheme because it has eight data lines but otherwise performs as the MC68000 does. This CPU is much more powerful than typical 8-bit microprocessors in terms of both its speed of operation and its processing capability.

2.2.2 Operand Length

As an integer number, the m-bit operand transferred between the processor and other elements of the system, such as memory, can represent 2^m numbers in the decimal range 0 to 2^m-1. An 8-bit value allows only 256 values in the range 0 to 255. The MC68000 data word of 16 bits allows 65,536 values. Internally, the MC68000 CPU can operate on data values 8, 16, or 32 bits in length. This is true for all of the members of the 68000 family, including the MC68008.

The significance of operand length is revealed by examining the numerical range. Generally, the design, the intended purpose, and the efficiency of a processor can be surmised from the length of the largest operand that can be processed, at least for general-purpose processors. The length of all the information transferred to and from the processor, including the binary-coded instruction patterns, are multiples of the m-bit operand. Because these m-bit quantities allow 2^m different combinations to be assigned to the bit pattern, there is a possibility of 2^m instructions, data values, or other entities encoded in some desirable way. In itself, the m-bit length is not a great limitation. By proper programming, a quantity can be represented as two or more m-bit values combined to form multiple-length entities.

2.2.3 Addressing Range

Another characteristic of a microprocessor that determines its capability is the *addressing range* of the processor. Each instruction and data value transferred via the data lines of a processor is located by an address that designates the exact position of the item in memory. The addressing range determines the size of the largest program or the maximum number of data values the processor may address. If k bits are used for an address, the processor can address 2^k separate locations.

As stated previously, the MC68000 can address 2^{24} byte locations. Except for 32-bit microprocessors, the address length of k bits is longer than the data bus size of m bits for most processors. For example, for the MC68000, $k = 24$ and $m = 16$. A 32-bit address, as in the MC68020, is considered sufficient for almost all applications because the CPU can address over 4.2 *billion* locations. Many 8-bit processors employ a 16-bit address length, allowing 2^{16} or 65,536 locations to be referenced, with each location considered to contain an 8-bit value representing a data value, an instruction, or another item. Today this addressing range is considered inadequate or at least inconvenient for many of the applications to be satisfied with 16-bit microprocessors.

Table 2.1 lists characteristics of typical microcomputers with 8-bit, 16-bit, and 32-bit data bus sizes. The decimal range of a number for the given operand length is

Table 2.1 Characteristics of Typical Microcomputers

Characteristic	8-bit Family	16-bit Family	32-bit Family
Decimal range of data	0 to 255	0 to 65,535	0 to 4,294,967,295
Number of address bits	14 to 16	16 to 24	32
Typical memory length (1K = 1,024 bytes)	16K to 64K	64K to over 16 million bytes (16MB or 16,384K)	Any size up to 4,194,304K

shown in the first row. Then follows the number of address signal lines and the memory length.

Memory length is often given as a multiple of 1,024 (2^{10}) locations, which is termed 1K of memory. Thus, a 16-bit address would allow 65,536 locations, or 64K of memory. For memories of greater capacity, the number of locations is defined in *megabytes* (MB) with 1,048,576 bytes called 1 MB of memory. The maximum memory capacity of the 32-bit processors is often stated as 4 *gigabytes* (GB).[1]

Example 2.1 _____

One rough measure of the processing power of a CPU is determined by multiplying the width or number of signal lines of the address bus by the width of the data bus by the clock frequency in millions of cycles per second (MHz). Thus, using the 16.7-MHz version of the MC68020, the product is

$$32 \times 32 \times 16.7 = 17,100$$

By way of comparison, for the 12.5-MHz version, the MC68000 has a product of

$$24 \times 16 \times 12.5 = 4,800$$

The ratio indicates a 3.56 times improvement in performance for the MC68020. This is not an unreasonable estimate as determined by direct comparison of the times required to execute various programs on an MC68000-based system versus the times needed for a computer with an MC68020 as its central processor. However, due to other features of the MC68020, the improvement in processing power is typically greater than that indicated by the calculation.

Characterization by CPU Type.   The MC68000 is considered a member of the 16-bit family of processors, and the MC68020, MC68040 or MC68060 are members of the 32-bit family. These families of processors are thus divided according to the number of data signal lines that are available to transfer data and instructions from memory. Another classification of a microprocessor or a microcontroller is based on the type of CPU without consideration of the number of data and address signal lines. This represents the view of the programmer rather than that of a hardware designer.

In fact, all of the members of the 68000 family can operate internally on 32-bit data and addresses. Therefore, even the MC68000 is technically a 32-bit CPU because its instructions can process byte, word, or longword data.

Although the CPU32 of the MC68332 can process 32-bit data and addresses internally, the microcontroller transfers 16-bit data and uses 24 bits to represent an address for external memory. Thus, the MC68332 can address 2^{24} different memory locations, each containing a byte of information. This is the same capability as the MC68000. The designation CPU32 recognizes that the CPU can process 32-bit operands.

Example 2.2 _____

As an illustration of the use of the internal system bus, assume that a processor with 16 address lines and 16 data lines is to execute an instruction. The instruction is 32 bits long. It is stored in two 16-bit words at locations 10 and 12 of memory as shown in Table 2.2. The hypothetical

---

[1] Strictly speaking, the prefixes kilo-, mega-, and giga- mean 10^3, 10^6 and 10^9, respectively. When referring to memory capacity, these are frequently taken to mean 2^{10}, 2^{20}, and 2^{30}, respectively.

Table 2.2 Memory Contents for CLR 100 Instruction.

Memory Address (Decimal)	Memory Contents (Binary)
10	0100 0010 0111 1000
12	0000 0000 0110 0100
14	(Next instruction)
•	•
•	•
•	•

Note: The even locations contain one 16-bit word that is considered as two 8-bit (byte) locations.

instruction written in mnemonic form is CLR 100. This instruction sets the contents of memory location 100 to zero, which is represented as a 16-bit string of {0}'s in location 100. The first 16-bit word of the instruction contains the operation and indicates that the address of the location involved follows in the next word. The address 100 is stored in binary at location 12.

Figure 2.7 shows the state of the signal lines with increasing time as the processor executes the instruction. The instruction fetch requires two read cycles by the CPU to determine the operation and clear the location. For each read operation, the memory responds by presenting the contents of the addressed location on the data lines. There is a slight delay while the data signal lines change to the new value. The CPU then decodes the CLR instruction and executes it by writing zeros into location 100 of memory, via the data signal lines, after the control signal to write is given. The next instruction is then fetched and executed.

The 16-bit memory locations are given even addresses in this example rather than consecutive values because processors such as the MC68000 can address 8-bit operands in memory. Thus, word location 10 contains two bytes at locations 10 and 11. Each instruction consists of one or more 16-bit words.

The timing sequence of Figure 2.7 is typical for a processor executing a simple instruction such as the clear instruction (CLR). Although the functional operation is similar to that shown, the execution of this instruction by the MC68000 would in fact be more complicated. This is discussed briefly in Chapter 4, where the instruction prefetch characteristic of the MC68000 is

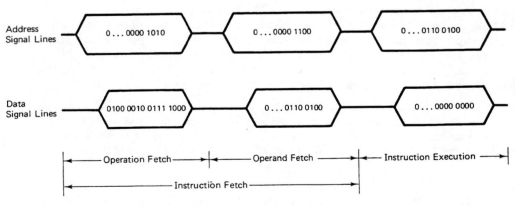

Figure 2.7   States of signal lines with increasing time.

considered. Actually, the MC68000 CPU would first read the contents of location 100 before zeros are written there when CLR is executed. However, the preceding description of the operation is adequate for our present purposes.

EXERCISES

2.2.1 It was mentioned that the Motorola MC68008 was similar to the MC68000 processor, except that it had 8 data signal lines instead of 16. In what situations might the MC68008 be useful?

2.2.2 What is the width of the data bus of the MC68020 and the MC68040?

2.2.3 What is the largest program (in bytes) for a microprocessor with:
   (a) 16 addressing lines
   (b) 20 addressing lines
   (c) 24 addressing lines
   (d) 32 addressing lines

2.2.4 The MC68000 can address byte (8 bits), word (16 bits), or longword (32 bits) operands. For each operand length, draw a diagram showing the organization of an 8-byte buffer in memory starting at location 1,000. Label the bit positions and addresses. A buffer is an area of memory that holds data temporarily during I/O transfers.

2.2.5 How many bits (address lines) are necessary to address a memory with:
   (a) 4,096 locations
   (b) 65,536 locations
   (c) 16,777,216 locations

2.2.6 The MC68000 uses a 24-bit address and can address each 8-bit byte in memory. How many bits for addressing would be needed to access the same memory space if the MC68000 could address only 16-bit words?

2.3 THREE VIEWS OF THE MICROPROCESSOR

Modern microprocessors such as the MC68000 are incorporated into computer systems to control the overall operation of the system. These processors are capable of directing system activity by executing programs and performing the necessary I/O functions. Furthermore, the modern processors separate the processing into supervisor and user states or modes, allow memory management and protection, and detect various types of errors. These capabilities are not common for earlier processors of the 8-bit class.

Programming features of the 16-bit processors include a general and powerful instruction set, as well as a number of different ways to reference an operand in memory, called *addressing modes*. Many of these processors provide the capability to support special programming techniques and debugging aids.

Interaction of the processor and the other hardware elements of the system is via the system bus, where transfers of control signals, addresses, and data occur. The capability and flexibility of the processors in this regard are determined by the functions of the signal lines from the processor. For example, sophisticated I/O and interrupt capabilities alleviate the need to provide a great deal of special hardware in a complex system. Most 16-bit processors provide the state of the processor and other relevant information to external circuits as the processor operates, a feature that simplifies the hardware design.

The features of 16-bit processors that help meet system, programming, and interfacing requirements of complex systems are presented briefly in this section. Although the material is somewhat general, it forms a base for understanding many of the characteristics of the 68000 family processors. Figure 2.8 summarizes three different views of a microprocessor. The system designer, assembly language programmer, and the interface designer each focus on different aspects of the processor, so these views may coincide or even conflict in some instances. Obviously, the assembly language programmer and the interface designer have the same goals in debugging a prototype system, but their approach to producing a correct product may differ considerably.

2.3.1 System Design

The system designer is concerned with the overall operation of the system, including its performance and reliability. The designer also determines the memory usage and how the memory areas occupied by the operating system will be protected if read/write memory is used. This protection is vital if the system is used for developing software, in which addressing errors and runaway programs may exist. When many peripheral units are attached to the system, the design of the I/O portion of the system is critical to assure proper coordination between programs and hardware during data transfers. Table 2.3 summarizes the aspects of system design covered in this section.

System Characteristics.   One important criterion used to measure the performance of a microcomputer is its speed of operation while executing a given program. A key element, although not the only one, in determining the speed of operation is the maximum rate at which the CPU executes instructions. As discussed in Section 1.2, the MC68000 is produced in versions providing a wide selection of operational speeds. The MC68000L12, for example, can execute 12.5 million processor cycles per second, whereas the MC68000L8 executes only 8 million. These relative speeds are useful in comparing the performance of different systems based on different versions of the processor.

To keep errors in applications programs from affecting the overall operation of the system (or each other), the MC68000 provides two modes of processor execution, the *supervisor mode* and the *user mode.* Programs executing in the supervisor mode have full control of the processor and system functions. As expected, the operating system executes in the supervisor mode. This includes all of its routines to handle I/O and interrupts. A *routine* is usually a short program segment intended to accomplish one specific operation such as transferring a data value.

Typically, applications programs execute in the user mode. In this mode, certain processor instructions and perhaps certain memory areas are inaccessible to the applications program. The selection of the mode for various programs is determined by the system designer, and the transitions between the two modes are carefully controlled.

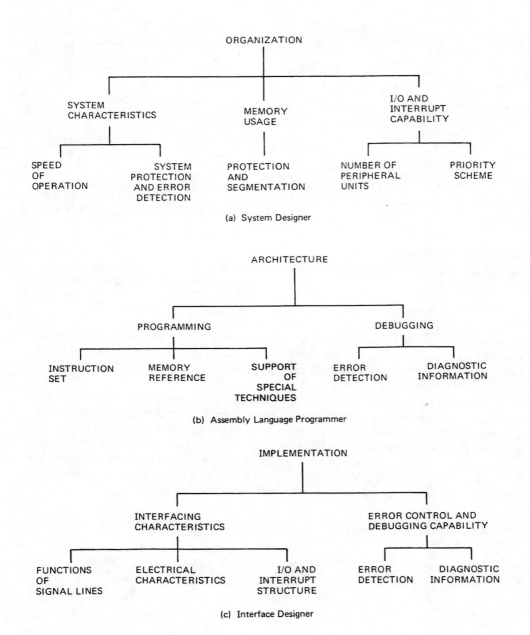

Figure 2.8 Three views of the microprocessor.

Table 2.3   System Considerations

Characteristic	Purpose
System	
Speed of operation	One measure of system performance
System protection (supervisor versus user modes)	To prevent user programs from interfering with the operating system
Error detection	To detect and isolate various errors through traps and interrupts
Memory usage and protection	
Separation of programs and restricted access to certain memory areas	To prevent user programs from interfering with memory allocated to the operating system or other user programs
I/O and interrupt capability	
Number of peripheral units allowed and interrupt priority	To determine the I/O capability of the system (e.g., number of devices and response times for data transfer)

Figure 2.9 shows an example of system operation versus time in which control is passed to a user program and returned to the operating system. The return to the supervisor mode may be caused by program completion, an error detected by the CPU, or an interrupt.

Processors such as the MC68000 allow certain errors occurring during program execution in either mode to be detected and trapped. The trap mechanism passes control of the processor from the program causing the trap to a routine of the system software that processes the error. The execution of an illegal instruction is an example of an operation that causes a trap. An illegal instruction is a machine language instruction with a bit pattern not recognized by the CPU. In the MC68000, an attempt to divide by zero in an arithmetic operation will also cause a trap.

Figure 2.9   CPU operation showing transitions between modes.

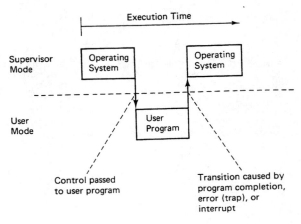

Certain hardware errors can be detected by using the interrupt mechanism. The system could be designed to process an interrupt caused by power failure, for example. The interrupt routine would typically cause the CPU to save information that allows a program to be restarted where it was interrupted after power returns.

Interrupts may occur at any time, asynchronous with program execution. Traps, on the other hand, occur only as a result of execution of program instructions.

Memory Usage and Protection. Because the MC68000 has the capability to address 2^{24} byte locations, very few systems are limited by a lack of memory space. In most systems, both the operating system and the applications programs can be held in memory without conflict. The system designer determines the allocation of memory by specifying the required number of locations for each program. Particularly in those systems used for program development or multiuser applications, a method must be available that prevents an executing program from accessing (reading from or writing to) any memory locations not assigned to that program. This protection is usually provided to the operating system's memory space to prevent access by programs executing in the user mode.

In MC68000-based systems, the memory can be protected in this way by memory management circuits in a chip called a memory management unit (MMU). The MC68000 indicates the type of access as supervisor or user via three of its control signal lines. Simultaneously, memory management circuits compare the memory location being addressed to the valid range for the mode assigned to the program. A violation is indicated by an interrupt from the memory management circuitry to the CPU. Thus, if the CPU provides the mode and address for each memory access as the MC68000 does, the protection of memory areas is easily accomplished. None of the earlier microprocessors of the 8-bit class provided this feature.

The MC68010 processor and an MMU can be combined to provide the hardware support for both memory protection and virtual memory. A virtual memory system has an advantage in being a system with a relatively small physical memory space that can be used with programs of any size, at least theoretically. Several references in the Further Reading section of this chapter discuss virtual memory systems with the MC68010 and the MC68000. Processors such as the MC68030, MC68040 and MC68060 have memory management features suitable for use by an operating system with virtual memory capability.

I/O and Interrupt Capability. The system designer will specify the number and type of peripheral devices necessary to satisfy the requirements of an application. Subsequent design of the I/O subsystem becomes complicated when many devices are attached because different units have different time requirements to complete data transfers. For example, a line printer is much slower in printing characters than a disk unit is in storing them. Due to these timing variations, the system is typically designed so that each device can issue an interrupt request via several of the control signal lines from the interface to the CPU.

An interrupt request is issued when the device is ready to receive or transmit data or when an error condition is detected. Because the interrupt originates outside of the CPU, the external device determines when the interrupt occurs. From the point of view of an executing program, the interrupt causes a break in normal program execution until the interrupt routine completes the data transfer or other processing. Such routines execute in the supervisor mode in an MC68000-based system. When the interrupt processing is complete, control returns to the program that was interrupted. In time, the transition between a user program and the interrupt occurs, as shown in Figure 2.9.

The interrupt mechanism is a primary determinant of the I/O capability of a system when a number of peripheral devices are attached to the microcomputer. In processors such as the MC68000, multilevel interrupt circuitry is part of the CPU. There are seven possible interrupt levels arranged in priority so that a higher-priority interrupt will interrupt a routine executing due to a lower-level interrupt request. In theory, seven interrupt routines could be in various states of execution at the same time in an MC68000-based system.

At each interrupt level, a number of devices could be given the same priority. External circuitry is required, in this case, to resolve conflicts if two or more devices interrupt on the same level at the same time or if several interrupts are pending (waiting for the interrupt routine at this level to complete). The Motorola CPU is capable of handling up to 192 devices distributed according to the system requirements across its seven interrupt levels. However, such a configuration would require extensive hardware design to control the devices that could interrupt at the same CPU interrupt level.

2.3.2 Assembly Language Programming

The ease with which a program may be created, debugged, and tested to satisfy a specific application depends largely on the characteristics of the processor, rather than on the development software, if the program is written in assembly language. Editors, assemblers, and other development aids vary in quality and efficiency, but a good development system cannot make up for an inadequate processor.

Processors of the MC68000 class are adequate for many applications because of their powerful instruction sets, numerous addressing modes, and other special features not typically found in the 8-bit microprocessors. Table 2.4 summarizes several characteristics of a microprocessor that are used to determine the ability of the processor to satisfy the programming requirements of sophisticated software.

The term *computer architecture* sometimes refers to the complete set of characteristics of a computer system that are important to the programmer. A description of the architecture includes a definition of the overall organization of the system and a complete discussion of the programming characteristics of the CPU. The processor's instruction set and addressing modes constitute two of the most important characteristics of the CPU for this description.

Programming. The instruction set of a microprocessor is the collection of all the machine language instructions available to the programmer. Each instruction can be described by its operation or function and the number and type of operands it manipulates. For example, the binary addition instruction of the MC68000 adds two signed

Table 2.4   Programming Considerations

Characteristic	Purpose
Programming	
Instruction set	Determines efficiency and ease of programming
Addressing modes	Indicates the ability to allow creation of data structures in memory
Special instructions	Convenient for creating modern sophisticated programs and systems
Debugging	
Error detection and diagnostic information	Useful in isolating certain errors and determining cause

integers. The MC68000 also allows subtraction, multiplication, and division of two such integers.

The 68000 family CPUs have instructions to add and subtract decimal numbers. Decimal numbers are represented in memory as coded bit sequences in a representation called binary-coded decimal. Thus, the MC68000 has a fairly complete set of instructions to operate on numerical operands. Chapter 3 discusses the types of operands allowed with MC68000 instructions. Chapter 7 presents the complete set of arithmetic instructions that are available in the MC68000.

In contrast, earlier 8-bit microprocessors had a more restricted set of instructions for arithmetic operations. Divide and multiply instructions were not available, for example. Routines based on the operations of addition and subtraction were created to accomplish the tasks, such as performing multiplication by repeated addition. In most cases, the equivalent instructions of the 16-bit processors are more powerful and efficient than those of the 8-bit class of processors. The powerful instructions of the MC68000 simplify assembly language programming and increase the speed of execution of equivalent programs. One such example was given in Section 2.1 when data word length was discussed. The MC68000 can perform arithmetic operations on operands of 8 bits, 16 bits, or 32 bits in length. When 8-bit processors had equivalent instructions, the operand length was typically only 8 bits.

The number and type of instructions, including their operands, are examined for their capability and flexibility when comparing processors. If the operands are held in memory, the address of an operand may be specified in an instruction as a 24-bit integer by the MC68000. This method of directly addressing each operand is usually called *absolute addressing*. Other methods of addressing operands are possible, and these addressing schemes are called addressing modes. The MC68000, for example, has 14 distinct addressing modes.

MC68000 instructions must specify not only the operation to be performed, but also the addressing mode to be used to refer to each operand addressed by the instruction. The CPU calculates the hardware address of the memory location for each operand as the instruction executes. In terms of the number of addressing modes, the MC68000 system is more like a large computer than a microcomputer because 8-bit microprocessors usually have restricted addressing capability. For example, a variety of data structures, such as lists and arrays, can easily be created in memory with the addressing modes of the MC68000.

In addition to having a powerful and flexible general-purpose instruction set, the MC68000 has a number of special instructions. These instructions are of great value in certain applications, of which only a few are introduced here. For example, the MC68000 has a set of bit manipulation instructions that allow operations on individual bits within an 8-bit or 32-bit operand. The selected bits can be tested, set (to {1}), or cleared (to {0}) by the bit instructions. Such operations are important when the status of a device is indicated or set as a binary value (i.e., representing ON or OFF status).

Modern programming techniques dictate that programs be modular for ease of debugging and testing. Each module performs a concisely defined function, and a complete program is created by linking the modules together. The MC68000 has a number of instructions to support modular programming, including instructions to invoke subroutines (modules). Additionally, the MC68000 allows parameters to be easily transferred between modules via its LINK instruction. This instruction combines several operations that normally require a short program segment on some other processors.

A number of instructions of the MC68000 are useful for controlling system operation. The TRAP instruction is one example. When a TRAP instruction is executed in a user mode program, control is returned to the supervisor mode. This is useful in invoking operating system routines from a user program. Another instruction, the Test and Set (TAS) operand, allows several processors to share a common memory area in a multiprocessor application without danger of simultaneous access to an addressed memory location in the area.

These examples of the MC68000's capabilities as a programmable processor only begin to indicate its power and versatility. Many of the remaining chapters are devoted to exploring its instruction set and related concepts in more detail.

Debugging. The MC68000 incorporates several features that aid in the debugging and testing of programs. The mechanism that detects an error is a trap that occurs when an instruction causing an error executes. As mentioned previously, an illegal instruction or an attempt to divide by zero during program execution causes a trap.

Certain addressing errors and arithmetic conditions can also cause traps. When trapping occurs, control is passed to a specific routine of the operating system. The routine performs any required processing for the type of error detected. Information about the status of the processor and the address of the offending instruction are saved when a trap occurs. This facilitates diagnosis of the problem.

Trapping various error conditions serves to detect errors during program development and protect the system from unpredictable operation if these errors occur after the product is complete. Later chapters discuss the program actions that might be taken when errors causing traps are detected.

Advanced Processors. The processors developed from the MC68000 had a number of design goals with respect to the programming capabilities. These included enhancing the instruction set by adding new and more powerful instructions. Also, the number of addressing modes was increased in most of the later processors. Today, such CPUs are called CISC (Complex Instruction Set Computer) to distinguish them from processor designs that use a simplified instruction set. The differences in processor designs are discussed in Chapter 4.

One criterion was that the advanced processors must execute the machine language code of the previous members of the 68000 family. Thus, any user mode program using the MC68000 instruction set will execute unchanged on the MC68008, MC68010, MC68020, MC68030, MC68040, and MC68060. A supervisor mode program may require some modification, as discussed later in Chapter 4.

2.3.3 Interface Design

The interface designer is concerned with the implementation of the computer system in designing and debugging interfaces. The functional and electrical characteristics of the processor signal lines determine the design of the circuitry that connects the processor to external devices. The functional aspects of the signal lines determine their purpose. The electrical characteristics include the timing properties of the signals, the voltage levels, and other details of importance to circuit designers. When interface design is discussed in this textbook, the emphasis is on the functional approach rather than the electrical details. This subsection considers the inferfacing characteristics and hardware debugging features of the MC68000 as listed in Table 2.5.

Interfacing. Figure 2.10 shows a simplified diagram of the MC68000 that illustrates the major classes of signal lines connected to the system bus. These signal lines are physically connected to the processor by the signal pins of its integrated circuit package.

Twenty-three signal lines are used to address a memory word, and two of the five I/O transfer control lines indicate whether an 8-bit (one byte) or 16-bit (one word) location is selected. These signal lines allow an addressing range of 2^{24} bytes or 2^{23} words. Chapter 13 presents the actual details of addressing using the CPU signal lines.

Table 2.5 Interfacing Considerations

Characteristic	Purpose
Interfacing	
Availability of processor mode to external circuits	Allows indication of supervisor or user mode to external circuits for memory management or other purposes
Bus control	Allows shared bus systems
I/O capability and priority interrupt scheme	Determines the design of interfaces to meet specific requirements
Error control and debugging	
Hardware error detection and diagnostic information	Aids in debugging or recovery from error
CPU-generated error signal	Indicates a faulty CPU or system error

Figure 2.10 Signal lines of the MC68000.

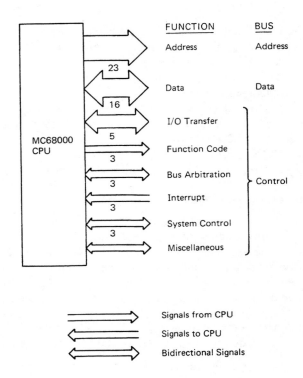

The 16 data lines transfer data values in either direction, as indicated by the double arrows. Such signal lines are said to be *bidirectional.* Four of the five I/O transfer control signals initiate processor reads or writes from the addressed location, while one line allows the external circuitry to acknowledge the processor's request. For normal data transfers, the address bus, data bus, and the five I/O transfer control lines are involved. The three function code lines indicate the processor mode and are used by external circuits to determine whether a supervisor program or a user program is making the transfer request. The function code lines also indicate when the CPU is processing an interrupt request.

The three control lines designated for bus arbitration allow an external circuit to take control of the CPU bus by issuing a request to the CPU. When acknowledged by the MC68000, the processor is electrically isolated from the bus. Three interrupt lines accept requests from external circuits and the requests are encoded into priority levels. Three other signals for system control are used to detect external errors or to allow the CPU to indicate that it has failed. Other miscellaneous input signals include the clock signal and connections for power and grounding.

Because the MC68000 and similar microprocessors are often incorporated into complex systems requiring memory protection and a large number of peripheral units, the processor is designed to accommodate such requirements. Figure 2.11 shows a microcomputer system with the signals for memory management, interrupt requests, and error indications separately indicated.

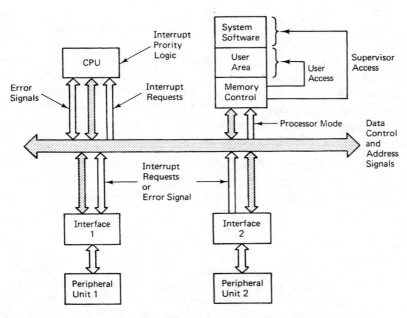

Figure 2.11   Microcomputer system bus structure.

As shown, the mode of the processor (supervisor or user) is used by the memory control circuits to address the correct area of memory. If a program in the user mode attempts to read or write in the supervisor area of memory, an error signal would be generated by the memory control circuitry and appropriate action can be taken by an error routine executed by the CPU.

The bus arbitration control lines are not shown explicitly in Figure 2.11. These three lines determine which device on the system bus will be the *master* and control I/O transfers and similar operations. Such arbitration is required when several processors share the same bus or when any device other than the CPU is capable of initiating I/O transfers on the bus. The DMA chips discussed in Chapter 1 are devices of this type.

The interrupt request signals shown in Figure 2.11 are processed by the interrupt circuitry of the CPU. The CPU determines the priority of the interrupt and passes control to the interrupt routine corresponding to the highest-priority peripheral unit that is requesting service. Upon completion of the interrupt routine, control passes back to the program that was interrupted.

Error Control and Debugging.   Various error signals can be generated by the interface circuitry (as shown in Figure 2.11) or perhaps even by the CPU itself. The MC68000 can issue an error signal when it detects that the system cannot continue to operate correctly due to a critical external failure or to a CPU failure. This feature can be vital in a multiprocessor system, where one faulty processor must be isolated when other processors in the system detect an error that could have a system-wide effect. Upon detecting failure, the CPU indicates the problem on one of the three system control signal lines. It then ceases to process instructions and another device or CPU must take control of the system.

If a hardware error is indicated, the CPU saves information in memory about the condition of the system when the error occurred. The information defines the status of the processor, the type of operation in progress, and similar data at the time the error occurred. This information can be used in certain cases to allow the system to recover from a hardware error during operation. Of course, such information is also valuable to the interface designer in debugging the hardware.

Other Processors.   The MC68000, MC68HC000 and MC68010 have the signal lines shown in Figure 2.10. The MC68HC001, MC68EC000 and MC68008 have slightly different signal pin arrangements. The advanced 32-bit processors have completely different signal lines.

Example 2.3 ——————————————————————————————————

Many systems include an external timing circuit that monitors the time taken for I/O transfers. Timers used in this manner are usually called *watchdog* timers. The MC68000 may use a watchdog timer to determine if an external device fails to respond to an I/O transfer request.

When an I/O request is made to an external device by the CPU, a timing circuit in the system counts the elapsed time until the device acknowledges the request via the I/O transfer control lines. If no acknowledgment occurs within a specified time (usually several microseconds), an error signal from the watchdog timing circuit would be placed on one of the system control lines to the processor. The processor's error-handling routine would then determine the next action. The processor could retry the I/O transfer or indicate a system failure on another system control line.

The system integration module of the M68300 family of microcontrollers described in Chapter 1 provides a hardware watchdog timer that monitors I/O transfers. The timer can be programmed to indicate an error after a specified number of clock cycles if an external device does not respond.

_____

EXERCISES

2.3.1 Why are interrupts useful in a system in which the I/O units vary greatly in data transfer rates?

2.3.2 List some features of the MC68000 instruction set that enable the programmer to use modern data structures and programming techniques.

2.3.3 The three Function Code lines in Figure 2.10 indicate whether the CPU is in supervisor or user mode or is arbitrating an interrupt request. What use might an interface designer make of these control lines? Consider address space protection and control.

2.3.4 How might an operating system be protected from being corrupted by an application routine? Give details of the memory use, hardware devices, and processing required.

2.3.5 As the number of interrupt priority levels of a processor increases, does the hardware design become easier or more difficult? How does the software design change? Explain your answers.

2.3.6 What are the three methods by which control is transferred from user-mode execution to supervisor-mode execution?

2.3.7 One of the following statements is false. Which one is it? Explain why each statement is true or false.
  (a) The MC68000 provides support for virtual memory.
  (b) The MC68000 provides support for modular programming.
  (c) The MC68000 provides support for calls to privileged operations.
  (d) The MC68000 provides support for multiprocessor hardware designs.

REVIEW QUESTIONS AND PROBLEMS

2.1 Discuss the differences between a trap and an interrupt. Include both hardware and software considerations.

2.2 What are the possible consequences of allowing the CPU to execute illegal instructions?

2.3 The MC68000 requires 44 clock cycles to respond to an interrupt. At 10 MHz, each clock cycle is one-tenth of a microsecond, and the response time is 4.4 microseconds. Assume that three interrupt routines require the following total execution time after the interrupt is recognized:
  (a) R1 = 20 microseconds
  (b) R2 = 30 microseconds
  (c) R3 = 20 microseconds
The priority levels are such that R3 has the highest priority and R1 has the lowest. What is the possible range of time in which each routine can be executed when the corresponding interrupt occurs?

2.4 Each instruction of the MC68000 is at least 16 bits in length and occupies one word in memory. An absolute address can occupy 16 bits or 32 bits in memory. What are the possible lengths, in words, of MC68000 instructions, including their operand addresses? Instructions can have none, one, or two operands.

2.5 Explain the use of the supervisor and user modes. Describe the programs that might execute in each mode, and explain why.

Design Problems and Further Exploration

Some of these problems may require reference to Manufacturer's literature or other references.

2.6 What interfacing features of the 68000 are intended to allow creation of a complex system with several peripheral devices, memory protection, and bus sharing? State the purpose of each feature you list.

2.7 Assume that no two devices in a system should have the same interrupt priority level. Assign interrupt priority levels to the following peripheral devices or chips and explain your choices:
  (a) Memory management unit
  (b) Disk controller
  (c) Keyboard
  (d) Printer
  (e) Serial communications chip
  (f) Time-of-day clock
  (g) Power failure detector

2.8 Which routines should run in supervisor mode and which in user mode? Explain your choices for routines to do the following:

(a) Binary to ASCII conversion

(b) Read from disk

(c) Write to disk

(d) Accept a character from the operator's keyboard

(e) Validate data entry

(f) Find day of the week

(g) Clear the display

(h) Output print line to printer

(i) Format text file for printer

2.9 Compare and contrast the Motorola 68000 CPU bus with the Intel 80386 CPU bus.

2.10 How do other processors (e.g. Motorola 68040, Intel 80386) implement multilevel interrupts?

Design of an Instrument

The design problem in the Review Questions of Chapter 1 required a block diagram for a proposed product such as an instrument for measurements. Continue that problem for the product you chose but with more detail based on the material in Chapter 2. Be more specific as to the bus structure and other specifications for the product.

FURTHER READING

The reader should refer to the latest issue of the computer journals to keep abreast of the rapidly changing microcomputer area. The latest publications from Motorola should be consulted for specific details about their product lines, because even the journal articles tend to be outdated quickly when specific product information is needed. A useful reference for the MC68000, MC68008, MC68010, and variations of the processors is the *8-/16-/32-Bit Microprocessor User's Manual* (M68000UM/AD) available from Motorola.

A number of articles have been written about the MC68000 family in computer journals. Three of historical interest are listed here.

The MC68010 is treated in some detail in the article by MacGregor and Mothersole referenced below. Its capability in a virtual memory system is described by two designers from Motorola.

MC68000

DeLaune, J., and T. Scanlon. "Supporting the MC68000." *Mini-Micro Systems* 13, no. 8 (August 1980): 95–102.

Kister, Jack, and I. Robinson. "Development System Supports Today's Processors and Tomorrow's." *Electronics* 53, no. 3 (January 31, 1980): 81–88.

Stritter, Edward, and T. Gunter. "A Microprocessor Architecture for a Changing World: The Motorola 68000." *IEEE Computer* 12, no. 2 (February 1979): 43–52.

MC68010

MacGregor, Douglas, and Davis S. Mothersole. "Virtual Memory and the MC68010." *IEEE MICRO* 3, no. 3 (June 1983): 24–39.

CHAPTER 3

Representation of Numbers and Characters

The digital computer has the capability of storing and processing information of interest to the programmer. The information is stored in memory as sequences of binary digits that are processed by the CPU. For example, the machine language instructions discussed in Chapter 2 represent information that controls the operation of the computer system. *Programs,* consisting of these instructions, operate on other binary sequences of *data* that represent information that has been stored for processing. This chapter explores the storage methods commonly used to represent numbers and characters for MC68000-based computer systems.

Numbers that are interpreted as positive or negative integers or fractions may be represented in memory in many ways. The most common number system used to represent numbers in microcomputers is the *two's-complement* system, which represents signed numbers as binary (base 2) values. The MC68000 provides instructions for addition, subtraction, multiplication, and division of these binary numbers. These two's-complement numbers form a fundamental *data type* for the MC68000.

Decimal numbers can also be added and subtracted with instructions of the MC68000 if the decimal values are coded into binary by a scheme called *binary-coded decimal* (BCD). To incorporate both positive and negative numbers, the *ten's-complement* system is used.

Many engineering and scientific applications require a large range for numbers that are represented in a *floating-point* format. This binary equivalent of scientific notation, which uses a mantissa and an exponent to represent a number, is employed and sometimes required in MC68000-based systems. No MC68000 instructions are available to manipulate these numbers directly, so floating-point routines must be provided to perform arithmetic operations when required. However, the MC68020 CPU with its floating-point coprocessor has floating-point instructions, so that programming will be greatly simplified. The floating-point format used in the MC68020 system is discussed in this chapter.

Text is stored in memory by assigning a specific bit pattern to each character in the alphabet. The *ASCII code* is the most popular code used to represent characters in microcomputer systems. As with floating-point numbers, any processing of the

ASCII-coded characters is via software routines because the MC68000 has no instructions that operate specifically on characters.

The basic characteristics of the data types commonly used in MC68000 systems are discussed in this chapter. Machine instructions to manipulate the data types and various other programming considerations are discussed in later chapters. In particular, arithmetic operations are treated in detail in Chapter 7.

3.1 NUMBER REPRESENTATION

This section discusses the representation of positive and negative integers and fractions. A general formulation with the base or radix r for each number representation is presented and then applied to the discussion of binary values with $r = 2$ and other bases as appropriate. The generalized presentation is useful for conversion of numbers from one base to another and techniques of numerical analysis. Representation of binary numbers in the sign-magnitude, one's-complement, and two's-complement systems are presented. Decimal representations for the nine's-complement and ten's-complement systems are also presented.

3.1.1 Nonnegative Integers

A nonnegative integer in base r is written in positional notation as

$$N_r = (d_{m-1}\, d_{m-2} \cdots d_0)_r \tag{3.1}$$

where each digit d_i has one of the distinct values $[0, 1, 2, \ldots, r - 1]$ and m represents the base 10 or decimal number of digits in the integer. Thus the number 324 would have $d_0 = 4$, $d_1 = 2$, and $d_2 = 3$ in Equation 3.1 and could be written as

$$N_{10} = 324_{10}.$$

For numbers in base 10, the subscript is omitted if no confusion could result from its omission. The form specified by Equation 3.1 is generally referred to as *positional* notation. The position of the digit starting from the rightmost digit represents a power of the base r; that is, 324 represents 4 ones (4×10^0), 2 tens (2×10^1), and 3 hundreds (3×10^2). Mathematically, the value of the number is calculated as

$$N_r = d_{m-1}r^{m-1} + d_{m-2}r^{m-2} + \cdots + d_1r + d_0$$
$$= \sum_{i=0}^{m-1} d_i r^i \tag{3.2}$$

in which the digits are restricted in value such that $0 \le d_i \le r - 1$. Thus the number 324 can be calculated as

$$324 = 3 \times 10^2 + 2 \times 10^1 + 4 \times 1.$$

The arithmetic operations in Equation 3.2 could be carried out in any number base and this equation is frequently used to determine the decimal equivalent of a number in another base.

Table 3.1 Digits in Various Number Systems

Number System	Base r	Digits
Hexadecimal	16	0, 1, 2, 3, 4, 5, 6, 7, 8, 9, A, B, C, D, E, F
Decimal	10	0, 1, 2, 3, 4, 5, 6, 7, 8, 9
Octal	8	0, 1, 2, 3, 4, 5, 6, 7
Binary	2	0, 1

Table 3.1 lists the range of possible values of the digits in the hexadecimal, decimal, octal, and binary number systems. The decimal system is, of course, used primarily for ordinary arithmetic by human beings, and the binary system is used for computer arithmetic. Octal and hexadecimal representations are convenient for writing long binary numbers. For example,

$$01011010_2 = 5A_{16} = 132_8.$$

The decimal value of the number represented in these bases is obtained from Equation 3.2 by converting the base r digits to decimal equivalents such that

$$N = 5 \times 16^1 + 10 \times 1$$
$$= 1 \times 8^2 + 3 \times 8^1 + 2 \times 1$$
$$= 90_{10}.$$

The hexadecimal digits [A, B, . . . F] represent the decimal numbers [10, 11, . . . 15] in the conversion from hexadecimal to decimal.

Example 3.1

Consider the largest m-digit positive integer in positional notation,

$$N_r = ((r - 1)(r - 1) \cdots (r - 1)),$$

as in $(1111 \ldots 1111)_2$ or $(9999 \ldots 9999)_{10}$ with m digits each. The sum of Equation 3.2 indicates that the decimal value is

$$N = (r - 1) \sum_{i=0}^{m-1} r^i$$

which is an easily summed geometric series. The result is $r^m - 1$. For example, an 8-bit binary number has a maximum value of $2^8 - 1$, or 255.

Positive Fractional Values. The positional representation defined by Equation 3.1 is valid for integers only. If a fraction is to be represented, a radix point in the base r is used to separate the integer from the fractional part of the number. The radix point is called the *binary point* in base 2 and the *decimal point* in base 10. Thus 324.14 has the value

$$3 \times 10^2 + 2 \times 10^1 + 4 \times 1 + 1 \times 10^{-1} + 4 \times 10^{-2}.$$

In general, a k-digit positive fraction is written with a leading radix point as

$$.(d_{-1} \, d_{-2} \cdots d_{-k})_r \tag{3.3}$$

with the value

$$.n_r = d_{-1}r^{-1} + d_{-2}r^{-2} + \cdots + d_{-k}r^{-k} \tag{3.4}$$

where the negative subscript for the digits indicates the appropriate negative power of r.

Internally, the processor performs arithmetic on integers without taking into account the position of the radix point. It is then possible to interpret the internal value of a fraction by writing $.n_r$ in the form

$$.n_r = r^{-k} \times (d_{-1}d_{-2} \cdots d_{-k}) \tag{3.5}$$

with the value in parentheses treated as an integer value. For example, the number $.1000_2$ (0.5_{10}) can be written as

$$2^{-4} \times (1000.)_2 = 2^{-4} \times 8$$

both of which have the value 0.5, as expected. The scaling factor r^{-k} has the effect of shifting the radix point k positions to the left. Thus $.n_r r^k$ may be used internally as an integer operand and the final result scaled by r^{-k}. For example, the addition of the binary values $.1000$ (0.5_{10}) and $.0100$ (0.25_{10}) can be accomplished as

$$
\begin{array}{r}
1000. \times 2^{-4} \\
+\underline{0100. \times 2^{-4}} \\
1100. \times 2^{-4}.
\end{array}
$$

The machine addition results in 1100_2 or 12 decimal, and the programmer must apply the scale factor to obtain the correct arithmetic result:

$$12 \times 2^{-4} = 0.75.$$

In addition and subtraction, each scaled value must have the same scaling factor. The choice of the scaling factor may cause the radix point to be at the right of the number (integer), at the left (fraction), or anywhere within the number. Thus the value 0.5 in four-digit binary could be written as a fraction

$$.1000_2$$

as an integer with scaling 2^{-4} as

$$(1000.)_2 \times 2^{-4}$$

or as a mixed quantity scaled arbitrarily. For example:

$$(10.00)_2 \times 2^{-2}.$$

When the radix point is fixed for a particular problem and the programmer must take the scaling into account, the system is called a *fixed-point* representation. All integer operations with the MC68000, such as addition or subtraction, assume that the scaling factor is 2^0. Therefore, the binary point is on the right. The importance of Equation 3.5 is that both for analysis and for machine operations, a fractional value may be treated as an integer during all the intermediate steps of a computation. The appropriate scale factor can be applied as the last step when the actual numerical value is desired.

Example 3.2

The first example showed that the largest m-digit integer for unsigned integers has the value $r^m - 1$. Thus the largest 16-bit (binary) integer

$$1111\ 1111\ 1111\ 1111_2$$

has the value

$$2^{16} - 1 = 65,535$$

in decimal representation. The largest 16-bit fraction

$$.1111\ 1111\ 1111\ 1111_2$$

has the decimal value

$$2^{-16} \times 65,535 = 0.99998474.$$

This is obtained by scaling the 16-bit fraction as

$$2^{-16} \times (2^{16} - 1) = 1 - 2^{-16}$$

and performing the arithmetic on a calculator with a sufficient number of decimal places.

EXERCISES

3.1.1.1. Convert the binary number

$$0100.0110_2$$

to decimal.

3.1.1.2. What is the decimal value of

$$1111\ 1111\ .\ 1111\ 1111\ 1111\ 1111_2$$

to five decimal places in the fraction?

3.1.1.3. Compute the decimal value of the following numbers.
  (a) 130_9
  (b) 120_5
  (c) 0.7632_8
  (d) $F00A_{16}$

3.1.1.4. If

$$111_x = 31_{10}$$

what is the base x?

3.1.1.5. Compute the largest integer representable in a 32-bit computer word. Give the answer as a decimal value.

3.1.2 Representation of Signed Numbers

The positive integers, including zero, can be conveniently represented as shown in Section 3.1.1. However, to represent the complete set of integers, which includes positive integers, zero, and negative integers, a notation for negative values is necessary. In ordinary arithmetic, a negative number is represented by prefixing the magnitude (or absolute value) of the number with a minus sign. Thus, -5 is a negative integer with a magnitude of 5. For hand calculations, the use of separate symbols to indicate positive $(+)$ and negative $(-)$ numbers is convenient. Computer arithmetic circuits to manipulate positive and negative integers are also simplified if one of the digits in the positional notation of a number is used to indicate the sign of the integer. Two such possible representations of signed integers are *sign-magnitude* notation and *complement* notation. In both notations, the most significant digit on the left in the positional form of the number indicates the sign. Negative fractions can also be represented in either of these systems. For fractions, the sign digit is written to the left of the radix point.

The binary arithmetic instructions of the MC68000 operate directly only on integers in two's-complement notation if signed integers are considered. Integers in other binary notations or fractions must be manipulated by programs designed for that purpose. The treatment of decimal numbers by the CPU is discussed in Section 3.2 although the mathematical representation of decimal numbers is first introduced in this section.

Sign-Magnitude Representation. The sign-magnitude representation of a number in positional notation has the form

$$N_r = (d_{m-1}d_{m-2} \cdots d_1 d_0)_r \tag{3.6}$$

where the sign of the number is indicated by the most significant (leftmost) digit:

$$d_{m-1} = \begin{cases} 0 & \text{if } N_r \geq 0 \\ r-1 & \text{if } N_r < 0 \end{cases} \tag{3.7}$$

Thus, using the sign-magnitude representation, 1011_2 and 9003 are four-digit negative numbers in the binary and decimal systems, respectively. The magnitude of the number, written $|N_r|$, is

$$|N_r| = \sum_{i=0}^{m-2} d_i r^i \tag{3.8}$$

where only the first $m-1$ digits from the right are considered. The positive version of a number differs from the negative only in the sign digit, and the digits $(d_{m-2}d_{m-3} \cdots d_1 d_0)$ indicate the magnitude. According to the definitions and Equation 3.8, the four-digit number $0011_2 = 3$, and $1011_2 = -3$. The number of digits, including the sign digit, in the representation must be specified or confusion could result.

For example, 1011_2 in an eight-digit representation is assumed to be $0000\ 1011_2$, which has the decimal value 11. For binary values, a negative fraction in sign-magnitude notation has a leading digit of 1 followed by the fractional part. Thus 1.100_2 is the number -0.5.

Example 3.3 _____

The number 16 is written in a 16-bit binary system as

$$0000\ 0000\ 0001\ 0000_2.$$

The number -16 has the sign-magnitude representation

$$1000\ 0000\ 0001\ 0000_2.$$

Complement Representation. Most microprocessors, including the MC68000, have arithmetic instructions that operate on negative numbers represented in a _complement_ number system.[1] In these systems, positive numbers have the same representation as in sign-magnitude notation, but the negative numbers are formed by computing the complement of the number according to the rules of the specific system being used. The two most common complement systems are the _radix-complement_ and the _diminished radix-complement_ systems. The general theory of these systems will be presented first. Then the two's and ten's complements of numbers will be discussed as examples of radix-complement numbers. The one's and nine's complements are examples of diminished radix-complement systems for base 2 and base 10, respectively.

In general form, the radix complement of an m-digit number is computed mathematically as

$$N'_r = r^m - N_r \tag{3.9}$$

where N'_r is the radix complement of the base r number N_r. In the machine computation, only m-digit values can be represented. If any operation produces a result that requires more than m digits, the higher-order digit is ignored. This is taken to be an out-of-range condition. A machine error of this type is called _overflow_.

The two radix-complement systems used with the MC68000 instructions are the two's-complement and the ten's-complement systems. The two's complement of a number N_2 given by Equation 3.9 using 2 as the base r is

$$N'_2 = 2^m - N_2. \tag{3.10}$$

Thus the four-digit two's-complement form of -1 is

$$N'_2 = 2^4 - 1 = 1\ 0000 - 0001 = 1111_2.$$

---

[1] Complement representations have an advantage over sign and magnitude notation because the sign digit does not have to be treated in a special way during addition and subtraction. This simplifies the arithmetic circuits of the CPU somewhat, as is discussed in several references in the Further Reading section of this chapter.

If the number and its complement are added:

$$\begin{array}{ll} \quad 0001 & N \\ +1111 & N' \\ \hline 1\ 0000 \end{array}$$

the result is 0 to four digits, as expected. In a two's-complement system, negative values always have a leading digit of 1 and positive values have 0 as the leading digit. Thus $+4 = 0100_2$ and $-4 = 1100_2$. If addition is performed in a four-digit representation on two positive numbers, the result must not be greater than 7 or overflow occurs. This limits the range of the m-bit positive numbers to the decimal value $2^{m-1} - 1$. In the case of 4-bit numbers, the addition of $4 + 5$ in binary yields

$$\begin{array}{l} \quad 0100 \\ +0101 \\ \hline 1001_2 \end{array}$$

which is a negative number in two's-complement notation. The magnitude as derived from Equation 3.9 would be

$$N_2 = 2^4 - 1001 = 1\ 0000 - 1001 = 0111_2$$

which is +7 in decimal, clearly an error. In the MC68000, an indication is given when such an overflow occurs and the programmer must make provisions in the program for such occurrences.

In the ten's-complement system, the ten's complement of a number N is formed as

$$N' = 10^L - N \tag{3.11}$$

in an L-digit representation. For four digits, -1 is represented as

$$N' = 10^4 - 1 = (10,000 - 1) = 9999.$$

The MC68000 has arithmetic instructions to operate on numbers represented in the ten's-complement system.

The diminished radix complement is computed as

$$\overline{N}_r = r^m - N_r - 1 \tag{3.12}$$

which is one less than the radix-complement value computed by Equation 3.9. The diminished radix complement, or simply complement as it is usually called, is the one's complement for binary values and the nine's complement for decimal numbers. The four-digit decimal value 0002 has the complement

$$\overline{N}_r = (10^4 - 0002) - 1 = 9999 - 0002 = 9997. \tag{3.13}$$

From the complement, the radix complement is formed by adding 1, as a comparison of Equations 3.12 and 3.9 shows.

Table 3.2   Complement Systems

Value	One's	Two's	Value	Nine's	Ten's
7	0111	0111	4999	4999	4999
6	0110	0110	4998	4998	4998
5	0101	0101	•	•	•
4	0100	0100	•	•	•
3	0011	0011	•	•	•
2	0010	0010	0002	0002	0002
1	0001	0001	0001	0001	0001
0	0000	0000	0000	0000	0000
−0	1111	—	−0000	9999	—
−1	1110	1111	−0001	9998	9999
−2	1101	1110	−0002	9997	9998
−3	1100	1101	•	•	•
−4	1011	1100	•	•	•
−5	1010	1011	•	•	•
−6	1001	1010	•	•	•
−7	1000	1001	−4999	5000	5001
−8	—	1000	−5000	—	5000

Table 3.2 lists the radix complement for four-digit binary and decimal numbers. The one's- and nine's-complement values are also presented for comparison. Notice that in the nine's- and ten's-complement notation, the negative values have leading digits in the range 5 through 9. The sign digit is not unique as it is in the case of two's-complement negative values.

Example 3.4 _____

The radix complement of a number is easily computed by complementing each digit [subtracting it from $(r - 1)$] and adding 1 to the result formed from the complemented digits. Thus the value -2 is represented as follows in various four-digit systems:

2's complement: $-2 = (1111 - 0010) + 1 = 1110_2$
10's complement: $-2 = (9999 - 0002) + 1 = 9998$
16's complement: $-2 = (FFFF - 0002) + 1 = FFFE_{16}$

Example 3.5 _____

For a fraction of length k digits, the radix complement is computed by complementing each digit and adding r^{-k} (not 1) to the result. Thus the number

$0101.01_2 = 5.25$

has as its complement the value

$1010.10_2.$

Its radix or two's-complement representation would be

$$
\begin{array}{r}
1010.10 \\
+\quad\underline{.01} \\
1010.11_2
\end{array}
$$

where the value 2^{-2} is added to the one's complement of the number since $k = 2$. Similarly, the fraction 0.01_2 has complement 1.10_2 and two's complement 1.11_2.

Number Range.   The range of integers (or fractions) for a given number system is specified by the smallest and largest values that can be represented. For positive integers represented in m digits, for example, there are r^m possible values with a numerical range of 0 to $r^m - 1$. For the 8-bit representation of a positive binary integer, the range is 0 to $2^8 - 1$, or 0 to 255. The range of signed integers in an m-digit representation still allows r^m values, but one-half of these values are negative numbers.

The maximum positive integer in sign-magnitude, one's-complement, or two's-complement representation is

$$(0111 \ldots 111)_2$$

where $(m - 1)$ 1's are shown. The maximum decimal value is thus $2^{m-1} - 1$, considering the discussion in Example 3.1. In an 8-bit representation, the largest positive number is $2^7 - 1$, or 127. In sign-magnitude notation, the most negative value is

$$(1111 \ldots 111)_2 = -(2^{m-1} - 1).$$

The most negative one's-complement number $(100 \ldots 00)_2$ has the same decimal value. Both of these systems allow a positive and a negative value of zero since the "positive" zero

$$(000 \ldots 000)_2$$

has negative values of $(1000 \ldots 000)_2$ and $(1111 \ldots 111)_2$ in the sign-magnitude and one's-complement notations, respectively. In the two's-complement notation, however, only one value of zero is allowed since the two's complement of the number 0 is the same value to m bits. Since there are a total of 2^m values for each m-bit representation, the two's-complement notation allows one more negative value than the others.

The two's complement of the positive value

$$(011 \ldots 111)_2$$

is the negative number

$$(100 \ldots 001)_2 = -(2^{m-1} - 1)$$

for m bits. The integer

$$(100 \ldots 000)_2$$

must then represent -2^{m-1} with no positive counterpart.

The m-digit ten's complement allows 10^m values in the range

$$-10^m/2 \quad \text{to} \quad 10^m/2 - 1$$

as from -5000 to $+4999$ in the four-digit representation shown in Table 3.2. The positive values are

$$0, 1, 2, \ldots, 499 \ldots 99$$

for m digits and the negative values are represented as

$$999 \ldots 999, 999 \ldots 998, \ldots, 500 \ldots 001, 500 \ldots 000$$

with values $-1, -2, \ldots$. The nine's-complement representation of the negative numbers has a negative 0 and consequently one less nonzero negative value than the ten's-complement notation allows, that is, a range from

$$-10^m/2 + 1 \quad \text{to} \quad 10^m/2 - 1.$$

In this case, the magnitude of the number is restricted so that

$$|N| \leq 10^m/2 - 1$$

which causes the most significant digits of 0, 1, 2, 3, or 4 to indicate a positive number and the digits 5, 6, 7, 8, or 9 to indicate a negative value.

Example 3.6

The largest positive number for an m-digit radix-complement number in base r is

$$(1/2) \times r^m - 1$$

because there are $(1/2) \times r^m$ positive integers, including zero. Thus the two's-complement maximum value is

$$(1/2) \times 2^m - 1 = 2^{m-1} - 1$$

while the ten's-complement number has the maximum value of

$$(1/2) \times 10^m - 1$$

as indicated in the previous discussion. The four-digit binary number allows values up to $+7$ in the two's-complement system, while the four-digit decimal value has a largest positive value of 4999.

Example 3.7

Applying the formulas for the most negative and most positive numbers in m-bit representations gives the following results:

Representation	Most Negative	Most Positive
Sign-magnitude	$-2^{m-1} + 1$	$2^{m-1} - 1$
One's complement	$-2^{m-1} + 1$	$2^{m-1} - 1$
Two's complement	-2^{m-1}	$2^{m-1} - 1$

For a 16-bit representation, the sign-magnitude and one's-complement numbers range from $-32,767$ to $32,767$, and the two's-complement numbers range from $-32,768$ to $32,767$.

If a signed, binary fraction is represented as

$$(b_0.b_{-1} \cdot \cdot \cdot b_{-(k-1)})_2$$

the range is determined by scaling by $2^{-(k-1)}$. For example, the 8-bit fraction in two's-complement representation has the range -1 to $1 - 2^{-7}$. The binary values in this range are

1.000 0000	(-1)
1.000 0001	
•	
•	
•	
0.000 0000	(0)
•	
•	
•	
0.111 1110	
0.111 1111	$(1 - 2^{-7})$.

EXERCISES

3.1.2.1. Find the two's-complement representation of the following numbers.
   (a) -0647_{16} to 16 bits
   (b) -11_{10} to 16 bits
   (c) -00101.110_2 to 8 bits

3.1.2.2. The most negative two's-complement number is $100 \ldots 0_2$ for m bits. What value results when the two's complement of the number is taken?

3.1.2.3. In the two's-complement notation, the sign bit has the weight -2^{m-1} for an m-bit integer. Determine the procedure to extend the m-bit number to $2m$ bits for (a) a positive number and (b) a negative number. This is called *sign extension.*

3.1.2.4. Represent the given numbers in the notation specified.
   (a) Nine's complement of 653.72 with five digits
   (b) -223_{16} in sign-magnitude form with four hexadecimal digits
   (c) $-3/8$ in one's-complement form with 8 bits, including the sign bit

3.1.2.5. Determine the range of numbers for the sign-magnitude, one's-complement, and two's-complement forms for an m-bit representation if
   (a) $m = 8$
   (b) $m = 16$
   (c) $m = 32$

3.1.2.6. If the largest positive number in a four-digit, ten's-complement representation is limited to 999 (i.e., three digits), determine the corresponding range and representation of the negative numbers. Note that negative numbers always begin with 9 as the sign digit when the magnitude of the numbers is limited in this way.

3.1.3 Conversions Between Representations

The number systems of most interest to computer users are the binary, octal, decimal, and hexadecimal systems. Although the internal machine representation in microcomputers is binary, the other representations are important for the convenience of the user. In this regard, the conversions of numbers in other bases to decimal, and vice versa, are frequently required.

Conversion between arbitrary number bases is sometimes necessary, although in computer work, the conversions between binary, octal, and hexadecimal systems are of greatest importance. Fortunately, conversions between these bases are straightforward.

Conversion to Decimal for Positive Numbers. The number N_r in base r can be represented in decimal as

$$N.n = D_{m-1}r^{m-1} + \cdots + D_0 + D_{-1}r^{-1} + \cdots + D_{-k}r^{-k} \qquad (3.14)$$

where the D_i are the equivalent values in the base 10 of the digits in base r. The number is converted by multiplying each digit by the appropriate power of r and adding each result to the sum. The number is designated here as $N.n$ to emphasize the fact that it has an integral as well as a fractional part.

Example 3.8 _____

To convert $11\ 1110_2$ to decimal, the value is computed as a series from Equation 3.2 or 3.14 with the result

$$N = 1 \times 2^5 + 1 \times 2^4 + 1 \times 2^3 + 1 \times 2^2 + 1 \times 2^1 + 0$$
$$= 32 + 16 + 8 + 4 + 2$$
$$= 62.$$

Example 3.9 _____

The value of 0.502_8 in decimal is

$$0.n = 5 \times 8^{-1} + 0 + 2 \times 8^{-3}$$
$$= 0.6250 + 0.003906250$$
$$= 0.62890625_{10}$$

as determined by Equation 3.4 or 3.14.

_____

Conversions from Decimal to Any Number Base. To convert a decimal number by hand to a number in another base, it is convenient to work in the decimal system with the series representation of the number. Conversion of a positive integer is accomplished by repeated division by the new radix using successive remainders as digits in the new system. A fraction is converted by repeated multiplication by the radix, with the resulting integer parts of the products taken as digits of the result.

Example 3.10 _____

To convert 3964_{10} to octal, the number is repeatedly divided by 8, as follows:

$$3964/8 = 495 + (4/8)$$
$$495/8 = 61 + (7/8)$$
$$61/8 = 7 + (5/8)$$
$$7/8 = 0 + (7/8).$$

The remainders of each division, represented by the numerators of each fraction, are digits in the resulting answer. The order of these digits is the reverse of the order in which they were obtained. Thus, in the example above:

$$3964_{10} = 7574_8.$$

Example 3.11 _____

The number 0.78125_{10} is converted to a hexadecimal fraction as follows:

$$0.78125 \times 16 = 12.0 + 0.5$$
$$0.5 \times 16 = 8.0 + 0.$$

The result is therefore $0.78125_{10} = .C8_{16}$.

To understand the theory of these conversions, write Equation 3.14 in the form

$$N = ((\cdot \cdot \cdot((d_{m-1}r + d_{m-2})r + d_{m-3})r + \cdot \cdot \cdot + d_1)r + d_0)$$

for the integer part and set it equal to the decimal value it represents. Then divide N by the base desired. Using 8 as the base in Example 3.10, the first remainder is the octal value d_0. Continuing to divide the whole numbers (quotients) by the base successively yields d_0, d_1, d_2, . . . , d_{m-1}, in that order. Try the fractional part of Equation 3.14 in a similar manner with negative powers of r to see how the value in Example 3.11 was computed.

Conversion of a Number from Any Base to Another.   A number written in base r_1 can be converted to a number in base r_2 by performing the arithmetic operations in a base other than decimal. For human computation a more acceptable method is to convert the selected number to decimal from base r_1 and then convert the result from decimal to base r_2.

Example 3.12 _____

Converting 112_3 to base 5 is accomplished by converting 112_3 to decimal:

$$1 \times 3^2 + 1 \times 3^1 + 2 = 9 + 3 + 2 = 14_{10}.$$

Then the decimal number 14 is converted to base 5 in the form

$$14/5 = 2 + (4/5)$$
$$2/5 = 0 + (2/5)$$

or $14_{10} = 24_5$. Thus $112_3 = 24_5$.

Conversion of Positive Numbers with Bases That Are Powers of 2. If the relationship between a number base r_1 and another base r_2 is of the form

$$r_2 = r_1^L$$

where L is a positive or negative integer, conversion between the bases is particularly simple using positional notation. In particular, because the binary, octal, and hexadecimal bases are related as

$$16 = 2^4$$
$$8 = 2^3$$

the conversion from binary to octal or binary to hexadecimal requires only the grouping of the binary digits by threes or fours, respectively. The conversion from octal to binary or hexadecimal to binary requires that each octal or hexadecimal digit be replaced by its binary equivalent.

Conversion between octal and hexadecimal numbers is best accomplished by using the binary representation as an intermediate step, because the bases 8 and 16 are not related.

Example 3.13 ——

Conversion of the binary number $1011\ 0111.0010\ 1_2$ to octal requires grouping the digits by threes, starting with the least significant or rightmost binary digit for the integer portion but starting with the most significant or leftmost digit for the fractional portion. Thus the conversion proceeds as

$$(010)\ (110)\ (111).\ (001)\ (010) = 267.12_8$$

where extra binary digits of zero were added at each end to yield legitimate octal digits before conversion.

——

Conversion of Negative Numbers from Binary to Decimal. The conversion of a positive binary number to decimal is easily achieved by the power series method shown previously. The conversion of negative numbers in one's-complement or two's-complement notation can also be achieved in this manner when the sign bit is given the proper decimal value or weight. As an example, consider the following negative numbers in 8-bit two's-complement representation and their decimal equivalents:

$$1111\ 1111_2 = -1$$
$$1111\ 1110_2 = -2$$
$$\cdot$$
$$\cdot$$
$$\cdot$$
$$1000\ 0001_2 = -127$$
$$1000\ 0000_2 = -128.$$

By associating the leading digit with -2^7 (-128) and adding the positive positional value of the remaining digits, the proper decimal value results. Thus the decimal value of an 8-bit negative number in two's-complement notation is

$$-2^7 + d_6 \times 2^6 + d_5 \times 2^5 + \cdot \cdot \cdot + d_0$$

when the negative number in its positional form is

$$(1d_6d_5d_4 \cdot \cdot \cdot d_0)_2.$$

By inspecting the positive values, the general case for both positive and negative two's-complement numbers may be derived to compute the decimal equivalent as

$$N = -d_{m-1} \times 2^{m-1} + \sum_{i=0}^{m-2} d_i \times 2^i \tag{3.15}$$

with $d_{m-1} = 0$ for a positive value or $d_{m-1} = 1$ when the number is negative. From this equation, the 8-bit number $1000\ 0010_2$ has the decimal value

$$N = -2^7 + 0 + \cdot \cdot \cdot + 1 \times 2^1 + 0 = -126$$

with sign bit $d_7 = 1$. The number $0000\ 0010_2$ with $d_7 = 0$ has the value

$$-0 \times 2^7 + 0 + \cdot \cdot \cdot + 1 \times 2^1 + 0 = +2.$$

In essence, Equation 3.15 represents a compact notation for computation of the decimal value of an m-bit number in the two's-complement system. The decimal weights of the leading digit for one's-complement and two's-complement integers and fractions are given in Table 3.3. The magnitude of a sign-magnitude number is simply multiplied by +1 or −1 according to its sign.

Example 3.14

(a) The integer $1000\ 0011_2$ in sign-magnitude notation has the value

$$(-1)\ (0 \times 2^6 + \cdot \cdot \cdot + 1 \times 2^1 + 1) = -3_{10}$$

because the magnitude is multiplied by −1.

(b) The number $1111\ 1001_2$ in one's-complement notation has the value

$$(1 - 2^7) + (1 \times 2^6 + 1 \times 2^5 + 1 \times 2^4 + 1 \times 2^3 + 1)$$
$$= -127 + 121 = -6_{10}$$

using the weight for the leading bit shown in Table 3.3.

(c) The two's-complement number $1111\ 1001_2$ has the value

$$-2^7 + (1 \times 2^6 + 1 \times 2^5 + 1 \times 2^4 + 1 \times 2^3 + 1)$$
$$= -128 + 121 = -7_{10}.$$

Table 3.3 Values of the Sign Bit

	Weight in Decimal	
Representation	Integer (m bits)	Fraction (k bits with sign)
One's complement	$1 - 2^{m-1}$	$2^{-(k-1)} - 1$
Two's complement	-2^{m-1}	-1

Example 3.15

(a) The fraction $1.001\ 0000_2$ in sign-magnitude notation has the value

$$(-1)\ (0 \times 2^{-1} + 0 \times 2^{-2} + 1 \times 2^{-3}) = -0.125_{10}.$$

(b) The fraction $1.100\ 1111_2$ in one's-complement notation has the value

$$(2^{-7} - 1) + 1 \times 2^{-1} + 1 \times 2^{-4} + 1 \times 2^{-5} + 1 \times 2^{-6} + 2^{-7} \times 1)$$
$$= -0.375_{10}.$$

EXERCISES

3.1.3.1. Convert the following numbers as indicated:
    (a) 1024_{10} to binary
    (b) 53000_{10} to hexadecimal
    (c) $FFFF\ FFFF_{16}$ to decimal
    (d) 35_{10} to base 5

3.1.3.2. Convert the repeating octal fraction $(0.333\ .\ .\ .)_8$ to decimal.

3.1.3.3. Show that adding 1 to the complement form of the positive number N yields the radix-complement representation $r^m - [N]$ for an m-digit number.

3.1.3.4. Convert the fraction $1.111\ 1111_2$ in two's-complement notation to its decimal value.

3.1.3.5. Using two's-complement representation, represent numbers in the range -2, $-1\frac{3}{4}$, . . . , $1\frac{1}{2}$, $1\frac{3}{4}$.

3.2 BINARY-CODED DECIMAL

Many microcomputers provide instructions to perform arithmetic operations on data considered as decimal numbers. This is convenient for business processing and in representing data that are inherently decimal in nature. Thumb-wheel switches, for example, may present an output as a decimal digit encoded in binary to represent a selected digit on the switch. Many displays are designed to receive encoded decimal digits and display the result in decimal.

One decimal coding system is the *binary-coded decimal* or BCD system. In the BCD system, the first 10 binary numbers correspond to decimal digits. This is sometimes called the "natural" binary-coded decimal system.

This section discusses the binary-coded decimal system and various conversion operations with BCD numbers. The use of ten's-complement notation is convenient to represent negative BCD values, although other representations are possible. Only ten's complement is discussed here because it is the method assumed for addition and subtraction of signed decimal values performed by the MC68000 processor.

3.2.1 BCD Representation of Positive Integers

In many applications, particularly those involving financial transactions, a true representation of decimal numbers in a machine that operates on binary digits is desirable. Because any decimal digit can be represented by four binary digits, it is natural to

Table 3.4 Binary-Coded Decimal Values

BCD	Binary
0	0000
1	0001
2	0010
3	0011
4	0100
5	0101
6	0110
7	0111
8	1000
9	1001

select a binary code in which four binary digits are used for each decimal digit. Such a code is shown in Table 3.4, which lists the binary-coded decimal (BCD) representation. The possible binary values 1010_2 through 1111_2 are not used since the decimal values 10 through 15 require two BCD digits for their representation.

Each decimal digit has the value

$$D_i = b_{i3} \times 2^3 + b_{i2} \times 2^2 + b_{i1} \times 2^1 + b_{i0} \tag{3.16}$$

where b_{ij} is the jth binary digit in the representation of the ith decimal digit. The value of the L-digit BCD number is calculated in decimal as

$$N = D_{L-1} \times 10^{L-1} + D_{L-2} \times 10^{L-2} + \cdot \cdot \cdot + D_0 \tag{3.17}$$

where each D_i is formed as shown in Equation 3.16. The decimal value 95, for example, would be coded into binary as

$$1001\ 0101_2$$

and it is stored in this form. Using Equation 3.16, we have

$$D_0 = 1 \times 2^2 + 1 = 5$$

and

$$D_1 = 1 \times 2^3 + 1 = 9.$$

Numbers in BCD can be added or subtracted by MC68000 instructions that perform decimal arithmetic. Therefore, the programmer does not need to be concerned with the internal representation. It is only when conversions between BCD numbers and other representations are required that the internal format must be considered.

Example 3.16 ———————————————————————

The binary number

$$0001\ 0111\ 0011\ 1001$$

has the BCD value

$$N = 1 \times 10^3 + 7 \times 10^2 + 3 \times 10^1 + 9 = 1739.$$

The 16 binary digits encoded as positive BCD numbers have a range of only $0 \leq N \leq 9999$.

Example 3.17 _____

Microprocessors that perform arithmetic operations on 8-bit (byte) and longer operands may allow such operations on "packed" BCD integers, as the MC68000 does. In this representation, two BCD digits are contained in each 8-bit value rather than storing each BCD digit in a separate byte (called *unpacked notation*).

The BCD number 3475 may be treated for the purposes of machine calculation as either packed BCD or unpacked as follows:

Decimal Value		Memory (binary)
Packed	34	0011 0100
	75	0111 0101
Unpacked	03	0000 0011
	04	0000 0100
	07	0000 0111
	05	0000 0101

In the unpacked representation, the single digit is shown in a byte location with a leading 0 since the MC68000 and most other processors address memory locations containing 8 bits. The unpacked format is typically used when algorithms to perform multiplication or division of BCD numbers are employed.

EXERCISES

3.2.1.1. Show the internal machine representation (binary) of the following positive numbers in packed BCD format.
   (a) 07
   (b) 13
   (c) 99

3.2.1.2. Determine the decimal values of the following positive numbers coded in BCD format.
   (a) $0001\ 1001\ 0111\ 0000_2$
   (b) $0001\ 1111_2$

3.2.1.3. Assuming that the representation is packed BCD, compute the decimal range for the positive BCD representation of numbers using
   (a) 8 bits
   (b) 16 bits
   (c) 32 bits

3.2.1.4. Describe the test required to assure that an out-of-range condition is detected when two positive BCD integers are added (subtracted). Assume that the maximum length is L decimal digits for each BCD integer. The MC68000 has built-in hardware (condition codes) for detecting these conditions.

3.2.2 Conversion between BCD and Binary

When the machine representations as binary sequences are required, arithmetic in base 2 instead of base 10 to convert between BCD values and binary numbers, or vice versa, is convenient with the MC68000 because it has multiply and divide instructions to perform the binary arithmetic. Thus the conversion of the positive BCD number

$$(D_{L-1}D_{L-2} \cdots D_0)$$

can be accomplished by first writing Equation 3.17 in the form:

$$N = (\cdots ((D_{L-1}) \times 10 + D_{L-2}) \times 10 + \cdots + D_1) \times 10 + D_0.$$

Converting all the digits to binary yields an equation useful for machine implementation. The binary value of the BCD number is then

$$N_2 = (\cdots ((D_{L-1} \times 1010_2 + D_{L-2}) \times 1010_2 + \cdots + D_1) \times 1010_2 + D_0$$

where 1010_2 is 10 decimal and the digits D_i are used in their 4-bit binary form. First the most significant digit is multiplied by 1010_2, then the next most significant digit is added and the sum multiplied by 1010_2, and so on, until the last digit D_0 is added. The result is an m-digit binary representation of the BCD number. Programming the conversion equation is simple using the MC68000 instruction set. BCD values are sometimes converted to binary for machine processing since the CPU has an extensive set of instructions that operate on binary numbers but relatively few to manipulate BCD integers.

When it is necessary to convert binary numbers to BCD representation, the conversion is accomplished by repeated division by 10 (binary 1010). Each remainder is a BCD digit starting with the low-order digit. This is frequently done before internal binary numbers are output for display as a decimal value.

Example 3.18

Using base 2 arithmetic, 99 in BCD is converted to binary as

$$N_2 = 1001 \times 1010 + 1001 = 0110\ 0011.$$

Example 3.19

The binary number $0010\ 0100_2$ is converted to the machine representation of the BCD equivalent using binary arithmetic for division as follows:

$$\frac{0010\ 0100}{1010} = 0011 + (0110)$$

$$\frac{0011}{1010} = 0 + (0011).$$

Here the remainders in parentheses represent the binary sequence

0011 0110

which is interpreted as 36 as a decimal equivalent.

EXERCISES

3.2.2.1. Convert $1000\ 0000_2$ to BCD by repeatedly dividing by 1010_2 in binary.

3.2.2.2. Convert the BCD number 509 to binary using both base 10 and base 2 arithmetic.

3.2.2.3. Consider the multiplication of numbers in BCD representation. Devise an algorithm to multiply a multidigit BCD value by a single-digit multiplier, assuming that only binary multiplication is possible and the digits are in unpacked form. Further assume that a BCD addition instruction is available to sum the partial results.

3.2.3 Negative BCD Integers

The MC68000 has instructions to perform arithmetic on BCD numbers represented in ten's-complement notation. As described in Section 3.1, the ten's complement of a decimal number N is formed by

$$N' = 10^L - N \tag{3.18}$$

when L digits are used to represent the number. For hand calculation, the ten's complement is easily formed by complementing the number digit by digit (nine's complement) and adding 1 to the result.

Example 3.20 _____

The ten's complement of 1319 in a five-digit representation is

$$N' = 100,000 - 1319 = (99,999 - 1319) + 1 = 98,681.$$

The BCD representation in memory would be

$1001\ 1000\ 0110\ 1000\ 0001_2$.

Example 3.21 _____

The ten's complement of a number can also be formed by subtracting it from 0 and ignoring the high-order borrow. Thus the ten's complement of 98,681 is

$$0 - 98,681 = 01319$$

as shown. The MC68000 instruction NBCD (negate decimal) performs this operation to form the ten's complement of a number.

EXERCISES

3.2.3.1. Determine the machine representation (binary) of the following signed BCD numbers with a word length of 16 bits.

    (a) 124

> **(b)** -1
> **(c)** -1000
> **(d)** 5024

3.2.3.2. What is the range of a signed BCD number that can be represented by m binary digits for
(a) $m = 8$
(b) $m = 16$
(c) $m = 32$
when the positive value is restricted to the maximum value of $10^L - 1$ for L decimal digits?

3.2.3.3. Show that the nine's complement of a BCD digit can be formed by adding 6 and then forming the one's complement of the result in binary notation.

3.3 FLOATING-POINT REPRESENTATION

The representation for numbers that we considered previously assumed that the radix point was located in a fixed position, yielding either an integer or a fraction as the interpretation of the internal machine representation. The programmer's responsibility would be to scale numerical operands to fit within a selected word length and then unscale the results to obtain the correct values. Of course, the radix point is not actually stored with the number, but its position must be remembered by the programmer. This method of representation is called *fixed-point*.

In practice, the machine value is limited to a finite range that is determined by the number of binary digits used in the representation. For a 32-bit word, the range of signed fixed-point integers is about $+2^{31}$ or $+10^{11}$. Thus the limited range of fixed-point notation is a drawback for certain applications. Furthermore, arithmetic units operating on fixed-point numbers generally have no capability of rounding results. As discussed in several references in the Further Reading section for this chapter, this limits the usefulness of fixed-point notation in scientific computing.

To overcome many of the limitations of fixed-point notation, a notation that is the counterpart of scientific notation is used for numbers in digital systems. The *floating-point* notation represents a number as a fractional part times a selected base raised to a power. In the machine representation, only the fractional part and the value of the exponent are stored. The decimal equivalent is written as

$$N.n = f \times r^e \tag{3.19}$$

where f is the fraction or *mantissa* and e is a positive or negative integer called the *exponent*. The choice for the base is usually base 2, although base 16 is sometimes used.

A number of choices are presented to the designer of a floating-point format. This applies whether the arithmetic operations are carried out by the CPU or its coprocessor directly or by a software package containing routines for floating-point arithmetic. The number of formats is bewildering and only recently has an attempt been made to standardize floating-point arithmetic for computers.

The IEEE standard has been adopted by Motorola for a number of their products. This IEEE standard describes precisely the data formats and other aspects of floating-point arithmetic required to provide consistent operation of a program even when it is executed on different computer systems.

3.3.1 Floating-Point Formats

The typical floating-point format stores the fraction and the exponent together in an m-bit representation. The choice for a fixed-length floating-point format is commonly 32 or 64 bits, referred to as single precision and double precision, respectively. Extended formats with $m > 64$ are occasionally used when greater range or precision is required.

Once the length of the floating-point representation is chosen, a number of choices for both the length and the format of the fraction and exponent are possible. Because either or both could be negative as well as positive, a signed fraction and a signed exponent are required. Finally, the interpretation of the bits within the floating-point representation depends on the placement of the fraction and exponent.

Many floating-point formats employ a sign-magnitude representation for the fraction. The most significant bit of the word is reserved for the sign, and this facilitates testing for a positive or negative number. The fraction is generally normalized to yield as many significant digits as possible. Thus, in a base r system, the most significant digit is in the leftmost position in the fraction. For nonzero numbers in the binary system, the leftmost digit will be a 1. As the arithmetic unit or the program shifts the digits in the fraction during arithmetic operations, the exponent is adjusted accordingly. When normalized as a base 2 value, the magnitude of the fraction is

$$0.5 \leq |f| < 1 \tag{3.20}$$

unless the number is zero. The number of digits reserved for the fraction represents a compromise between the precision of the fraction and the range of the exponent. A typical single-precision format (32 bits) might contain an 8-bit exponent and a 23-bit fraction, excluding sign.

An exponent could be represented in two's complement or any other notation that allows signed values. A different alternative, which permits the exponent to be represented internally as a positive number only, is to add an offset value. This value is often called an *excess*. For this format, positive bias or offset is added to all exponents such that the number is read as

$$N.n = fr^{e'-N_b} \tag{3.21}$$

where e' is the actual value of the stored exponent and N_b is the positive offset. For an L-bit exponent in a binary base, the positive number added is usually of the form

$$N_b = 2^{L-1} \tag{3.22}$$

although the IEEE standard format discussed later uses a value of $N_b - 1$.

Example 3.22 _____

One IBM floating-point format has the following representation for a 32-bit word:

    (a) Sign as the most significant bit (leftmost bit)
    (b) Next 7 bits as exponent with excess-64 or 40_{16} with radix 16
    (c) Next 24 bits as fraction in base 16.

Thus, +1.0 is represented as

$$(0.1)_{16} \times 16^1$$

for which the stored exponent becomes 41_{16}. The internal representation is

$$4110\ 0000_{16}.$$

Example 3.23 _____

One PDP-11 (Digital Equipment Corporation) floating-point format uses the following conventions in a 32-bit word:

    (a) Sign as the most significant bit (leftmost bit).
    (b) Next 8 bits as exponent in excess-128 notation with radix 2.
    (c) Next 23 bits as fraction. These 23 bits are derived from a 24-bit fraction always normalized (i.e., the leftmost bit will be a 1 in a nonzero number). The most significant bit of the normalized fraction is not stored since it is always 1.

The representation of +12 is thus

$$0.11_2 \times 2^{132-128}$$

and is represented internally as

$$0\ 1000\ 0100\ 1000\ 0000\ 0000\ 0000\ 0000\ 000_2.$$

Notice that the leading bit of the fraction is a 1 and it is not stored with the floating-point number. This has been termed a *hidden bit* and would be restored by a floating-point hardware processor when the value is processed.

EXERCISES

3.3.1.1. Discuss the various factors that influence the choice of the exponent length, mantissa length, and choice of radix in a floating-point number. Compute the various ranges for a choice of an L-bit exponent in excess notation, a k-bit fraction, and a total length of m bits.

3.3.1.2. Express 1/32 in binary floating-point format using an 8-bit excess-128 exponent and a 24-bit fraction with the leading bit implied. The order in the word from left to right is sign, exponent, and then fraction (PDP-11).

3.3.1.3. Convert the numbers indicated to a 32-bit floating-point representation with the characteristics: the leading bit (bit 31) is the sign of the number; the next 9 bits are the exponent in excess-256 notation and following bits are 22 bits of fraction; and a negative number is represented as the integer's two's complement of the positive floating-point number.

(a) +16.0

(b) −1.0

3.3.2 Standard Floating-Point Format

Although the Motorola MC68000 processor does not provide floating-point instructions, many Motorola products support the standard floating-point format proposed by the IEEE. These products include software routines, routines in read-only memory, and a coprocessor chip to support floating-point arithmetic in the MC68020 systems. (The MC68040 and MC68060 have a floating-point unit on the chip.)

The basic format allows a floating-point number to be represented in single or 32-bit format as

$$N.n = (-1)^S 2^{e'-127}(1.f) \tag{3.23}$$

where S is the sign bit, e' is the biased exponent, and f is the fraction stored normalized without the leading 1. Internally, the exponent is 8 bits in length and the stored fraction is 23 bits long. A double-precision format allows a 64-bit representation with an 11-bit exponent and a 52-bit fraction.

Various features of these floating-point formats are presented in Tables 3.5 and 3.6. Other extended-precision formats are presented in the references in the Further Reading section of this chapter.

Example 3.24 _____

The numbers +1.0, +3.0, and −1.0 have the following representations in the standard 32-bit format:

(a) Since $1.0 = 1.0 \times 2^0$, the internal value is $3F80\ 0000_{16}$.

(b) Since $3.0 = 1.5 \times 2^1$, the exponent is 128 and the fraction is 1.100_2 without the leading 1. Thus the internal value is $4040\ 0000_{16}$.

(c) Since $-1 = -1.0 \times 2^0$, the result requires the sign bit to be 1 in the representation $BF80\ 0000_{16}$.

_____

Table 3.5 IEEE Standard Floating-Point Notation

	Single	Double
Length in bits		
Sign	1	1
Exponent	8	11
Fraction	23 + (1)	52 + (1)
Total	$m = 32 + (1)$	$m = 64 + (1)$
Exponent		
Max e'	255	2047
Min e'	0	0
Bias	127	1023

Note: Fractions are always normalized and the leading 1 (hidden bit) is not stored.

Table 3.6   Internal Format by Bit Number

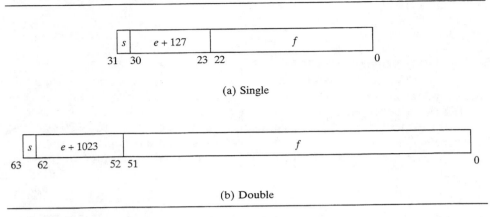

(a) Single

(b) Double

EXERCISES

3.3.2.1. Write the internal machine representation for the following numbers in the standard floating-point single-precision format.

    (a) 0.5

    (b) −0.5

    (c) 2^{-126}

3.3.2.2. What is the decimal value of the largest positive number that may be represented in single-precision standard format if the biased exponent is limited to a maximum of 254 when a valid number is being represented? (The value 255 is reserved for special operands.)

3.3.2.3. Express the following numbers in the internal representation using standard double-precision floating-point format.

    (a) 7.0

    (b) −30

3.4   ASCII REPRESENTATION OF ALPHANUMERIC CHARACTERS

Because m binary digits can represent 2^m distinct states, it is possible to assign a meaning to each possible combination of the m digits to produce a code that represents alphabetic, numeric, or other information. One of the major uses of codes is to allow human-readable input or output data to be manipulated internally by the computer. These internal binary representations are seldom desired as output except possibly for debugging purposes. Thus most computer systems will have routines (or hardware) to convert internal binary codes to readable form, and vice versa.

The code used by Motorola to represent alphanumeric characters is the American Standard Code for Information Interchange (ASCII) code, which is given in Appendix I of this text. The 7-bit codes shown in the appendix are the hexadecimal values of the ASCII characters as the values would be stored in memory. For example, the ASCII value "1" would be stored as 31_{16} or $0011\ 0001_2$.

Numbers, letters, and special characters recognized by the assembler and other Motorola system programs are stored internally in the ASCII code as 8-bit values in which one byte is used to represent one character. On output to devices such as CRT terminals or line printers, the ASCII code is transmitted unchanged. If the internal values are in binary or BCD, they must be converted to ASCII format before output for external devices that require ASCII. References in the Further Reading section give a number of examples of such conversions useful to the assembly language programmer. Programs for this purpose are given in Chapter 7 of this book.

A number of other codes are available for computer applications, each with special characteristics and advantages. Several of the references for this chapter discuss codes in more detail. Mackenzie, in particular, gives a comprehensive discussion of many of the commonly used codes and a complete presentation for the ASCII code.

Example 3.25 _____

The ASCII character string 'INPUT' has the following internal representation:

49 4E 50 55 54

where each two digits (in hexadecimal) represent one alphabetic character.

_____

EXERCISES

3.4.1. How many bytes of memory are required to store the text of a textbook with 100,000 words if the text is stored in ASCII code? Assume five characters per word. If an MC68000 system memory can store 2^{24} bytes, what percentage of the memory is used by the text?

3.4.2. Convert the following text into ASCII code (internal hexadecimal representation):

"THE MOTOROLA MC68000"

3.4.3. Show the machine representation of the number 255 in the following ways:
  (a) Binary
  (b) Binary-coded decimal
  (c) ASCII

3.4.4. Define the method and convert the following as directed:
  (a) 45 (BCD) to binary
  (b) 45 (BCD) to ASCII
  (c) −45 (BCD in ten's-complement notation) to ASCII assuming a three-digit BCD number with sign

3.4.5. Can a single binary variable be used to represent the Morse code?

FURTHER READING

The books by Stein and Munro, Hwang, and Sterbenz listed below present a highly mathematical view of number representations. Sterbenz concentrates on floating-point notation. The articles in *IEEE Computer* are recommended for information on the IEEE standard format.

Weller presents a number of conversion techniques between data types of use to the assembly language programmer. The design of computer hardware and the use of various number representations for arithmetic operations is discussed by Abd-Alla and Meltzer. Mackenzie discusses the details of a large number of codes for representation of characters.

Abd-Alla, Abd-Elfattah M., and Arnold C. Meltzer. *Principles of Digital Computer Design,* Vol. 1 (Englewood Cliffs, N.J.: Prentice Hall, 1976). (Chapter 4 presents a number of codes.)

Coonen, Jerome T. "An Implementation Guide to a Proposal Standard for Floating-Point Arithmetic" *IEEE Computer,* 13, no. 1 (January 1980): 68–79.

Hwang, Kai. *Computer Arithmetic.* (New York: Wiley, 1979).

IEEE Computer 14, no. 3 (March 1981). (Several articles in this issue discuss the proposed floating point standard.)

Mackenzie, Charles E. *Coded Character Sets: History and Development.* (Reading, Mass.: Addison-Wesley, 1980).

Stein, Marvin L., and William D. Munro. *Introduction to Machine Arithmetic.* (Reading, Mass.: Addison-Wesley, 1971).

Sterbenz, Pat H. *Floating-Point Computation.* (Englewood Cliffs, N.J.: Prentice Hall, 1974).

Weller, Walter J. *Assembly Language Programming for Small Computers.* (Lexington, Mass.: Lexington Books, 1975).

CHAPTER 4

Introduction to
the M68000 Family

The characteristics of the MC68000 as a programmable processor are introduced in this chapter. These characteristics are of interest primarily to the programmer. For the most part, the terminology follows Motorola's literature. The discussion applies to the MC68000 and other members of the M68000 family, including the MC68008, MC68HC000, MC68HC001, MC68EC000, and MC68010, except where differences are noted.

Assembly language programmers are interested in the characteristics of the MC68000 as a programmable processor. The descriptions in Table 4.1 of the processor's register set, instruction set, addressing modes, and other features are important for this purpose. To a large extent, these characteristics define the power and the flexibility of the processor in meeting the requirements of a programming application. This chapter is organized into sections that generally follow the order of topics listed in Table 4.1. The features in Table 4.1 are called *architectural features* of the processor if they effect only software design. Programs that are compatible will execute on various processors that share these features whatever the differences in hardware implementation.

This chapter first introduces the M68000 family of processors with emphasis on their general features. The design of the CPU as a Complex Instruction Set Computer (CISC) is also discussed. Subsequent sections then describe the register set, instruction set, and addressing modes, all of which define the programming model of the processor. The machine language of the MC68000 is also presented to illustrate the correspondence between the bit patterns of the instructions and the operation of the processor. The chapter concludes with a presentation of the organization of memory in an MC68000 system. Many of the features covered in this chapter are summarized briefly in the Appendices.

4.1 THE PROCESSORS

One advantage to studying the MC68000 in detail is that its characteristics form the basis for the entire M68000 family of CPUs. As previously discussed in Chapter 1, the CPUs that follow the MC68000 include the CPU32 of Motorola's microcontrollers,

Table 4.1 Features of the M68000
Family CPU

Feature	Use
Register set	Data, address, and special purpose registers for on-chip storage
Instruction set	Fifty-six basic instructions for data movement, arithmetic and logical operations, and program control
Addressing modes	Fourteen modes to designate operands in memory
Machine language	Defines the binary format of the instruction set
Memory organization	Specifies the format of instructions and operands in memory

the CPU of the integrated processors, and other processors as powerful as the MC68060.

The 32-bit CPUs contain the basic registers, instruction set and addressing modes of the MC68000 but add other special registers, new instructions, and enhanced addressing modes. One design criterion for the advanced processors required that they execute user mode programs of the MC68000 without modification.

This section presents the M68000 family of processors that are essentially identical to the MC68000 from the programmer's point of view. All the 16-bit family processors are listed in Table 4.2 with the differences shown. Recall that reference to the *M68000 family* refers to these processors. The *68000* family includes both these 16-bit processors and the more advanced 32-bit processors and microcontrollers with the CPU32.

The MC68000 was the first implementation of the family architecture. This CPU has a 16-bit data bus and a 24-bit addressing capability. Internally, the CPU has 32-bit address and data busses. The MC68008 has the basic instruction set, but with an 8-bit data bus and a limited addressing range.

Table 4.2 M68000 Family
of Processors

Processor	Features
MC68000	First member of the 16-bit M68000 family
MC68HC000	Low-power version
MC68HC001	Selectable 8-bit or 16-bit data bus
MC68EC000	Embedded controller version with 8-bit or 16-bit data bus
MC68008	8-bit data bus and reduced addressing capability
MC68010	Enhanced version

Other processors in the family provide features such as a low-power requirement compared to the MC68000. Several versions offer a choice of an 8-bit or 16-bit data bus. However, the architecture of these processors is identical and independent of these special features. For example, the same program will execute on processors with different CPU speeds of operation and external bus structure, even if the processors are fabricated with a different semiconductor technology.

The MC68000 uses the Complementary Metal Oxide Semiconductor (CMOS) technology. An MC68HC000 is fabricated with the high-speed CMOS process (HCMOS), which scales down the elements compared to the standard MOS process, and thus increases the speed and reduces the power consumption for each transistor in the CPU. As a historical note, the original MC68000 was said to contain approximately 68,000 transistors.

The MC68010 has a few extra instructions and other features not available in the other processors. These differences are explained in the appropriate section of this textbook.

4.1.1 Processor Design

As with many microprocessors, the circuit-level operation of the MC68000 is controlled by a microprogrammed sequence of instructions stored in two ROM sections within the CPU chip. As machine language instructions from external memory are fetched and decoded, they, in turn, cause the execution of a microprogram. This microprogram controls all activity on the external signal lines and all the data transfers or operations within the CPU during the execution of that instruction. Because the microprogram cannot be modified except by creating a new chip with a different microcode, the user is rarely concerned with the details of the processor operations at this level. However, understanding the microcode is the only way to determine exactly what the processor is doing in response to an instruction.

One aspect of the microprogram that might affect the design of a complex system is the processor's prefetch operation. When one instruction is fetched by the CPU, the word in memory following the instruction word is also fetched as the instruction is being decoded and executed. This means that the processor signal lines are addressing and reading from a new location before the previous operation is finished. If hardware is being designed or debugged based on the cycle-by-cycle operation of the processor, it may be necessary to consider the microprogrammed sequence.

References in the Further Reading section discuss the prefetch feature and other aspects of the microprogram for the MC68000. Also, a brief discussion of the prefetch is presented in the Appendix. The exact details of the microcoded sequence for a particular instruction can be obtained by contacting Motorola.

Example 4.1 ───

According to Motorola designers, the advanced processors, such as the MC68020, use microprogrammed CPUs to simplify design, modification, and testing of the product line. A two-level microcode was chosen as a compromise between execution efficiency and the size of the control store (ROM) on the CPU chip. In this scheme each machine language instruction is emulated by a sequence of micro-instructions. These micro-instructions in turn address more detailed "nano-instructions" that directly control the execution unit and other sections of the CPU.

One important benefit of the microprogrammed processors to the manufacturer is that different members of the family can execute the same instruction if the microcode is common to the processors in the family. The MC68020 thus executes the machine-language instructions of the 16-bit MC68000 without modification. Of course, the converse is not necessarily true because the MC68020 has additional instructions that are not present in the MC68000.

Finally, microprogrammed instructions are provided in the MC68020 ROM for testing the CPU. These test routines can test logic paths and circuits in the CPU at a level that is inaccessible by machine language instructions. Such testing is required during production of the CPU to assure that the chip is functioning properly.

4.1.2 Instruction Set Design

As technology advanced in the 1970s, it became possible to implement in the CPU features traditionally associated with software. Such features include instructions to perform complicated operations and sophisticated addressing modes. For example, instructions for multiplication, division, and Binary Coded Decimal arithmetic are included in the instruction set of all the 68000 family processors. These operations were normally created by software routines in earlier 8-bit CPUs.

The M68000 family instructions can be divided into categories as follows:

Data movement

Arithmetic operations

Logical operations

Bit manipulation and shift operations

Program control

System control

All of the pertinent instructions in these categories will be studied later. However, by the range of these instructions, it is evident that one of the CPU designer's goals was to reduce the gap between the instructions available in a high-level compiler language and the instructions of the CPU. An important purpose of the powerful microcoded instructions in the CPU was to reduce reads and writes to external memory to obtain instructions. The microcoded instructions could execute much faster than instructions fetched from external memory.

This approach to instruction set design gave rise to the notion of a *Complex Instruction Set Computer* (CISC). This designation was put forth to distinguish CPUs such as those in the 68000 family from another approach to CPU design that emphasized a simplified instruction set with fewer but possibly faster executing instructions.

4.1.3 Processor Type (CISC Versus RISC)

Studies in the 1970s of the use of CPU instruction sets caused some designers to conclude that many of the complex instructions of the then-current processors were not being used by programmers and compilers. To simplify CPU design and increase instruction execution speed, several organizations proposed an architecture for processors that came to be called *Reduced Instruction Set Computers* (RISC). Some authors call machines with this architecture *Reusable Instruction Set Computers*. Table 4.3 lists some of the differences between RISC and CISC processors.

In a RISC processor, each instruction is of the same length, usually four bytes (32 bits). This simplifies instruction fetching and decoding and allows the instruction to be executed in one clock cycle. The 68000 family CPUs have instructions that vary in length from two bytes (16-bits) to 10 bytes in the MC68000 with even longer instructions for the MC68020 and other advanced processors.

The RISC processors have a *load and store* architecture. The instructions are executed by first fetching operands from memory and loading them into registers. Then, arithmetic or other operations are performed on the operands in one clock cycle, because register-to-register operations are very fast compared to memory accesses. Finally, the results are stored back in memory. Each fetch, execution, and store operation requires one instruction. In CISC processors, some instructions perform the complete process with one instruction. This means that many CISC instructions access memory as well as perform arithmetic operations. The register set of a CISC processor is typically much smaller than that of a RISC processor, so memory accesses are sometimes necessary to store data temporarily.

RISC processors have a small number of simple instructions compared to CISC processors. This makes assembly language programming more tedious and difficult compared to CISC programming and most users of RISC machines program in a high-level language. The compiler must be written to take advantage of the RISC features. Also, the limited number of instructions may lead to larger programs for the RISC systems. Some defenders of CISC architecture have called the processors *Complete Instruction Set Computers* to indicate that the CISC CPU has a large set of powerful instructions. The desire to minimize the use of memory for instructions was a factor

Table 4.3 RISC Versus CISC Architecture

RISC	CISC
Fixed instruction size, one-cycle execution	Variable instruction size and execution times
Load/store instructions	Memory operations possible
Large number of registers (32 or more)	Limited register set
Small instruction set	Large and complete instruction set
Difficult assembly language programming	Relatively simple assembly language programming
Compiler optimization necessary	No special compiler requirements
Hardware design of CPU	Microcoded CPU

when CISC designers originally incorporated single instructions with multiple operations.

RISC machines seem to allow faster execution than CISC machines for many tasks, although the programs may be larger and the compiler or assembly language program is more complicated. What then is the main advantage of RISC versus CISC? In the CISC processors with a large number of instructions and addressing modes, microinstructions (microcode) are used to implement the instructions. This requires a large number of transistors and logic elements in the CPU for the microcode. In contrast, the RISC CPU with fewer instructions is created with hard-wired circuits sometimes called *custom logic*. This method requires far fewer circuit elements and eliminates the overhead associated with microcode due to the time necessary to interpret the instructions and branch within the microcode. Because the instruction circuits require fewer components, additional functions, such as a floating-point unit, can be implemented on a RISC processor with the same chip size as a CISC processor.

Example 4.2 _____

IBM, Motorola, and Apple have produced a new RISC processor family called the Power PC. Apple has chosen this family as the CPU for the new generation of its personal computers. Previously, Apple computers used M68000 family CPUs in their computers. The first version, the Power PC 601, contains 2.8 million transistors on a silicon chip (die) 120 mm^2 in area. One of its important competitors is the Intel Pentium, which contains 3.1 million transistors on a chip that is 262 mm^2 in area.

The chip size of the original MC68000 is conveniently measured in square mils, 1 mil being one-thousandth of an inch (.0254 mm). For comparison, the chip size was about 246 by 280 mils, which is approximately 68,000 square mils in area (44.4 mm^2). By coincidence, the chip also contained about 68,000 transistors.

The Power PC has 64 registers for integer and floating-point operations, while the Pentium has sixteen. One other feature of the Power PC is that it uses separate busses for data and instructions, which is commonly called *Harvard architecture* to distinguish it from the single bus structure of most CISC processors. The purpose is to transfer instructions and data to the CPU at the fastest possible rate, assuming that the CPU can process information faster than the memory can supply it in a single bus system. The single bus systems, such as the MC68000-based systems discussed in Chapter 2, are said to have a *Von Neumann* architecture.

It is interesting that many of the advanced processors of the M68000 family use RISC-like techniques to increase the speed of execution. Most of these features are not apparent to the programmer because the architecture of the processors remains the same when programming applications. Several references listed at the end of the chapter discuss the RISC versus CISC issue. The Motorola *User's Manuals* for the MC68030, MC68040, and MC68060 describe the architectures in detail.

EXERCISES

These questions can best be answered by reference to other textbooks or journal articles listed in the Further Reading section at the end of this chapter.

4.1.1. The MC68020 is considered a CISC processor. This implies that the instruction set contains relatively complex instructions. Another approach to CPU design employs RISC principles. Here the instructions are simpler and perhaps more efficient. By reference to appropriate literature, compare the two approaches to CPU design and give examples of processors using each method.

4.1.2. The 68000 family of microprocessors are microprogrammed. Compare the design of a CPU using microprogrammed control with a design employing custom (hard-wired) logic circuits. Consider the initial design and testing, as well as speed of operation and flexibility when design changes are needed.

4.2 THE MC68000 REGISTER PROGRAMMING MODEL

Just as the memory is used to store instructions and data associated with a program, the CPU contains storage elements called *registers,* which hold information needed for the instruction currently being processed. A register consists of one or more storage cells, each containing 1 bit of information. The *length* of the register is defined as the number of bits that may be stored or read simultaneously. A CPU contains a large number of registers, most of them used for specific purposes within the CPU. A few of these registers, called *programmable registers,* are available to the machine language or assembly language programmer via the processor's instruction set. In the MC68000, the programmable registers are the general-purpose registers, the program counter, and the status register. The basic register set of the MC68010 is the same although several new registers have been added.

As described in Chapter 2, the MC68000 allows programs to be divided into user-mode programs and supervisor-mode programs. Critical routines of the operating system execute in the supervisor mode. These are usually routines that control the operation of the hardware in some way. Its routines have a higher privilege than those of a user-mode program. These supervisor routines can execute instructions that are forbidden to the user mode. Programs in the user mode are also restricted from modifying the contents of certain processor registers, such as the supervisor stack pointer and the status register of the CPU. A supervisor-mode program can use any processor register when it executes.

This section first describes all of the processor registers that are used for general-purpose programming. These are the data registers, seven of the nine address registers, the program counter, and the condition code register. Following those descriptions, the system stack pointers and the status register are covered. Programs in both the user mode and the supervisor mode use the same general-purpose registers.

4.2.1 The MC68000 CPU General-Purpose Register Set

The general-purpose registers of a processor hold addresses or data values being manipulated by an instruction. The *program counter* holds the address of the next instruction to be fetched from memory. The *status register* contains bits that indicate the state or status of the processor and information about the results of arithmetic or similar operations. In the MC68000, the general-purpose registers are divided into eight *data registers* and nine *address registers.* Two of these address registers are used as system stack pointers. This CPU has a 32-bit program counter and a 16-bit status register.

Figure 4.1 shows the MC68000 register set and a simplified diagram of the internal and external transfer paths of the processor. The address, data, and control signal lines connect to the system bus as explained in Chapter 2. Internally, the processor contains programmable registers to hold values that are treated as data or addresses. These values can be transferred to the arithmetic and logic unit (ALU) for arithmetic computations. Data values held in the data registers of the processor can be transferred

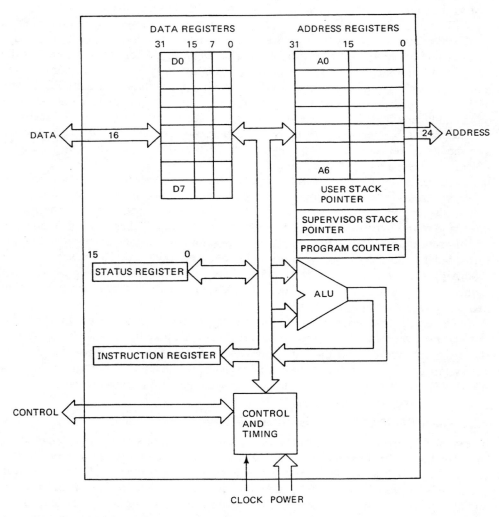

Figure 4.1 The MC68000 register set and transfer paths.

to and from memory or between the CPU and peripheral devices via the data signal lines. Values in the address registers can be placed on the address bus to reference locations in memory. These operations are controlled by machine language instructions that are fetched from memory via the data signal lines and then decoded in the instruction register. The instruction register is not part of the programmable register set of the processor because it cannot be referenced directly by MC68000 instructions.

The programmable register set of any processor may be defined in terms of the features useful to a programmer. The number, type, and length (in bits) of the registers and their connection via internal and external data paths determine the basic capability of the processor to execute instructions using these registers. In general, an increased number of available registers will simplify programming at the assembly language level. Additionally, program execution speed is increased if registers are used to hold operands because operations between registers require less processing time than operations that reference external devices or memory.

Registers are also used to hold addresses of data items stored externally to the processor. These addresses can easily be changed during program execution by modifying the contents of the registers. Modern programming techniques favor this method of indirect register addressing, rather than allowing the program to modify itself by changing an address that is part of an instruction. In many processors, including the MC68000, the registers holding the address of an operand may be modified by the addition (or subtraction) of the contents of other registers called *index registers.*

Two of the address registers in Figure 4.1 are designed as the *user stack pointer* and the *supervisor stack pointer.* These stack pointers are used by the CPU to save return addresses and similar information on the stack during operations such as subroutine calls. In the user mode, only the user stack pointer (USP) is active and available to a program or the CPU. In the supervisor mode, the CPU uses the supervisor stack pointer (SSP) to store return addresses after an interrupt or subroutine call. However, it can also read or modify the USP. This use of two separate stack pointers assures that user-mode programs cannot alter the supervisor stack area in memory, and thus information for the operating system is protected.

The program counter of the MC68000 is used to address 2^{23} word (16-bit) locations in memory—over 8 million locations. The full 32-bit addressing capability of the program counter is not currently available in the MC68000 because there are a limited number of address signal lines.

The CPU also contains a status register that holds logical variables that indicate the status of the CPU and the interrupt system, as well as the results of arithmetic operations. Instructions are available to test or change the contents of the status register as a program executes.

Figure 4.2 shows the programmable register set of the MC68000 and indicates the numerical designation of the bits in each register. Table 4.4 lists the registers according to their primary use and also gives the symbolic notation used to refer to a particular register. The assembly language programmer, for example, would use this notation to designate a register in an assembly language statement. The fact that either the USP or SSP can be referred to as A7 causes no confusion because the stack pointer

Figure 4.2 Programmable register set of the MC68000.

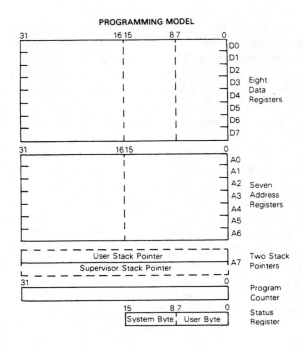

PROGRAMMING MODEL

	User Stack Pointer	Two Stack
	Supervisor Stack Pointer	Pointers

Eight Data Registers

Seven Address Registers

Program Counter

Status Register

Table 4.4 Register Usage and Symbolic Notation

Register	Symbolic Notation	Usage
Data	D0	Accumulator
	D1	Buffer register
	•	Index register
	•	Temporary storage
	•	
	D7	
Address	A0	Indirect addressing
	A1	Stack pointer
	•	Index register
	•	
	•	
	A6	
System stack pointer	A7 or SP	
User SP	A7 or USP	User subroutine calls
Supervisor SP	A7 or SSP	Interrupt processing or subroutine calls (supervisor mode)
Program counter	PC	Instruction addressing
Status register	SR	System status, condition codes

referenced is determined by the mode (supervisor or user) of the processor as explained later.

In order to be consistent with the register transfer notation defined in the Appendix, the contents of a register or memory location are designated here by enclosing the item in parentheses. Thus (D2) means the contents of data register D2. When selected bits of an operand are designated, the bit numbers are enclosed in brackets, with the beginning and ending bit number separated by a colon if consecutive bits are specified. For example, bits 0 through 7 of data register D1 are indicated by (D1)[7:0]. An operand designated by ⟨operand⟩ means that any valid operand, as determined in the discussion, may be substituted into the expression. The designation ⟨Dn⟩, for example, means that any data register can be specified (i.e., any one of D0, D1, . . . , D7).

4.2.2 Data Registers

The MC68000 has eight registers designated as data registers. The registers are referred to symbolically by number as Dn, where $n = 0, 1, . . . , 7$. The internal bus structure of the CPU allows a byte operand (Dn)[7:0], a word operand (Dn)[15:0], or a longword operand (Dn)[31:0] to be manipulated in the data register selected. Because three lengths are possible, the processor instructions that reference a data register must indicate the operand length. Only the corresponding bits of the specified register are modified by that instruction. The portion of the register involved may be used as an accumulator, a storage register, a buffer register, or as an index register.

As *accumulators,* the data registers hold operands of the specified length and allow arithmetic, logical, and other operations. These registers are also used temporarily to store operands generated in other registers of the processor. The data registers act as *buffer* registers when data values are transferred in or out of the processor via the data signal lines. For the MC68000 with its 16 data signal lines, the possible transfers include an 8-bit or 16-bit quantity in a single transfer or a 32-bit value in two transfers of 16 bits each.

A data register can be used as an *index* register whose contents are added to the value in an address register to form an address of an operand. The power of this addressing capability is that the index can be modified by any processor instruction that operates on a data register, so the index value may be changed in very sophisticated ways during program execution. This usually occurs within a program loop.

Example 4.3 _____

Figure 4.3 shows a data register of the MC68000 with the bits designated from 0 on the right to 31 on the left. The sign bit of a byte-length two's-complement number would be bit 7, as indicated. Any access of the register specifying a byte operand would affect only bits 0 through 7; the remaining bits would be unchanged. Similarly, the sign bit of a word operand is bit 15 and that of a longword operand is bit 31, as shown in the figure.

4.2.3 Address Registers

The MC68000 has nine address registers, which accept word or longword values only. Seven of the address registers, symbolically designated A0, A1, . . . , A6, are shared by programs in either the supervisor or the user mode, but only one of the two

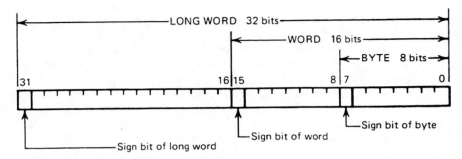

Figure 4.3 Data register format.

remaining address registers can be referenced as A7 by a program operating in a specific mode. These two registers are the system stack pointers. The other seven registers available to a program can also be used to address "private" stacks defined by the programmer.

The primary use of the address registers is to hold the address of an operand in memory. Because an address register is 32 bits in length, an address may range up to

$$2^{32} = 4,294,967,296$$

locations. In the MC68000, however, only the lower 24 bits may be output via the address and control signal lines of the processor. Thus, when accessing memory, an address register of the MC68000 should be considered to contain a 24-bit address.

Private Stacks. In the MC68000, a *stack* consists of a set of contiguous memory locations addressed by a register designated as a *stack pointer.* Items stored on the stack are retrieved in reverse order, reminiscent of a push-down stack of cafeteria plates. The stack is accessed to store or retrieve data from one end only in a last-in-first-out (LIFO) manner.

Functionally, a stack pointer contains a value, designated (SP), that is used as an address to point to the top of the stack. The processor uses the value in the stack pointer to address a location from which the processor reads or to which it writes a data item. Information stored on the stack is said to be *pushed* on the stack by a processor write cycle. The item of information is retrieved by a processor read cycle that is called a *pop* or *pull* of the item.

Each address register of the MC68000 can be used as a stack pointer by a program using one of the addressing modes appropriate for stack operations. In these addressing modes, the value in the address register being used as a stack pointer is automatically changed by the proper amount after each push or pop of a data item. The data values in the stack may be either byte, word, or longword values if any of the registers designated as A0, A1, . . . , A6 are used as stack pointers.

Index Registers. Another use of an address register is as an index register. The index value is added to the contents of another address register to compute an effective address of an operand in memory. The same usage was defined for a data register in the preceding section. Thus the MC68000 allows both address and data registers to serve as index registers.

Example 4.4 _____

Figure 4.4(a) shows the format of an address register of the MC68000. The address can be either 16 bits or 32 bits in length. However, the corresponding longword address would be only 24 bits in length when output to memory.

An indirect memory reference is shown in Figure 4.4(b). The address held in A1 in the example points to an operand at hexadecimal location 1000. Thus the effective address of the operand is designated as the contents of A1, or (A1).

Indexed addressing, shown in part (c) of the figure, allows the sum of two values to be used to determine the operand location. The maximum 24-bit address of a byte location in memory is FFFFFF in hexadecimal.

Figure 4.4   MC68000 address register usage.

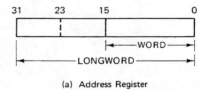

(a) Address Register

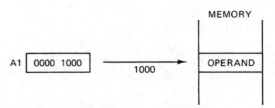

(b) Indirect Memory Reference

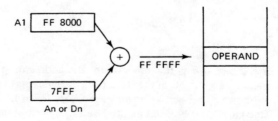

(c) Indexed Memory Reference

Note: All values are in hexadecimal.

Example 4.5 _____

The *bottom* of the stack is the first item pushed onto the stack and the top of the stack is the last added. Removing (popping) an item is done from the top. A stack may grow from lower addresses to higher memory addresses, or it may grow "down" in memory. These two cases

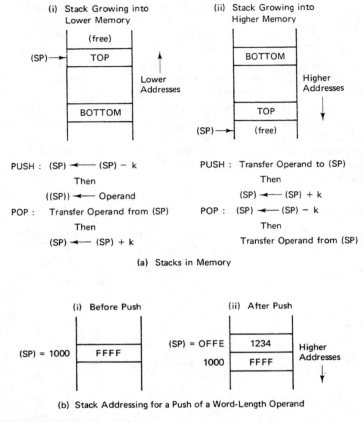

(i) Stack Growing into Lower Memory

(ii) Stack Growing into Higher Memory

PUSH : (SP) ←—— (SP) − k

Then

((SP)) ←—— Operand

POP : Transfer Operand from (SP)

Then

(SP) ←—— (SP) + k

PUSH : Transfer Operand to (SP)

Then

(SP) ←—— (SP) + k

POP : (SP) ←—— (SP) − k

Then

Transfer Operand from (SP)

(a) Stacks in Memory

(i) Before Push

(ii) After Push

(b) Stack Addressing for a Push of a Word-Length Operand

Figure 4.5 Stack operation and growth.

are shown in Figure 4.5(a), where the addressing required is also shown. Our notation ((SP)) means the contents of the stack (i.e., the contents of the location addressed by the stack pointer). When the stack grows down in memory toward lower addresses, (SF) points to the last item (top) and must be decremented before a push. After a pop, (SP) must be incremented to point to the top again. The opposite is true for a stack that grows upward in memory. The increments (or decrements) are 1, 2, or 4 depending on whether the size of each item on the stack is byte, word, or longword. If an instruction specifies the *predecrement* mode of the MC68000, the CPU automatically subtracts k (k = 1, 2, or 4) from (SP) before the stack pointer value is used as an address. The *postincrement* mode is used to add k to (SP) after use. These addressing modes are discussed in more detail in Section 4.4.

Figure 4.5(b) shows the stack contents before and after a push of the hexadecimal value 1234 onto a word (16-bit) stack. Before the push, the stack pointer contains the hexadecimal address 1000 in the example and the value ((SP)) is FFFF. To push the value, the stack pointer is first decremented by 2 and then used as the address of the stack location for the data value. Therefore, the item is pushed to location 0FFE.

4.2.4 System Stack Pointers

The MC68000 uses stacks to store information when a subroutine call is made by a program or when an *exception* occurs during system operation. These exceptions as defined by Motorola for the MC68000 include traps, interrupts, and several error conditions recognized by the CPU. When any one of these events occurs, the normal flow of control through sequential instructions in memory is altered. Control is passed to the instructions associated with the subroutine, trap, or interrupt until its specific task is completed. Then control is normally returned to the next instruction in the preempted sequence. To allow control to be returned, the CPU stores or saves the information on the appropriate system stack. For a subroutine call, only the value of (PC) needs to be saved and restored after the subroutine completes. When an exception occurs, the contents of the PC and the status register (SR) are saved as well as other information according to the requirements of the exception. The saving and restoring operations are automatic and require no programmer intervention.

The MC68000 separates the stack location in memory into a user stack area and a supervisor stack area. In the user mode, the system stack is the user stack and the user stack pointer is used to address this stack. The initial location of this user stack, before it is used for saving and restoring information, is assigned by the operating system. When the CPU operates in the supervisor mode, the system stack is a supervisor stack. The SSP addresses the supervisor stack areas in memory. A user mode program can alter the user stack pointer, but it cannot access the supervisor stack pointer. Therefore, even if a user-mode program mishandles its stack and causes a serious error, the supervisor stack is not affected. The operating system needs only to manipulate its own stack properly to assure that the system operates correctly. It should also be noted that the contents of the general-purpose data registers or address registers are not saved on the system stack automatically in response to an exception. Program instructions must be used to save the contents of these registers if necessary.

System Stack Operations. Any time the MC68000 executes a program, only one system stack is being used. Its stack pointer will be designated SP. In the user mode, SP indicates the user stack pointer. The supervisor stack pointer is explicitly designated as SSP or as SP in a supervisor-mode program. No confusion should result in discussing the stack pointers or in programming because the mode of operation of the CPU determines which SP is being referenced.

Subroutine Calls. Figure 4.6(a) shows the flow of control when a subroutine call is made. The program has executed a Jump to Subroutine (JSR) or Branch to Subroutine (BSR) instruction described in Chapter 6. The CPU saves the 32-bit contents of the program counter on the active system stack and transfers control to the subroutine by loading the starting address in the PC. When the subroutine completes by executing a Return from Subroutine (RTS) instruction, the CPU reloads the PC with the value that was saved on the stack. The value is the address of the instruction following the JSR or BSR instruction. If the subroutine call were made in the user mode, the user stack is

Figure 4.6 Flow of control during program execution.

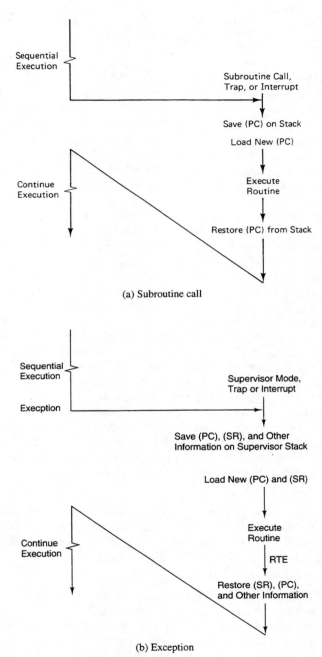

Sequential
Execution

Subroutine Call,
Trap, or Interrupt

Save (PC) on Stack

Load New (PC)

Continue
Execution

Execute
Routine

Restore (PC) from Stack

(a) Subroutine call

Sequential
Execution

Supervisor Mode,
Trap or Interrupt

Execption

Save (PC), (SR), and Other
Information on Supervisor Stack

Load New (PC) and (SR)

Continue
Execution

Execute
Routine

RTE

Restore (SR), (PC),
and Other Information

(b) Exception

employed. In the supervisor mode, the return address is saved on the supervisor stack. Example 4.6 describes the use of both stacks for a subroutine call and an interrupt.

Exception Processing. When an exception is recognized by the MC68000, the transition to the exception handling routine is controlled automatically by the CPU, as shown in Figure 4.6(b). Any exception causes the CPU to change to the supervisor mode, regardless of the mode that was active when the exception occurred. When the exception is recognized, the (PC) and (SR) are saved on the supervisor stack. Some exceptions require that even more information be saved. In any case, enough information about the preempted program is available to allow it to resume execution after the exception routine has completed. The last instruction in the exception routine (RTE) causes the information to be restored so that control is returned to the original program.

4.2.5 Supervisor Versus User Mode

The difference between the supervisor mode and the user mode is one of privilege that concerns control of the CPU itself and perhaps external devices. Programs executing in the user mode are restricted from executing certain instructions and cannot access the supervisor stack pointer. The user-mode programs may also be restricted from controlling certain elements in the computer system or in the regions of memory they may access for reading or writing operands. These system restrictions must be enforced by external circuitry, not by the CPU. The CPU indicates its mode via several of its control signal lines to allow external circuits to determine the proper action. For example, external circuits can be used to restrict memory access for a user-mode program. If such access is attempted, the memory-management circuits signal the MC68000 CPU and cause an error exception. The supervisor program then determines the appropriate action, such as preventing the offending program from executing further. These system considerations are considered in more detail in Chapters 10, 11, and 13.

Transitions Between Modes. Figure 4.7 shows the method of transition between modes. The user mode can be entered only by having a supervisor program change the mode to the user mode. Return to the supervisor mode is automatic when the MC68000 processes an exception. The exceptions as defined by Motorola include traps, interrupts, and several error conditions recognized by the CPU. The occurrence of any exception causes the appropriate routine to execute in the supervisor mode.

Privilege Distinctions. Table 4.5 indicates the difference between the supervisor and user modes. The modes share the general-purpose register set and the program counter. However, a user mode program cannot execute all of the instructions of the MC68000.

Figure 4.7 Transition between modes.

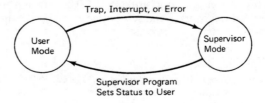

Table 4.5   Supervisor Versus User Mode

	Supervisor	**User**
Register usage	D0–D7, A0–A6, PC, SSP	D0–D7, A0–A6, PC, CCR
Stack pointer	SSP	USP
Instructions	All	Restricted set
Entered by:	Exception processing	Supervisor program changing mode to user

In particular, instructions used to change the mode or control certain system operations are not executable by programs operating in the user mode. These are called privileged instructions and if an attempt is made to execute one by a user mode program, an exception occurs.

In the supervisor mode, the active stack pointer is designated as the SSP. The system stack pointer in the user mode is the USP.[1] Each stack accommodates only word or longword data elements that can be pushed onto the stack using the predecrement addressing mode to reference the stack pointer. Thus, the stacks extend into memory locations with lower addresses as items are pushed onto them. At any time after a stack access, (SP) points to the last location used. A push operation causes (SP) to be decremented by 2 for a word operand or by 4 for a longword operand before the item is stored. After an item on the stack is retrieved, (SP) is increased by 2 or 4, as required.

The system stack is used to save the contents of the program counter automatically during subroutine calls. The 32-bit contents are saved and eventually restored after the subroutine completes. During interrupt or trap processing, however, both the contents of the program counter and that of the status register are saved on the supervisor stack, as indicated in Table 4.6. Two error conditions (address error or bus error) cause seven 16-bit words to be saved on the supervisor stack to enable the programmer to investigate the cause of the error. In addition to the (PC) and (SR), the contents of the instruction register and other data are saved. These exceptions are discussed in Chapter 11.

Example 4.6 _____

Figure 4.8 shows the contents of the stacks used during the execution of a user-mode program. If the program calls a subroutine, the return address is pushed on the stack using (USP) as the address that is decremented by 4 to accommodate the 32-bit (PC). If an interrupt occurs during subroutine execution, the return address within the interrupted subroutine and the contents of the status register are saved using (SSP). The return from the interrupt processing followed by the eventual return from the subroutine leave (USP) and (SSP) as they were initially.

_____

[1] In Motorola literature, the designations for USP and SSP are sometimes given as A7 and A7′, respectively. In this text, the acronyms will be used.

Table 4.6 Stack Usage in Supervisor Mode for the MC68000

Activity	Items Saved on Stack
Subroutine call	(PC) [31:0]
Interrupt or trap	(PC) [31:0]
	(SR) [15:0]
Address error or bus error	(PC) [31:0]
	(SR) [15:0]
	Instruction register [15:0]
	Access address [31:0]
	Access information [15:0]

Notes:
1. Access address is the address that was being accessed when the error occurred.
2. Access information:
   [2:0]: function code
   [3:3]: 0 = instruction, 1 = not an instruction
   [4:4]: 0 = write, 1 = read
3. The MC68010 has a slightly different stack format, as described in the text.

Other Processors. Exception processing with the MC68010 is slightly different in that at least four 16-bit words are saved on the supervisor stack after an exception is recognized. After an interrupt or trap, the first word pushed indicates the stack length and other information, followed by (PC) and (SR) as for the MC68000. Thus for interrupts and traps, four words are saved, including the (PC) and (SR). An address error or bus error causes a total of 29 words to be placed on the stack. This is to allow instruction continuation if the problem can be corrected by the supervisor program. Instruction continuation allows a virtual memory system to be implemented with the MC68010.

The contents of the system stack after an exception is processed may not be the same for different processors of the 68000 family. This is the reason that Motorola only guarantees that user mode programs written for the MC68000 are compatible for execution on the advanced processors.

4.2.6 Program Counter

In the 68000 family of processors, the program counter internally is 32 bits in length, although only (PC)[23:0] can be used for addressing an instruction location in an MC68000 system. This allows an addressing range of over 8 million words in memory. Because the MC68000 may have instructions occupying as little as one word or as many as five words in memory, the program counter is incremented by the proper amount automatically as the current instruction is executed.

The program counter can be modified by the programmer to change the control sequence in several ways. The normal sequence of execution can be directly altered by a program instruction that causes a jump or branch in the program. In this case, no return address is saved by the CPU because control is not returned to the instruction following the jump or branch instruction. The jump instruction thus loads the program

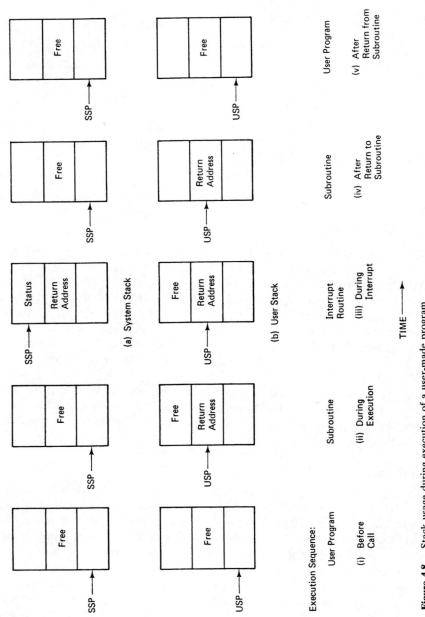

Figure 4.8 Stack usage during execution of a user-made program.

98

counter with the new address and destroys its previous contents. In contrast, a subroutine call or exception causes the current value of the program counter to be saved on the system stack before its contents are changed to the address of the new routine to be executed. The last instruction in a subroutine or exception routine must be an instruction (e.g., Return) to restore the contents of the program counter.

4.2.7 Status Register

During execution, a program can be described at any instant in time by its state. The state description contains all the information necessary to stop, then restart, the program. Specifying the instructions and data values, including the contents of the programmable register set, constitutes a complete state description. Assuming that the instructions and data values in memory are not altered if a program is suspended from execution temporarily, only the contents of the register set used by the suspended program must be saved in temporary locations. The contents of these registers must be restored before the program can execute again.

In the MC68000, this register set would include selected address and data registers, the program counter, and the status register. Important information about the state of the program is contained in the status register, including its mode, the interrupt status, and the arithmetic or logical conditions that were obtained after the last instruction was executed. While a program is suspended, the "temporary" storage locations for all of this information is usually the stack.

Figure 4.9 shows the 16-bit status register of the MC68000. The entire contents of the status register (SR) can only be modified by a program in the supervisor mode.

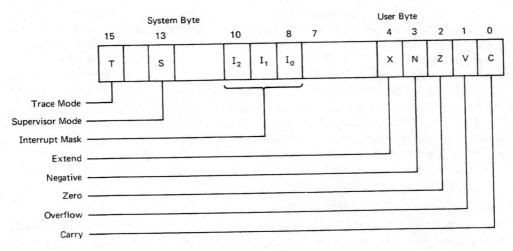

Notes:
(1) Conditions stated are true when the corresponding bit = {1}.
(2) The user byte portion of the status register is referred to as the condition code register (CCR).

Figure 4.9   MC68000 16-bit status register.

The upper 8 bits are sometimes called the *system byte* because these bits determine the state of the CPU. The low-order 8 bits are called the Condition Codes Register (CCR) and can be read or modified by programs executing in either the supervisor or the user mode. In the figure, each bit is considered separately, except the interrupt mask bits SR[10:8] which are taken together as a 3-bit binary number indicating the interrupt level. The other bits are considered individually, with a {1} indicating that particular condition is true.

Thus, SR[13] = {1} indicates that a supervisor mode program is executing, and SR[13] = {0} indicates a user mode program. In practice, the operating system sets SR[13] = {0} and modifies other bits as necessary when control is passed to the user mode program. If an exception condition then occurs, the CPU operating in the supervisor mode saves the current (PC) and (SR) on the stack. The exception condition automatically changes the S bit to {1} and causes the other bits of the status register to be modified as required. As control is returned to the user program, the previous contents of the status register and program counter are pulled from the system stack and restored. When the previous contents of the program counter are restored, the user program continues execution where it left off. Thus control of the CPU mode is determined by the setting of the S bit, which provides a simple and efficient means of switching between modes.

The operating system can cause each instruction to be executed individually by setting the trace mode bit, SR[15], to {1}. After each instruction, a trap is caused and control passes to the trap-handling routine, which is typically designed to aid in debugging a user-mode program.

An interrupt request is issued by an external device using three signal lines of the CPU control bus that indicate the priority level for the device. The internal interrupt system of the MC68000 is controlled by three corresponding bits in the status register. These bits act as an interrupt *mask* for the seven-level priority-interrupt system of the MC68000. The decimal value of the interrupt mask is 0 through 7, with priority levels in ascending order from 1 to 7. Setting these bits with a value from 1 to 6 disables or masks interrupts at the level indicated and those levels below in priority. The CPU will ignore all interrupt requests that are disabled. A zero ({000}) value for the interrupt mask means that every interrupt level is enabled.

The level 7 interrupt is referred to as a *nonmaskable interrupt* and cannot be disabled. If an interrupt occurs at a given level below level 7, the mask bits are automatically set to that level to prevent additional interrupts from that level or a level below it from being received.

4.2.8 Condition Code Register (CCR)

The user byte or CCR of the status register contains condition codes, which are single-bit variables indicating the results of arithmetic or logical operations. These are set automatically by many of the instructions of the MC68000. For example, if an addition results in a zero sum, the Z bit is set to {1}. The other bits in the status register have the meanings shown in Tables 4.7 and 4.8.

Table 4.7 Interpretation of Condition Codes

Name	Symbol	Meaning
Extend	X	Used in multiple-precision arithmetic operations; in many instructions it is set the same as the C bit
Negative	N	Set to {1} if the most significant bit of an operand is {1}
Zero	Z	Set to {1} if all the bits of an operand are {0}
Overflow	V	Set to {1} if an out-of-range condition occurs in two's-complement operations
Carry	C	Set to {1} if a carry is generated out of the most significant bit of the sum in addition; set to {1} if a borrow is generated in subtraction

Table 4.8 Interpretations of System Status

Name	Symbol	Meaning
Trace	T	Set to {1} if the trace mode is being used (single-instruction stepping)
Supervisor	S	Set to {1} if the processor program is in the supervisor mode
Interrupt mask	I0, I1, I2	Coded interrupt level; interrupts at level indicated and below will not be recognized (levels 1–6); level 7 is not maskable

Example 4.7

The 8-bit operand {1XXX XXXX} would cause N = {1} if tested for a negative value. The setting of the other bits designated by X has no effect on the test. Depending on the program application, the interpretation might be a negative two's-complement number, an unsigned binary number greater than or equal to 128, or a BCD number greater than or equal to 80.

Example 4.8

The status register contents 0700_{16} indicates the user mode with all interrupts below level 7 masked (disabled). All the condition codes are {0}. Only a level 7 interrupt will be acknowledged because it is nonmaskable. If such an interrupt occurs, the (SR) will be set to 2700_{16} during interrupt processing, indicating that the processing occurs in the supervisor mode. Upon completion of the interrupt routine, control returns to the user mode program with (SR) = 0700_{16}.

EXERCISES

4.2.1. Determine the status if the system status register contains the following hexadecimal values.
   (a) 0400

      (b) 2000

      (c) 0004

      (d) A000

4.2.2. Show the contents of the system status register after a level 4 interrupt is accepted. Assume that the status register initially contained 0 for each bit.

4.2.3. What registers must be initialized before the processor can execute a program? Consider a supervisor mode program. What registers must the operating system initialize before control is passed to a user-mode program?

4.2.4. Show the contents of the supervisor stack if a program is interrupted by a level 1 interrupt when $(PC) = 101C_{16}$. If the level 1 interrupt routine is itself interrupted by a level 2 interrupt when $(PC) = 200C_{16}$, show the changes to the system stack. Assume that initially $(SR) = 0$ and $(SSP) = 8000_{16}$.

4.3 INTRODUCTION TO THE MC68000 INSTRUCTION SET

The instruction set for the MC68000 determines the operations that are available to perform data transfer, arithmetic processing, and control program flow. Each complete MC68000 instruction consists of the following:

    (a) an operation code determining the operation to be performed;

    (b) a designation of the length of the operand or operands;

    (c) specification of the locations of any operands involved by indicating an addressing mode for each.

Figure 4.10 lists the instruction set for the MC68000 in alphabetical order. Each mnemonic represents the operation code. A letter is used to indicate a length of byte (B), word (W), or longword (L) for 8-bit, 16-bit, and 32-bit operands, respectively. For example, the symbolic instruction to add two 16-bit operands would be

ADD.W   X,Y

where X and Y designate the locations of the operands.

Instructions for the MC68000 can be classified by type or by the number of operands. The number of operands for an instruction determines whether it is classified as a single-address or double-address instruction. Classification by type groups the basic operations allowed by the processor into categories, such as those for data movement or those for arithmetic operations.

A processor instruction set is sometimes separated into types in order to compare it to instruction sets of other processors. After comparison, a processor with an extensive arithmetic set might be chosen over one with less capability to support mathematical programming, for example. The division into types is also convenient for coding instructions, because this grouping allows the programmer to select the best instruction to perform an operation of a particular type. The Exchange instruction (EXG), listed in Figure 4.10, for example, exchanges the contents of two MC68000 registers and is more efficient for this purpose than several other data movement instructions that could accomplish the same result.

Figure 4.10 Instruction set for the MC68000. (Courtesy of Motorola, Inc.)

Mnemonic	Description
ABCD	Add Decimal with Extend
ADD	Add
AND	Logical And
ASL	Arithmetic Shift Left
ASR	Arithmetic Shift Right
Bcc	Branch Conditionally
BCHG	Bit Test and Change
BCLR	Bit Test and Clear
BRA	Branch Always
BSET	Bit Test and Set
BSR	Branch to Subroutine
BTST	Bit Test
CHK	Check Register Against Bounds
CLR	Clear Operand
CMP	Compare
DBcc	Test Condition, Decrement and Branch
DIVS	Signed Divide
DIVU	Unsigned Divide
EOR	Exclusive Or
EXG	Exchange Registers
EXT	Sign Extend
JMP	Jump
JSR	Jump to Subroutine
LEA	Load Effective Address
LINK	Link Stack
LSL	Logical Shift Left
LSR	Logical Shift Right
MOVE	Move
MOVEM	Move Multiple Registers
MOVEP	Move Peripheral Data
MULS	Signed Multiply
MULU	Unsigned Multiply
NBCD	Negate Decimal with Extend
NEG	Negate
NOP	No Operation
NOT	Ones Complement
OR	Logical Or
PEA	Push Effective Address
RESET	Reset External Devices
ROL	Rotate Left without Extend
ROR	Rotate Right without Extend
ROXL	Rotate Left with Extend
ROXR	Rotate Right with Extend
RTE	Return from Exception
RTR	Return and Restore
RTS	Return from Subroutine
SBCD	Subtract Decimal with Extend
Scc	Set Conditional
STOP	Stop
SUB	Subtract
SWAP	Swap Data Register Halves
TAS	Test and Set Operand
TRAP	Trap
TRAPV	Trap on Overflow
TST	Test
UNLK	Unlink

The basic types of instructions for the MC68000 are those of data movement, arithmetic and logical operations, program control, and processor or system control. These categories are expanded in Chapter 5, but for our purposes at present, only a few instructions representing several types will be discussed. For convenience, the instruction set is also presented in Appendix IV.

4.3.1 The CLEAR Instruction

The CLEAR (CLR) instruction is considered a single–address arithmetic instruction that has the symbolic form

CLR.X   ⟨EAd⟩

where X is B, L, or W and the ⟨EAd⟩ is the effective address of the destination. Zeros are transferred to the portion of the destination location specified by the operation, as shown in Table 4.9. If the destination location originally contained all ones, executing the CLR.X instruction causes the designated portion of the location to be cleared. The operation can be defined as

(EAd)[X] < - - 0[X]

which is read: "The contents of location EAd of length X is replaced with zeros."

The description of the CLR instruction of the MC68000 is shown in Figure 4.11. This summary, taken from the Motorola *User's Manual* presents the characteristics of the instruction in several ways. The *operation* indicates that the destination location is replaced by 0. Motorola refers to this notation as Register Transfer Language (RTL), which is summarized in Appendix IV. The assembler recognizes the CLR mnemonic and converts it and the effective address of the destination location to machine language. The valid destinations, in terms of the possible addressing modes, are listed in the table accompanying the description of the instruction. These modes are discussed in Section 4.4. Other important characteristics, such as the effect on the condition codes and the machine language format, are also given.

There are a number of ways to specify the location to be cleared by the CLR instruction. The method is chosen from among the eight valid addressing modes shown in the table at the bottom of the figure. The length of the operand at the destination location is called its size and is specified as byte (8 bits), word (16 bits), or longword (32 bits), with the corresponding symbolic designation B, W, or L, respectively. For example, the assembler recognizes

CLR.B   D1

Table 4.9 Operation of the CLR Instruction

Instruction	Contents of the Destination After Instruction Executes
CLR.B ⟨EAd⟩	FFFF FF00
CLR.W ⟨EAd⟩	FFFF 0000
CLR.L ⟨EAd⟩	0000 0000

Note: Destination contains FFFF FFFF before each instruction executes.

CLR

CLR   Clear an Operand   **CLR**

Operation: 0 → Destination

**Assembler
Syntax:**   CLR <ea>

Attributes: Size = (Byte, Word, Long)

Description: The destination is cleared to all zero bits. The size of the operation may be specified to be byte, word, or long.

Condition Codes:

X	N	Z	V	C
—	0	1	0	0

N   Always cleared.
Z   Always set.
V   Always cleared.
C   Always cleared.
X   Not affected.

Instruction Format:

15	14	13	12	11	10	9	8	7	6	5 4 3	2 1 0
0	1	0	0	0	0	1	0	Size		Effective Address Mode	Register

Instruction Fields:

Size field — Specifies the size of the operation:
00 — byte operation.
01 — word operation.
10 — long operation.
Effective Address field — Specifies the destination location. Only data alterable addressing modes are allowed as shown:

Addressing Mode	Mode	Register	Addressing Mode	Mode	Register
Dn	000	register number	d(An, Xi)	110	register number
An	—	—	Abs.W	111	000
(An)	010	register number	Abs.L	111	001
(An) +	011	register number	d(PC)	—	—
−(An)	100	register number	d(PC, Xi)	—	—
d(An)	101	register number	Imm	—	—

Note: A memory destination is read before it is written to.

Figure 4.11   Description of the CLR instruction. (Courtesy of Motorola, Inc.)

as the instruction to clear 8 bits of register D1. To describe this operation precisely, our notation will be

$$(D1)[7:0] <-- 0$$

indicating that bits 0 through 7 of register D1 are cleared. The replacement symbol $(<--)$ will mean that the operand on the left is replaced by the value on the right. After the instruction executes, the contents of the destination locations are equal to zero, which is indicated as

$$(D1)[7:0] = 0$$

Example 4.9

Several examples of the CLR instruction are given below. The addresses for the destination locations in memory are indicated as decimal values. This conforms with the assembly language notation to be discussed in Chapter 5. A word (16-bit) location in memory consists of two consecutive bytes.

Instruction Symbolic Form	After Execution
CLR.B D1	$(D1)[7:0] = 0$
CLR.W D1	$(D1)[15:0] = 0$
CLR.B 1000	$(1000) = 0$
CLR.W 1000	$(1000) = 0$
	$(1001) = 0$

4.3.2 The MOVE Instruction

The fundamental data movement instruction for the MC68000 is the MOVE instruction, which is a double-address instruction written in symbolic form as

MOVE.X   ⟨EAs⟩,⟨EAd⟩

where X = B, W, or L specifies the length or size of the operand. A copy of the source operand of length X in location ⟨EAs⟩ is transferred to the destination location ⟨EAd⟩, leaving the source location unchanged. Both the source and destination operands are treated as though they are of length X. The MOVE instruction copies the source operand into bits $[7:0]$, $[15:0]$, or $[31:0]$ for the transfer of a byte, word, or longword, respectively. The effective addresses, ⟨EAs⟩ and ⟨EAd⟩, are computed by the processor according to the specification of the addressing mode. They may indicate processor registers or memory locations. Figure 4.12 shows the operation of the MOVE instruction for the three operand lengths. In each case, the destination and source locations contain 32-bit values, but only the specified portion of the operand is copied from the source to the destination location.

Example 4.10

A number of examples of the MOVE instruction are shown in the following summary. In each case, the source location, data register D2, contains the hexadecimal value 0FFF 0105 before each instruction executes. The destination register D1 contains 1000 0000 in 32 bits before execution.

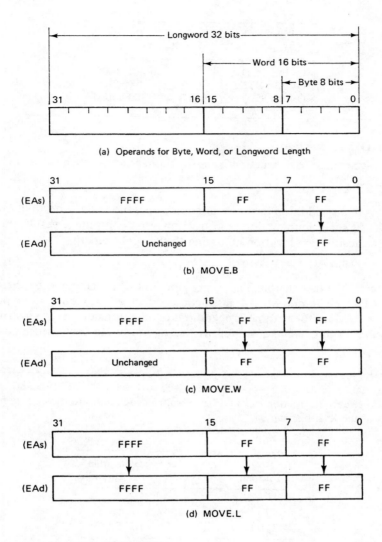

(a) Operands for Byte, Word, or Longword Length

(b) MOVE.B

(c) MOVE.W

(d) MOVE.L

Note: Source operand is FFFF FFFF$_{16}$.

Figure 4.12 Operation of the MOVE instruction.

Instruction	Destination After Execution
MOVE.B D2,D1	1000 0005
MOVE.W D2,D1	1000 0105
MOVE.L D2,D1	0FFF 0105

4.3.3 The ADD Instruction

An important arithmetic instruction is the ADD instruction. It has the form

ADD.X   ⟨EAs⟩,⟨EAd⟩

and performs binary addition between the source operand and the destination operand of length X. In such double-address instructions that compute a result, the result is stored in the destination location according to the replacement

(EAd)[X] < - - (EAs)[X] + (EAd)[X]

by execution of the instruction. The source operand is not changed by these instructions. The ADD, CLR, and MOVE instructions discussed previously are typical of the MC68000 instructions for arithmetic operations or data transfer. In such instructions, an operand location and length, as well as the operation to be performed, must be specified.

Example 4.11

Assume that data register D2 contains the hexadecimal value of 0FFF 0105 and D1 contains 1000 0001 before each instruction executes. The results stored in D1 are shown here for addition of operands of the length specified.

Instruction	Destination After Execution
ADD.B D2,D1	1000 0006
ADD.W D2,D1	1000 0106
ADD.L D2,D1	1FFF 0106

4.3.4 Other Instruction Types

Program control instructions may modify the flow of control in a program by changing the value in the program counter and thereby causing a new sequence of instructions to be executed. For example, the Jump instruction

JMP   ⟨EA⟩

causes program control to be transferred to the instruction contained in the location designated by the effective address ⟨EA⟩. The Branch instruction

BRA   ⟨disp⟩

adds a displacement value to the contents of the program counter at the time the instruction is executed. This causes program control to be transferred within the range allowed by the value ⟨disp⟩, which is either a positive or negative integer. Both the BRA (Branch Always) and JMP instructions cause unconditional transfer of control. Other branch instructions may or may not cause a branch, depending on conditions set by an arithmetic operation. For example,

BGT   ⟨disp⟩

causes a branch if the result was greater than zero. The condition is indicated by the setting of the condition code bits in the status registers. These program control instructions are discussed in detail in Chapter 6.

Instructions that control the operation of the processor or the system are generally reserved for programs operating in the supervisor mode. The instruction STOP, for example, causes the processor to discontinue fetching and executing instructions. Such instructions are discussed in more detail in Chapter 10.

EXERCISES

4.3.1. Describe the contents of each byte location affected by the instruction CLR.L 1000.

4.3.2. Before each instruction given executes, assume the following hexadecimal contents of D1, D2, and longword location 1000.

(D1) = 0601
(D2) = 0805
(1000) = 1913

Determine the results of executing each of the following instructions.
(a) MOVE.B   D1,D2
(b) MOVE.B   D1,1001
(c) CLR.W   1000
(d) ADD.B   D1,D2
(e) ADD.W   1000,D1

4.4   ADDRESSING MODES FOR THE MC68000

The *addressing modes* of a CPU determine the ways in which a processor can reference an operand held in one of its registers or in memory. For each operand, the addressing mode specifies how the processor is to locate or calculate the actual address of the operand. The actual address is called the effective address and is determined when the instruction referencing the operand is executed.[2]

---

[2] When the CPU directly addresses the memory, an effective address is the actual or physical hardware address. In many sophisticated systems, special circuitry called memory-mapping circuitry is employed. This circuitry then computes the physical address that corresponds to the CPU address. The physical addresses involved depend entirely on the design of the system and are independent of programming references to operands in memory.

The broad categories of addressing for the MC68000 include direct addressing, indirect addressing, and addressing relative to the program counter. A special immediate mode is also provided. Table 4.10 defines these basic modes for the MC68000.

In the table, the category of addressing mode and the effective address that results from instruction execution are listed. The symbolic designation is the reference to the given addressing mode recognized by a Motorola assembler. An absolute address is considered to be a decimal value unless it is preceded by "$" to indicate a hexadecimal value. The symbol "*" in a symbolic instruction references the current value of the program counter and the symbol "#" preceding a number indicates immediate addressing.

Instructions of the MC68000 can specify one or two operands in the manner described in Section 4.3. The CLR instruction, for example, may specify the destination by any of the modes indicated in Table 4.9 except immediate. The MOVE or ADD instructions require two operands, and both the source and the destination addressing modes must be specified. A number of examples in this section show how the basic addressing modes are specified symbolically. The discussion presented in this section is limited to those modes shown in Table 4.10, which represent only 8 of the 14 possible addressing modes for the MC68000. A more detailed study of the MC68000 addressing modes is given in Chapter 5 after assembly language programming is introduced.

4.4.1 Direct Addressing

The *direct* addressing modes of the MC68000 include register addressing and absolute addressing. In either case, the location or address of an operand is specified directly as part of the instruction, so that no calculation of an effective address by the CPU

Table 4.10 Basic Addressing Modes

Type	Effective Address	Symbolic Designation
Direct		
Register	EA = Rn	D0, D1, . . . , D7; A0, A1, . . . , A7
Absolute	EA = ⟨address⟩	⟨decimal address⟩ or $⟨hexadecimal address⟩
Indirect		
Address register	EA = (An)	(A0), (A1), . . . , (A7)
Predecrement	An = An − k,EA = (An)	−(An)
Postincrement	EA = (An),An = An + k	(An)+
Relative with displacement	EA = (PC) + ⟨disp⟩	* + ⟨disp⟩
Immediate	None	#⟨data⟩

Notes:

1. Rn refers to any register Dn or An.

2. For the predecrement and postincrement modes, k is 1, 2, or 4 for byte, word, or longword operations, respectively.

3. Angle brackets ⟨⟩ imply that the indicated value must be specified.

is necessary. For the register modes, the operand is in one of the address or data registers. In the absolute mode, the operand is in memory at a location designated by a positive integer representing its address. This address is not related to the length or size of the operand except that word and longword operands must be addressed at even locations in memory.

The basic format for the CLR instruction using register addressing is

CLR.⟨X⟩   ⟨Dn⟩

where the operand of length X is cleared in register Dn, which is written specifically as one of D0, D1, . . . , D7. Thus the instruction

CLR.W   D2

clears the low-order 16 bits of register D2. The MOVE instruction requires two operands and has the form

MOVE.⟨X⟩   ⟨Dm⟩,⟨Dn⟩

as, for example,

MOVE.W   D1,D2

which copies (D1)[15 : 0] into (D2)[15 : 0].

An absolute address can be specified as a decimal or hexadecimal integer in an instruction. For example, the instruction

MOVE.W   10000,D1

transfers 16 bits from word location 10000 to (D1)[15 : 0]. According to the conventions of Motorola assemblers, the symbolic form for the same location in hexadecimal would be

MOVE.W $2710,D1

because the value 2710_{16} corresponds to 10000 and the $ indicates hexadecimal.

Example 4.12 _____

Being the simplest addressing schemes, the direct addressing modes were used in the preceding section to introduce important processor instructions. For example, the instruction

CLR.W   1000

specifies the *absolute* address 1000 as the destination. The address is stored with the instruction in memory. The instruction

MOVE.W   1000,D1

employs absolute addressing for the source location and register direct addressing for the destination. An instruction such as

MOVE.W   A1,D1

transfers the 16-bit address in A1 to the data register.

4.4.2 Indirect Addressing

In the MC68000, *indirect* addressing means the use of the contents of an address register as the address of an operand in memory. The contents are used as a pointer to reference the location. For example, if the instruction

MOVE.W   (A1),D1

is executed when (A1) = 1000, the 16-bit value in memory word location 1000 would be copied into (D1)[15:0]. To modify the address referenced in memory, the address register can be changed by any instruction that operates on the contents of address registers. This ability to modify the pointer in very flexible ways allows a programmer to address values in sophisticated data structures in memory. A simple example is the stack structure discussed in Section 4.2.3.

In fact, the stack structure is so common in modern programs that the MC68000 has two indirect addressing modes that are used primarily with stacks. These stacks can be created and used by employing the address register indirect with postincrement or predecrement addressing modes. To add data to a stack that grows from high memory to low memory, for example, the instruction

MOVE.W   D1, − (A1)

transfers a word from D1 to the stack after the stack pointer (A1) is decremented by 2 (bytes) to point to the next free memory location. A data item could be retrieved with the instruction

MOVE.W   (A1) + ,D2

which pops the word from the stack addressed by A1 and copies it to (D2)[15:0]. After the transfer, A1 is incremented by 2. The push operation, for the downward-growing stack, employs the predecrement addressing mode using A1 as the stack pointer. The pop requires the postincrement mode for the source addressing mode. Any address register of the MC68000 can be used as a stack pointer. The source location in the push or the destination location in the pop operation can be a memory location or any register of the MC68000.

Several other indirect addressing modes are provided by the MC68000. An address register containing an indirect address may be indexed by another address register or a data register. Variations on indirect addressing are described in more detail in Chapter 5.

4.4.3 Relative Addressing

A program counter *relative* address is an address that the CPU calculates by adding a displacement to the value in the program counter. The calculated effective address is then

$$EA = (PC) + \langle disp \rangle$$

where the displacement value $\langle disp \rangle$ is specified in the instruction. The displacement is a positive or a negative integer, so the referenced location can be higher or lower in memory relative to the instruction using this addressing mode.

An example of relative addressing is indicated by the instruction

BRA   * + 10

which, when executed, would cause a branch 10 byte locations ahead of that indicated by the program counter.[3] In the case of the BRA instruction, the value in the program counter is changed to the new address to point to the next instruction six word locations farther up in memory from the location of the BRA instructions. Relative addressing can also be used to address data values in memory.

Because the program counter contents act as a pointer to the instruction currently executing, the displacement value indicates the distance between the operand referenced in the relative mode and the instruction itself. If the program is moved in memory, the relative references in the program are still correct. When the memory references used by a program are relative, the program is said to be position-independent. Such programs are discussed in Chapter 9. Programs with absolute references to memory locations cannot be moved unless the absolute addresses are changed to indicate the new locations.

The MC68000 also allows program counter relative addressing with indexing. In this mode, the effective address is calculated as the contents of the PC, plus a displacement value, plus the contents of an index register. Such variations on relative addressing are discussed in Chapter 5.

4.4.4 Immediate Addressing

The *immediate* addressing mode is used to specify a constant 8, 16, or 32 bits long. The constant is included in the instruction in memory. For example, the instruction

ADD.W   #5,D1

adds 5 to the value in $(D1)[15:0]$. The instruction

MOVE.B   #'A',(A1)

moves the ASCII value "A" into the byte addressed by A1 in memory. The assembler recognizes the source addressing modes in these two examples as immediate. Of course, the immediate mode is never allowed as a destination mode because the destination location must be alterable (writable).

---

[3] When the branch is taken, the value in the PC is the address of the BRA instruction +2. The BRA instruction requires two word locations in memory. Thus, the BRA instruction shown causes a branch to an instruction six word locations higher in memory than the first word of the BRA instruction itself.

Example 4.13 ———————————————————————————

The symbolic instruction

MOVE.L #'1234',D1

causes the contents of D1 to be replaced with the hexadecimal value 3132 3334. Similarly, the instruction

MOVE.W   #$F0,D1

has the effect

$(D1)[15:0] < -- F0_{16}$

An instruction with an immediate operand as a destination such as

MOVE.B   1000,#10000

would be illegal and cannot be assembled.

EXERCISES

4.4.1. Using hexadecimal values for all of your answers, determine the operation and locations affected by each of the following instructions.
   (a) MOVE.W   1000,2000
   (b) MOVE.W $1000,D1
   (c) MOVE.B   1000,D1
   (d) CLR.L $FFFFFC

4.4.2. Compare the operation of the following instructions when $(A1) = 1000$ and $(1000) = FFE0_{16}$ before each execution.
   (a) MOVE.W   A1,D1
   (b) MOVE.W   (A1),D1
   (c) MOVE.W   1000,D1
   (d) MOVE.W   #1000,D1

4.4.3. Determine the contents of the destination in hexadecimal after each instruction executes.
   (a) MOVE.W   #'AB',D1
   (b) MOVE.W   #$C1,D1
   (c) MOVE.W   #1000,D1

4.4.4. Using only the instructions and techniques discussed thus far, write the symbolic instructions to store the low-order word of D1 into memory locations 1001 and 1002, that is, after execution $(1001) = (D1) [15:8]$ and $(1002) = (D1) [7:0]$. (Remember that word-length operands must start at even locations in memory and that they occupy 2 bytes.)

4.5 MACHINE LANGUAGE FOR THE MC68000

The instructions that the processor executes are coded in machine language format in memory. These instructions can be created by a programmer coding these binary sequences directly. More likely, an assembler program translates assembly language

Bits 15 through 12	Operation		Bits 15 through 12	Operation
0000	Bit Manipulation/MOVEP/Immediate		1000	OR/DIV/SBCD
0001	Move Byte		1001	SUB/SUBX
0010	Move Long		1010	(Unassigned)
0011	Move Word		1011	CMP/EOR
0100	Miscellaneous		1100	AND/MUL/ABCD/EXG
0101	ADDQ/SUBQ/Scc/DBcc		1101	ADD/ADDX
0110	Bcc/BSR		1110	Shift/Rotate
0111	MOVEQ		1111	(Unassigned)

Figure 4.13 Operation codes. (Courtesy of Motorola, Inc.)

statements written in symbolic notation into machine language instructions. The assembly language programmer rarely works directly with the machine language program, but a knowledge of the machine language formats is necessary for a full understanding of the capability of the processor.

The machine language instructions for the MC68000 consist of from one to five 16-bit words in memory. The first word is the operation word, which contains the operation code (op code), as well as the size or length and the addressing modes for any operands, if necessary. For most instructions, the op code is contained in bits 12 through 15 of the first word. Various combinations of these 4 bits yield 16 different op codes.

The meaning of each of these is defined in Figure 4.13. The remaining 12 bits in the operation word are used to further define the operation to be performed. Additional extension words for the machine language instructions may contain immediate data or absolute addresses for source or destination operands. A short absolute address (16 bits) requires one extra word, and a long address (32 bits) requires two. The format of the machine language instruction is shown in Figure 4.14. The extension words follow the operation code at higher memory addresses.

The formats for single- and double-address instructions are discussed in this section. As in previous sections of this chapter, the CLR, ADD, and MOVE instructions are used for specific examples. Every instruction for the MC68000 is described in Appendix IV.

15	14	13	12	11	10	9	8	7	6	5	4	3	2	1	0
Operation Word (First Word Specifies Operation and Modes)															
Immediate Operand (If Any, One or Two Words)															
Source Effective Address Extension (If Any, One or Two Words)															
Destination Effective Address Extension (If Any, One or Two Words)															

Figure 4.14 Machine language instruction formats. (Courtesy of Motorola, Inc.)

4.5.1 Single-Address Instructions

The operation word for a single-address instruction is shown in Figure 4.15(a). Bits [15:6] define the operation and bits [5:0] designate the addressing mode. The effective address field is itself divided into mode and register subfields of 3 bits each.

For an addressing mode employing a register, the register number (0-7) is given in the register subfield. In this case, the mode subfield specifies whether direct or indirect addressing is used and the variations shown in Figure 4.16 apply. Absolute, relative, or immediate addressing modes have a fixed encoding for the entire 6-bit field.

As an example, the instruction

CLR.B   D1

specifies the destination D1 by direct register addressing. The machine language format is shown in Figure 4.15(b). For data registers, the mode in the effective address field is {000} and the register number is {001}. The bit pattern in bits [15:8] specifies the CLR instruction. The operand size is a byte in this example and is indicated by 00 in bits [7:6]. Because only register addressing is used, the instruction requires only one word of memory.

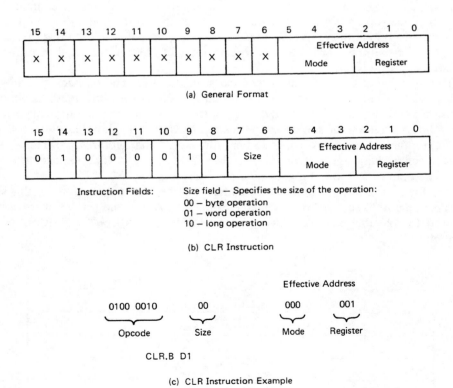

Figure 4.15   Single-address instructions. (Courtesy of Motorola, Inc.)

Addressing Mode	Mode	Register
Data Register Direct	000	register number
Address Register Direct	001	register number
Address Register Indirect	010	register number
Address Register Indirect with Postincrement	011	register number
Address Register Indirect with Predecrement	100	register number
Address Register Indirect with Displacement	101	register number
Address Register Indirect with Index	110	register number
Absolute Short	111	000
Absolute Long	111	001
Program Counter with Displacement	111	010
Program Counter with Index	111	011
Immediate or Status Register	111	100

Figure 4.16   Effective address encoding. (Courtesy of Motorola, Inc.)

A number of other single-address instructions, such as NEG (negate), NOT (one's complement), and NBCD (negate decimal), have the same general format as the CLR instruction. Other instructions with single operands or those with no operands can vary considerably in their machine language format from that shown for the CLR instruction. Instructions to control the processor can have no address specification but use a unique 16-bit operation word with fixed format. For example, the STOP instruction has the single hexadecimal word 4E72 as its op code.

4.5.2 Double-Address Instructions

When an instruction uses two operands, the addressing modes for both the source and destination must be specified in the operation word. If each double-address instruction encoded the addressing modes into 6 bits each, as shown previously, and specified length (byte, word, or longword) for each operand using 2 bits, a total of 14 bits of the 16-bit operation word would be taken; thus only 2 bits would remain for the op code. Because 4 bits are always used for the op code, flexibility in addressing for double-address instructions must be limited further to provide a full set of instructions. A comparison of the MOVE instruction and the ADD instruction shows the approach taken by the designers of the MC68000.

The MOVE Instruction.   The format of the MOVE instruction is shown in Figure 4.17(a) for MOVE.B, MOVE.L, and MOVE.W with bits [13:12] of the op code specifying the length. The source addressing mode is specified as before for single-address instructions. However, the destination addressing mode for MOVE reverses the mode/register encoding as shown. As an example, the format for the instruction

MOVE.W   D1,D3

is illustrated in Figure 4.17(b).

The ADD Instruction.   The ADD instruction has the format shown in Figure 4.18(a). It requires the source or destination operand to be held in one of the data registers of the processor. The symbolic form of the ADD instruction is either

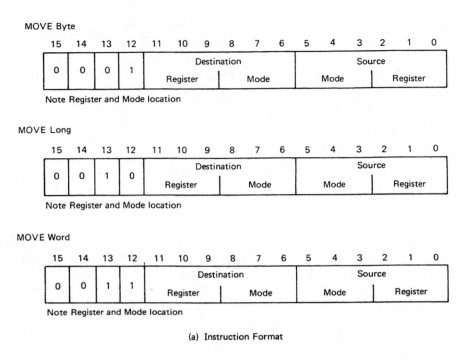

(a) Instruction Format

(b) MOVE.W D1,D3

Figure 4.17 MOVE instruction format. (Courtesy of Motorola, Inc.)

ADD.X   ⟨EAs⟩,⟨Dn⟩

or

ADD.X   ⟨Dn⟩,⟨EAd⟩

with X = B, W, or L, as before. In the first case, Dn specifies the destination for the result of the addition. The "op mode" bits [8:6] determine the length X and specify the destination as Dn. In the second instruction, the location specified by ⟨EAd⟩ is the destination and Dn is the source, so the op mode changes. An example is shown in Figure 4.18(b).

Many other double-address instructions restrict the source or destination location so that it can be only a processor register. Thus memory-to-memory operations are not allowed except with the MOVE instruction. The MOVE instruction is therefore the most flexible of the MC68000 instructions with respect to its allowed addressing modes.

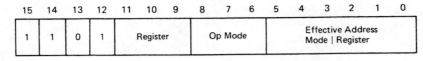

Instruction Fields:

Register field — Specifies any of the eight data registers
Op-Mode field —

Byte	Word	Long	Operation
000	001	010	$(<Dn>) + (<ea>) \rightarrow <Dn>$
100	101	110	$(<ea>) + (<Dn>) \rightarrow <ea>$

(a) ADD Instruction Format

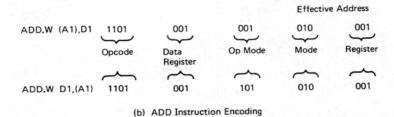

(b) ADD Instruction Encoding

Figure 4.18   ADD instruction format. (Courtesy of Motorola, Inc.)

Example 4.14 _____

Figure 4.19 shows several examples of the machine language and assembly language forms for the CLR, ADD, and MOVE instructions. The hexadecimal values to the left of the instruction represent the machine language code. Any immediate value or absolute address is held in memory words following the operation word.

4.5.3 Compatibility with MC68000 Code

The 16-bit MC68000 instructions have the same bit patterns as the corresponding instructions of the more advanced processors in the M68000 family. Thus, the machine-language or object code for a user-mode program written for the MC68000 will execute unaltered on processors such as the MC68020, MC68040, MC68060 and the microcontrollers containing the CPU32. This compatibility also extends to the supervisor mode except for certain operations that manipulate the data contained in the supervisor stack. This slight incompatibility is due to the fact that the stack formats of saved information are not the same for the various processors.

However, with few exceptions, software developed for the 16-bit Motorola family will execute correctly on the other processors in the family. Programs will execute faster because the 32-bit processors have a faster instruction execution time for most instructions. One drawback to executing unmodified MC68000 programs on the 32-bit CPUs is that those processors have more efficient instructions for certain operations.

```
abs.  rel.   LC    obj. code    source line
----  ----   ----  ---------    -----------
   1     1   0000                         TTL      'FIGURE 4.19'
   2     2   0000                         LLEN     100
   3     3   0000            *
   4     4   8000                         ORG      $8000
   5     5   8000                         OPT      P=M68000
   6     6   8000            *
   7     7   8000  421D                   CLR.B    (A5)+
   8     8   8002  4258                   CLR.W    (A0)+
   9     9   8004  42B8 0568              CLR.L    $0568
  10    10   8008  4239 0002              CLR.B    $00020000
  10         800C 0000
  11    11   800E            *
  12    12   800E  D800                   ADD.B    D0,D4
  13    13   8010  D378 308E              ADD.W    D1,$308E
  14    14   8014  0642 0030              ADD.W    #$30,D2
  15    15   8018            *
  16    16   8018  141D                   MOVE.B   (A5)+,D2
  17    17   801A  3401                   MOVE.W   D1,D2
  18    18   801C  33C1 0003              MOVE.W   D1,$300E8
  18         8020  00E8
  19    19   8022  1CFC 002D              MOVE.B   #'-',(A6)+
  20    20   8026  23F9 0002              MOVE.L   $00022000,$14000
  20         802A  2000 0001
  20         802E  4000
  21    21   8030            *
  22    22   8030                         END
  22 lines assembled
```

Figure 4.19 Program examples of instruction formats for the CLR, ADD, and MOVE instructions.

Thus, if the programs are not rewritten, they will not take advantage of the enhanced instruction set of the advanced processors.

EXERCISES

4.5.1. Write the symbolic statements necessary to add two values in memory together and store the results in a third location.

4.5.2. Assume that (A1) = $1000 and ($1000) = $0010 before the execution of each instruction listed. Determine the resulting action of each instruction.
    (a) CLR.B $1000
    (b) CLR.W   (A1)
    (c) MOVE.W   A1,(A1)
    (d) MOVE.W $1000,D1
    (e) MOVE.W   #$1000,D1
    (f) MOVE.B   (A1),D1
The values are hexadecimal numbers for addresses and contents.

4.5.3. Translate the following machine language statements, given in hexadecimal, into the assembler language (symbolic) equivalent.

    (a) 4241

    (b) 200B

    (c) 103C 002E

4.5.4. Write the machine language instruction for the following symbolic instructions.

    (a) CLR.W   D0

    (b) MOVE.L   A0,D0

    (c) ADD.B   D0,D5

4.6   THE MC68000 AND MEMORY ORGANIZATION

A simplified diagram of a MC68000 system is shown in Figure 4.20, which illustrates the 24 signal lines for addressing and 16 signal lines for data transfer or instruction fetching by the processor.[4] The MC68000 is considered a byte addressing processor and each address indicates a byte (8 bits) location in memory or the address of a location associated with an interface. The range of possible addresses is called the addressing space of the processor. This space for the 24-bit address of the MC68000 is shown in Figure 4.21.

    The system designer can allocate the addressing space for programs, data, or I/O interfaces as necessary, but certain conventions are required for products that use the MC68000. The lower 1,024 (400_{16}) locations are reserved by the MC68000 processor for use as addresses (called vectors by Motorola). These addresses point to routines for servicing interrupts or processing traps. As such, they indicate the starting address of operating system routines that process exceptions in the supervisor mode. In MC68010 systems, one or more vector tables can be present and each located anywhere in memory.

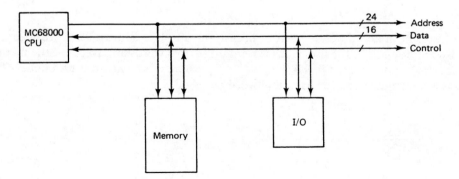

Figure 4.20   The MC68000 and memory.

---

[4] Depicting a 24-bit address bus represents the functional view. Chapter 13 presents the actual signal lines of the MC68000.

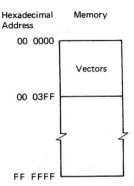

Figure 4.21   Address space for byte data.

Because the MC68000 can address byte, word, or longword operands, the physical organization of memory into bytes as shown in Figure 4.21 may be confusing when word or longword operands are addressed. The programmer must be aware of the relationship between the physical organization of memory into bytes and the operand length specified in an instruction. For byte-length operands, the physical address directly identifies the byte addressed. When a word or longword operand is specified in an instruction, the address identifies 2 or 4 bytes in memory, respectively.

4.6.1 Memory Organization and Addressing

The physical byte configuration of MC68000 memory can be logically organized into words as shown in Figure 4.22. Although each byte in the word can be addressed by the processor, instructions and word operands must be referenced only at even addresses and, therefore, occupy locations designated $n, n + 2, n + 4, . . .$, where n is an even integer. An attempt to address a word operand at an odd boundary, as in the instruction

MOVE.W $1001,D1

would result in an addressing error because the absolute address $1001 resides at an odd-word boundary. If a longword is referenced, two words are used and the memory addresses are designated $n, n + 4, n + 8, . . .$, where n is again an even integer. Thus 4 bytes are transferred for each longword accessed.

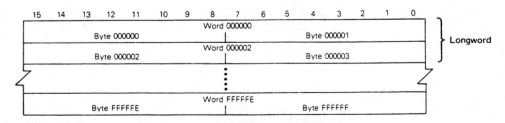

Figure 4.22   Memory organization by address. (Courtesy of Motorola, Inc.)

4.6.2 Data Organization in Memory

Data or addresses are stored in memory as illustrated in Figure 4.23. Within a byte, bit 0 is the rightmost and bit 7 is the leftmost bit. For integer data, bit 0 represents the least significant digit for either 8-bit (byte) or 16-bit (word) values. A longword is stored with the high-order 16 bits in location n and the low-order portion at location $n + 2$, where n is an even integer. For example, the hexadecimal value 0200 0100 is stored with (0200) at the lower memory address and (0100) at the next word location. If the value is an address, the address is stored in the same way. Binary-coded decimal (BCD) values are stored two digits per byte, as shown.

Example 4.15 _____

Table 4.11 shows a number of items stored in memory at the locations specified. In each case, the address and its contents are hexadecimal. The instructions CLR and MOVE require one word for their operation word. The MOVE instruction also requires an additional word to indicate the short absolute address $1000. The long address $200F6 is stored as shown, with the most significant word appearing first in memory. For example, a return address saved on the system stack would be stored in this manner. The word location $2000 contains $2001 in the figure, but each individual byte could be addressed. Thus the byte address $2000 contains $20 and the byte address $2001 contains $01, as shown. The instruction

MOVE.B $2000,D1

would cause (D1)[7:0] = $20. The transfer

MOVE.W $2000,D1

results in (D1)[15:0] = $2001.

Finally the location $2008 contains the decimal value 1021 stored as a BCD number. The two low-order digits are stored in byte location $2009 and the two high-order digits are in location $2008. MC68000 instructions that operate on multidigit BCD numbers require this format for BCD data storage.

Table 4.11   Example of Memory Contents

Memory Address	Contents	Meaning	
Increasing			
1000/1001	42 83	CLR.L	D3
1006/1007	11 C0	MOVE.B	D0,$1000
1008/1009	10 00		
100A/100B	00 02	Address	$200F6
100C/100D	00 F6		
	•		
	•		
	•		
2000/2001	20 01	Byte data	
	•	(2000) = 20	
	•	(2001) = 01	
	•		
2008/2009	10 21	BCD 1021_{10}	

Note:
Except for the BCD value 1021, all numbers are hexadecimal.

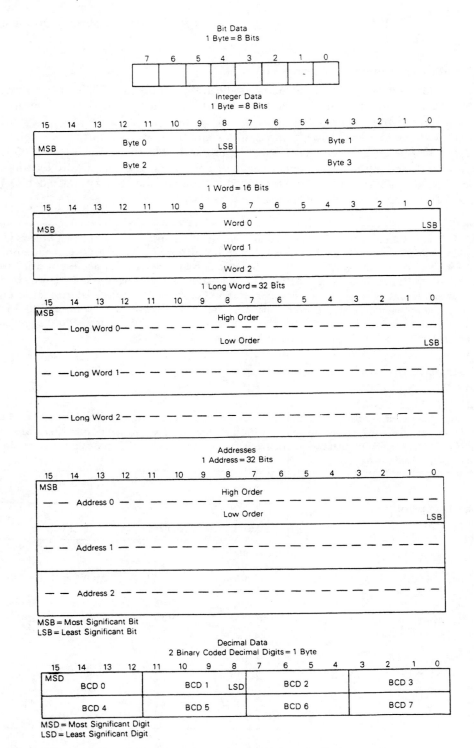

Figure 4.23 Data organization in memory. (Courtesy of Motorola, Inc.)

Table 4.12 Alternative Memory Storage

Memory Address	Big Endian (Motorola)	Little Endian (Intel)
$1000	00	F6
$1001	02	00
$1002	00	02
$1003	F6	00

4.6.3 Other Memory Organizations (Big- and Little-Endian)

Incompatibility of hardware and software is one of the most frustrating things about dealing with computers produced by different manufacturers. In fact, compatibility between members of the 68000 family was one of the most important reasons given in Chapter 1 for choosing a family of processors. Although attempts have been made to standardize various aspects of different computer systems such as the floating-point notation and standard busses previously discussed, most of the components of interest to the assembly language programmer or interface designer are not compatible. When designing an interface between different computer systems this lack of compatibility may even apply to the storage of data in memory.

Notice that in Example 4.15 the MC68000 convention is that the most significant unit of data or an address is stored at the lowest memory address. This is called *big-Endian* storage. The Intel family of 80x86, such as the 80486 processor, take a *little-Endian* approach and store the least significant unit at the lowest address.[5] Table 4.12 indicates the storage of the value $000200F6_{16}$ in locations $1000 through $1003 in both cases.

The storage standard can be imposed by the processor or by the system bus. For a 32-bit transfer, the VME bus stores byte 0 of data being transferred at the lowest memory address as expected by the M68000 family processors. The NuBus is another standard bus that stores the least significant byte at the lowest address. However, Apple incorporates the NuBus in many of their computers that contain M68000 family CPUs. If values are exchanged between computers or between a bus and a computer with incompatible storage, the conversion can be accomplished by the interface (hardware) or by a program.

EXERCISES

4.6.1. Determine the decimal number of bytes, words, or longwords the MC68000 can address.

4.6.2. Show how the following numbers or characters are stored in memory if each starts at hexadecimal location 1000. The data and formats are as follows:

(a) 10,203,040 (BCD)

---

[5] The terms big-Endian and little-Endian are taken from Jonathan Swift's satirical story, *Gulliver's Travels*.

(b) 0200 00FC (hexadecimal)

(c) 'ABCD'(ASCII)

4.6.3. The program counter contained 0002 FFF0$_{16}$ before it was transferred into memory starting at location 1002$_{16}$. What are the memory contents in each byte of the memory area where (PC) is stored?

REVIEW QUESTIONS AND PROBLEMS

Multiple Choice

4.1. The data registers of the MC68000 are used as
    (a) accumulators and buffer registers
    (b) index registers
    (c) both (a) and (b)

4.2. The MC68000 stack is
    (a) First In First Out (FIFO)
    (b) Last in First Out (LIFO)
    (c) Circular

4.3. Which of the following is *not* an exception?
    (a) trap
    (b) interrupt
    (c) subroutine call

4.4. The Status Register value of $0404 indicates
    (a) user mode, interrupts masked at level 4 and below, Z = 1
    (b) interrupt level 4 enabled, Z = 1
    (c) Trace T0 on, Z = 1

4.5. If (D1) = $FFFFFFFF, the instruction CLR.B D1 yields
    (a) $00000000
    (b) $FFFFFF00
    (c) $FFFF0000

4.6. If (A1) = $0000 1000 and (D1) = $FFFF1005, the instruction MOVE.W D1, − (A1) yields
    (a) ($1000) = $FFFF1005
    (b) ($1000) = $1005
    (c) ($0FFE) = $1005

4.7. Let (D2) = $0FFF0105 and (D1) = $10000001. The result of ADD.W D2,D1 is
    (a) $1000 0006
    (b) $1000 0106
    (c) $0FFF 0106

4.8. The contents of D1 after the instruction MOVE.W 10000,D1 is
    (a) the number 10000
    (b) the contents of location 10000
    (c) D1[W] = 16 bits from word location 10000

4.9. The addressing mode for the source operand in the instruction MOVE.W (A1)+,D1 is
    (a) postincrement
    (b) register indirect
    (c) stack addressing

In 4.10 and 4.11, (A1) = 1000 and (1000) = $FFE0.

4.10. The result of the instruction MOVE.W A1,D1 is
  (a) (D1) = 10000000
  (b) (D1)[W] = 1000
  (c) (D1)[W] = (1000)

4.11. The result of the instruction MOVE.W (A1),D1 is
  (a) (D1)[W] = $FFE0
  (b) (D1)[W] = 1000
  (c) (D1) = $0000FFE0

4.12. The length of the instruction CLR.B $00020000 is
  (a) two words
  (b) three words
  (c) four words

4.13. True or false, object code for the MC68000 will execute on the MC68332.
  (a) True
  (b) False

4.14. The address $0002 0005 is loaded into memory at location 1002. The $05 is in byte location
  (a) 1003
  (b) 1004
  (c) 1005

Essay Questions

4.15. What are the advantages and disadvantages of separate address and data registers?

4.16. Describe the operation of the system stack pointer A7 as follows:
  (a) Describe the use of the system stack in the user and supervisor mode.
  (b) How does the stack grow in memory as items are pushed?
  (c) Show the addressing modes for manipulating data on the system stack.

4.17. Explain the term *effective address.* How is it used for M68000 family of processors?

Design Problems and Further Exploration

4.18. Consider the design of the instruction set of the MC68000. Why must some instructions be limited in their addressing flexibility compared to the MOVE instruction? For example, the ADD instruction requires that one of the operands be held in a data register.

4.19. Describe some of the factors that the designers must have considered when selecting the instructions and addressing modes for the MC68000. (This problem is considered in several articles listed in the Further Reading section of this chapter.)

4.20. The MC68000 object code for user-mode programs executes on MC68020-based systems. The assembler source code, however, is not necessarily compatible. When is source-code compatibility desirable? When is object-code compatibility advantageous?

FURTHER READING

The articles and textbooks listed here treat various aspects of the design of modern computer systems and CPU chips. The articles by Johnson, Kuban, and the Motorola Semiconductor Group deal specifically with the Motorola family of products. In particular, Johnson's article compares CISC and RISC architectures.

Johnson, Thomas L. "The RISC/CISC Melting Pot." *Byte* 12, no. 4 (April 1987): 153–160.

Kuban, John R., and John E. Salick. *Testing Approaches in the MC68020.* Motorola Publication AR225. (Austin, TX.: Motorola Semiconductor Products Inc.).

"MC68000 Microprogrammed Architecture." Motorola Publication AR235. (Austin, TX.: Motorola Semiconductor Group).

Tredennick's textbook explains the design of microprocessors using the IBM micro/370 and the MC68000 as examples. He was one of the designers of the MC68000 CPU. The textbook by Stone gives a detailed analysis of many of the features of modern microprocessors.

Tredennick, Nick. *Microprocessor Logic Design* (Bedford, Mass.: Digital Press, 1987).

Stone, Harold S. *High-Performance Computer Architecture* (Reading, Mass.: Addison-Wesley, 1987).

The textbook by Mazidi presents a complete description of Intel microprocessors and includes a discussion of RISC versus CISC designs.

Mazidi, Muhammad Ali, and Janice G. Mazidi. *The 80x86 IBM PC & Compatible Computers, Volume II.* (Englewood Cliffs, N.J. Prentice Hall, 1995).

The article by Stritter and Gunter is of historical interest and it covers many of the topics discussed in this chapter.

Stritter, Edward, and T. Gunter. "A Microprocessor Architecture for a Changing World: The Motorola 68000." *IEEE Computer* 12, no. 2 (February 1979): 43–52.

CHAPTER 5

MC68000 Assembly Language and Basic Instructions

The brief introduction in Chapter 4 to the machine language of the MC68000 should indicate the complexity involved in machine language programming. The extensive instruction set combined with the variety of addressing modes for many instructions would preclude efficient coding in machine language except for the simplest of programs. In assembly language, instructions and addresses are designated by symbolic names that the assembler program translates into the appropriate binary code. Motorola has defined a standard assembly language for the M68000 family of processors. The rules for the language specify the instruction mnemonics, symbolic addressing references, and the format for each assembly language statement. These conventions are generally followed by other suppliers of assemblers for the MC68000. Differences in assemblers must be resolved by reference to the *User's Manual* for a particular assembler.

This chapter begins in Section 5.1 with a discussion of techniques for program development. Section 5.2 introduces the assembly-language instructions for the MC68000. The emphasis is on standard features common to all assemblers although the program examples in this chapter were created with Motorola development software. Section 5.2 also describes the more sophisticated capabilities available with some assemblers. Section 5.3 describes each addressing mode of the MC68000. Section 5.4 summarizes the addressing modes and their use.

Notation. In this chapter, a hexadecimal number in the text itself is preceded by a "$". Otherwise, numerical values are decimal. However, assembler listings and outputs from monitor sessions use hexadecimal values for addresses of memory locations and their contents. No preceding symbol is used to indicate hexadecimal notation by these programs. However, the assembly language statements created by the programmer require the form $hhhh to indicate a hexadecimal number. As an immediate value, the hexadecimal number $hhhh would be written in a program as #$hhhh.

When an address register is used as the destination location in an instruction of an example, the instruction variations ADDA, MOVEA, SUBA, and so on, are used. Instruction variations for immediate (16 to 32 bits) and quick immediate (range 1–8) are referenced as ADDI, ADDQ, SUBI, SUBQ, and so on. Most assemblers recognize instruction variations without explicitly defining the suffix. The programs explicitly use the variations for clarity. For convenience, the assembly language instruction set for the MC68000 is summarized in the Appendix. The variations of the arithmetic instructions are presented in Chapter 7. Chapter 9 presents instruction variations that reference an address register as the destination.

5.1 SOFTWARE DEVELOPMENT

Software development consists of problem analysis, software design, and program coding, followed by debugging and testing. Appropriate documentation should be provided at each stage. The programming activities are shown in simplified form in Figure 5.1, which emphasizes the cyclic or iterative nature of the process. The editor program is used to create an assembly language *source* program, which is translated by the assembler.

At this stage in development, the assembler *listing* is used to find errors in the source program. The listing gives the assembly-language source program and the machine language equivalent if no errors are detected by the assembler. Errors in the source program are indicated on the listing.

Once the program is free of assembly errors, an *object module* is produced. The object module contains the translated assembly-language statements and other information to allow the program to be loaded into the target machine's memory for execution. Thus, the object module also requires processing by another program before the machine-language code for the CPU is produced. The processed object module is typically called a *load module*. A *binary* load module contains the machine–language program and any data in binary code ready to be loaded at the starting address of the program in memory.

Execution of the code is controlled during debugging by a program called a *debugger.* The debugger program allows the user to cause instruction-by-instruction execution and to display intermediate results after each instruction completes. Errors in the design of the program may be detected at this stage. To correct the errors, the source program must be reedited and reassembled. The loading and debugging of a program is often accomplished using a *monitor* that is a utility program stored in a ROM of the target MC68000 system. A typical monitor program will be described in Section 5.1.

Practical Considerations. The details involved in executing the development software (editor, assembler, debugger) vary greatly with different systems. Also, the source and object programs are normally stored as disk files on the development system disk. The *User's Manual* or operating system manual for a particular system will describe the procedure required to create, store, and execute programs.

In practice, the object program may require processing by another program, called a linkage editor, before it is loaded into memory for execution. The distinction between assembly, loading, and execution operations is discussed later in this chapter.

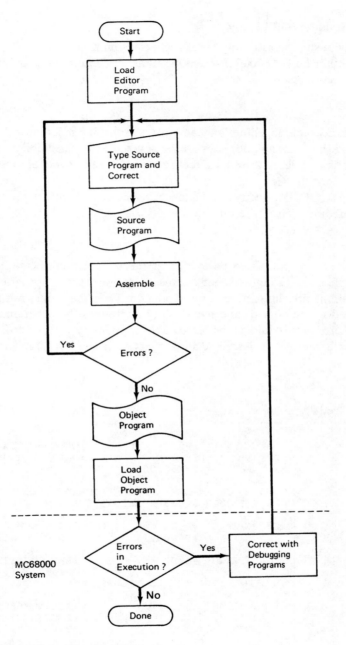

Figure 5.1 Programming a microcomputer.

5.1.1 The Assembler and Listing

As noted previously, the assembly process checks source statements for errors. Each statement is either an MC68000 instruction, an assembler directive, or a comment. A symbolic instruction such as

ADD.W D1,D2

becomes an executable machine-language instruction. The mnemonic ADD, the operand size, and the operand addresses are recognized by the assembler and converted to binary machine code. Assembler directives, on the other hand, are instructions for the assembler, not the CPU.

The Origin (ORG) directive in Figure 5.2 specifies where the program is to be loaded into memory. Thus, the directive

ORG $8000

specifies that the program is to be loaded at the hexadecimal location $8000. This directive specifies an absolute starting address in memory for the program. An object module containing absolute addresses is produced. Thus, the program is not intended to be relocated in memory to another starting address as the program is loaded. In contrast, the program is said to be *relocatable* in memory if the starting address is assigned after assembly as described in Subsection 5.1.2. For simplicity, the ORG

```
abs. rel.    LC   obj. code    source line
---- ----    ----  ---------    -----------
   1    1    0000                       TTL     'FIGURE 5.2'
   2    2    0000                       LLEN    100            ;LINE LENGTH
   3    3    8000                       ORG     $8000          ;ORIGIN IN MEMORY
   4    4    8000                       OPT     P=M68000
   5    5    8000               *
   6    6    8000               * ADD FOUR 16-BIT NUMBERS STORED IN LOCATIONS $9000
   7    7    8000               *    THROUGH $9006.  RETURN THE SUM IN D1[15:0].
   8    8    8000               *
   9    9    8000 7200          INIT MOVE.L  #0,D1             ;ZERO SUM
  10   10    8002 227C 0000          MOVEA.L #$9000,A1         ;ADDR OF FIRST NUMBER
  10         8006 9000
  11   11    8008 7404               MOVE.L  #4,D2             ;SET COUNTER TO 4
  12   12    800A               *
  13   13    800A               * DEFINE LOOP TO ADD VALUES
  14   14    800A               *
  15   15    800A D259          LOOP ADD.W   (A1)+,D1          ;SUM NUMBERS
  16   16    800C 5342               SUB.W   #1,D2             ;DECREMENT COUNTER
  17   17    800E 66FA               BNE     LOOP              ;TILL (D2)=0
  18   18    8010               *
  19   19    8010 1E3C 00E4          MOVE.B  #228,D7
  20   20    8014 4E4E               TRAP    #14               ;RETURN TO MONITOR
  21   21    8016               END
  21 lines assembled
```

Figure 5.2   Typical assembly listing.

directive is used to specify the absolute load address of the program examples in this chapter. The Option (OPT) directive directs the assembler to assemble code for the MC68000. This particular assembler, discussed later, can assemble code for a number of the 68000 family processors.

The title (TTL) directive names the program. The Line Length (LLEN) directive determines the format of the assembler listing. Comments for the convenience of the programmer may also appear in the source program. These comments are ignored by the assembler and simply printed on the listing.

Example Listing.   Figure 5.2 shows a typical MC68000 listing in the same format as example listings to be presented in this chapter. The first two columns present the absolute (abs.) and relative (rel.) line numbers, respectively. The next column is the hexadecimal value of the location counter at each instruction. The location counter keeps track of instruction locations during assembly much as the program counter does during program execution. If the program shown were loaded beginning at location $8000, the program counter would change just like the location counter in Figure 5.2.

The column labeled "obj.code" in Figure 5.2 is the machine-language translation of the assembly-language program that is part of the object module. The machine language consists of an operation word for each instruction followed by the value of any extension words required for the instruction. Any value assigned by an assembler directive, such as the origin (ORG) statement shown, also appears in this column. The machine language translation is followed to the right by the source program statement that generated it. In our examples, the in-line comments are preceded by a semicolon. An entire line may be treated as a comment if an asterisk (*) is the first character of the line.

The simple program in Figure 5.2 sums four 16-bit integers in locations $9000, $9002, $9004, and $9006. The result is stored in (D1)[15:0]. The three directives LLEN, ORG, and END define the width of the listing, the origin, and the end of the program, respectively. INIT and LOOP are labels attached to a particular line in the program so that the line can be referred to symbolically from elsewhere in the program. The labels have values assigned by the location counter.

In the program, INIT indicates the location of the first instruction. LOOP defines the start of a sequence of instructions that is repeated four times during execution to sum the values. This iteration, or "loop," is terminated when the value in D2 reaches zero.

The last two instructions return control to our monitor program. These instructions apply only to the TUTOR monitor used on the authors' evaluation module for the MC68000. The reader should refer to the manual describing the procedure for the particular monitor being used. Each monitor will have a different method of returning control to it after the program finishes execution.

The END statement is a directive to the assembler that defines the end of the program being assembled. It must be the last directive in the program. Because the program assembled correctly, it is only necessary to load it at location $8000 and execute it. This is accomplished in our examples by use of the monitor program as described later.

In Figure 5.2, the numbers (immediate values) that appear in instructions must be preceded by the number symbol (#). However, the numerical values for the assembler directives do not require the designation as immediate values. Such assembler conventions must be determined by reference to the *User's Manual* for the particular assembler being used.

5.1.2 Cross-Assembly and Linking

Figure 5.3 shows the host computer system and software used by the authors for program development with cross-assemblers. The IBM PC is used for software development. Once the assembly language program is corrected and ready to be executed, the load module can be transferred to the MC68000-based system for execution and debugging. Alternatively, the execution can be simulated with a simulator program

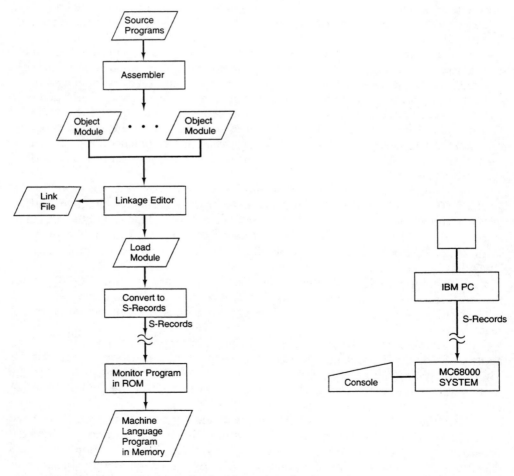

Figure 5.3   Hardware and software involved in program development.

on the host computer. These programs execute under the control of the operating system of the host computer being used.

The first step is to create the source program and translate it with the assembler into a *relocatable* object module. This object module generally does not contain absolute addresses referencing the MC68000 system memory, but rather relocatable addresses that will be assigned by the *linkage editor* program. A linkage editor serves to combine several object modules into a *load module*. The link file lists the names and other information for the modules being linked. All of the programs are stored as disk files on the host computer's disk unit.

In the transfer of the load module from the host computer to the MC68000 system, the load module is treated as a file of text in ASCII. In response to the operator's request, the monitor program in the ROM of the MC68000-based development system causes the transfer from the host. This monitor converts the transmitted object module from the form of S-records (ASCII) into machine-language instructions and loads the executable instructions into memory. The S-record format is used by Motorola to transfer programs and data between computers. The host computer only needs to have the file transfer program to respond to the monitor's request.[1] Once the machine-language program is in memory, the monitor program is used to execute the program and perform debugging functions, as described in Subsection 5.1.3.

Example 5.1 _____

The load module produced by the cross-software is not directly executable on a MC68000-based system. This module is stored as a disk file on the host computer's disk unit. The file actually contains ASCII text in a special format that Motorola calls *S-records*. This format was created to allow files to be transmitted between computer systems via serial communication lines. The general format is shown in Figure 5.4(a). A specific example for the assembly language program of Figure 5.2 is given in Figure 5.4(b). The S0 record indicates the first record of a series of S-records. Each following record (S1), (S2), or (S3) contains the type, the hexadecimal length of the record in bytes, the hexadecimal starting address at which the program segment is to be loaded, and finally, the instructions or data. The length is the number of character pairs following the length specification.

There are 27 ($1B) bytes or character pairs in the S2 record of Figure 5.4(b). These values without the checksum characters are stored in memory starting at address $8000. The last two characters represent a check sum for verifying the correctness of the record. A termination record (S9) indicates the end of the object module. The checksum $FC represents the one's complement of the sum of the hexadecimal values ($03) in the S9 record, excluding the type specification itself.

_____

Motorola Assembler and Development Systems.   The particular assembler used in our examples is a Motorola cross-assembler designated MASM (M68000-family-structured assembler). The assembler converts the assembly program to an object module. A

_____

[1] A number of file transfer programs are available. For example, KERMIT and PROCOMM+ are programs that allow an IBM PC to emulate a terminal and perform file transfers. PROCOMM+ is a program available from Datastorm Technologies, Inc.

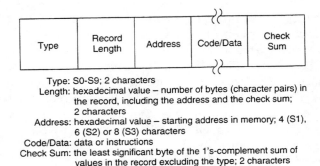

Type: S0-S9; 2 characters
Length: hexadecimal value – number of bytes (character pairs) in
the record, including the address and the check sum;
2 characters
Address: hexadecimal value – starting address in memory; 4 (S1),
6 (S2) or 8 (S3) characters
Code/Data: data or instructions
Check Sum: the least significant byte of the 1's-complement sum of
values in the record excluding the type; 2 characters

(a) Format definition

```
S0060000612E6FFB
S11B80007200227C000090007404D259534266FA1E3C00E44E4E4E7193
S9030000FC
```

(b) Sample object module in S-record format

Figure 5.4   S-record format.

listing is produced in the form shown in Figure 5.2. If only one program is to be
assembled and loaded into memory without linking, the assembler object module is
converted to a load module containing S-records by a program called HEX. This
module is suitable for loading by the monitor program of the MC68000-based system.
When several object modules must be linked, a link editor program and then HEX
must be used to create the load module.

Several evaluation systems were used to load and test the assembly language
programs in this textbook. One of the systems was the Integrated Development Platform from Motorola described in Chapter 1. The other was the SBC68K single-board
evaluation module available from Arnewsh, Inc. Both systems have monitor programs
stored in ROM and RAM to store programs and data loaded during software development. The Arnewsh board uses the popular TUTOR monitor, which will be described next.

5.1.3 TUTOR Monitor Program

In general, any monitor or debugging program will allow the user to load a program, initialize memory locations and processor registers, including the program counter, and then execute the program. This monitor session is carried on interactively via an operator's terminal. From the terminal, values can be changed as required during execution of the program for the purpose of testing. Monitors also allow the contents of memory locations and registers to be displayed by the user. This is one of the primary debugging aids available to the assembly language programmer. Finally, the monitor allows the user to stop and start execution of the program. One or more selected instructions can be executed and the results displayed. If errors are found, the program is usually modified and then reassembled, linked, and loaded for another debugging session.

The monitor program used for most of the examples in this book is called TUTOR. It provides aid for three fundamental operations:

(a) loading a program;
(b) executing a program;
(c) debugging a program.

The monitor itself resides in read-only memory, which is part of the evaluation module.

Load modules must be loaded from the disk unit of a development system or from another computer if cross-development software is being used. Each system has a different procedure and TUTOR or other monitors will have features to allow program loading. These features are not discussed here because they depend so completely on the system being used. However, many of the TUTOR capabilities are provided by most other monitors.

Program Development and Debugging Using TUTOR.   Table 5.1 lists a few of the commands for the TUTOR monitor. The monitor "prompt" on the screen

TUTOR 1.32⟩

indicates that the monitor (version 1.32) is ready to accept commands. The operator enters the two letter command followed by any parameters and a carriage return (CR) to invoke a command. Any value displayed by the monitor in response to a command is not preceded by the prompt in the examples in this chapter. Also, all addresses and memory contents are given in hexadecimal unless otherwise indicated. Therefore, hexadecimal values entered by the operator do not use the "$" designation that is required by the assembly language notation.

Processor registers are designated by their symbolic names: A0, A1, . . . , A6 for address registers and D0, D1, . . . , D7 for data registers. The other registers of immediate interest are program counter (PC), status register (SR), user stack pointer (US), and supervisor stack pointer (SS).

The monitor commands in Table 5.1 are divided into those that allow communication with a host computer and those that are used for program debugging. In the former category, the load (LO) command is used to transfer a file from the host to

Table 5.1   TUTOR Commands

Commands	Meaning
General Format	
TUTOR 1.32⟩ ⟨command⟩ (CR)	Command format
Communication with host computer and help	
LO	Load S-records from host
HE	List TUTOR commands
Initialize register or memory contents	
MM ⟨address⟩[;DI]	Memory modify (sequential)
MS ⟨address⟩ ⟨value⟩	Memory set (one location)
.Rx	Display or change a register value
Display registers or memory contents	
MD ⟨address⟩[;DI]	Memory display
DF	Register display
Execute and trace	
BR ⟨address⟩[;⟨count⟩]	Insert a breakpoint at ⟨address⟩
NOBR	Delete breakpoints
GO [⟨address⟩]	Execute a program at ⟨address⟩
GT ⟨address⟩	Go till ⟨address⟩
TR [⟨count⟩]	Trace
TT ⟨address⟩	Trace till ⟨address⟩

Notes:
1. ⟨ ⟩ indicates any valid selection.
2. All values for data and addresses are hexadecimal.
3. Registers are selected by their assembly-language mnemonic.
4. [;DI] indicates an optional value; in this case, assemble or disassemble (DI).

the target computer. The name of the file to be transferred may be part of the command to the host in the LO command. The exact procedure depends on the file transfer program executing on the host computer. The help (HE) command displays all of the TUTOR commands.

For debugging, the contents of a register or a memory location can be initialized, modified, or displayed by the appropriate command in Table 5.1. The MM command allows changes to the values stored in memory, but the MD command simply displays the values beginning at the address specified. The command .Rx is used to modify the contents of a particular register when Rx is replaced by the register designation, such as .PC, to change the program counter value. Display formatted registers (DF) displays the contents of all of the CPU registers in an easily readable form.

A number of commands control the execution of a program. Typically, for testing, the program counter is set to the first instruction address of the program and the GO command is used to execute the code for the whole program. During a debugging session, single instructions or a group of instructions might be executed if errors exist in the program. For example, the trace (TR) command sets the trace mode of the CPU. In the trace mode, one instruction is executed and control is returned to the monitor program. The monitor program displays the contents of the CPU registers after tracing.

The TR command with a value count = N traces N instructions and then returns control to the monitor. Breakpoints can be set at various locations in the program with the BR command. If a count is given with the BR command, the program will execute the instruction at the breakpoint for the specified number of times before control is returned to the monitor. This is useful for debugging loops in the program.

The monitor also has a "one-line" assembler, which converts a single assembly-language statement into machine language at the address specified. It is an option for the memory modify (MM) command. Such a one-line assembler is of limited use in program development. For example, no labels are allowed.

The reverse process of assembly is called *disassembly*. When the contents of memory locations containing machine language instructions are displayed and disassembled, the TUTOR monitor will display the corresponding assembly language statements in mnemonic form. This can be used to verify the program in memory without translating the machine-language code when the memory display (MD) command is used.

Example 5.2

Use of the TUTOR monitor for program loading, execution, and debugging is shown in Figure 5.5b. An assembly-language listing of the program is included as Figure 5.5a for convenience. The file of S–records was created using the Motorola cross-assembler and HEX programs. Before the object module can be loaded into memory, it is necessary to establish communication with the host computer. This part of the session is not shown in the figure because it depends on the file transfer program being used. Once this communication to the host is established, the object module may be loaded by invoking the monitor again with LO1 (load command). LO1 directs the monitor to load a program from I/O port 1, which is the serial connection to the host computer.

The host responds to the LO1 command by transmitting the S–record file. The program is loaded by the monitor in memory starting at location $8000. It is now possible to begin testing the program. As part of the initialization, which is not shown, registers D0–D7 and A0–A6 were initialized with zero values. The user stack pointer (US) was loaded with the initial value $1000. Before execution, the code loaded at location $8000 is disassembled. The address $9000 in the first instruction is given as its decimal value (36864). Next, the program counter is initialized to $8000. Then the register contents are displayed to show the initial values.

Our simple program adds four 16-bit numbers in memory locations $9000 through $9006 and accumulates the sum in (D1)[15:0]. The address of the first number is loaded into A1 by the MOVEA instruction, a variation of the MOVE instruction. To test the program, four test values are loaded into memory as shown in Figure 5.5b using the MM (memory modify) command. Values of 1, 2, 3, and 4 are used. The last value is followed by a period to tell the monitor that no more locations are to be initialized. The values are displayed (MD) to verify their correctness.

Then, the GT command causes execution from the starting location until the instruction at $8010. The final display in Figure 5.5b shows that the value in D1 is indeed the hexadecimal

```
abs. rel.   LC   obj. code    source line
---- ----   ----  ----------  -----------
   1   1    0000                        TTL      'FIGURE 5.5(a)'
   2   2    0000                        LLEN     100              ;LINE LENGTH
   3   3    8000                        ORG      $8000            ;ORIGIN IN MEMORY
   4   4    8000                        OPT      P=M68000
   5   5    8000              *
   6   6    8000              * ADD FOUR 16-BIT NUMBERS STORED IN LOCATIONS $9000
   7   7    8000              *    THROUGH $9006.  RETURN THE SUM IN D1[15:0].
   8   8    8000              *
   9   9    8000 7200         INIT     MOVE.L   #0,D1            ;ZERO SUM
  10  10    8002 227C 0000             MOVEA.L  #$9000,A1        ;ADDR OF FIRST   NUMBER
  10        8006 9000
  11  11    8008 7404                  MOVE.L   #4,D2            ;SET COUNTER TO 4
  12  12    800A             *
  13  13    800A             * DEFINE LOOP TO ADD VALUES
  14  14    800A             *
  15  15    800A D259         LOOP     ADD.W    (A1)+,D1         ;SUM NUMBERS
  16  16    800C 5342                  SUB.W    #1,D2            ;DECREMENT COUNTER
  17  17    800E 66FA                  BNE      LOOP             ;TILL (D2)=0
  18  18    8010             *
  19  19    8010 1E3C 00E4             MOVE.B   #228,D7
  20  20    8014 4E4E                  TRAP     #14              ;RETURN TO MONITOR
  21  21    8016                       END
  21 lines assembled
```

(a)   (continued)

Figure 5.5 A monitor session: (a) assembly listing, (b) monitor results.

sum ($A) of the values added. The counter (D2) has been decremented to zero and the address register A1 was incremented by 2 each time through the loop to point to the next operand in memory. The command GO causes the value 228 to be loaded into D7 and the TRAP #14 instruction returns control to the monitor.

Debugging can continue with different values in memory and the (PC) reset to $8000. More extensive testing would reveal a flaw in the program because the sum in register D1 is not tested to see if it overflows the register length (16 bits) being used. Techniques to test for this situation are discussed in Chapters 6 and 7.

5.1.4 Resident Assembler

The Motorola Resident Assembler executes under the control of a Motorola operating system. It translates source statements written in assembly language into relocatable object code. One or more object modules can be linked by Motorola's linkage editor program and stored on the disk unit of the computer. Subsequently, the complete module can be loaded into the memory of the system. Thus, the procedure for program development and debugging is essentially the same as that described previously in Section 1.2 for the cross-software development. The difference is that the *resident* development software associated with assembly language programming executes directly on the 68000-based system used for program execution. To use the resident software, a disk storage unit with its associated hardware and software must be part

```
TUTOR  1.32> LO1                                                      ;LOAD S-RECORDS
S0060000612E6FFB
S11B80007200227C000090007404D259534266FA1E3C00E44E4E4E7193
S9030000FC

TUTOR  1.32> MD 8000;DI                                               ;DISPLAY MEMORY
008000    7200                 MOVEQ.L  #0,D1                         ;WITH DISASSEMBLY

TUTOR  1.32>
008002    227C00009000         MOVE.L   #36864,A1
008008    7404                 MOVEQ.L  #4,D2
00800A    D259                 ADD.W    (A1)+,D1
00800C    5342                 SUBQ.W   #1,D2
00800E    66FA                 BNE.S    $00800A
008010    1E3C00E4             MOVE.B   #228,D7
008014    4E4E                 TRAP     #14

TUTOR  1.32> .PC 8000                                                 ;SET (PC)=$8000

TUTOR  1.32> DF                                                       ;DISPLAY REGISTER
PC=00008000 SR=2704=.S7..Z.. US=00001000 SS=0000077A                  ;VALUES
D0=00000000 D1=00000000 D2=00000000 D3=00000000
D4=00000000 D5=00000000 D6=00000000 D7=00000000
A0=00000000 A1=00000000 A2=00000000 A3=00000000
A4=00000000 A5=00000000 A6=00000000 A7=0000077A
--------------------008000    7200                 MOVEQ.L  #0,D1

TUTOR  1.32> MM 9000;W                                                ;MODIFY MEMORY
009000    FFFF ?0001                                                  ;INITIALIZE DATA
009002    FFFF ?0002
009004    FFFF ?0003
009006    FFFF ?0004

TUTOR  1.32> MD 9000                                                  ;DISPLAY DATA
009000    00 01 00 02 00 03 00 04  FF FF FF FF FF FF FF FF

TUTOR  1.32> GT 8010                                                  ;EXECUTE PROGRAM
                                                                      ;AND VERIFY RESULTS
PHYSICAL ADDRESS=00008010
PHYSICAL ADDRESS=00008000

AT BREAKPOINT
PC=00008010 SR=2704=.S7..Z.. US=00001000 SS=0000077A
D0=00000000 D1=0000000A D2=00000000 D3=00000000
D4=00000000 D5=00000000 D6=00000000 D7=00000000
A0=00000000 A1=00009008 A2=00000000 A3=00000000
A4=00000000 A5=00000000 A6=00000000 A7=0000077A
--------------------008010    1E3C00E4             MOVE.B   #228,D7

TUTOR  1.32> GO
PHYSICAL ADDRESS=00008010

TUTOR  1.32>
```

(b)

Figure 5.5 *(continued)*

of the 68000-based computer. The operating system is used to load the program into memory from the disk unit and execute it. The command sequence to accomplish this depends entirely on the characteristics of the operating system being used.

EXERCISES

5.1.1. The single-step and breakpoint features of a debugger are similar. Discuss the use of each technique and compare them.

5.1.2. Consider the use and advantages or disadvantages of software development with the following features:
   (a) a monitor program with a one-line assembler and disassembler;
   (b) cross-assembler and simulator executed on a host computer;
   (c) disk-based system with a resident assembler;
   (d) assembler with linking loader versus absolute loader. An absolute loader cannot relocate programs.

5.1.3. Compare the testing and debugging procedure using a high-level language with the techniques available to the assembly-language programmer. Use the simple program shown previously in this section as a specific example.

5.2 ASSEMBLY LANGUAGE CHARACTERISTICS

The source program statements, as processed by the assembler, consist of strings of ASCII characters combined to form symbols. These symbols are constructed according to the rules of the language. Each statement consists of four *fields:* label, operations, operand(s), and comments. The fields are separated by spaces or other delimiters according to the *format* required. For example, the statement

```
INIT    MOVE.L    #0,D1    ;ZERO SUM
```

consists of the label INIT; a mnemonic instruction MOVE, which represents the operation; an operand field "#0,D1"; and a comment. In this case, the delimiter between fields is a blank or space character. At least one space is required to separate the fields. However, multiple spaces can be used. This is referred to as a *free-field format.*

Typically, the assembler first scans each source statement to determine that the formatting and symbol usage are valid. An error results if the format is incorrect or if the operation is not either a processor instruction or an assembler directive.

This subsection divides the discussion of MC68000 assembly language into two parts. The first deals with the construction of source statements representing executable instructions for the processor. The second covers assembler directives, which control the way the assembler itself operates. The characteristics discussed in this section are necessary for the creation of useful programs, although most assemblers have many additional capabilities.

5.2.1 Statement Formats

The source statements processed by the assembler must follow a precise format defining the order and relationship of the elements in the statement. The MC68000 assemblers recognize source statements composed of the following fields:

(a) Label field
(b) Operation code or directive field
(c) Operand(s) field
(d) Comment field

Each assembly language statement consists of these elements separated by spaces.

Table 5.2 shows the format of a general assembly language statement with optional fields enclosed in brackets. If a "*" is encountered in the first column of a statement, the entire statement is a comment. If another character is encountered in the first column, the symbol is considered to be a label, which must have from one to eight alphanumeric characters. In most assemblers, the first character of a symbol must be a letter (A–Z), although different assemblers have other conventions. If no label is used, the first column must contain a blank (space) if other fields are present.

The next field encountered is interpreted as an instruction mnemonic or assembler directive. For example, in a statement without a label such as

MOVE.W   D1,D2   ;COMMENT

the instruction mnemonic must start in column 2 or beyond. A space must precede the operands and the comment. The semicolon is not needed before the comment but is used here to enhance readability.

Labels.   The label is optional for most instructions and directives. When one is used, it represents an address. The label is assigned the value of the location counter when the label is encountered. In statements such as

HERE   MOVE.W   D1,D3

the label HERE defines the location of the instruction in memory after the program is loaded and can be used to define the beginning of a program segment for later reference.

Table 5.2   Assembly Language Format

Label Field	Operation Code and Directive Field	Operand(s) Field	Comment Field
[⟨LABEL⟩]	⟨op code⟩ or ⟨directive⟩	[⟨operand 1⟩[,⟨operand 2⟩]]	[⟨comment⟩]

Notes:
1. An asterisk in column 1 indicates a comment line.
2. Angle brackets indicate any valid symbol.
3. Square brackets indicate an optional field.

Example 5.3

Figure 5.6 shows several examples of labels used as addresses. As discussed previously, the program adds four values to form a sum. The program starts at location $8000 and initializes the sum to zero when it executes. The symbol INIT is a label associated with the first statement and has the value $8000. Another program (not shown) could use the instruction

JMP   INIT

to begin execution of this segment. The label LOOP locates the first instruction of a repeated sequence of instructions. The loop is executed four times until the counter (D2) is decremented to zero. This label is simply for the addition sequence and would not be referenced by instructions other than the BNE instruction in the loop.

Operation Codes. The second field in the source statement must contain an instruction mnemonic or assembler directive. When the operands require a length to be specified, a length specification is included as part of the instruction field. The length specification is preceded by a period ".", which is appended to the operation code, and consists of B, W, or L to specify byte, word, or longword, respectively. For example, the instruction

MOVE.W   D1,D2

defines word-length operands.

```
abs. rel.   LC    obj. code    source line
---- ----   ----  ---------    -----------
   1    1   0000                      TTL      'FIGURE 5.6'
   2    2   0000                      LLEN     100         ;LINE LENGTH
   3    3   8000                      ORG      $8000       ;ORIGIN IN MEMORY
   4    4   8000                      OPT      P=M68000
   5    5   8000               *
   6    6   8000               * ADD FOUR 16-BIT NUMBERS STORED IN LOCATIONS $9000
   7    7   8000               *    THROUGH $9006.  RETURN THE SUM IN D1[15:0].
   8    8   8000               *
   9    9   8000 7200          INIT MOVE.L   #0,D1          ;ZERO SUM
  10   10   8002 227C 0000          MOVEA.L  #$9000,A1      ;ADDR OF FIRST NUMBER
  10        8006 9000
  11   11   8008 7404               MOVE.L   #4,D2          ;SET COUNTER TO 4
  12   12   800A               *
  13   13   800A               * DEFINE LOOP TO ADD VALUES
  14   14   800A               *
  15   15   800A D259          LOOP ADD.W    (A1)+,D1       ;SUM NUMBERS
  16   16   800C 5342               SUB.W    #1,D2          ;DECREMENT COUNTER
  17   17   800E 66FA               BNE      LOOP           ;TILL (D2)=0
  18   18   8010               *
  19   19   8010 1E3C 00E4          MOVE.B   #228,D7
  20   20   8014 4E4E               TRAP     #14            ;RETURN TO MONITOR
  21   21   8016                    END
  21 lines assembled
```

Figure 5.6   Use of labels as addresses.

Table 5.3   MC68000 Instruction References

Operand	Format	Typical Reference or Operand	Example
None	OPR	External device	RESET
Implied	OPR	PC, SP, or SR	NOP, TRAPV, RTS
Immediate	OPR ⟨value⟩	Processor control or instructions requiring a value	TRAP, STOP
Single	OPR ⟨address⟩	Relative address	BRA
		Instruction address	JMP
		Operand address	CLR, NEG
Double	OPR ⟨value⟩,⟨destination⟩ OPR ⟨source⟩,⟨destination⟩	Immediate value to destination or double address	ADD, MOVE

Notes:
1. OPR is any valid operation code.
2. Minor variations from the formats shown are possible.

Operands.   Location of the operands for the instructions are accessed according to the addressing mode for each operand. The general formats for operands are shown in Table 5.3. A few instructions require no operands. Others refer to the processor registers implicitly and their execution may cause the program counter, stack pointer, or status register values to be modified. The TRAP and STOP instructions require an immediate value in the form of a decimal or hexadecimal number.

Most instructions require operands to be specified by addressing modes. Single-address instructions contain the specification of one operand. Double-address instructions contain two operands, which are separated by a comma in the operand field. Processor address or data registers are designated symbolically by the letter A or D followed by the register number. Thus the instruction

MOVE.W   A1,D1

designates A1 as the source register and D1 as the destination. Indirect addressing is specified by enclosing the address register symbol in parentheses, as in the instruction

MOVE.W   (A1),D2

which causes the word in the location pointed to by (A1) to be copied into D2. The addressing modes are described in detail in Section 5.3.

Example 5.4 _____

Figure 5.7 is an assembler listing showing a number of statements to illustrate the specification of operands. The RESET and RTS (Return from Subroutine) instructions require no operands. A TRAP instruction must have the trap number specified as an immediate value. Such instructions have unique requirements for operand specification.

```
abs. rel.    LC   obj. code      source line
---- ----    ----  ---------      -----------
   1    1   0000                          TTL       'FIGURE 5.7'
   2    2   0000                          LLEN      100
   3    3   8000                          ORG       $8000
   4    4   8000                          OPT       P=M68000
   5    5   8000              *
   6    6   8000              * MISCELLANEOUS INSTRUCTIONS
   7    7   8000              *
   8    8   8000 4E70                RESET
   9    9   8002              *
  10   10   8002 4E4F                TRAP      #15
  11   11   8004              *
  12   12   8004 4E75                RTS
  13   13   8006              *
  14   14   8006              * SINGLE ADDRESS
  15   15   8006              *
  16   16   8006 4241                CLR.W     D1                 ;DATA REG. DIRECT
  17   17   8008 4278 1000           CLR.W     $1000              ;ABSOLUTE
  18   18   800C 4251                CLR.W     (A1)               ;INDIRECT
  19   19   800E 4261                CLR.W     -(A1)              ;PREDECREMENT
  20   20   8010 4259                CLR.W     (A1)+              ;POSTINCREMENT
  21   21   8012 4269 0002           CLR.W     2(A1)              ;INDIR. WITH DISP.
  22   22   8016 4271 1002           CLR.W     2(A1,D1.W)         ;INDIR. WITH INDEX
  23   23   801A 4271 A802           CLR.W     2(A1,A2.L)
  24   24   801E              *
  25   25   801E              * DOUBLE ADDRESS (SOURCE ADDRESS SPECIFIED)
  26   26   801E              *
  27   27   801E 3401                MOVE.W    D1,D2              ;DATA REG. DIRECT
  28   28   8020 3409                MOVE.W    A1,D2              ;ADDR REG. DIRECT
  29   29   8022 3438 1000           MOVE.W    $1000,D2           ;ABSOLUTE
  30   30   8026 3411                MOVE.W    (A1),D2            ;INDIRECT
  31   31   8028 3421                MOVE.W    -(A1),D2           ;PREDECREMENT
  32   32   802A 3419                MOVE.W    (A1)+,D2           ;POSTINCREMENT
  33   33   802C 3429 0002           MOVE.W    2(A1),D2           ;INDIR. WITH DISP.
  34   34   8030 3431 1002           MOVE.W    2(A1,D1.W),D2      ;INDIR. WITH INDEX
  35   35   8034 343C 0005           MOVE.W    #5,D2              ;IMMEDIATE
  36   36   8038 3439 0000           MOVE.W    *+8,D2             ;RELATIVE
  36        803C 8040
  37   37   803E              *
  38   38   803E 1E3C 00E4           MOVE.B    #228,D7
  39   39   8042 4E4E                TRAP      #14                ;RETURN
  40   40   8044                     END
  40 lines assembled
```

Figure 5.7 Examples of operand specification.

Single-address instructions, such as CLR, require one address in the operand field. The address may be specified by any addressing mode valid for the particular instruction. Several examples are shown for specifying the operand for the CLR instruction.

The double-address MOVE instruction is shown with various addressing modes used to specify the source operand. The destination is a register in each example, although the MOVE instruction does allow other addressing modes for the destination operand.

Expressions as Operands. The assembler recognizes certain symbols in the operand field. A symbol may designate an absolute address, an immediate value, or any other valid operand. An *expression* is a combination of symbols, constants, algebraic operators, and parentheses, which the assembler evaluates to determine the address or value of the operand.

To specify a constant value, sometimes called a *literal,* the immediate addressing mode is used. For most assemblers, the constants can represent either numbers or ASCII characters. These constants are the simplest form of expressions and are specified using the definitions in Table 5.4. A numerical constant can be any decimal or hexadecimal value that can be represented as an 8-bit, 16-bit, or 32-bit integer. The size specification of the instruction determines the appropriate length. A decimal number is defined by a string of decimal digits and a hexadecimal number is defined by a dollar sign ($) followed by a string of hexadecimal digits. Thus the instruction

MOVE.W   #$2000,D1

defines the 16-bit hexadecimal value 2000 as the immediate source operand.

ASCII literals consist of up to four ASCII characters enclosed in apostrophes. For example, the string 'ABCD' is recognized by the assembler and converted into the ASCII code

41   42   43   44

which occupies 4 bytes when stored in memory. Longer character strings cannot be used as a literal value in an expression because the size is limited by the 32-bit registers of the MC68000.

Table 5.4   Assembler Symbols for Expressions

Symbolic Format	Interpretation
$⟨Number⟩	Hexadecimal number
⟨Number⟩	Decimal number
'⟨String⟩'	ASCII string of characters
#⟨Number⟩	Immediate operand
#'⟨String⟩'	Immediate operand
In expressions	
+	Add
−	Subtract
*	Multiply
/	Divide
( )	Grouping

The operators for addition, subtraction, and multiplication can be used in an expression. The result of any arithmetic operation is a 32-bit integer value. The use of the unary minus $(-)$ is recognized as a means to define negative numbers. For example, the immediate value -1 is computed in the instruction

MOVE.W   #−1,D1

and is stored as $FFFF with the instruction. An equivalent specification is

MOVE.W   #$FFFF,D1

where the value $FFFF is the two's-complement number in 16 bits.

Example 5.5 ──

The short program segment in Figure 5.8 shows various uses of labels and expressions. The labels START and ENDLP serve to define the values of the beginning and end of a group of statements in the example. The MOVEA and MOVE instructions initialize the address and counter, respectively, when the program is executed. When the addition is complete, the result is in (D1). The segment length is calculated as 14 bytes (seven words) in (D3). The length in words in (D4) is half of the length in bytes. If the machine language instructions were moved in memory without reassembly, statements referencing START as a source address or operand would be in error. The length calculations, however, would be correct.

The instruction itself at location START is moved into D5 when the source operand in a MOVE instruction specifies the label only. The immediate form #START selects the address.

The final MOVE instruction transfers the ASCII string 'DONE' to D3. Note that the immediate mode must be indicated or the value would be interpreted as an address. A suitable I/O routine could be used to print the string to indicate completion of the program.

──

5.2.2 Assembler Directives

The mnemonic symbols for instruction op codes and for the various addressing modes are part of an internal *symbol table* used by the assembler to translate the source statements into machine language. The user-defined symbols, such as labels, are used to reference instructions or data within the assembly language program. The assembler automatically keeps track of locations and offsets associated with the machine language program.

The use of symbolic forms as addresses of instructions or as operands is of valuable assistance in writing assembly language programs. This is one of the principal advantages of assembly language over machine language. However, most assemblers aid the programmer in other ways by providing *assembler directives* that are actually instructions to the assembler rather than for the processor. The action caused by each directive occurs only when the source program is being assembled. The major categories of directives are for assembly control, symbol definition, data definition and storage allocation, and listing control, as shown in Table 5.5.

Assembly Control.   The location counter of the assembler normally begins with the value $0000 to indicate the location of the first executable instruction. This counter is increased by the appropriate amount as each instruction is assembled. If the machine

```
abs.  rel.   LC    obj. code    source line
----  ----   ----  ---------    -----------
  1    1     0000                              TTL    'FIGURE 5.8'
  2    2     0000                              LLEN   100
  3    3     8000                              ORG    $8000
  4    4     8000                              OPT    P=M68000
  5    5     8000               *
  6    6     8000               *   USE OF LABELS AND EXPRESSIONS
  7    7     8000               *
  8    8     8000
  9    9     8000 227C 0000     START          MOVEA.L #$9000,A1          ;FIRST ADDRESS
  9          8004 9000
 10   10     8006 7404                         MOVE.L  #4,D2              ;COUNTER
 11   11     8008               *
 12   12     8008               *    THE SUM IS ADDED TO THE INITIAL VALUE IN D1
 13   13     8008               *
 14   14     8008 D259          LOOP           ADD.W   (A1)+,D1           ;ADD 4 NUMBERS
 15   15     800A 5382                         SUBQ.L  #1,D2
 16   16     800C 66FA                         BNE     LOOP
 17   17     800E 4E71          ENDLP          NOP
 18   18     8010               *
 19   19     8010               *   LENGTH OF PROGRAM IN BYTES
 20   20     8010               *
 21   21     8010 760E                         MOVE.L  #(ENDLP-START),D3     ; (D3)=14
 22   22     8012               *
 23   23     8012               *   LENGTH OF PROGRAM IN WORDS
 24   24     8012               *
 25   25     8012 7807                         MOVE.L  #(ENDLP-START)/2,D4   ; (D4)=7
 26   26     8014               *
 27   27     8014               *   CONTENTS OF START (THE INSTRUCTION) TO D5
 28   28     8014               *
 29   29     8014 2A39 0000                    MOVE.L  START,D5              ; (D5)=227C 0000
 29          8018 8000
 30   30     801A               *
 31   31     801A               *   ADDRESS OF START TO A2
 32   32     801A               *
 33   33     801A 247C 0000                    MOVEA.L #START,A2             ; (A2)=8000
 33          801E 8000
 34   34     8020               *
 35   35     8020               *   ASCII STRING TO D6
 36   36     8020               *
 37   37     8020 2C3C 444F                    MOVE.L  #'DONE',D6            ; (D6)=444F 4E45
 37          8024 4E45
 38   38     8026               *
 39   39     8026 1E3C 00E4                    MOVE.B #228,D7
 40   40     802A 4E4E                         TRAP    #14
 41   41     802C                              END
 41 lines assembled
```

Figure 5.8   Uses of labels and expressions.

Table 5.5 Assembler Directives

Directive and Format			Meaning
Assembly control			
ORG ⟨expression⟩			Origin
END			End source
Symbol definition			
⟨label⟩	EQU	⟨expression⟩	Equate value of ⟨label⟩
Data definition and storage			
[⟨label⟩]	DC.⟨1⟩	⟨value(s)⟩	Define constant(s)
[⟨label⟩]	DS.⟨1⟩	⟨number⟩	Reserve storage
Listing control			
LLEN ⟨N⟩			Line length
LIST			List (default)
NOLIST			No listing
SPC ⟨N⟩			⟨N⟩ blank lines
PAGE			Next page

Notes:
1. Square brackets indicate an optional field.
2. ⟨1⟩ denotes B, W, or L.

language program were loaded into memory at location $0000 and executed, the program counter would follow the same sequence as the location counter as each instruction is executed in turn.

Loading programs at location $0000 is not possible in MC68000 systems because the lowest addressed area in memory is reserved for MC68000 vectors. This restriction does not apply to MC68010 systems. Use of the ORG directive allows the programmer to define the starting value of the location counter and, consequently, the first address of the program in memory. In the previous examples in this chapter the ORG directive was used to indicate that hexadecimal location $8000 was to be used to store the first machine language statement in a program.

The format of the ORG directive is

ORG   ⟨expression⟩

in which the ⟨expression⟩ has the same meaning as previously defined. When the directive is encountered, the location counter is "loaded" with the value much as a JMP instruction changes the contents of the program counter (PC). The ORG directive can appear anywhere in the source program and can be used, for example, to divide the program into instruction and data sections. This is particularly useful if the instructions are to be held in a read-only memory and data are held in a writable memory at another starting location.

Another assembly control directive is the END directive, which is always the last source statement in a program. It causes the assembler to stop its top-to-bottom scan of the program. Any source statements after the END directive are not processed by the assembler.

Symbol Definition. An EQU directive is used to equate a number to a symbol. The value may represent an address or a constant. The format is

⟨label⟩   EQU   ⟨expression⟩

where ⟨label⟩ is assigned the value of the expression when the statement is assembled. The expression may contain a label if it has been defined previously in the program. Thus the statement

TTYOUT   EQU $7FFF

assigns the value $7FFF to the symbol TTYOUT. The intent might be to define the address of an output buffer using a mnemonic term. Then a statement such as

MOVEA.L   #TTYOUT,A1

transfers the number $7FFF to the address register A1. If the immediate mode is not used for the source, the contents of the location are transferred. The instruction

MOVE.B   D1,TTYOUT

would move a byte from D1 to the location $7FFF, which might be the location of an output buffer, for example.

An important advantage of the EQU directive is evident when a value is defined that is referenced several times within a program. If the address TTYOUT needs to be changed in a subsequent assembly, reassembly with a new EQU set to the correct address would change the value throughout the program. This might be necessary if the program is executed on several systems, each with different buffer locations.

The EQU directive can also be used to give mathematical constants useful names, as in the statement

ONEK   EQU   1024

which defines "ONEK" as 1024 decimal. As another example, MAXMEM could be equated to the maximum memory space available for a system. The value could be changed, if necessary, when the program is reassembled on a new system. The only drawback is that the value defined by the EQU directive is known only to the assembler and does not exist in memory. Therefore, it cannot be changed without reassembly.

Data Directives. The two directives Define Constant (DC) and Define Storage (DS) are available to initialize values in memory and reserve space in memory, respectively. The DC directive is similar to a DATA statement in FORTRAN, in which the variables defined are assigned initial values. The DS directive is similar to the DIMENSION statement, which reserves space for variables but assigns no values to them.

The DC directive causes the assembler to store specified values in the location or locations associated with the location counter value at the time the DC directive is encountered during assembly. When the machine language program is loaded into memory, the locations involved have the initial values specified. For example, the statement

INITV   DC.W   20

causes the decimal value 20 to occupy a word at location INITV. However, if the program is executed more than once and the value at INITV changes between executions, any program statement depending on the initial value of 20 may not yield the correct results when executed. Thus, the DC directive should never be used to initialize a value that may be modified after the program is loaded into memory if multiple program executions depend on the initial value. A better approach to initialization of values is to reserve space for the values with a DS directive and then initialize the values with executable instructions.

    Both the DC and DS directives require a length specification (B, W, or L). The length specification determines whether bytes, words, or longwords are reserved. Thus the directive

DS.W $10

reserves 16 words in memory. The length of each constant defined by the DC directive is determined by the size specification. For example, the directive

DC.W   LABEL+1

will store the address of LABEL plus 1 in a word location.

Listing Control.   The last group of directives in Table 5.5 indicates a few of the options available to format the listing produced by the assembler. The LLEN directive determines the number of characters in the printed lines. LLEN 72 is typically used for CRT units, but longer lines with more than 72 characters may be used for line printers. The SPC directive causes the specified number of blank lines to appear on the printout to enhance readability. Other directives, such as PAGE, are usually offered to format the listing. The PAGE directive causes an advance to the top of a new page each time it is encountered. The page length depends on the printer being used and is generally set as a parameter in the operating system. One other directive (not shown in Table 5.5) is useful for MC68000 assemblers. The G directive instructs the assembler to list the contents of every location initialized by a DC directive. Otherwise, only the first locations have their values listed when the operand is a string of characters.

    Generally, each option for listing has a default value for which the opposite can be specified. The NOLIST directive in Table 5.5, for example, causes the statements following it to be omitted from the listing. Its opposite, the LIST directive, is the default value and need not be specified unless the NOLIST option is to be reversed. Thus the sequence

LIST
(segment I)

NOLIST
(segment II)
LIST
(segment III)
END

lists segments I and III of a program but not segment II.

Example 5.6 ———————————————————————————

Figure 5.9 illustrates a program employing a number of assembler directives. The first three directives set the line length, cause printing of the values defined by the DC directives, and set the origin to $8000, respectively. The EQU directives define the constant ONE and also the starting address of the program for the second ORG directive.

The first ORG directive defines the area for data storage beginning at location $8000. The executable program begins at location $8100, as specified by the directive

ORG   PROGRAM

The program clears the data locations between COMMON and the last location used for data. The DC directives define a number of constants in the data area. Note that

INITADD   DC.L   INITDT

initializes INITADD with the address of INITDT. The instruction

MOVE.L   #INITDT,INITADD

would accomplish the same thing, but only as the program executes. A total of 55 bytes are reserved by the various DS directives and

DS.B   55

would accomplish the same results. However, it is assumed that reference to the individual blocks (COMMON, HEXVAL, BYTES) is required in another program segment not shown. The final

EVEN

directive aligns the next location address on a word (even) boundary.

———

5.2.3 Advanced Features of Assemblers

Most modern assemblers provide a number of features to improve program structure and readability. Most of these features have always been present in assemblers for large computers but were unavailable in assemblers for many microcomputer–based systems until recently. They are introduced only briefly here, to acquaint the reader with these useful techniques. Anyone attempting an extensive development project using assembly language would be well advised to explore these capabilities for the

```
abs. rel.   LC    obj. code    source line
---- ----   ----  ---------    -----------
   1    1   0000                              TTL    'FIGURE 5.9'
   2    2   0000                              LLEN   100
   3    3   8000                              ORG    $8000
   4    4   8000                              OPT    P=M68000
   5    5         0000 0001    ONE            EQU    1                      ;A CONSTANT
   6    6         0000 8100    PROGRAM        EQU    $8100                  ;STARTING ADDR
   7    7   8000              *
   8    8   8000              *    DATA AREA
   9    9   8000              *
  10   10   8000 0A05 0700    INITDT         DC.B   10,5,7,0               ;BYTES-DECIMAL
  11   11   8004 0000 000A                   DC.L   10,5,7                 ;LONGWORDS
  11   11   8008 0000 0005
  11   11   800C 0000 0007
  12   12   8010 FF10 AF00                   DC.B   $FF,$10,$AF,0 ;BYTES HEX
  13   13   8014 0000 00FF                   DC.L   $FF,$20,$AE   ;LONGWORDS
  13   13   8018 0000 0020
  13   13   801C 0000 00AE
  14   14   8020 4142 4344                   DC.B   'ABCDEFGH'    ;BYTES ASCII
  14   14   8024 4546 4748
  15   15   8028 0000 0041                   DC.L   'A','BC'      ;LONGWORDS
  15   15   802C 4243
  16   16   802E             *
  17   17   802E 0000 8000    INITADD        DC.L   INITDT                 ;ADDRESS INPUT
  18   18   8032              COMMON         DS.W   10                     ;10 WORDS
  19   19   8046              HEXVAL         DS.W   $10                    ;16 WORDS
  20   20   8066              BYTES          DS.B   3                      ;3 BYTES
  21   21   806A                             EVEN                          ;EVEN BOUNDARY
  22   22   806A             *
  23   23   806A             *    THE LENGTH IS COMPUTED AS THE LAST DATA LOCATION
  24   24   806A             *    MINUS THE FIRST DATA LOCATION:   (BYTES+3)-COMMON
  25   25   806A             *
  26   26         0000 0037    LENGTH         EQU    (BYTES+3-COMMON)  ;NUM. OF BYTES
  27   27   806A             *
  28   28   8100                              ORG         PROGRAM
  29   29   806A             *
  30   30   8100             *    CLEAR COMMON VALUES
  31   31   8100             *
  32   32   8100 123C 0037    BEGIN          MOVE.B  #LENGTH,D1             ;COUNTER
  33   33   8104 227C 0000                   MOVEA.L #COMMON,A1            ;ADDR OF FIRST WORD
  33        8108 8032
  34   34   810A             *
  35   35   810A 12FC 0000    LOOP           MOVE.B  #0,(A1)+              ;CLEAR COMMON AREA
  36   36   810E 5301                        SUBQ.B  #ONE,D1
  37   37   8110 66F8                        BNE     LOOP
  38   38   8112             *
  39   39   8112 1E3C 00E4                   MOVE.B  #228,D7
  40   40   8116 4E4E                        TRAP    #14
  41   41   8118                             END
  41 lines assembled
```

Figure 5.9   Use of directives.

particular computer being used. Most of the features are invoked by including assembler directives in the program. In some cases, the linkage editor is needed to complete the operations.

Table 5.6 lists a number of capabilities of certain assemblers that go beyond the translation of each assembly language statement into a machine language instruction. They are listed in alphabetical order in the table but otherwise have no priority of importance. Their use depends on the application.

The technique of *conditional assembly* allows a programmer to select certain portions of the program to be assembled or not according to certain logical conditions. For example, the conditional-assembly specification (directive)

IFEQ ⟨expression⟩
⟨assembly language statements⟩
ENDC

causes the statements between the IFEQ and ENDC directives to be assembled if the value of the ⟨expression⟩ is equal (EQ) to zero. Otherwise, these statements are ignored by the assembler. Other conditions might include a specification of not equal, greater than, or less than zero. The purpose is to allow one program to take care of several variants of the same basic problem. The choice is made only when the program is assembled. Each time a different segment of code is to be assembled, the ⟨expression⟩ must be changed appropriately and the program reassembled.

The ability to reference *external variables* in an assembly language program is vital when a number of modules are to be linked together to form a load module. With this capability, a program may reference programs or variables by name in another object module without a specification of the exact address of the referenced entity. In one module, for example, the directive

XDEF Y

defines the variable Y. Another module would include the statement

XREF Y

Table 5.6   Advanced Features of Assemblers

Type	Purpose or Definition
Conditional assembly	To allow selective assembly
External references	References to variables or programs in another object module
Library	A collection of object modules taken as a unit
Macro instructions	A group of assembly language statements referenced by name

to indicate that Y is defined elsewhere. The linkage editor resolves all external references and concatenates separate object modules into one contiguous, relocatable object module.[2] One or more of these modules can also consist of a "library" of modules.

A *library* in this context is considered a group of object modules ready to be linked with the module that references programs or data in the library. For example, a mathematical library is convenient in certain applications. It might consist of routines to perform functions such as calculating the roots of equations. An assembly language program would simply reference the routines by name. The linking program would search the library and add the proper modules to form the complete load module before program loading and execution occurs. A library is created and maintained (modified by having modules added or deleted) by a program that is sometimes called a "librarian." This program works in conjunction with the assembler and linkage editor programs to provide the necessary information to allow the linkage editor to link the appropriate modules together.

Assemblers allow programming to be more modular and convenient by providing a macroassembler to assemble *macro instructions* or *macros*. The macro instruction can be viewed as the name of a sequence of assembly language instructions. In this regard it is similar to a subroutine in a high-level language. An example of a macro structure might be

```
MAC1 MACRO
⟨assembly language statements⟩
ENDM
```

where MAC1 is a label. The statements in the body of the macro instruction are included in the assembly language program each time the instruction MAC1 is encountered. Values can also be passed into the macro instruction in a manner similar to a subroutine call with various arguments.

EXERCISES

5.2.1. Write a routine to reserve a 20-word block of memory for storage and then initialize it to the successive values 1 through 20 upon execution.

5.2.2. Find the errors in the following program to add four 16-bit numbers in locations $2000 through $2006.

```
ORG       $1000
MOVE.L    $2000,A1
MOVE.L    4,D2
ADD.W     (A1)+,D2
SUB.W     #ONE,D2
BNE       LOOP
```

---

[2] The exact procedure depends on the system software being used. It differs slightly between computer systems.

```
ONE DC.W          1
    JMP           MACSBUG
    END
```

5.2.3. Determine the values (if any) created by the following directives.
   (a) DC.B 'N IS'
   (b) DC.B 20
   (c) HERE EQU *
   (d) DS.L 1
   (e) DC.L LABEL+2

5.2.4. Assume that EQU directives have been used to assign START the value $1000 and END the value $2000. Find the value computed by the assembler for the following expressions.
   (a) START-2
   (b) END-START
   (c) (END-START)/2
   (d) (END-START)/3
   (e) 2*END

5.2.5. Discuss how each of the advanced features of the assembler allows programs to be created that are easy to read, debug, and modify (maintain).

5.2.6. Give some examples that show where conditional assembly would be convenient. As a specific example, consider several models of the same basic computer with different memory capacity and disk units with slightly different characteristics. Assume an assembly language program assigns memory space and determines when memory is full. Other routines handle I/O transfer with the disk unit. Describe the structure of such assembly language routines. How would they be changed when they are made ready for execution on different models of the computer?

5.2.7. Compare the use of subroutines and macro instructions. Consider memory use as well as programming convenience and efficiency. A subroutine is sometimes called a *closed* routine, while a macro is called an *open* routine.

5.2.8. Consider a large programming project that involves the development and debugging of a number of assembly language modules, perhaps written by different programmers. Discuss various problems that might arise if the development software included only a nonrelocating assembler with no advanced features. By comparison, how does a relocating assembler with advanced features and a linkage editor facilitate program development?

5.3 ADDRESSING MODES FOR THE MC68000

The different addressing modes of a processor determine the variety of ways that an operand or its address can be referenced by an instruction. Generally speaking, processors employed in sophisticated applications require a large number of addressing modes in order to be effective and efficient. The MC68000 allows 14 different modes, which classifies it among the most powerful microprocessors in this regard. The basic addressing modes were introduced in Chapter 4, where the operation of the CPU and its machine language were emphasized. In the present sections, all of the modes are discussed using the MC68000 assembly language notation.

The general classification of addressing modes includes register direct, absolute, indirect, relative, and immediate addressing. In each class, a number of variations are available for the MC68000, as indicated in Table 5.7. For each addressing mode, the processor addressing circuitry computes the effective address of any operand specified in an instruction when that instruction is fetched and decoded. The table defines how the effective address, EA, is calculated for each addressing mode.

Table 5.7 MC68000 Addressing Modes

Mode	Effective Address Calculation
Register direct addressing	
Data register direct	$EA = Dn$
Address register direct	$EA = An$
Absolute data addressing	
Absolute short	$EA = $ (next word)
Absolute long	$EA = $ (next two words)
Indirect addressing	
Address register indirect	$EA = (An)$
Indirect with postincrement	$EA = (An); (An) \leftarrow (An) + N$
Indirect with predecrement	$(An) \leftarrow (An) - N; EA = (An)$
Indirect with displacement	$EA = (An) + \langle d_{16} \rangle$
Indirect with index	$EA = (An) + (Xn) + \langle d_8 \rangle$
Relative addressing	
PC relative with offset	$EA = (PC) + \langle d_{16} \rangle$
PC relative with index	$EA = (PC) + (Xn) + \langle d_8 \rangle$
Immediate data addressing	
Immediate	$DATA = $ next word(s)
Quick immediate	Inherent data
Implied addressing	
Implied register	$EA = $ SR, USP, SP, or PC

Notes:
1. EA   Effective address
   An   Address register
   Dn   Data register
   Xn   Address or data register used as index register
   SR   Status register
   PC   Program counter
   $\langle d_8 \rangle$ 8-bit offset (displacement)
   $\langle d_{16} \rangle$ 16-bit offset (displacement)
   ( )   Contents of
   ←   Replaces

2. $N = 1$ for byte, 2 for words, and 4 for longwords. If An is the stack pointer and the operand size is byte, $N = 2$ to keep the stack pointer on a word boundary.

3. The designation Ri is also used to indicate a register used for indexing. Motorola literature also uses Xi or ix for an index register as in Appendix IV.

The classification of addresses according to mode is convenient for the programmer to determine the ways in which operands can be referenced. Certain addressing modes are allowed and others forbidden to specific instructions. Motorola literature further classifies addressing into categories to simplify the discussion of those modes available for an instruction. As defined in Table 5.8, the addressing categories are *data, memory, control,* and *alterable,* referring to the operand characteristics. The significance of the categories is discussed later in this section. The table also lists the assembler syntax for each addressing mode and the effective address encoding (mode/register) used as part of the machine language format.

5.3.1 Register Direct and Absolute Addressing

In both the register direct and absolute addressing modes, the location of an operand is specified explicitly in the instruction. An MC68000 *direct address* must refer to one of the processor data or address registers. The contents of the register is the operand. An *absolute address* is a 16-bit or 32-bit memory address. In this case the contents of the location is the operand. A 16-bit address is termed "short" and a 32-bit address is considered "long."

Register Direct Addressing.　In effect, the processor registers represent a high-speed memory within the CPU and operations between registers require no external memory references. The two modes for register direct addressing are *data register direct* and *address register direct.* The effective address calculated is

EA = Dn

Table 5.8　Addressing Categories and Assembler Syntax

Addressing Mode	Mode	Register	Addressing Categories				Assembler Syntax
			Data	Mem.	Cont.	Alter.	
Data register direct	000	Reg. no.	X	—	—	X	Dn
Address register direct	001	Reg. no.	—	—	—	X	An
Address register indirect	010	Reg. no.	X	X	X	X	(An)
Address register indirect with postincrement	011	Reg. no.	X	X	—	X	(An)+
Address register indirect with predecrement	100	Reg. no.	X	X	—	X	−(An)
Address register indirect with displacement	101	Reg. no.	X	X	X	X	d(An)
Address register indirect with index	110	Reg. no.	X	X	X	X	d(An,Ri)
Absolute short	111	000	X	X	X	X	NNN
Absolute long	111	001	X	X	X	X	NNNNNN
Program counter with displacement	111	010	X	X	X	—	d(PC)
Program counter with index	111	011	X	X	X	—	d(PC,Ri)
Immediate	111	100	X	X	—	—	#NNN

which specifies data register Dn. The address register direct mode calculates the effective address as

EA = An

when An specifies the nth address register. The instruction referring to a data register can specify a byte (B), word (W), or longword (L) length for the operand. Thus the instruction

CLR.⟨X⟩   D2

clears register D2 for the length ⟨X⟩ = B or W or L. This is the typical format for a single-address instruction using a data register as the destination. An address register is considered to hold an address 16 or 32 bits long. The MC68000, however, uses only a 24-bit address to reference memory. Instructions are not available for byte operations using address registers. For example, the instruction

MOVE.⟨X⟩   A1,D1

will allow a size specification of only ⟨X⟩ = W or ⟨X⟩ = L. In addition to the size restriction, other limitations are placed on the use of an address register as the destination location.

The MC68000 allows manipulation of addresses by a group of instructions which use an address register as the destination. These instructions normally take the suffix "A" in assembly language notation. The instruction

MOVEA.L   A1,A2

transfers the 32-bit contents of A1 to A2, for example. The Move Address (MOVEA) instruction performs the same function as the MOVE instruction, but is designed to treat operands as addresses (i.e., positive numbers), indicating memory locations. Thus the data registers and address registers of the MC68000 are not considered equivalent for many operations. The differences are presented in detail in Chapter 9, where instructions that operate on addresses are considered.

Absolute Short Addressing.   The absolute short or 16-bit address for an operand in memory is contained in an extension word to the operation word for an instruction. The address is actually converted to a 32-bit address by extending the sign bit (bit 15) of the short address. Figure 5.10(a) shows the calculation. A 24-bit address is used to address memory with leading zeros or ones in bits [23:16], replicating the most significant bit of the extension word. The extended address is considered to be a positive 24-bit number.

Example 5.7 _____

Figure 5.11 shows the effect of using the short addressing mode. If the hexadecimal address is between 0 and $7FFF, the low 32K bytes of memory can be addressed. Short addresses $8000 and above are sign-extended to 24 bits, resulting in addresses in memory between $FF8000 and

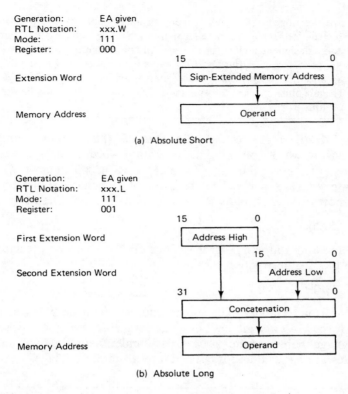

Generation: EA given
RTL Notation: xxx.W
Mode: 111
Register: 000

Extension Word

15   0

Sign-Extended Memory Address

Memory Address

Operand

(a) Absolute Short

Generation: EA given
RTL Notation: xxx.L
Mode: 111
Register: 001

First Extension Word

15   0

Address High

15   0

Second Extension Word

Address Low

31   0

Concatenation

Memory Address

Operand

(b) Absolute Long

Figure 5.10   Absolute addressing calculations. (Courtesy of Motorola, Inc.)

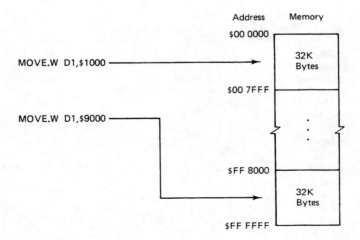

                        Address   Memory

                        $00 0000

MOVE.W D1,$1000   32K Bytes

                        $00 7FFF

MOVE.W D1,$9000

                        $FF 8000

                        32K Bytes

                        $FF FFFF

Figure 5.11   Effect of absolute short addressing.

$FFFFFF that are the upper 32K bytes of memory addressable by the MC68000. Thus the designers of the instruction set treated these segments of memory at the extremes in a special way. Normally, the system designer specifies the lowest segment of memory for system parameters. In fact, the MC68000 processor uses the first 1024 bytes for its vectors, which define the starting addresses for trap and interrupt routines. The highest segment is reserved for I/O interfaces in many systems. The absolute short addressing mode allows efficient access to fixed locations in either of these regions.

Absolute Long Addressing. As shown in Figure 5.10(b), a long address is formed from two extension words following the operation word of an instruction. Such an address can span the entire addressing space of the MC68000 memory. The magnitude of the absolute address specified in the assembly language statement determines whether the short or long absolute addressing mode is used. The instruction

MOVE.W $12000,D1

requires absolute long addressing for the source operand because the address requires more than 16 bits. An instruction such as

MOVE.B $3FFF,$12000

specifies both absolute modes. The source address could be given as $0000 3FFF, which would force most assemblers to use the long absolute mode. Reference to the assembly language manual of the particular assembler being used will determine how these addresses are handled and translated into machine language.

Example 5.8 _____

Table 5.9 shows a few of the addressing modes just discussed as they would be used in instructions. The table also lists the machine language statements as they are stored in memory. The instructions specifying register modes are one word in length and are the most efficient for storage

Table 5.9   Register and Absolute Addressing

Machine Language (hexadecimal)	Instruction		After Execution
2406	MOVE.L	D6,D2	(D2) = (D6)
4240	CLR.W	D0	(D0) [15:0] = 0
2640	MOVEA.L	D0,A3	(A3) = (D0)
1038 2001	MOVE.B	$2001,D0	(D0) [7:0] = ($2001)
21FC 0002 14AA 0024	MOVE.L	$214AA,$24	($24) = ($214AA)
21F8 0534 0528	MOVE.L	$0534,$0528	($0528) = ($0534)

and execution. As noted previously, the mnemonic MOVEA is used as the operation code when the destination is an address register.

In the absolute modes, the address is independent of the operand length except that word or longword operands must be addressed at even locations. The instruction

MOVE.L $0534,$0528

copies the 32-bit value in locations $0534 through $0537 into four bytes beginning at address $0528 since longword operands are specified. However, both addresses are short absolute addresses.

5.3.2 Indirect Addressing

Five indirect addressing modes are available for the MC68000 to reference an operand that is part of a data structure in memory. All five modes use an address register to hold the basic address, which can then be modified in various ways to compute the effective address of a specific operand. Unlike the absolute addresses that are defined when the program is assembled, the indirect address is computed as the program executes. Address register indirect, register indirect with displacement, and register indirect with indexing are discussed in this subsection. The postincrement and predecrement indirect modes are discussed in Section 5.3.3.

Address Register Indirect.   Any of the eight address registers of the MC68000 can be used to address indirectly an operand in memory. The effective address is calculated as

$$EA = (An)$$

when the nth register is designated. The contents of the selected address register are used as the operand address when an instruction using this mode executes. To load the register, an instruction such as

MOVEA.L   #⟨addr⟩,A1

transfers the address ⟨addr⟩ to A1. Then, reference to (A1), as in the instruction

MOVE.W   (A1),D1

would move the 16-bit value contained at location ⟨addr⟩ into D1. The specification of the address used as the source in the MOVEA instruction can be by any of the addressing modes, including the immediate mode just shown. For example, the instruction

MOVEA.L $1000,A1

moves the contents of the longword at location $1000 into A1.

Figure 5.12(a) shows the calculation of the operand address in this indirect mode. The assembler recognizes indirect addressing when (An) is used to designate a source or destination operand. In this text, (An) denotes the contents of register An and ((An)) indicates the operand, which is the contents of the contents of An. In Motorola's

Figure 5.12 Indirect addressing. (Courtesy of Motorola, Inc.)

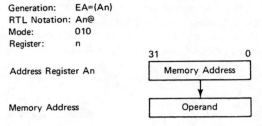

```
Generation:    EA=(An)
RTL Notation: An@
Mode:          010
Register:      n
```

(a) Address Register Indirect

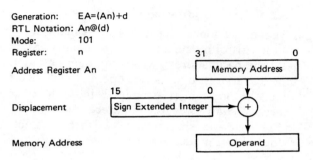

```
Generation:    EA=(An)+d
RTL Notation: An@(d)
Mode:          101
Register:      n
```

(b) Indirect with Displacement

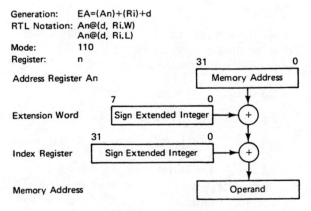

```
Generation:    EA=(An)+(Ri)+d
RTL Notation: An@(d, Ri.W)
              An@(d, Ri.L)
Mode:          110
Register:      n
```

(c) Indirect with Index

Register Transfer Language (RTL), the indirect address is indicated as An @. This notation is defined in Appendix IV.

Example 5.9

The address held in an address register can be modified in a number of ways during program execution. For example, the instruction Add Address has the format

ADDA.⟨X⟩ ⟨EAs⟩,⟨An⟩

to add the value in location ⟨EAs⟩ to the value in An. The length is restricted to W or L for 16 and 32-bit values, respectively. The address

⟨An⟩ ← ⟨An⟩ + ⟨EAs⟩

is generated when the ADDA instruction is executed. An operand referenced using An in a subsequent instruction would cause the new location to be addressed. Figure 5.13 shows a possible sequence.

The value $2000, representing the contents of location $1000, is first moved into A1. This address $2000 might point to the first word in a data structure in memory. Adding the constant $100 using the ADDA variation of ADD causes A1 to point to location $2100, an offset of 100 locations (hexadecimal) from the beginning of the data structure. Using A1 indirectly in the MOVE instruction transfers the value in the location $2100 to the designated location at $3000.

Indirect with Displacement. The address register indirect mode just discussed can be used in a variety of ways to address operands in memory. Using the address manipulation instructions of the MC68000 (MOVEA, ADDA, and others), the address in register An can be modified as desired when the program executes. In certain applications, it is not desirable to modify the address in the address register. This could be the case when the register is used to point to a segment of memory, perhaps the first address of an array or table. A selected element in the data structure could be addressed by adding an offset to the value in the address register. The address register thus holds the starting address and the offset or displacement specifies the relative position of the element within the data structure. When the offset is a fixed value that is known

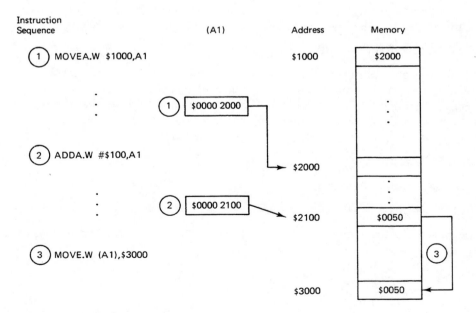

Figure 5.13   Examples of indirect addressing.

at assembly time, the *indirect with displacement* addressing mode of the MC68000 can be used to reference the operand.

The effective address for this mode is calculated, as shown in Figure 5.12(b), by the equation

$$EA = (An) + \langle d_{16} \rangle$$

where $\langle d_{16} \rangle$ is a 16-bit signed integer.[3] The range of this 16-bit offset is $-32{,}768$ to $+32{,}767$ bytes in memory. This displacement or offset is stored as an extension word of the machine language instruction. The assembler uses the indirect with displacement mode in instructions such as

MOVE.W $\langle d_{16} \rangle (An),\langle EAd \rangle$

where $\langle d_{16} \rangle$ is specified as a hexadecimal ($XXXX) or decimal (XXXX) constant. For example,

MOVE.W $20(A1),D1

transfers 16 bits to register D1 from the location 32 bytes beyond the address in A1. Note that the immediate symbol "#" is not needed because the assembler recognizes $\langle d_{16} \rangle$ as an offset and not a possible address.

Indirect with Indexing.   When the offset from a base address must be varied during program execution, the indirect with indexing mode may be used. The MC68000 provides indexed addressing that allows an effective address to be computed as the sum of three components: a value in an address register, plus a value in an index register, plus a displacement. The index register is any of the address or data registers of the MC68000. The basic calculation, shown in Figure 5.12(c), is

$$EA = (An) + (Ri)\, [X] + \langle d_8 \rangle$$

where Ri is any An or Dn and $\langle d_8 \rangle$ is an 8-bit signed integer. Notice that the displacement is restricted compared with the range of displacement for the indirect with displacement mode. The designation (Ri) [X] is used to indicate that the contents of Ri can be used as a sign-extended 16-bit integer or as a 32-bit value (i.e., X = W or L for word or longword, respectively). The designation

$\langle d_8 \rangle (An,Ri.W)$

indicates a word-length index and

$\langle d_8 \rangle (An,Ri.L)$

specifies the entire contents of the index register Ri. Thus the assembly language instruction

MOVE.W   2(A1,D1.W),D2

---

[3] In this context the subscript denotes the length of the integer, *not* the number base.

moves 16 bits into D2 from the address given by:

$$EA = (A1) + (D1) [15:0] + 2$$

If the value in the index register Ri is a hexadecimal number from $8000 through $FFFF, the value is considered negative. Thus the range of the 16-bit index in decimal is from −32,768 to +32,767. Use of the long value in Ri results in an index that can span the entire addressing space. This mode requires one extension word to the instruction in memory.

Example 5.10 _____

Assume that a list of words is stored in memory starting at location $1000. If the first four 16-bit entries in the list are used to determine the length of the list and similar information, the first data item can be accessed by adding a displacement of 8 bytes to the starting address. Consecutive items in the list can be accessed by indexing. This scheme is shown in Figure 5.14 using A1 as the base address and D1 as an index register. The first execution of the instruction at label LOOP addresses location $1008.

_____

5.3.3 Predecrement and Postincrement Addressing

In many programming applications, it is necessary to access data stored in consecutive memory locations. If the length of each item is a byte, word, or longword, the indexing can be accomplished using the MC68000 *indirect with predecrement* or *indirect with postincrement* addressing. In these modes, the contents of an address register are changed automatically as the instructions using the modes execute. Thus, no time is wasted increasing index values by program instructions as with the indirect with indexing addressing mode. Perhaps the most important use for these modes is to manipulate operands in a stack in memory as discussed in Chapter 4. However, many other data manipulation operations are simplified through the use of these modes.

The effective address for each mode is calculated as shown in Figure 5.15. In the postincrement mode, the address in An is used before An is incremented by 1, 2, or 4 for byte, word, or longword operands, respectively. This mode allows a program to address consecutive values stored at increasingly higher addresses in memory. In the predecrement mode, An is decremented first, then used as a pointer to a memory location.

The assembler recognizes the predecrement mode as −(An) and the postincrement as (An)+. In the instruction

MOVE.W   −(A1),(A2)+

a word is moved from location (A1)−2 to location (A2). If the instruction is executed again, the source address is one word lower in memory and the destination one word higher than the original values. This is typical usage in program loops in which the

```
        MOVEA.L  #$1000,A1        ;START OF LIST
        CLR.L    D1               ;ZERO INDEX
          .
          .
          .
LOOP MOVE.W   8(A1,D1.L),D2   ;MOVE ITEM TO D2
          .
          .
          .
        ADD.W    #2,D1            ;NEXT WORD

   (BRANCH TO LOOP IF NOT DONE)
```

(a) Program Segment

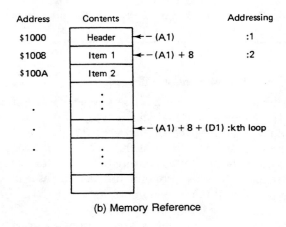

(b) Memory Reference

Figure 5.14   Indirect addressing with indexing.

autoindexed instruction is repeatedly executed until the condition to quit looping is met.

These modes are also used to move blocks of data from one memory segment to another when statements such as

MOVE.W   (A1)+,(A2)+

are used in a loop. Here (A1) designates the first block and (A2) the second. The addressing in this case is equivalent to using

MOVE.W   (A1),(A2)
ADD.L   #2,A1
ADD.L   #2,A2

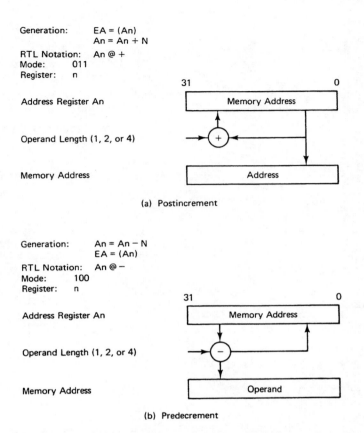

Generation: EA = (An)
               An = An + N
RTL Notation: An @ +
Mode: 011
Register: n

Address Register An

Operand Length (1, 2, or 4)

Memory Address

(a) Postincrement

Generation: An = An − N
               EA = (An)
RTL Notation: An @ −
Mode: 100
Register: n

Address Register An

Operand Length (1, 2, or 4)

Memory Address

(b) Predecrement

Figure 5.15   Predecrement and postincrement addressing. (Courtesy of Motorola, Inc.)

where both the source and destination addresses are incremented by 2 after the transfer because the operands are 2 bytes in length.

Example 5.11 _____

The simple program segment in Figure 5.16 moves a 32-byte block of data from location $2000 to location $3000, but in reverse order. The order of the data is reversed by starting the transfer from the first byte of the source block using postincrement addressing to the last byte of the destination block with predecrement addressing. The byte at $2000 is moved to $301F, the byte at $2001 is moved to $301E, and so on. Register D1 is used as a counter for the loop and registers A1 and A2 contain the addresses of the blocks.

_____

5.3.4 Relative Addressing

The relative addressing modes of the MC68000 allow an operand address to be calculated with respect to the value in the program counter. The *program counter relative with displacement* addressing mode allows a 16-bit integer value to be added to the

```
abs. rel.    LC   obj. code    source line
----  ----   ----  ---------   -----------
   1     1   0000                              TTL     'FIGURE 5.16'
   2     2   0000                              LLEN    100
   3     3   8000                              ORG     $8000
   4     4   8000                              OPT     P=M68000
   5     5   8000              *
   6     6   8000              *
   7     7   8000              *          MOVE 32 BYTES FROM $2000 TO $3000
   8     8   8000              *                  REVERSING THEIR ORDER
   9     9   8000              *
  10    10   8000 123C 0020               MOVE.B  #32,D1          ;SET COUNTER TO 32
  11    11   8004 227C 0000               MOVEA.L #$2000,A1       ;SET UP ADDRESSES
  11         8008 2000
  12    12   800A 247C 0000               MOVEA.L #$3020,A2       ;   FOR TRANSFER
  12         800E 3020
  13    13   8010              *
  14    14   8010 1519         LOOP       MOVE.B  (A1)+,-(A2)     ;MOVE NEXT BYTE
  15    15   8012 5301                    SUBI.B  #1,D1           ;DECREMENT COUNT
  16    16   8014 66FA                    BNE     LOOP            ;CONTINUE UNTIL
  17    17   8016              *                                 ;   COUNT = 0
  18    18   8016              *
  19    19   8016 1E3C 00E4               MOVE.B #228,D7
  20    20   801A 4E4F                    TRAP    #15             ;RETURN TO MONITOR
  21    21   801C                         END
  21 lines assembled
```

```
TUTOR  1.32> MD 2000                                    ;DATA
002000    00 01 02 03 04 05 06 07  08 09 0A 0B 0C 0D 0E 0F
TUTOR  1.32>
002010    10 11 12 13 14 15 16 17  18 19 1A 1B 1C 1D 1E 1F


TUTOR  1.32> DF
PC=00008000 SR=2704=.S7..Z.. US=00001000 SS=00000774
D0=00000000 D1=00000000 D2=00000000 D3=00000000
D4=00000000 D5=00000000 D6=00000000 D7=00000000
A0=00000000 A1=00000000 A2=00000000 A3=00000000
A4=00000000 A5=00000000 A6=00000000 A7=00000774
-----------008000    123C0020              MOVE.B  #32,D1


TUTOR  1.32> GT 8016
PHYSICAL ADDRESS=00008016
PHYSICAL ADDRESS=00008000


AT BREAKPOINT
PC=00008016 SR=2704=.S7..Z.. US=00001000 SS=00000774
D0=00000000 D1=00000000 D2=00000000 D3=00000000
D4=00000000 D5=00000000 D6=00000000 D7=00000000
A0=00000000 A1=00002020 A2=00003000 A3=00000000
A4=00000000 A5=00000000 A6=00000000 A7=00000774
-----------008016    1E3C00E4              MOVE.B  #228,D7


TUTOR  1.32> MD 3000 20·                                ;RESULTS
003000    1F 1E 1D 1C 1B 1A 19 18  17 16 15 14 13 12 11 10
003010    0F 0E 0D 0C 0B 0A 09 08  07 06 05 04 03 02 01 00
```

Figure 5.16   Examples of postincrement and predecrement addressing.

170

value in the PC to compute the effective address. If the displacement is not fixed but must be altered during program execution, the *program counter relative with index* mode may be used. The effective address is the sum of the address in the PC, an 8-bit integer that is sign-extended, and the contents of the index register. The calculation of the effective address for these two modes is shown in Figure 5.17(a). Each mode requires one extension word for the displacement in the instruction and the value in the program counter is the address of the extension word when the effective address is computed.

The assembler automatically treats any branch instruction (BRA, BGT, etc.) as an instruction using relative addressing with displacement. The distance between the branch instruction and the destination is computed as an offset from the contents of the program counter. A forced reference to the (PC) occurs in statements such as

ADD.W   *+$10,D1

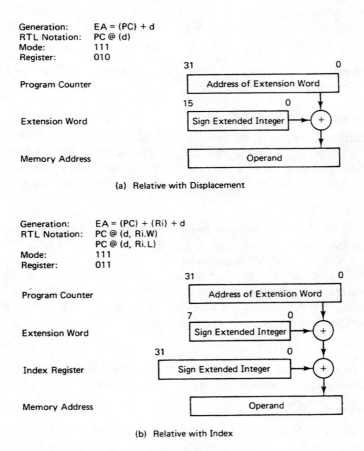

(a) Relative with Displacement

(b) Relative with Index

Figure 5.17 Relative addressing. (Courtesy of Motorola, Inc.)

where the operand is in a location eight words (16 bytes) past the location of the extension word of the ADD instruction. The operand is added to the contents of D1 [15:0]. The symbol "*" indicates the location counter to the assembler and the instruction is assembled using PC relative with displacement addressing. In this example instruction, the operand is nine words or 18 bytes past the location of the operation word of the ADD instruction.

The relative with index addressing mode causes the effective address to be computed as

$$EA = (PC) + (Ri) [X] + \langle d_8 \rangle$$

where the length of the index register is either 16 bits or 32 bits, indicated by $X = W$ or L, respectively. The fixed displacement is a signed 8-bit value. It is contained in an extension word in the instruction, as shown in Figure 5.17(b).

The relative modes of addressing are described in more detail in Chapter 9, particularly when position-independent coding is discussed. It should be noted that different assemblers may reference the relative addressing modes differently. One Motorola assembler (the resident assembler) allows references such as

ADD.W $10(PC),D1

which forces the relative mode for the address of the source. The * + $10 reference would also be recognized by this assembler.

5.3.5 Immediate and Implied Addressing

The immediate addressing mode for the MC68000 allows byte, word, or longword values to be used as constants in an instruction. The byte or word-length values require one extension word, and the long value adds two words to the instruction in memory. The assembler recognizes the immediate mode in instructions such as

MOVE.W   #50,D2

by the symbol "#" and the instruction moves 50 (decimal) into the low-order word of D2. As defined previously, hexadecimal values are specified with "$" and ASCII characters are enclosed within apostrophes.

Some MC68000 instructions do not require an operand to be specified and other instructions affect processor registers without explicitly referencing them in the instructions. These *implicit* references can be to the program counter, the status register, or the stack pointer. For example, the instruction

RTS

which returns from a subroutine to the calling program uses the stack pointer to retrieve the return address from the stack. The instruction

BSR   ⟨address⟩

does not explicitly reference the stack pointer or the program counter, but modifies both during its execution. These types of instructions have an addressing mode that is implied.

5.3.6 Addressing Categories

The individual addressing modes of the MC68000 are further characterized into four groups of *addressing categories*. These categories refer to the characteristics of the operand being addressed. As defined in Figure 5.18, the categories are *data, memory, control,* and *alterable.* The data category includes every mode but address register direct addressing. This is logical because the operand in an address register is assumed to be an address, not a data item, in this context. Except for register direct addressing, the other addressing modes can all refer to an operand in memory. However, all the memory references are not considered alterable. An immediate value is obviously not alterable. Also, relative addressing modes do not refer to an alterable operand. This means that the relative modes cannot be used to define destination addresses that are to be written (altered). For example, a MOVE instruction requires an alterable destination, which excludes the immediate and relative modes according to Figure 5.18.

The programmer uses the category information to determine which addressing modes are allowed for particular instructions. For example, according to the description found in the Motorola *User's Manual* or Appendix IV, the CLR instruction allows only data-alterable addressing modes for the destination operand. This excludes the address registers and relative addresses as destinations according to Figure 5.18. Thus the instruction

CLR   A1

Effective address modes can be categorized by the ways in which they may be used. The following classifications will be used in the instruction definitions.

Data   If an effective address mode can be used to refer to data operands, it is considered a data-addressing effective address mode.

Memory   If an effective address mode can be used to refer to memory operands, it is considered a memory-addressing effective address mode.

Alterable If an effective address mode can be used to refer to alterable (writable) operands, it is considered an alterable-addressing effective address mode.

Control   If an effective address mode can be used to refer to memory operands without an associated size, it is considered a control-addressing effective address mode.

Addressing Mode	Mode	Register	Addressing Categories				Assembler Syntax
			Data	Mem	Cont	Alter	
Data Reg Dir	000	reg no.	X	—	—	X	Dn
Addr Reg Dir	001	reg no.	—	—	—	X	An
Addr Reg Ind	010	reg no.	X	X	X	X	(An)
Addr Reg Ind w/Postinc	011	reg no.	X	X	—	X	(An)+
Addr Reg Ind w/Predec	100	reg no.	X	X	—	X	−(An)
Addr Reg Ind w/Disp	101	reg no.	X	X	X	X	d(An)
Addr Reg Ind w/Index	110	reg no.	X	X	X	X	d(An, Ri)
Absolute Short	111	000	X	X	X	X	XXX
Absolute Long	111	001	X	X	X	X	XXXXXX
Prog Ctr w/Disp	111	010	X	X	X	—	d(PC)
Prog Ctr w/Index	111	011	X	X	X	—	d(PC, Ri)
Immediate	111	100	X	X	—	—	#XXX

Figure 5.18   Addressing categories. (Courtesy of Motorola, Inc.)

is not allowed. In the discussion of the instructions in later chapters, the allowed addressing will be described in terms of these categories.

A few observations on the categories in Figure 5.18 should be noted:

(a) Instructions that require data-alterable addressing cannot use address registers as operands.
(b) The relative addressing modes reference operands that are not alterable (writable) by MC68000 instructions.
(c) Address register indirect addressing is allowed for all instructions that take operands. The same comment applies to the other modes that are considered to encompass all four categories.

EXERCISES

5.3.1. List some reasons why operands addressed relative to the program counter can be read and manipulated but their locations cannot be written.

5.3.2. The MC68000 does not allow the CLR instruction to specify an address register as a destination. Is the restriction on using CLR with address registers a problem? How does a program address location 0?

5.3.3. Discuss the various addressing modes in terms of values to be defined before assembly and those that can be defined or changed during program execution. (Assume that the program will not change its own instructions in memory.)

5.3.4. Discuss the advantages and disadvantages of the absolute short addressing mode compared to the absolute long addressing mode.

5.3.5. If (A1) = $1000, determine the operand address for the following instructions.
   (a) CLR.B $FFFF (A1)
   (b) MOVE.B   (A1)+,D1
   (c) MOVE.W   −(A1),D1

5.3.6. What do these instructions accomplish?
   (a) MOVE.L   4 (A0),(A0)
   (b) MOVE.W $9000,D1
   (c) MOVE.B   0(A0,D1),D1

5.3.7. Compare the absolute short addressing mode with the indirect displacement mode when the displacement value is $8000.

5.4 SUMMARY OF ADDRESSING MODES

The MC68000 has a number of addressing modes to reference an operand in a register, in the instruction itself, or in memory. Each instruction of the MC68000 allows certain addressing modes to define the location of operands (if any). With a few restrictions, the addressing modes can be combined with any operation to form a complete instruction. These restrictions will be treated in detail when particular instructions are encountered in subsequent chapters. In this section we concentrated on the use of various addressing modes and their assembly language format.

Table 5.10 briefly summarizes the addressing modes and lists possible uses for a particular mode. In the register direct modes, the operand is held in an address register or a data register. For the immediate mode, also called the "literal" mode, the operand is defined in memory as part of the instruction in machine language. These basic modes are sufficient for many programs in which operands are not part of a data structure in memory, but are treated individually as in mathematical calculations or data transfers. The absolute addressing modes are used to reference fixed addresses in memory, usually hardware related, such as for I/O devices. When the data structure in memory is more complex, an indirect addressing mode is usually more convenient to address an operand. Such data structures are treated in Chapter 9.

EXERCISES

5.4.1. Let a program have the instruction sequence:

```
ADD.W     FIRST(PC),D1
  •
  •
  •
ADD.W     FIRST(PC),D1
  •
  •
  •
END
```

Is the machine-language code for the two ADD instructions the same? Would the code be the same if (PC) were replaced by (A1)?

Table 5.10   Use of Addressing Modes

Mode	Example Use
Register direct	Mathematical or data transfer operations on operands held in registers
Immediate	Increment or initialize values in registers or memory
Absolute	Reference to hardware-related locations
Indirect	Operations on data values in memory
Predecrement or postincrement	Operations on values in arrays, stacks and queues
Indirect with displacement	Reference to an operand in memory from a base address
Indirect with index	Reference to an operand in a table of values in memory
PC-relative addressing	Create position-independent references to operands or addresses

REVIEW QUESTIONS AND PROBLEMS

In questions 5.1 through 5.14, choose the correct answer to the multiple choice question.

5.1. In an assembler, the instruction addresses are counted by the
   (a) program counter
   (b) location counter
   (c) line number

5.2. The Motorola S–records consist of
   (a) the object module
   (b) executable code
   (c) the ASCII version of the load module

5.3. In a single-board computer, the monitor resides in
   (a) ROM
   (b) RAM
   (c) the disk

5.4. The assembler directive DS.W $10
   (a) reserves 10 words in memory
   (b) reserves 16 words in memory
   (c) defines a constant to be $10

5.5. Which instruction initializes the 32-bit value of D1 to 5?
   (a) MOVE.L 5,D1
   (b) MOVE.W #5,D1
   (c) MOVE.L #5,D1

5.6. Macro instructions are used to
   (a) allow selective assembly
   (b) call a sequence of instructions by name
   (c) reference variables outside of a program

5.7. The MC68000 CPU has how many addressing modes?
   (a) 14
   (b) 18
   (c) 16

In 5.8 and 5.9, assume that ($1000)=$2000 and ($2100)=50.
5.8. The instruction MOVEA.W $1000,A1 results in
   (a) (A1)=$1000
   (b) (A1)=$2000
   (c) (A1)=$2100

5.9. Following the instruction in 5.8, the instruction MOVE.W ($100,A1),$3000 loads the value into location $3000 of
   (a) $2100
   (b) $50
   (c) $1100

5.10. In the instruction ADD.W $D1,$2F000 the source and destinations addressing modes are
   (a) absolute and immediate
   (b) register direct and immediate
   (c) register direct and absolute

5.11. In the instruction ADD.W $20(A1,D1),(A2) the source and destinations addressing modes are
   (a) register direct and register indirect
   (b) indirect with indexing and register indirect
   (c) indirect with indexing and register direct

5.12. In the instruction ADD.W #50,D2 the source and destinations addressing modes are
   (a) immediate and register direct
   (b) absolute and register direct
   (c) register direct and immediate

5.13. If (A1)=$0000 1000, the location addressed by CLR.B $FFFE(A1) is
   (a) $10FFE
   (b) $FFFE 1000
   (c) $FFFE

5.14. Which is *not* an assembler directive?
   (a) DC.W
   (b) END
   (c) STOP

Design Problems and Further Exploration

5.15. A number of development systems are available for the MC68000. Select several systems and compare them for both software and hardware development and debugging. Reference to manufacturers' literature is encouraged.

FURTHER READING

Barron's short but lucid text describes in detail the operation of an assembler. The text by Wakerly covers many members of the 68000 family. Other such references were cited in Chapter 4.

Barron, D. W. *Assemblers and Loaders,* 2nd ed. (New York: American Elsevier, 1972).

Wakerly, John F. *Microcomputer Architecture and Programming. The 68000 Family.* (New York: Wiley, 1989).

CHAPTER 6

Data Transfer, Program Control, and Subroutines

Several chapters are devoted to discussion of the assembly language instruction set of the MC68000. The purpose of these chapters is to analyze each instruction in detail and illustrate its use in assembly language programs. These discussions begin in this chapter. The instructions and their variations are separated into categories based on the operation performed. The instructions listed in Table 6.1 for *data transfer, program control,* and *subroutine use* are discussed in this chapter. Instructions for arithmetic, logical, and similar operations are presented in subsequent chapters.

The MOVE instruction is the primary instruction in the data transfer category. It does not have many restrictions on the location and length of operands that can be transferred between the CPU and memory. Two variations of this instruction are also listed in Table 6.1. They are distinguished from the MOVE instruction by adding a letter suffix, Q or M, to form MOVEQ or MOVEM. The MOVEQ "quick" form is a one-word instruction to load a data value into a data register. The MOVEM instruction is a variation that allows the contents of a selected group of registers to be transferred to or from consecutive memory locations. These two variations may be more efficient than the MOVE instruction for some purposes.

Instructions for program control are used to define the flow of control within a program. The BRA and JMP instructions cause unconditional transfer of control. The Bcc and DBcc instructions branch when certain conditions, which are defined by the condition codes, are met. The CMP instruction and its variations, which are listed in Table 6.1, are discussed in this chapter because they set the condition codes based on the values of the operands.

Subroutines can be called with the BSR or JSR instructions. The execution of a RTR or RTS instruction in the subroutine causes control to be returned to the calling program. The saving and restoring of the return address is handled automatically by the CPU when the call and return instructions are executed in the proper sequence.

In this chapter the instructions are defined in terms of the assembler syntax, operand length, valid addressing modes, and the effect on the condition codes. The entire instruction set is presented in Appendix IV.

Table 6.1   Selected Instructions

_____

Data transfer

MOVE	Move
MOVEQ	Move Quick (immediate)
MOVEM	Move Multiple Registers
EXG	Exchange Registers
SWAP	Swap Data Register Halves

Program control

Unconditional

BRA	Branch Always
JMP	Jump

*Conditional branch
and compare*

Bcc	Branch Conditionally
DBcc	Test Condition, Decrement, and Branch
CMP	Compare
CMPI	Compare Immediate
CMPM	Compare Memory
TST	Test

Subroutine

BSR	Branch to Subroutine
JSR	Jump to Subroutine
RTR	Return and Restore (CCR)
RTS	Return from Subroutine

_____

6.1   DATA TRANSFER

The instructions MOVE, MOVEQ, MOVEM, EXG, and SWAP represent data transfer instructions. The MOVE instruction is the most flexible and, consequently, the most frequently used. The quick variation, MOVEQ, transfers an 8-bit data value to a designated data register. MOVEM has the letter suffix M designating "multiple." It is used to save the contents of a selected group of registers in memory or to restore their contents from memory. The EXG and SWAP instructions transfer data between and within registers.

Table 6.2 summarizes the instructions covered in this section. For each instruction, the listings show the assembler syntax, the valid operand lengths, the possible addressing modes, and the condition codes affected by the instruction. When a number of modes are possible for the source or destination operand, the addressing modes are defined in terms of the addressing categories introduced in Chapter 5. The data-alterable category excludes address register direct, relative, and immediate addressing. Control-alterable addressing prohibits register direct, postincrement, predecrement, relative, and immediate addressing. The control category does not include the register direct, postdecrement, predecrement, or immediate modes.

Table 6.2 Instructions for Data Transfer

Instruction	Syntax	Operand Length (bits)	Addressing Modes		Condition Codes Affected
			Source	Destination	
Move	MOVE.⟨l⟩ ⟨EAs⟩,⟨EAd⟩	8, 16, 32	All[2]	Data alterable	N, Z V = {0}, C = {0}
Move Quick	MOVEQ #⟨d₈⟩,⟨Dn⟩	32	Immediate (sign-extended)	Dn	N, Z V = {0}, C = {0}
Move Multiple Registers	MOVEM.⟨l₁⟩ ⟨list⟩,⟨EA⟩	16 or 32	Register list	Control alterable or predecrement	None
	MOVEM.⟨l₁⟩ ⟨EA⟩,⟨list⟩	16 or 32	Control or postincrement	Register list	None
Exchange Registers	EXG ⟨Rx⟩,⟨Ry⟩	32	Rx	Ry	None
Swap Register Halves	SWAP ⟨Dn⟩	16	(Dn)[31:16] ↔ (Dn)[15:0]	—	N, Z V = {0}, C = {0}

Notes:
1. ⟨EA⟩   effective address
   ⟨l⟩   B, W, or L
   ⟨l₁⟩   W or L
   ⟨Rn⟩   Any Dn or An
   ⟨list⟩ register list
2. If the operand size is 1 byte, ⟨An⟩ cannot be a source.

6.1.1 The MOVE Instruction

The MOVE instruction transfers data between registers, between registers and memory, or between different memory locations. The format is

MOVE.⟨l⟩   ⟨EAs⟩,⟨EAd⟩

in which ⟨l⟩ = B, W, or L, indicating 8-bit, 16-bit, or 32-bit operands, respectively. Only the portion of the destination location specified by ⟨l⟩ is replaced in the operation

(EAd) [l] ← (EAs) [l]

where [l] designates the appropriate bits affected. If An is specified as a source operand, the length must be a word or longword (W or L) because byte operations on address registers are not allowed. The destination addressing mode must be data-alterable, which excludes address registers and relative addresses. The instruction variation MOVEA transfers values to address registers.

The value transferred by the MOVE instruction is treated as a signed integer of the specified length. Condition codes N and Z are set as a result of the operation and V and C are set to {0}. Thus tests on the operand for negative or zero values are possible immediately following a MOVE instruction. For example, the sequence

```
MOVE.W      LENGTH,D1
BEQ         DONE
```

causes a branch to the instruction at DONE if the contents of location LENGTH were zero because BEQ (Branch if Equal Zero) branches when Z = {1}. Otherwise, program execution continues with the next instruction. This sequence could be used, for example, to manipulate a data table of given length. If the length held in location LENGTH is zero, the instructions to manipulate the data values are skipped and control is passed to the instruction labeled DONE.

The instruction

```
MOVE.B      #$FF,D1
```

sets Z = {0} because the immediate value is not zero, but also sets N = {1} because the 8-bit value is considered negative.

The MOVE instruction is used in most of the example programs in this book, so specific examples of its use are not given here. Instead, variations of the MOVE instruction are presented in the next subsection and the MOVE instruction is compared with each variation.

6.1.2 Variations of MOVE

The MOVEQ instruction is a one-word instruction that has the symbolic form

MOVEQ   #⟨d₈⟩,⟨Dn⟩

where $\langle d_8 \rangle$ is an 8-bit constant. The 8-bit value is sign-extended to 32 bits and transferred to Dn. This instruction is very efficient when loading a data register with a constant in the decimal range −128 to +127. It executes in only four machine cycles, which is why it is called "quick." Standard MOVE instructions require more cycles for all transfers except register-to-register transfers. The condition codes are affected by the MOVEQ instruction exactly as they are for the MOVE instruction, allowing a test for zero or negative.

The instruction to clear a register

MOVEQ   #0,D1

has the same effect as CLR.L D1 because both instructions affect the full 32 bits of the register. There is no basic difference in efficiency or operation between these instructions.

Another variation of the MOVE instruction is MOVEM, which transfers data between processor registers and memory locations. The instruction has the symbolic form

MOVEM.$\langle l_1 \rangle$ $\langle$list$\rangle$,$\langle$EA$\rangle$

where the registers in the $\langle$list$\rangle$ are to be moved to memory locations beginning at address $\langle$EA$\rangle$; the low-order word or the whole register contents is moved if $\langle l_1 \rangle$ = W or L, respectively. The syntax for the $\langle$list$\rangle$ is shown in Table 6.3(a). Thus the instruction

MOVEM.W   D0/D1/A2–A4,$1100

transfers the low-order contents of D0, D1, A2, A3, and A4 to locations $1100, $1102, $1104, $1106, and $1108, respectively. To restore the register values, the format is

MOVEM.$\langle l_1 \rangle$ $\langle$EA$\rangle$,$\langle$list$\rangle$

where $\langle l_1 \rangle$ and $\langle$list$\rangle$ have the same meaning as before.

When the destination for a MOVEM instruction is memory, only control-alterable addressing modes or the predecrement mode are allowed. This excludes register direct, postincrement, and program counter relative addressing for the destination $\langle$EA$\rangle$. In

Table 6.3   MOVEM Instruction

(a) Register list
   (1) Selected registers are separated by "/" (e.g., D1/D3/D4)
   (2) Consecutive registers are specified by "–" (e.g., D1–D4)

(b) Transfer order

Order	Type
D0–D7, A0–A7 into higher address	1. Register to memory (control-alterable)
A7–A0, D7–D0 into lower addresses	2. Memory to register Register to memory (predecrement)

the control modes, the data registers are stored in the first locations of the specified memory area followed by the address registers, regardless of the order specified in the register list. The predecrement mode is used to store the register contents on a stack growing downward in memory and the order of storage is reversed. That is, address registers are stored first, followed by data registers.

A transfer from memory to the registers allows only control-alterable modes or postincrement addressing for the source operands. The registers are transferred in the order D0, D1, . . . , A0, . . . , A7, regardless of the order in the list. The storage addresses are assumed to increase in memory. The postincrement addressing pops the designated registers off the stack. Table 6.3(b) summarizes the transfer order for the various cases.

Example 6.1

The MOVEM instruction is used in several ways in the program shown in Figure 6.1. The first MOVEM saves the low-order contents of D4, D3, D2, and D1 on the system stack. The second MOVEM transfers four words from locations $1100, $1102, $1104, and $1106 to registers D1, D2, D3, and D4, respectively. The numbers are added and the sum stored in location $1108.

```
abs. rel.   LC    obj. code    source line
---- ----   ----  ---------    -----------
   1    1   0000                       TTL    'FIGURE 6.1'
   2    2   0000                       LLEN   100
   3    3   8000                       ORG    $8000
   4    4   8000                       OPT    P=M68000
   5    5   8000               *
   6    6   8000               *
   7    7   8000               *        USE OF MOVEM INSTRUCTION
   8    8   8000               *
   9    9   8000  48E7 7800             MOVEM.L   D1-D4,-(SP)    ;SAVE REGISTERS
  10   10   8004               *                                ; ON STACK
  11   11   8004  4CB8 001E             MOVEM.W   $1100,D1-D4    ;LOAD DATA
  11        8008  1100
  12   12   800A               *
  13   13   800A  D441                  ADD.W     D1,D2          ;ADD THE NUMBERS
  14   14   800C  D642                  ADD.W     D2,D3
  15   15   800E  D843                  ADD.W     D3,D4
  16   16   8010               *
  17   17   8010  31C4 1108             MOVE.W    D4,$1108       ;SAVE THE RESULT
  18   18   8014               *
  19   19   8014  4CDF 001E             MOVEM.L   (SP)+,D1-D4    ;RESTORE REGISTERS
  20   20   8018               *
  21   21   8018  1E3C 00E4             MOVE.B    #228,D7
  22   22   801C  4E4E                  TRAP      #14            ;RETURN TO MONITOR
  23   23   801E               *
  24   24   801E
  25   25   801E                        END
  25 lines assembled
```

Figure 6.1   Use of MOVEM instruction.

The registers are then restored from the stack in the reverse order from that in which they were stored.

6.1.3 Internal Data Transfer

The instruction EXG exchanges the 32-bit longword in the source register with that in the destination register. The registers involved can be data registers, address registers, or an address register and a data register. The instruction format is

EXG   ⟨Rx⟩,⟨Ry⟩

where the contents of Rx and Ry are exchanged. Only 32-bit exchanges are allowed and the condition codes are not affected.

The EXG instruction removes the need for temporary storage of register contents when 32-bit operands in two registers must be interchanged. The use of the exchange instruction is illustrated in Example 6.2.

The SWAP instruction exchanges the low-order word with the high-order word in a single register. The instruction has the form

SWAP   ⟨Dn⟩

in which only a data register can be specified. The condition codes are set according to the full 32-bit result to test for negative or zero.

Example 6.2 ──

The program shown in Figure 6.2 illustrates two methods of exchanging the contents of two registers, A0 and D1. In the first method, the contents of register D1 are saved in a temporary location in memory. Then D1 can be loaded with the value in A0 and, finally, A0 can be loaded with the original contents of D1 from the temporary memory location. The second method uses the EXG instruction to accomplish the same operation in one instruction.

EXERCISES

6.1.1. Why must the MOVE instruction clear the condition codes C and V?

6.1.2. Write a routine to find the largest value of a set of 16-bit integers stored in consecutive locations in memory by comparing numbers in order and exchanging a smaller one for a larger one in a processor register until the register contains the largest one. Try it with and without use of the EXG instruction.

6.1.3. A stack with N words has its first element at location BOTTOM. Write the instructions to transfer the word values to another stack. Let both stacks grow down in memory.

6.1.4. What advantage does MOVEM have over a series of MOVE instructions to save register contents in memory?

```
abs.  rel.    LC    obj. code    source line
----  ----    ----  ---------    -----------
   1     1    0000                          TTL      'FIGURE 6.2'
   2     2    0000                          LLEN     100
   3     3    8000                          ORG      $8000
   4     4    8000                          OPT      P=M68000
   5     5    8000               *
   6     6    8000               *
   7     7    8000               *  EXCHANGE (D1) AND (A0)
   8     8    8000               *
   9     9    8000  23C1 0000               MOVE.L   D1,TEMP        ;SAVE CONTENTS OF D1
   9          8004 8016
  10    10    8006 2208                     MOVE.L   A0,D1          ;D1 := A0
  11    11    8008 2079 0000                MOVEA.L  TEMP,A0        ;A0 := SAVED D1
  11          800C 8016
  12    12    800E               *  USE OF EXG INSTRUCTION
  13    13    800E C388                     EXG      D1,A0          ;A0 <---> D1
  14    14    8010               *
  15    15    8010 1E3C 00E4               MOVE.B   #228,D7
  16    16    8014 4E4E                     TRAP     #14            ;RETURN TO MONITOR
  17    17    8016               *
  18    18    8016               TEMP       DS.L     1              ;TEMPORARY STORAGE
  19    19    8016               *                                 ;   LOCATION
  20    20    801A               *
  21    21    801A                          END
  21 lines assembled
```

Figure 6.2   Use of EXG instruction.

6.1.5. Think of possible uses of the SWAP instruction. Remember that the second most significant byte in a 32-bit register is not directly accessible. What type of instruction is needed to access any particular byte in a 32-bit register?

6.2   PROGRAM CONTROL

In every sophisticated program, it is necessary to select which sequence of instructions to execute based on the results of computations. Thus the flow of control will follow different paths through the program, depending on these computations. In some cases, the transfer of control is *unconditional.* For example, the statement

JMP   MACSBUG

serves to transfer control to the address MACSBUG. Other programs require *conditional* transfer to control, which is based on the results of an arithmetic or other operation. These operations set the condition codes of the MC68000 status register. The conditional branch instructions test these condition codes to determine whether a branch is required.

Two MC68000 instructions, BRA and JMP, cause unconditional transfer of control. The branch instruction BRA is limited in the range of memory locations that can be bypassed and allows only relative addressing to determine the branch distance or displacement in memory. The Branch Conditionally (Bcc) instruction allows one of 14 conditions to be specified. If the selected condition is met, program execution continues at the branch location designated in the instruction. The MC68000 also has a more complicated Test Condition, Decrement, and Branch (DBcc) instruction, which is useful for conditional branching in an iterative program structure. These branch instructions and those instructions that compare the test operands are discussed in this section. The setting of the condition codes and the MC68000 instructions that affect them are listed in Appendix IV.

6.2.1 Unconditional Branch and Jump

The instructions Branch Always (BRA) and Jump (JMP) cause unconditional transfer of control by changing the value in the program counter, as shown in Table 6.4. Any time a branch or jump occurs, the new instruction address must be on a word (even) boundary or an addressing error will occur. The JMP instruction differs from the BRA instruction because the jump can be to anywhere in memory, while the range of the BRA transfer is limited to a displacement of an 8-bit or 16-bit signed integer. Additionally, the jump address can be specified by different addressing modes, but branch addressing is always relative to the program counter.

BRA Instruction. The form of the BRA instruction is

BRA   ⟨disp⟩

which allows either an 8-bit or a 16-bit displacement. For an 8-bit displacement, the branch range is −126 byte locations to +129 byte locations from the BRA instruction location because the value (PC) is the current instruction location plus 2 when the new value is calculated. In the case of a 16-bit displacement, the range is −32,766 to +32,769 bytes. The value of the displacement is determined automatically by the assembler when the displacement is specified as a label in the form:

BRA   ⟨label⟩

If the displacement to the location specified by ⟨label⟩ exceeds the branch range, the JMP instruction must be used instead.

Table 6.4 Unconditional Branch and Jump

Syntax	Operation	Address Modes
BRA ⟨disp⟩	(PC) = (PC) + ⟨disp⟩	Relative
JMP ⟨EA⟩	(PC) = (EA)	Control modes

Notes:
1. In BRA ⟨disp⟩, ⟨disp⟩ is a signed 8-bit or 16-bit integer.
2. (PC) is the BRA instruction location + 2.
3. Condition codes are not affected.

JMP Instruction. The JMP instruction has the form

JMP   ⟨EA⟩

where EA specifies the location of the next instruction to be executed. Only control addressing modes are allowed, which eliminates register direct and autoindexing modes. An indirect jump in the form

JMP   (An)

takes the address of the next instruction from the contents of An. This indirect jump can be used to create jump tables, as shown in the following example.

Example 6.3 _____

The program segment shown in Figure 6.3 shows how a jump table can be created in memory. The table contains the addresses of specific sequences of instructions. The address of the proper sequence to execute is selected by a code that specifies the entry in the table. The address is then loaded from the table into an address register and an indirect jump is executed using the register contents. In this example, the code is simply the number of the entry.

```
abs. rel.   LC   obj. code    source line
---- ----   ----  ---------    -----------
   1    1   0000                         TTL       'FIGURE 6.3'
   2    2   0000                         LLEN      100
   3    3   8000                         ORG       $8000
   4    4   8000                         OPT       P=M68000
   5    5   8000               *
   6    6   8000               *   JUMP TABLE EXAMPLE
   7    7   8000               *      INPUT : (D0.W) = ENTRY NUMBER
   8    8   8000               *              (A0.L) = TABLE ADDRESS
   9    9   8000               *
  10   10   8000               *
  11   11   8000 2248                    MOVEA.L   A0,A1             ;SAVE STARTING ADDRESS
  12   12   8002 3200                    MOVE.W    D0,D1             ;SAVE ENTRY NUMBER
  13   13   8004
  14   14   8004 C2FC 0004     JMPTBL    MULU.W    #4,D1
  15   15   8008 2471 1000               MOVEA.L   0(A1,D1.W),A2     ;DEFINE JUMP ADDRESS
  16   16   800C 4ED2                    JMP       (A2)              ;TRANSFER CONTROL
  17   17   800E               *
  18   18   8010                         ORG       $8010
  19   19   800E               *
  20   20        0000 4000     SCHED     EQU       $4000             ;SET UP EXAMPLE ENTRY
  21   21        0000 4100     QUEUE     EQU       SCHED+$100        ; POINTS FOR TABLE
  22   22        0000 4200     DISP      EQU       QUEUE+$100
  23   23        0000 4300     TIMER     EQU       DISP+$100
  24   24   8010               *
  25   25   8010 0000 4000     TABLE     DC.L      SCHED             ;JUMP TABLE
  26   26   8014 0000 4100               DC.L      QUEUE
  27   27   8018 0000 4200               DC.L      DISP
  28   28   801C 0000 4300               DC.L      TIMER
  29   29   8020               *
  30   30   8020                         END
  30 lines assembled
```

Figure 6.3   Jump table creation.

The table starting address must be loaded into A0 and the entry number 0, 1, 2, . . . into D0 before execution of the program segment. Each address is assumed to be held in a long-word, so the entry number is multiplied by 4 to index into the table. The table has the following format:

Entry Number	Address	Memory Contents
0	(A0)	First address
1	(A0) + 4	Second address
.	.	.
.	.	.
.	.	.
$n - 1$	(A0) + 4 * (D0)	nth address

The initial values in A0 and D0 are first transferred to A1 and D1, respectively, so the original contents are saved. Multiplying the contents of D1 by 4 with an unsigned multiply instruction (MULU) gives the offset from the starting address. The contents are located and transferred to A2 using address register indirect with index addressing for the source operand in the instruction

MOVEA.L   0(A1,D1.W),A2

where the 16-bit value in D1 is the index value. The indirect jump address has now been determined. Then the transfer of control is caused by the JMP instruction.

After the jump, the original contents of the PC are lost, so no return is possible in the program as shown. To transfer control and return from a segment addressed by the table, a subroutine call can be used with indirect addressing to specify the beginning location.

6.2.2 Branch Conditionally

The Bcc instructions allow selection of a control path in a program based on conditions:

IF (condition is true)
    THEN (branch to new sequence)
    ELSE (execute next instruction)

The new sequence of instructions may be at either higher or lower memory addresses relative to the branch instruction. The assembly language format for the general conditional branch is

Bcc   ⟨label⟩

which uses program counter relative addressing. The displacement added to the value of the program counter to cause branching can be either an 8-bit or a 16-bit signed integer. Some assemblers accept Bcc.S to force an 8-bit displacement (short) when possible.

Figure 6.4 illustrates the operation of the Bcc instructions. Arithmetic operations as well as a number of other instructions set the condition codes based on the result of the particular operation. If the condition is true, the displacement value is added to the value in the program counter. At this time, the PC contains the conditional branch instruction address plus 2. The possible conditions are listed in Figure 6.5 together with the condition code settings that cause a branch. The instruction format indicates how the conditions are coded in machine language.

If an arithmetic instruction or a MOVE instruction is executed, the condition codes indicate the arithmetic conditions that apply to the destination operand. For instance, in the addition

ADD.$\langle l \rangle$ X,Y

the destination value, which can be of length $\langle l \rangle = B$, W, or L, is the sum. The contents of the destination, designated (Y), may represent a signed or unsigned integer. In the case of an unsigned integer, the result may have been zero (Z = {1}) or nonzero (Z = {0}). If the sum is too large, it is indicated by a carry (C = {1}). Signed arithmetic could yield an out-of-range condition (V = {1}) or a positive, zero, or negative sum. The possible tests for signed and unsigned integers are listed in Table 6.5.

After a MOVE instruction, the condition codes can be examined, but no out-of-range condition is possible. The condition codes C and V are cleared so that C = {0} and V = {0} after the transfer. As indicated in the table, except for the test for zero or nonzero, different condition tests apply to unsigned than signed integers. This

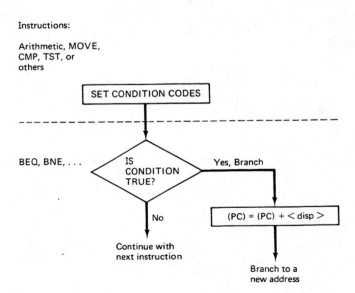

Figure 6.4   Operation of Bcc instructions.

CC	carry clear	0100	$\overline{C}$		LS	low or same	0011	$C + Z$
CS	carry set	0101	C		LT	less than	1101	$N \cdot \overline{V} + \overline{N} \cdot V$
EQ	equal	0111	Z		MI	minus	1011	N
GE	greater or equal	1100	$N \cdot V + \overline{N} \cdot \overline{V}$		NE	not equal	0110	$\overline{Z}$
GT	greater than	1110	$N \cdot V \cdot \overline{Z} + \overline{N} \cdot \overline{V} \cdot \overline{Z}$		PL	plus	1010	$\overline{N}$
HI	high	0010	$\overline{C} \cdot \overline{Z}$		VC	overflow clear	1000	$\overline{V}$
LE	less or equal	1111	$Z + N \cdot \overline{V} + \overline{N} \cdot V$		VS	overflow set	1001	V

Condition Codes: Not affected

Instruction Format:

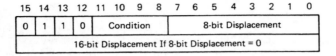

Notes:
(1) $\overline{X}$ as a condition code means $X = \{0\}$ is a TRUE condition, i.e., $\overline{X} = \{1\}$.
(2) "+" means LOGICAL OR.
(3) "·" means LOGICAL AND.

Figure 6.5   Conditional branch instructions.

section introduces the branching conditions required in both cases. Further discussion of the programming techniques for arithmetic operations is presented in Chapter 7.

Branch if Zero or Nonzero.   The conditional instructions BEQ and BNE are logical opposites. After a preceding instruction sets the condition codes, the instruction

BEQ   ⟨label⟩

Table 6.5   Conditional Tests for Integers

| | Conditions for Branch | |
Instructions	Unsigned	Signed
ADD.⟨l⟩ X,Y	$(Y) = 0$ BEQ	$(Y) = 0$ BEQ
or		
SUB.⟨l⟩ X,Y	$(Y) \neq 0$ BNE	$(Y) \neq 0$ BNE
or		
MOVE.⟨l⟩ X,Y		$(Y) \geq 0$ BPL
		$(Y) < 0$ BMI
Out of range for	$C = \{1\}$ BCS	$V = \{1\}$ BVS
arithmetic instructions	$C = \{0\}$ BCC	$V = \{0\}$ BVC

Note:
For the MOVE instruction with signed integers as operands, BGE, BLT, BGT, and BLE are also valid.

will branch if the result was zero. Branching on a nonzero condition would require

BNE   ⟨label⟩

Both instructions are valid for either signed or unsigned arithmetic and represent the only conditional branches that can be used with either.

Branches with Unsigned Integer Arithmetic. Unsigned arithmetic involves the positive integers and zero. Addresses should be treated as unsigned numbers. The only condition codes that should be tested after an addition or subtraction of unsigned integers are zero (Z) and carry (C).

The condition code C = {1} indicates that a carry occurred in addition because the sum was too large for the specified length of the operand. In subtraction, a carry bit set to {1} represents a borrow because the subtrahend was larger than the minuend. In either case, the result is not a valid unsigned integer. When arithmetic with unsigned integers is performed, the test for Branch on Carry Set (BCS) or Branch on Carry Clear (BCC) can be used to select between paths in the program.

The Bcc instruction is most often used to create loops in a program to perform the iterative parts of the algorithm being implemented. The Bcc instruction is used to exit the loop. For example, the simplest form of iteration occurs when the number of repetitions is known. In several previous programming examples, the termination of a loop relied on a counter value being decremented from the number of iterations required to zero. Then the BNE LOOP instruction following the decrement (SUB #1,Dn in most cases) instruction caused the looping to terminate when (Dn) = 0. If the branch instruction is the last statement in the loop, the loop is considered *post-tested* and has the general form:

REPEAT
     (body of loop)
UNTIL (count is zero)

These loops execute the instructions in the body of the loop at least once. The FORTRAN loop

     DO 10   I = 1,20
        (body of loop)
10 CONTINUE

is a loop of this type. Other types of loop structures are shown in examples elsewhere in the book.

Example 6.4 _____

The program shown in Figure 6.6 adds two arrays or vectors of unsigned 16-bit integers. The program computes X(I) + Y(I) and saves the sum in X(I) for each element in the arrays. The length of the arrays is input in D0 and the addresses of the arrays are input in A0 and A1. After each addition, the C condition code is checked. If the C bit is {1}, an overflow occurred and the low-order word of D0 will contain the value $FFFF.

In the program, (D1) is used as a counter and also as an index value into the arrays. After D1 is loaded with the word count and tested for zero, the value in D1 is multiplied by 2 to

```
abs. rel.    LC   obj. code    source line
---- ----    ----  ---------    -----------
   1    1   0000                        TTL    'FIGURE 6.6'
   2    2   0000                        LLEN   100
   3    3   8000                        ORG    $8000
   4    4   8000                        OPT    P=M68000
   5    5   8000          *
   6    6   8000          *
   7    7   8000          *  COMPUTE SUMS OF UNSIGNED INTEGER ARRAYS
   8    8   8000          *     INPUT : (A0.L) = X ARRAY OF 16-BIT NUMBERS
   9    9   8000          *             (A1.L) = Y ARRAY OF 16-BIT NUMBERS
  10   10   8000          *             (D0.W) = NUMBER OF ELEMENTS IN ARRAYS
  11   11   8000          *     OUTPUT: (D0.W) = $FFFF IF ERROR OCCURRED
  12   12   8000          *
  13   13   8000 48E7 60C0  SUM    MOVEM.L D1-D2/A0-A1,-(SP) ;SAVE REGISTERS
  14   14   8004 3200             MOVE.W  D0,D1            ;SET UP COUNTER
  15   15   8006 6700 0014        BEQ   ERROR             ;IF ZERO, EXIT
  16   16   800A          *                               ; WITH ERROR
  17   17   800A          *
  18   18   800A E349             LSL.W   #1,D1            ;COUNT BYTES
  19   19   800C 5541      LOOP   SUBQ.W  #2,D1            ;DECREMENT INDEX
  20   20   800E 6B00 0010        BMI   DONE              ;EXIT WHEN (D1) < 0
  21   21   8012 3431 1000        MOVE.W  0(A1,D1.W),D2    ;GET Y VALUE
  22   22   8016 D570 1000        ADD.W   D2,0(A0,D1.W)    ;ADD X := X + Y
  23   23   801A          *
  24   24   801A 64F0             BCC     LOOP             ;IF NO OVERFLOW, LOOP
  25   25   801C          *                               ;    ELSE
  26   26   801C 303C FFFF  ERROR  MOVE.W  #$FFFF,D0        ;SET STATUS TO ERROR
  27   27   8020          *
  28   28   8020 4CDF 0306  DONE   MOVEM.L (SP)+,D1-D2/A0-A1 ;RESTORE REGISTERS
  29   29   8024 1E3C 00E4        MOVE.B  #228,D7
  30   30   8028 4E4E             TRAP  #14                ; AND RETURN
  31   31   802A             END
  31 lines assembled
```

Figure 6.6   Overflow checking for unsigned arithmetic.

yield a byte count. The Logical Shift Left (LSL) instruction accomplishes the multiplication. In the loop, the counter and index in D1 are decremented by 2 and tested before each addition is performed. The Branch If Minus (BMI) instruction is used so the last addition is performed when (D1) = 0. Address register indirect with index addressing is used to select the values to be added. Notice that the last elements in the arrays are added first because (D1) begins in the loop with the byte count minus 2 as an index and decrements to zero. If no overflow occurs, the BCC instruction returns control to the first instruction in the loop. Otherwise, an error is indicated.

Branching with Signed Arithmetic.   If the values being manipulated are two's-complement numbers, the condition codes N, Z, and V are applicable. After arithmetic operations, V = {1} indicates an out-of-range condition. The instruction BVS branches when V = {1} and BVC branches when V = {0}. If the result is valid, the operand may be tested for a zero, nonzero, positive, or negative value. After a MOVE instruction, V = {0} and BGE, BLT, BGT, and BLE also perform valid tests according to the logic equations from Figure 6.5. When V = {0}, BGE has the same effect as BPL.

6.2.3 Branching After CMP or TST

The instructions Compare (CMP) and Test (TST) are used to set the condition codes based on operand values. Then conditional branch instructions can be used to direct the flow of control in the program. The instruction

CMP   X,Y

compares two operands by performing the computation

$$(Y) - (X)$$

to set the condition codes N, Z, V, and C. The instruction

TST   Y

evaluates one operand by performing the computation $(Y) - 0$, which always clears V and C but sets N and Z based on the result. Both of these instructions set the condition codes but do not modify the operands. The instructions discussed in this section are listed in Table 6.6. The CMP instruction has variations Compare Immediate (CMPI) and Compare Memory (CMPM), as shown.

The compare instruction

CMP.⟨l⟩   ⟨EA⟩,⟨Dn⟩

subtracts the source operand from the contents of the specified data register. The computation

$$(Dn) - (EA)$$

is performed without modifying the operand and the length ⟨l⟩ of each operand can be defined as B, W, or L. If the source is an address register, the operand length is restricted to word or longword.

The CMPI instruction compares an immediate value to an operand referenced by a data-alterable addressing mode. This excludes the address register direct and relative addressing modes. Byte, word, or longword operands are allowed. For example, the instruction

CMPI.B   #5,$2000

compares the value 5 with the contents of the byte at location $2000. Unlike the CMP instruction, the CMPI instruction may reference memory locations as the destination.

The CMPM instruction is used to compare sequences of bytes, words, or longwords in memory. Only the postincrement addressing mode is allowed for both operands. The instruction

CMPM.B   (A1)+,(A2)+

Table 6.6 Compare and Test Instructions

Instruction	Syntax	Addressing Modes		Operation	Condition Codes Affected
		Source	Destination		
Compare	CMP.⟨l⟩ ⟨EA⟩,⟨Dn⟩	ALL	⟨Dn⟩	(Dn) − (EA)	N, Z, V, C
Compare immediate	CMPI.⟨l⟩ #⟨d⟩,⟨EA⟩	d	Data alterable	(EA) − d	N, Z, V, C
Compare memory	CMPM.⟨l⟩ (Am)+,(An)+	(Am)+	(An)+	((An)) − ((Am))	N, Z, V, C
Test	TST.⟨l⟩ ⟨EA⟩	—	Data alterable	(EA) − 0	N, Z, C = {0}, V = {0}

Notes:
1. ⟨l⟩ denotes B, W, or L.
2. If An is the source for CMP, only word (W) or longword (L) operands are allowed.

compares the bytes addressed by A1 and A2 and then increments the addresses to point to the next bytes.

The TST instruction has the format

TST.⟨l⟩   ⟨EA⟩

where ⟨l⟩ can be B, W, or L and ⟨EA⟩ can be specified by all but the address register direct or relative addressing modes. The instruction sets the Z and N condition codes based on the value of the operand. However, V and C are always cleared.

The valid conditional branches after the TST instruction for an unsigned integer operand are BEQ and BNE. These conditions for a zero or nonzero value are also valid for signed integers. Because the TST instruction clears the C and V condition codes, a number of other tests for signed integers are also valid. After the instruction

TST   X

the conditional branch instructions BGT, BLT or BMI, BGE or BPL, and BLE can be used when the location X contains a signed integer.

When a Bcc instruction follows a CMP instruction in the sequence

CMP   X,Y
Bcc   ⟨label⟩

the comparison made is

IF (Y) "condition cc" (X)
    THEN branch to ⟨label⟩
    ELSE continue

For example, the instruction sequence

CMP.W   D1,D2
BGE   DONE

checks if the 16-bit value in D2 is greater than or equal to the 16-bit value in D1. If so, the branch is taken to DONE.

The TST instruction compares an operand with zero in a similar manner. Thus the sequence

TST.W   D2
BEQ   DONE

has the logic

IF (D2) [15:0] equals 0
    THEN branch to DONE
    ELSE continue

Table 6.7 lists the Bcc instructions that would cause a branch if executed after the CMP or TST instruction. The tests for Branch on Less Than (BLT) and Branch on Minus (BMI) as well as BGE and BPL are equivalent after a TST instruction because the overflow bit V is cleared.

Table 6.7 Bcc Instructions with CMP and TST

		Branch Condition	
Instruction	Result	Unsigned	Signed
CMP X, Y	(Y) = (X)	BEQ (Equal)	BEQ (Equal)
	(Y) ≠ (X)	BNE (Not Equal)	BNE (Not Equal)
	(Y) > (X)	BHI (High)	BGT (Greater Than)
	(Y) ≥ (X)	BCC (Carry Clear)	BGE (Greater or Equal)
	(Y) < (X)	BCS (Carry Set)	BLT (Less Than)
	(Y) ≤ (X)	BLS (Low or Same)	BLE (Less Than or Equal)
TST X	(X) = 0	BEQ	BEQ
	(X) ≠ 0	BNE	BNE
	(X) > 0	BNE	BGT
	(X) < 0	—	BLT, BMI
	(X) ≥ 0	—	BGE, BPL
	(X) ≤ 0	—	BLE

Notes:
1. In CMP X,Y; the destination is a data register.
2. TST sets C = {0} and V = {0}.
3. BMI (Branch on Minus) is the same as BLT, and BPL (Branch on Plus) is the same as BGE when V = {0}.

Both CMP and TST can be used with either signed or unsigned integer interpretations. However, the processor always performs the computation in two's-complement arithmetic. Therefore, the valid conditional branch instructions are different for the two interpretations of integers.

Example 6.5

The sample program in Figure 6.7 compares two tables of bytes addressed by A1 and A2. D1 contains the number of bytes in each table and D2 contains a status word or "flag" to indicate if the tables are the same. The flag is initialized to a default value of zero to indicate unequivalent values before the comparison begins. If all corresponding bytes of the two tables are equal, a value of +1 is entered into D2 by the last instruction before returning to the monitor program. If the number of bytes is zero, D2 will be left with the initialized zero value. Such a program is typically used to compare two strings of ASCII characters to see if they are equal.

Before comparison, a TST instruction determines if D1 has a nonzero byte count. Within the loop, the bytes addressed by A1 and A2 are compared and the addresses are automatically incremented. If any two bytes are not equal, the first BNE instruction causes a branch and the test ends. After each successful compare, the counter value in D1 is decremented. Looping continues until the counter reaches zero. If all the bytes are equal, D2 will be set to 1 by the instruction at label EQUAL. Before the program can be executed, A1, A2, and D1 must all be initialized with the proper input values.

Example 6.6

The program shown in Figure 6.8 compares 16-bit positive integers in an array previously stored in locations NUM1, NUM1 + 2, and so on, and leaves the largest one in D1. Register D3 is

```
abs. rel.   LC   obj. code   source line
---- ----   ----  ---------   -----------
   1    1   0000                           TTL   'FIGURE 6.7'
   2    2   0000                           LLEN  100
   3    3   8000                           ORG   $8000
   4    4   8000                           OPT   P=M68000
   5    5   8000             *
   6    6   8000             *
   7    7   8000             * COMPARE TWO TABLES OF BYTES
   8    8   8000             * INPUTS : (A1.L) = ADDRESS OF FIRST TABLE
   9    9   8000             *          (A2.L) = ADDRESS OF SECOND TABLE
  10   10   8000             *          (D1.B) = NUMBER OF BYTES IN THE TABLES
  11   11   8000             *                 (255 MAXIMUM)
  12   12   8000             *
  13   13   8000             * OUTPUT : (D2.W) = 0 : NOT EQUAL  ;STATUS
  14   14   8000             *                   1 : EQUAL
  15   15   8000             *
  16   16   8000 48E7 4060   COMPARE MOVEM.L D1/A1-A2,-(SP)  ;SAVE REGISTERS
  17   17   8004 4242                CLR.W   D2              ;SET FAULT STATUS
  18   18   8006 4A01                TST.B   D1              ;IF LENGTH IS ZERO
  19   19   8008 6700 0010            BEQ    DONE            ; THEN BRANCH TO DONE
  20   20   800C             *
  21   21   800C B509        LOOP    CMPM.B  (A1)+,(A2)+     ;ELSE IF TWO BYTES ARE
  22   22   800E             *                               ;     NOT EQUAL
  23   23   800E 6600 000A           BNE     DONE            ; THEN BRANCH TO DONE
  24   24   8012 5301                SUBQ.B  #1,D1           ;ELSE DECREMENT COUNTER
  25   25   8014 66F6                BNE     LOOP            ;     AND TEST NEXT VALUE
  26   26   8016             *                               ;     UNTIL (D1)=0.
  27   27   8016             *
  28   28   8016 343C 0001   EQUAL   MOVE.W  #1,D2           ;THEY ARE IDENTICAL
  29   29   801A             *                               ;  SET STATUS=1
  30   30   801A             *
  31   31   801A 4CDF 0602   DONE    MOVEM.L (SP)+,D1/A1-A2  ;RESTORE REGISTERS
  32   32   801E 1E3C 00E4           MOVE.B  #228,D7
  33   33   8022 4E4E                TRAP    #14             ; AND RETURN
  34   34   8024             *
  35   35   8024                     END
  35 lines assembled
```

Figure 6.7   Use of CMP and TST instructions.

used as a counter to determine when all the numbers have been tested. The count of numbers is assumed to be in D3 before the program executes.

Because only positive numbers are considered, Branch Lower or Same (BLS) is used to test if a value in D2 is smaller than the assumed maximum held in D1. If so, the loop is tested for completion. If the new value in D2 is larger than the "maximum" in D1, the register contents are exchanged.

6.2.4 DBcc Instruction

The Test, Decrement, and Branch (DBcc) instruction is a powerful instruction for control of loop structures. The basic format is

DBcc   ⟨Dn⟩, ⟨label⟩

```
abs. rel.   LC   obj. code    source line
---- ----   ----  ---------    -----------
   1    1   0000                        TTL    'FIGURE 6.8'
   2    2   0000                        LLEN   100
   3    3   8000                        ORG    $8000
   4    4   8000                        OPT    P=M68000
   5    5   8000            *
   6    6   8000            *
   7    7   8000            *  FIND THE LARGEST 16-BIT POSITIVE INTEGER IN TABLE NUM1
   8    8   8000            *
   9    9   8000            *  INPUTS :   (D3.W) = NUMBER OF INTEGERS TO EXAMINE
  10   10   8000            *             NUM1   = TABLE OF INTEGERS; 40 MAXIMUM
  11   11   8000            *  OUTPUTS :  (D1.W) = MAXIMUM INTEGER FOUND
  12   12   8000            *
  13   13   8000 48E7 3040            MOVEM.L D2-D3/A1,-(SP)  ;SAVE REGISTERS
  14   14   8004 227C 0000            MOVEA.L  #NUM1,A1       ;LOAD STARTING ADDRESS
  14        8008 8026
  15   15   800A 5343                 SUBQ.W   #1,D3         ;NUMBER OF INTEGERS-1
  16   16   800C 3219                 MOVE.W   (A1)+,D1      ;(D1)= FIRST INTEGER
  17   17   800E            *
  18   18   800E 3419       LOOP      MOVE.W   (A1)+,D2      ;GET NEXT INTEGER
  19   19   8010 B441                 CMP.W    D1,D2         ;COMPARE TO MAXIMUM
  20   20   8012 6300 0004            BLS      NEXT          ; IF LESS OR SAME
  21   21   8016            *                                ;    CHECK NEXT ONE
  22   22   8016 C342                 EXG      D1,D2         ; ELSE:EXCHANGE
  23   23   8018            *
  24   24   8018 5343       NEXT      SUBQ.W   #1,D3         ;DECREMENT COUNTER
  25   25   801A 66F2                 BNE      LOOP          ; LOOP TILL (D3)=0
  26   26   801C            *
  27   27   801C 4CDF 020C            MOVEM.L (SP)+,D2-D3/A1 ;RESTORE REGISTERS
  28   28   8020 1E3C 00E4            MOVE.B #228,D7
  29   29   8024 4E4E                 TRAP #14              ;AND RETURN
  30   30   8026            *
  31   31   8026            NUM1      DS.L 20               ;NUMBERS
  32   32   8026            *
  33   33   8076                      END
  33 lines assembled
```

Figure 6.8 Comparing positive integers.

which designates three parameters: the condition "cc", a data register, and a displacement. The displacement is represented here as a label in an assembly language program. The syntax and the various conditions for the instruction are shown in Figure 6.9.

The DBcc instruction can cause a loop to be terminated when either the specified condition, cc, is TRUE or when the count held in $\langle Dn \rangle$ reaches -1. Each time the instruction is executed, the value in Dn is decremented by 1. The flowchart in Figure 6.10 illustrates the conditions that cause the next instruction in sequence to be executed.

Fourteen of the logical conditions tested by DBcc are the same as those for the Bcc instruction discussed previously. However, the DBcc is sometimes called a "Don't Branch on Condition" instruction. If the condition is TRUE, no branch is taken, which is the opposite operation from that of the Bcc instruction. A loop structure with the DBcc terminating the loop has the logic

Test Condition, Decrement
and Branch

DBcc < Dn >, < disp >   (1)   Test "cc"
                                 (2)   If "cc" TRUE, Next Instruction
                                        ELSE
                                        (Dn) [15:0] = (Dn) [15:0] − 1
                                        IF (Dn) [15:0] = −1,
                                        · next instruction
                                        ELSE (PC) = (PC) + < disp >

Notes:
  (1) (PC) is instruction address +2.
  (2) < disp > is a signed 16-bit integer that is sign-extended.

(a) Syntax and Operation

CC	carry clear	0100	$\overline{C}$	LS	low or same	0011	$C + Z$
CS	carry set	0101	C	LT	less than	1101	$N \cdot \overline{V} + \overline{N} \cdot V$
EQ	equal	0111	Z	MI	minus	1011	N
F	false	0001	0	NE	not equal	0110	$\overline{Z}$
GE	greater or equal	1100	$N \cdot V + \overline{N} \cdot \overline{V}$	PL	plus	1010	$\overline{N}$
GT	greater than	1110	$N \cdot V \cdot \overline{Z} + \overline{N} \cdot \overline{V} \cdot \overline{Z}$	T	true	0000	1
HI	high	0010	$\overline{C} \cdot \overline{Z}$	VC	overflow clear	1000	$\overline{V}$
LE	less or equal	1111	$Z + N \cdot \overline{V} + \overline{N} \cdot V$	VS	overflow set	1001	V

Condition Codes: Not affected

Instruction Format:

15	14	13	12	11	10	9	8	7	6	5	4	3	2	1	0
0	1	0	1	Condition				1	1	0	0	1	Register		
Displacement															

(b) Conditions and Instruction format

Figure 6.9   DBcc instruction syntax and conditions.

REPEAT
      (body of loop)
UNTIL (condition)

For example, the sequence of instructions with the structure

LOOP . . .
      (body of loop)
      TST   X   ;Test for zero
      DBEQ   ⟨Dn⟩,LOOP   ;LOOP IF NOT ZERO

will continue to loop until the contents of location X are zero or until the count in
Dn has been exhausted. The TST instruction sets the condition codes based on the

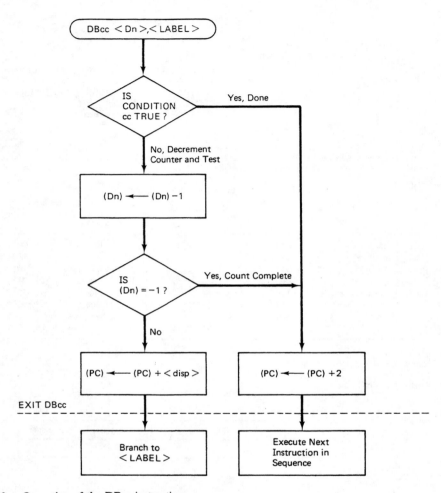

Figure 6.10 Operation of the DBcc instruction.

value (X). The DBcc tests the conditions and decrements ⟨Dn⟩ if the condition is false but does not affect the condition codes. When the specified condition is true or when ⟨Dn⟩ contains −1, the loop is terminated. If the example program above needed to loop (branch) on the X equal 0 condition, the opposite logical condition, DBNE, should be used. Then if X is zero, a branch is taken to LOOP; otherwise, execution continues to the next instruction.

In addition to the 14 testable conditions that are used by both DBcc and Bcc instructions, the DBcc has two other conditions, TRUE(T) and FALSE(F). The DBT instruction never branches. Its logical opposite, the DBF, always branches unless the count is exhausted. For the DBF instruction, no condition is tested. The DBF instruction replaces the "decrement and test for zero" sequence frequently used to terminate a loop.[1] There is one difference, however. Because the counter must reach −1 before looping stops, the initial value of the counter must be one less than the number of iterations required.

The register specified to hold the count contains a 16-bit integer with a decimal value between 0 and 65,535. Assuming that the register initially contained the integer value N and also that the condition cc is not true, then in a post-tested loop, the count would be N, N − 1, N − 2, . . . , 0 before the loop is exited; thus N + 1 iterations are executed. To loop N times, the counter register should contain the value N − 1 initially.

Example 6.7

The two short program segments shown in Figure 6.11 compare the use of the Bcc and DBcc instructions. Each program tests a table containing (N + 1) word-length operands to locate a nonzero entry. Before execution, address register A1 must contain the first address of the table and D1 contains N. In the first segment, a nonzero entry addressed by (A1) causes a branch to label DONE1. The address of the nonzero entry is (A1) −2 because postincrement addressing is used in the TST instruction. If no nonzero entries are found, (D1) is decremented until it reaches −1, upon which the BPL instruction terminates the loop. Thus, if (D1) = −1 after the program segment is complete, the table contains all zero values.

The instruction to test for a nonzero value and those to decrement and test the loop count in the first program can be replaced with a single DBcc instruction. The second segment in Figure 6.11 shows the instructions to test the (N + 1) locations for a nonzero value as before.

EXERCISES

6.2.1. Discuss the use of a jump table if the program is in read-only memory but the routines to be executed may have a different starting address in different systems.

6.2.2. Compare the use of indirect addressing in the following instructions.
   (a) JMP   (A2)
   (b) MOVEA   (A2),A1

6.2.3. If the instruction

   CMP.W ◂ I,J

   has been executed, specify the instruction that will perform the following.
   (a) branch to ZERO if I = J
   (b) branch to LESS if I < J
   (c) branch to MORE if I > J
   where ZERO, LESS, and MORE are statement labels.

---

[1] Some MC68000 assemblers use the form "DBRA" instead of the mnemonic DBF.

```
abs. rel.   LC    obj. code    source line
---- ----   ----  ----------   -----------
   1    1   0000                        TTL    'FIGURE 6.11'
   2    2   0000                        LLEN   100
   3    3   8000                        ORG    $8000
   4    4   8000                        OPT    P=M68000
   5    5   8000              *
   6    6   8000              *
   7    7   8000              *   COMPARISON OF DBCC AND BCC OPERATION
   8    8   8000              *     INPUTS  : (D1.W) = LENGTH OF TABLE - 1
   9    9   8000              *               (A1.L) = ADDRESS OF TABLE
  10   10   8000              *
  11   11   8000              *     OUTPUTS : (A1.L) = ADDRESS OF NONZERO ENTRY
  12   12   8000              *               (D1.W) = $FFFF : NO NONZERO ENTRIES
  13   13   8000              *
  14   14   8000              *   BCC OPERATION
  15   15   8000              *
  16   16   8000 4A59         LOOP1  TST.W   (A1)+
  17   17   8002 6600 0006           BNE     DONE1         ;DONE IF NONZERO
  18   18   8006              *                            ; ELSE:TEST NEXT VALUE
  19   19   8006 5341                SUBQ.W  #1,D1
  20   20   8008 6AF6                BPL     LOOP1         ;TEST TILL COUNTER IS -1
  21   21   800A 5589        DONE1  SUBQ.L  #2,A1         ;ADDRESS OF A NONZERO VALUE
  22   22   800C              *                            ; IF FOUND
  23   23   800C 1E3C 00E4           MOVE.B  #228,D7
  24   24   8010 4E4E                TRAP    #14
  25   25   8012              *
  26   26   8020                      ORG     $8020
  27   27   8012              *
  28   28   8020              *   DBCC OPERATION
  29   29   8020              *
  30   30   8020 4A59         LOOP2  TST.W   (A1)+
  31   31   8022 56C9 FFFC           DBNE    D1,LOOP2      ;LOOP IF VALUE IS ZERO
  32   32   8026              *                            ;   OR COUNTER IS >-1
  33   33   8026              *                            ; ELSE:GO TO NEXT INST.
  34   34   8026              *
  35   35   8026 5589        DONE2  SUBQ.L  #2,A1         ;ADDRESS OF FIRST NONZERO
  36   36   8028              *                            ;   VALUE
  37   37   8028              *
  38   38   8028 1E3C 00E4           MOVE.B  #228,D7
  39   39   802C 4E4E                TRAP    #14           ;RETURN TO MONITOR
  40   40   802E                      END
40 lines assembled
```

Figure 6.11   Comparison of Bcc and DBcc.

6.2.4. Rewrite the sequence

        LOOP . . .
                (body of loop)
                SUB.W #1,CNT
                BNE   LOOP

to use a DBcc instruction.

6.2.5. Write the assembly language program to implement the FORTRAN loop

        DO 10 I = 1,20
                (body of loop)
        10 CONTINUE

6.2.6. Write the assembly language program equivalent to the following:

```
      SUM = 0
      I = MAXVAL
10    SUM = SUM + I
      I = I − 1
      IF (I .NE. 0) GOTO 10
```

Assume that MAXVAL was defined previously. Test the program for the three values MAXVAL = 10, 1, and 0. Modify the program to take into account the case for MAXVAL = 0.

6.2.7. Since the MOVE and TST instructions always set the condition code $V = \{0\}$, show that BGE, BLT, BGT, and BLE are valid as branches after a MOVE or TST instruction when signed integers are moved. Also show that BGE is the same as BPL and that BLT is the same as BMI in this case.

6.3 SUBROUTINE USE WITH THE MC68000

The *subroutine* is a sequence of instructions that is treated as a separate program module within a larger program. The subroutine can be "called" or executed one or more times as the program executes. Generally, the subroutines associated with a program accomplish specific tasks, each of which represents a simpler procedure than that of the entire program. In fact, subroutines are called *procedures* in the Pascal language. Each module or single subroutine should be self-contained, that is, be testable by itself independent of the calling program. When the subroutine is called during execution of a program, its instructions are executed and control is then returned to the next instruction in sequence following the call to the subroutine.

The location of the first instruction of a subroutine is called its *starting address*. This must be defined in each program calling the subroutine. If the subroutine and the calling program are assembled at the same time, the subroutine starting address can be defined by a label at its first instruction. If the subroutine and calling program are not assembled together, the subroutine starting address must be explicitly defined in the call instruction.[2]

The MC68000 instructions to call and return from a subroutine are shown in Table 6.8. The Branch to Subroutine (BSR) and Jump to Subroutine (JSR) instructions perform calls. In each case, execution causes the longword address of the instruction following the call to be pushed onto the system stack. Execution then continues at the subroutine starting address. No other information is saved by the call, so it is the programmer's responsibility to preserve any register contents, including the contents of the status register if these are modified by the subroutine. These may be saved

---

[2] If a subroutine has several entry points, each address must be defined. Also, when independent programs are assembled separately, the external or "global" references are usually defined when the modules are linked together.

Table 6.8 Instructions for Subroutine Use

Instruction	Syntax	Operation	Comments
Branch to subroutine	BSR ⟨disp⟩	1. (SP) ← (SP) − 4; ((SP)) ← (PC) 2. (PC) ← (PC) + ⟨disp⟩	⟨disp⟩ is 8-bit or 16-bit signed integer
Jump to subroutine	JSR ⟨EA⟩	1. (SP) ← (SP) − 4; ((SP)) ← (PC) 2. (PC) ← (EA)	⟨EA⟩ is a control addressing type
Return and restore condition codes	RTR	1. (CCR) ← ((SP))[7:0]; (SP) ← (SP) + 2 2. (PC) ← ((SP)); (SP) ← (SP) + 4	(CCR) = (SR)[7:0]
Return from subroutine	RTS	(PC) ← ((SP)); (SP) ← (SP) + 4	—

Notes:
1. SP denotes the system stack pointer.
2. CCR is the condition code register, that is, (SR)[7:0].
3. PC is the program counter.

before the call and restored after the return, but a well-designed subroutine will save the values and restore them before returning. The latter approach is more reasonable because there are typically multiple calls to a single subroutine. Also, if subroutines are designed separately, the programmer designing the calling program may not be aware of what registers are modified by the subroutine unless good documentation is available. Therefore, our examples will show subroutines whose execution is *transparent* to the calling program except for modification of registers used to return values calculated by the subroutine.

Two return instructions are available for the MC68000. The Return and Restore (RTR) is used when the condition codes held in the condition code register (CCR) have been saved on the stack by the subroutine. Otherwise, Return from Subroutine (RTS) is used to simply load the program counter with the return address from the stack.

The instructions for calling and returning and their use with simple subroutine structures are discussed in this section. A more detailed look at subroutines is given in Chapter 9, where various methods of passing data between calling programs and subroutines is discussed. In the present discussion it is assumed that parameters are passed in processor registers.

6.3.1 Invoking Subroutines

The instructions BSR and JSR cause a transfer of control to the beginning address of a subroutine. In the statement BSR

BSR   ⟨label⟩

the ⟨label⟩ operand causes the assembler to calculate the displacement between the BSR instruction and the instruction identified by ⟨label⟩. This displacement is added during execution to the current contents of the program counter (the BSR location plus 2) to calculate the starting location of the subroutine. The displacement is stored as a 16-bit integer in two's-complement notation. The BSR operates in the same manner as the BRA instruction discussed in Section 6.2.1, except that the address of the instruction following the BSR is saved on the system stack. Similarly, the JSR is identical to the JMP instruction except for the saving of the return address. The addressing range of the JSR instruction is unlimited and any control addressing mode can be used. Thus the absolute modes, the relative modes, and the indirect addressing modes, except postincrement and predecrement, are allowed to specify the starting address. For example, the instruction

JSR   4(A5)

uses indirect with displacement addressing and causes a jump to the instruction four bytes past the address in A5. The return to the instruction following the BSR or JSR calling the subroutine is accomplished by executing a RTR or RTS instruction in the subroutine.

Figure 6.12 shows the general structure of the subroutine call. The call decrements (SP) by 4 and saves (PC) at location $7FFA. In the subroutine, the first statement saves the contents of the status register on the stack at location $7FFB just below (in memory) the two words containing the return address. The MOVEM.L instruction pushes all 32 bits of each register specified by the list onto the stack using predecrement

```
abs.  rel.   LC    obj. code     source line
----  ----   ----  ---------     -----------
  1     1    0000                          TTL      'FIGURE 6.12'
  2     2    0000                          LLEN     100
  3     3    8000                          ORG      $8000
  4     4    8000                          OPT      P=M68000
  5     5    8000               *
  6     6    8000               *    MAIN PROGRAM
  7     7    8000               *       ...
  8     8    8000  4EB9 0000              JSR      SUBR
  8          8004  800C
  9     9    8006               *       ...
 10    10    8006  1E3C 00E4              MOVE.B   #228,D7
 11    11    800A  4E4E                   TRAP     #14              ;RETURN TO MONITOR
 12    12    800C                                                  ; WHEN MAIN IS DONE
 13    13    800C               *
 14    14    800C               *
 15    15    800C               *    SUBROUTINE
 16    16    800C               *
 17    17    800C  42E7          SUBR     MOVE.W CCR,-(SP)          ;SAVE CONDITION CODES
 18    18    800E  48E7 FFFE              MOVEM.L D0-D7/A0-A6,-(SP) ;SAVE REGISTERS
 19    19    8012               *       ...
 20    20    8012  4E71                   NOP
 21    21    8014               *       ...
 22    22    8014  4CDF 7FFF              MOVEM.L  (SP)+,D0-D7/A0-A6 ; RESTORE REGISTERS
 23    23    8018               *
 24    24    8018               *    RETURN AND RESTORE CONDITION CODES
 25    25    8018               *
 26    26    8018  4E77                   RTR
 27    27    801A                         END
 27 lines assembled
```

Figure 6.12   Structure of a subroutine module.

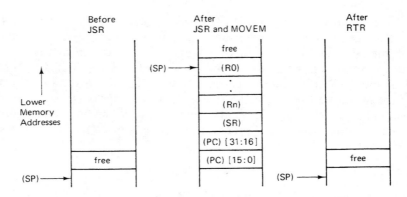

Note: Ri indicates an Address Register or a Data Register.

Figure 6.13 Stack usage during subroutine call.

addressing. Sixty bytes between $7FBC and $7FF7 are used. After processing, the registers are restored by the last MOVEM instruction before the return. The RTR instruction restores the condition codes to the status register, leaving the upper byte of SR unchanged before returning. If the condition codes had not been saved upon entry into the subroutine, the return would be through the RTS instruction. This instruction simply loads the return address from the stack into the program counter.

Figure 6.13 shows the stack contents before, during, and after the call to SUBR. The (PC) always occupies two words of the stack. The 16-bit contents of the status register are saved in the next (lower) word in memory. Then each register is saved when the MOVEM instruction is executed. The system stack pointer now points to the new TOP of stack, and the stack can be used by interrupt routines or during calls to other subroutines. After the subroutine has completed its processing, the return instruction should leave the stack pointer at its original value.

6.3.2 Program Structure

Large programs can be divided into a number of subroutines to accomplish specific tasks. This type of program structuring has the advantage of simplifying program testing and improving maintainability. The execution time is longer than that for a similar program without subroutines, due to the time required by the calling and return sequence for each subroutine. In most cases, the requirement for modularity is more important than minimizing execution time.

The possible structure of a program to read characters from a keyboard, convert them to BCD, and then process the numerical values is shown in Figure 6.14. The program begins by initializing values and then calls the first subroutine to read or input the characters as they are typed. After a string of characters is input, each is tested by a second subroutine. This test consists of checking each to see that it is a

Figure 6.14 An example program structure.

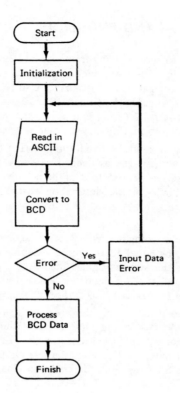

valid ASCII numeric character and converting valid characters to BCD. If an invalid character is detected, an error subroutine is called to display an invalid data message on the CRT screen. If no errors were detected, the BCD data can be processed further by additional subroutines.

EXERCISES

6.3.1. If the instruction

    BSR   SUB1

is located at address $3012 and the label SUB1 is at $3022, define the machine language instruction for

    BSR   SUB1

Refer to Appendix IV for the machine language format.

6.3.2. In the MC68000, what limits the number of subroutines that may be nested? A "nested" subroutine is one called by another subroutine.

REVIEW QUESTIONS AND PROBLEMS

6.1. Write a subroutine to copy a string from one address to another and return the length of the string (counted in bytes). Assume the strings are terminated by a null character ($00), such as the strings used in the C language. Use register A5 to pass the address of the source string and use register A6 to pass the address of the destination string. Return the string length in register D7 (don't count the null terminator). Make sure to restore any register that you need to modify in your routine (except D7, of course).

6.2. Write a subroutine to create a fixed length string by copying another string and truncating or padding with spaces as needed. Assume the strings are terminated by a null character ($00). Use register A5 to pass the address of the source string and use register A6 to pass the address of the destination string. Use register D7 to pass the length of the resulting string. Make sure to restore any register that you need to modify in your routine.

6.3. Write a subroutine to compare two character strings and determine whether one is less than, greater than, or equal to the other. Assume the strings are null-terminated. Use register A5 to pass the address of the first string and register A6 to pass the address of the second. Return -1 in D7 if the first is less than the second, return 0 in D7 if the strings are equivalent, and return +1 in D7 if the first is greater than the second.

6.4. Given a table of eight longword addresses, each pointing to a null-terminated character string, sort the addresses in the table in ascending sequence of string values. Do not move the strings themselves, just resequence the address table entries. Use the subroutine developed in problem 6.3 to determine the relative value of any two pairs of strings. (You might try a Bubble Sort algorithm described in Chapter 9.)

Use the following declarations to test your program.

```
S_TBL       DC.L S1
            DC.L S2
            DC.L S3
            DC.L S4
            DC.L S5
            DC.L S6
            DC.L S7
            DC.L S8
S1          DC.B 'String d, will not be last',0    ;7
S2          DC.B 'String 0, first',0               ;1
S3          DC.B 'String A',0                       ;2
S4          DC.B 'Strings end with $00',0           ;8
S5          DC.B 'String QED',0                     ;3
S6          DC.B 'String Z',0                       ;4
S7          DC.B 'String abc',0                     ;6
S8          DC.B 'String a',0                       ;5
```

6.5. Suppose you were designing a program that can be loaded anywhere in memory. Your design requires that the starting address of the program be identified at execution time. Using only the instructions discussed thus far (don't use LEA), write the initial series of instructions necessary to place this address in register A6.

REVIEW EXAMINATION (Chapters 1–6)

1. List ten aspects of a microcontroller family that are important to a product designer.

2. Which of the following features of the 68000 instruction set enable programmers to use modern programming techniques? Choose more than one.
   (a) Indexed addressing mode.
   (b) Integer multiply and divide.
   (c) Jump to subroutine, return from subroutine.
   (d) Supervisor and user modes.
   (e) Predecrement and postincrement.
   (f) Indirect addressing.
   (g) Long absolute addresses.
   (h) Branch on condition.

3. True or False: The only ways that a system running on the 68000 can switch from user mode to supervisor mode are a TRAP instruction is executed, an interrupt occurs that is not masked, an external reset occurs, or an exception occurs.

4. True or False: The only way that a system running on the 68000 can switch from supervisor mode to user mode is via a Return from Exception Handler (RTE) instruction, or by clearing bit #13 in the Status Register.

5. Which of the reasons is *least* correct? Motorola designers decided to use a microprogrammed design for the 68000 family in order to:
   (a) Simplify the design.
   (b) Simplify the testing.
   (c) Increase execution speed.
   (d) Help maintain upward compatibility within the processor family.

6. On which type of processor architecture will any given application run faster—CISC or RISC? Choose the best answer of those given below.
   (a) RISC
   (b) CISC
   (c) No difference
   (d) Answer cannot be determined by general architecture alone

7. Do the following conversions:
   (a) 1011.101_2 (unsigned) to decimal
   (b) 747_{10} to hexadecimal
   (c) sign extend $6C_{16}$ to 16 bits
   (d) -1809_{10} to 2's complement, 16 bits
   (e) 74 (two-digit BCD in ten's-complement notation) to ASCII digits with prefix sign character (show ASCII characters using hexadecimal notation)
   (f) IEEE standard floating point representation of -64.25

8. Write a subroutine to compare the word-length values in D0 and D1. Return the largest value in D0 and return -1 in D1 if D0 < D1, or $+1$ if D0 > D1, or 0 if D0 = D1.

9. Suppose there is an array of 1,000 bytes, each byte containing an unpacked BCD digit (value 0 through 9). Write an assembly program to read through the byte array and count the number of occurrences of each possible value. Store the result in a ten-word array, where word 0 contains the count of zeros, word 1 contains the count of ones, word 2 contains the count of twos, etc. Use the following statements to define the addresses of the two arrays:

```
DATA      EQU      $9000
RESULT    EQU      $9400
```

Use the DBcc instruction to control any loops. Make sure you initialize the RESULT array. Begin your program at $8000. HINT: Consider the "address register indirect with index" addressing mode.

10. Give the addressing modes (for *both* the source and destination operands) in the instruction
    (a) MOVE.W ($100,A1),$1000
    (b) MOVE.W −(A5),(A6)+
    (c) MOVE.W $7FE(PC),D4
    (d) MOVE.W $1000,D5
    (e) MOVEA.L #SOMEVAR,A0
        Note: SOMEVAR is a defined label.

11. Define the machine code for these instructions. Include both operation and operands.
    (a) MOVE.W #128,$7FFF00
    (b) MOVE.L −(A2),(A3,D1.W)

12. Show the user stack usage (addresses and description of contents) at the point where the following routine has reached the instruction labeled "LOOKHERE".

```
              ORG    $2000
MYPGM         BSR    MYSUBR
              STOP   $2000
MYSUBR        MOVE.W     CCR,−(SP)
              MOVEM.L    A2-A5/D0/D4-D6,−(SP)
LOOKHERE      NOP
              . . .
              MOVEM.L    (SP)+,A2-A5/D0/D4-D6
              RTR
              END
```

Note: Assume the initial stack pointer is $5000 and the routine is executing in user mode.

FURTHER READING

A number of works discuss assembly language programming in a manner that is useful to the MC68000 programmer. In particular, the book by Kane et al., gives a number of good examples of programs for the MC68000. Many of the textbooks written about the PDP-11 system can be applied to the study of the MC68000 if the differences between the processors are taken into account. For instance, the instruction

```
CMP    A,B
```

for the PDP-11 evaluates (A) − (B), which is the reverse of the MC68000. The texts by Eckhouse and Morris and by Lewis explain many of the aspects of assembly language programming for the PDP-11. Programs for the PDP-11 can be converted to MC68000, and vice versa, by a programmer familiar with both machines. The Further Reading section for Chapter 5 lists other references that provide more details about assembly language programming and the operation of assemblers.

Eckhouse, Richard H., Jr., and L. R. Morris. *Minicomputer Systems,* 2nd ed. (Englewood Cliffs, N.J.: Prentice Hall, 1979).

Kane, Gerald, Doug Hawkins, and Lance Leventhal. *68000 Assembly Language Programming.* (Berkeley, Calif.: Osborne (McGraw-Hill, 1981).

Lewis, Harry R. *An Introduction to Computer Programming and Data Structures Using MACRO-11.* (Reston, Va.: Reston, 1981).

CHAPTER 7

Arithmetic Operations

This chapter is concerned with arithmetic operations on numbers. Numerical data are used for identification of locations by their addresses and for representation of quantities such as temperatures or bank balances. In any case, the fundamental operations of addition, subtraction, multiplication, and division of numbers are essential in algorithms that are designed to process numerical data. Almost all programs incorporate some form of arithmetic operation.

The number system used in microprocessors is binary. The internal use of binary arithmetic and our external interpretation of the results as decimal numbers cause no problems if suitable conversion routines are available. Decimal values may be converted to ASCII code for input to the computer system and then be converted to binary for processing. The reverse conversions may be applied for output. A problem does arise in certain cases, however, due to the finite length of the machine representation. Stated another way, the length in bits of an integer value is a measure of the largest integer representable in a given format. Thus a 16-bit representation limits the magnitude of a positive integer to $2^{16} - 1$. Each operation, such as the addition of two 16-bit numbers, carries with it the danger of exceeding the maximum allowable value. The tests for such a condition and the means for extending the precision when necessary are two important aspects of the study of arithmetic operations using computers.

A review of the topics in Chapter 3 is suggested before or during the study of this chapter. Many of the mathematical details of binary and decimal arithmetic presented there are not repeated here. The setting of the condition codes resulting from execution of MC68000 instructions is summarized in Appendix IV.

7.1 SOME DETAILS OF BINARY ARITHMETIC

In the MC68000, the binary representation of operands can be 8, 16, or 32 bits in length. Numerically, positive integer values range from 0 to $2^m - 1$, where $m = 8$, 16, or 32. For signed integers, the range is -2^{m-1} to $+2^{m-1} - 1$ in two's-complement notation. In adding or subtracting m-bit integers, various out-of-range conditions can occur and these are indicated by condition code bits of the processor. The case of unsigned integers is treated differently from that for signed integers because the interpretation of the condition codes changes.

Consider the addition of two unsigned m-digit integers. For example, the 8-bit addition of 125 and 200 yields

Binary

$$
\begin{array}{r}
0111 \quad 1101 \\
+1100 \quad \underline{1000} \\
\hline
(1) \quad 0100 \quad 0101
\end{array}
$$

or an 8-bit result of 69 with a carry. This addition in the MC68000 would cause the carry condition code C to be set to {1} to indicate the overflow. Similarly, subtraction with a subtrahend larger than the minuend is indicated by a borrow out of the $(m + 1)^{st}$ place. In the MC68000, the borrow indication is also C = {1}. Thus, whenever the carry bit is {1} after addition or subtraction of unsigned integers, the m-bit result is incorrect and the programmer must decide on the appropriate action.

When integers in two's-complement notation are added or subtracted, the carry bit is ignored and the overflow (V) bit is checked for an out-of-range condition. For an m-bit representation, the positive range is only 0 to $2^{m-1} - 1$, so the addition of two positive numbers must not exceed that value. For example, the 8-bit addition of +125 and +127 yields

Binary

$$
\begin{array}{r}
0111 \quad 1101 \\
+0111 \quad \underline{1111} \\
\hline
1111 \quad 1100
\end{array}
$$

which is −4! The overflow condition, V = {1}, is also indicated. The problem that arises is that two positive numbers added together produced a negative result. Similarly, if the addition of two negative values produces a positive result, the V condition code is again set to {1}. This mathematical condition is termed *underflow* and means the result is too small (too negative) to be represented. However, the V or overflow bit indicates the error, and reference to out-of-range conditions in the machine is usually called "overflow." If the operands are of opposite sign, no out-of-range condition can occur. When subtraction is performed, the V bit set to {1} again indicates an erroneous result.

In the MC68000, division and BCD arithmetic can also yield out-of-range results. The conditions for these cases are covered in the appropriate sections of this chapter. In summary, the condition codes must be checked after an arithmetic operation to determine if an out-of-range condition occurred. Figure 7.1 indicates the procedure used to check for valid numbers. An error condition indicates that the m-bit representation is not valid. Either the result is rejected or a multiple-precision representation is required.

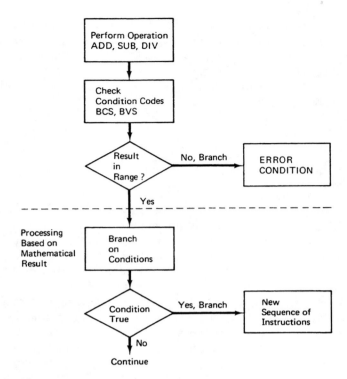

Figure 7.1 Testing for out-of-range conditions.

Conditional Tests. The Branch if Carry Set (BCS) following the arithmetic instruction causes a branch if a carry occurred during addition or a borrow occurred during subtraction of unsigned integers. Branch if Overflow Set (BVS) branches when an out-of-range condition occurs during arithmetic with two's-complement integers. It is recommended that such tests (or the opposite logic of BCC or BVC) be used before other conditional tests are executed.

When conditional branch instructions are executed after arithmetic instructions, the conditional tests described in Chapter 6 must be used with care. Consider the addition just shown of +125 and +127 for 8-bit signed numbers. The condition codes after the ADD instruction would be

$$Z = \{0\}, N = \{1\}, C = \{0\}, V = \{1\}$$

indicating nonzero, negative, no carry, and overflow, respectively. The result is in error mathematically, but the BMI instruction would cause a branch because it tests only for $N = \{1\}$. Worse yet, BGE would branch but BPL would not. BGE branches whenever N and V condition codes are the same, but BPL branches when $N = \{0\}$ without testing any other condition codes.

If no out-of-range condition is present, the conditional tests operate as shown in Figure 7.2(a). Here BPL and BGE are equivalent for valid signed integers, as are BMI and BLT. Figure 7.2(b) shows the conditions and the tests on the condition code.

Instruction	Branch Condition	Result
UNSIGNED ADD, SUB $C = \{0\}$	BEQ BNE	$X = 0$ $X \neq 0$
SIGNED ADD, SUB $V = \{0\}$	BEQ BNE BPL, BGE BLE BMI, BLT BGT	$X = 0$ $X \neq 0$ $X \geq 0$ $X \leq 0$ $X < 0$ $X > 0$

(a) Signed and Unsigned Tests

Mnemonic	Condition	Encoding	Test
T	true	0000	1
F	false	0001	0
HI	high	0010	$\overline{C} \cdot \overline{Z}$
LS	low or same	0011	$C + Z$
CC (HS)	carry clear	0100	$\overline{C}$
CS (LO)	carry set	0101	C
NE	not equal	0110	$\overline{Z}$
EQ	equal	0111	Z
VC	overflow clear	1000	$\overline{V}$
VS	overflow set	1001	V
PL	plus	1010	$\overline{N}$
MI	minus	1011	N
GE	greater or equal	1100	$N \cdot V + \overline{N} \cdot \overline{V}$
LT	less than	1101	$N \cdot \overline{V} + \overline{N} \cdot V$
GT	greater than	1110	$N \cdot V \cdot \overline{Z} + \overline{N} \cdot \overline{V} \cdot \overline{Z}$
LE	less or equal	1111	$Z + N \cdot \overline{V} + \overline{N} \cdot V$

(b) Condition Code Tests

Figure 7.2 Conditional branches for valid numbers.

The "encoding" specifies the condition in the machine language instruction. The instruction sequence in Figure 7.1 is followed in most of the examples in this chapter where the possibility of an out-of-range condition may occur.

EXERCISES

7.1.1. Compare the use of the Bcc instructions after arithmetic operations with their use after the MOVE, CMP, and TST instructions, as discussed in Chapter 6.

7.1.2. Show that BGE and BPL yield opposite results when overflow occurs.

7.1.3. Determine the result of subtracting −32,768 from 16,384 using 16-bit two's-complement arithmetic.

7.1.4. Show that subtraction can be accomplished with binary numbers by adding the two's complement of the subtrahend to the minuend.

7.2 ADDITION AND SUBTRACTION

Binary addition is performed on 8-, 16-, or 32-bit operands with the ADD instruction and its variations, as shown in Table 7.1. Similarly, byte, word, or longword operands can be subtracted with the SUB instruction. The NEG instruction forms the two's complement (negative) of the operand specified. These three instructions and their variations are used to perform basic arithmetic on binary integers. As seen in Table 7.1, each instruction has restrictions on the addressing modes allowed. In the operation of each instruction, which is defined in Table 7.2, the destination operand of specified length is replaced by the result. This is the sum for ADD, the difference for SUB, or the negative value when NEG is executed. The condition codes N, Z, V, and C are set according to the result. The X (extend) condition code is also set.

Addition. The assembly language format of the ADD instruction is

ADD.$\langle l \rangle$ $\langle EA \rangle, \langle Dn \rangle$

Table 7.1   Arithmetic Instructions ADD, SUB, and NEG

Syntax	Addressing Modes	
	Source	Destination
Addition or subtraction		
ADD.$\langle l \rangle$ $\langle EA \rangle, \langle Dn \rangle$	All[(1)]	Dn
SUB.$\langle l \rangle$ $\langle EA \rangle, \langle Dn \rangle$		
ADD.$\langle l \rangle$ $\langle Dn \rangle, \langle EA \rangle$	Dn	Memory alterable
SUB.$\langle l \rangle$ $\langle Dn \rangle, \langle EA \rangle$		
ADDI.$\langle l \rangle$ #$\langle d \rangle, \langle EA \rangle$	#$\langle d \rangle$	Data alterable
SUBI.$\langle l \rangle$ #$\langle d \rangle, \langle EA \rangle$		
ADDQ.$\langle l \rangle$ #$\langle d_3 \rangle, \langle EA \rangle$	#$\langle d_3 \rangle$[(2)]	Alterable[(3)]
SUBQ.$\langle l \rangle$ #$\langle d_3 \rangle, \langle EA \rangle$		
Negate		
NEG.$\langle l \rangle$ $\langle EA \rangle$	—	Data alterable

Notes:

·1. If the source effective address in the instructions ADD or SUB is an address register, the operand length is word or longword.

2. $\langle d_3 \rangle$ is a value between 1 and 8.

3. If An is a destination, only word or longword operations are allowed. In this case, the condition codes are not affected.

4. $\langle l \rangle$ denotes B, W, or L in all the instructions except as in notes 1 and 3.

5. Except as in note 3, all the condition codes are affected by the instructions.

Table 7.2 Arithmetic Instruction Operation

Instruction	Operation
ADD.⟨l⟩ ⟨EAs⟩,⟨EAd⟩	(EAd) [l] ← (EAs)[l] + (EAd)[l]
SUB.⟨l⟩ ⟨EAs⟩,⟨EAd⟩	(EAd) [l] ← (EAd)[l] − (EAs)[l]
NEG.⟨l⟩ ⟨EA⟩	(EA) [l] ← 0 − (EA)[l]

Notes:

1. ⟨EAs⟩ and ⟨EAd⟩ are the source and destination effective addresses, respectively.
2. ⟨l⟩ denotes B, W, or L.
3. [l] indicates corresponding bits in the operation.

when the destination operand is held in a data register. The source is specified as a byte, word, or longword operand (⟨l⟩ = B, W, or L) unless the source is an address register, in which case the length is restricted to word or longword operands. Thus the instruction

ADD.B   D1,D5

replaces (D5)[7:0] with the sum (D5)[7:0] + (D1)[7:0]. Only the specified length of the destination is affected.

The addition of the value in a data register to an operand in memory is also allowed. For example, the instruction

ADD.L   D1,(A1)

adds the 32-bit contents of D1 to the contents of the location addressed by A1. The destination operand cannot, however, be the contents of an address register or be addressed by program counter relative addressing.

Several variations of the ADD instruction shown in Figure 7.2 are Add Immediate (ADDI) and Add Quick (ADDQ). The immediate format can add an 8-, 16-, or 32-bit constant to a data-alterable location. This excludes an address register or a PC relative address for the destination. The ADDI instruction adds a specified constant to a destination location, while the ADD instruction operates only between registers and memory. Thus the instruction

ADDI.B   #20,(A1)

is used to add 20 to the byte addressed by A1 as an example.

The ADDQ instruction adds an immediate value between 1 and 8 to the specified destination location. Any alterable destination location is allowed, including an address register. Destinations addressed by the PC relative mode are prohibited. In the cases where An is the destination, the condition codes are not affected, but the whole register (32 bits) is changed. Arithmetic operations with addresses are considered in Chapter 9.

It should be noted that some assemblers select the appropriate variation of an ADD instruction without the programmer specifying it. For instance, the instruction

ADD   #1,D4

may be interpreted as an ADDI instruction or an ADDQ instruction by certain assemblers. The practice of letting the assembler choose the variation is not always wise if consistency in program documentation and machine language instruction length is desired.

Example 7.1

The subroutine shown in Figure 7.3 adds a column (vector elements) of 16-bit integers to form either a 16- or a 32-bit sum. Before the subroutine is called, D1 holds the number of integers to be summed and A1 contains the starting address of the column. If the length of the column is zero, D3 contains −1 after the subroutine executes. Otherwise, the low-order word of D3 contains zero if the sum is contained in 16 bits or +1 if the sum requires 32 bits. Each time a carry occurs in the addition of a new value to D2, 1 is added to the upper word of D2 to form the proper sum.

    The purpose of the program might be to use word (rather than longword) operations in subsequent processing if a 16-bit sum is the result. This would reduce the memory storage

```
abs. rel.   LC   obj. code   source line
----  ----  ----  ---------   -----------
  1    1    0000                     TTL      'FIGURE 7.3'
  2    2    0000                     LLEN     100
  3    3    8000                     ORG      $8000
  4    4    8000                     OPT      P=M68000
  5    5    8000            *
  6    6    8000            *  ADD 16-BIT UNSIGNED INTEGERS
  7    7    8000            *
  8    8    8000            *  INPUT  :  (D1.W) = NUMBER OF INTEGERS
  9    9    8000            *            (A1.L) = STARTING ADDRESS OF COLUMN
 10   10    8000            *               OF INTEGERS
 11   11    8000            *
 12   12    8000            *  OUTPUT :  (D3.W) = -1 : NUMBER OF INTEGERS IS ZERO
 13   13    8000            *                     0 : 16-BIT SUM IN D2 [15:0]
 14   14    8000            *                     1 : 32-BIT SUM IN D2 [31:0]
 15   15    8000            *            (D2.L) = SUM
 16   16    8000            *
 17   17    8000 48E7 4040  ADDUNS  MOVEM.L  D1/A1,-(SP) ;SAVE REGISTERS
 18   18    8004            *
 19   19    8004 4282               CLR.L    D2          ;SUM:=0
 20   20    8006 363C FFFF          MOVE.W   #-1,D3      ;SET STATUS TO ERROR
 21   21    800A 5341               SUBQ.W   #1,D1       ;SET UP LOOP COUNTER
 22   22    800C 6D00 0016          BLT      DONE
 23   23    8010 4243               CLR.W    D3          ;SET STATUS TO 16-BIT
 24   24    8012            *                            ; SUM
 25   25    8012 D459       LOOP    ADD.W    (A1)+,D2
 26   26    8014 6400 000A          BCC      ENDLP       ;IF NO OVERFLOW, THEN SKIP
 27   27    8018 0682 0001          ADDI.L   #$10000,D2  ; ELSE ADD 1 TO UPPER WORD
 27        801C 0000
 28   28    801E 7601               MOVEQ    #1,D3       ;SET STATUS TO 32-BIT SUM
 29   29    8020 51C9 FFF0  ENDLP   DBRA     D1,LOOP     ;CONTINUE
 30   30    8024 4CDF 0202  DONE    MOVEM.L  (SP)+,D1/A1 ;RESTORE REGISTERS
 31   31    8028 4E75               RTS
 32   32    802A                    END
```

Figure 7.3 Addition routine for 16- or 32-bit sums.

needed and be more efficient in time if memory accesses are made. A longword access to memory requires two read or write cycles by the processor because there are only 16 data signal lines.

Subtraction. The instructions SUB, SUBI, and SUBQ are exact counterparts of the addition instructions. They calculate the difference between the destination operand (minuend) and the source operand (subtrahend). This difference replaces the byte, word, or longword portion of the destination location as specified by the instruction.

If an address register is the destination in a SUBQ instruction, the immediate value can be a word or a longword operand. If the length is word, the immediate value is sign-extended to 32 bits before the subtraction is performed. For example, the instruction

SUBQ.W   #1,A1

has the effect of subtracting 1 from the entire contents of A1. Therefore, both ADDQ and SUBQ affect the 32-bit contents of a destination address register.

All the condition code bits are affected by a subtraction instruction unless the subtraction is made from an address register. The condition C = {1} indicates a borrow in unsigned subtraction. In signed arithmetic, N, Z, and V bits set to 1 indicate a negative number, zero, and an overflow condition, respectively.

Negation. The NEG instruction replaces a data-alterable destination location of specified length with the result of the calculation

0 − (destination)

thus forming the two's-complement value. As an example, if location $1000 contains the value 1, the instruction

NEG.B $1000

replaces the value with $FF. The only case in which an overflow can occur is when the value negated is -2^{m-1}, because this number has no positive equivalent in two's-complement notation.

Example 7.2

The subroutine shown in Figure 7.4 sums the differences between two columns or vectors of positive integers addressed by (A1) and (A2). The length of the columns is initially held in D1. If the length of the columns is zero or an overflow occurs, then (D4)[15:0] is set to 0 to indicate the error. Otherwise, the differences between corresponding entries in the columns are accumulated in the low-order word of D3. Once the summation is complete, the result is tested for positive or negative. If the result is negative, the NEG instruction is used to determine the magnitude of the number. A program using this routine must first check the status output in D4 for an error condition. If no error is indicated, the magnitude of the sum of the differences is held in the low-order word of D3. The program might then convert the magnitude to decimal ASCII and prefix the sign to the printed results.

```
abs. rel.   LC    obj. code    source line
---- ----   ----  ---------    -----------
   1    1   0000               |          TTL       'FIGURE 7.4'
   2    2   0000               |          LLEN      100
   3    3   8000               |          ORG       $8000
   4    4   8000               |          OPT       P=M68000
   5    5   8000               |*
   6    6   8000               |* SUM OF DIFFERENCES OF TWO COLUMNS OF 16-BIT
   7    7   8000               |* POSITIVE INTEGERS
   8    8   8000               |*
   9    9   8000               |* INPUT  : (D1.W) = LENGTH OF THE COLUMNS
  10   10   8000               |*          (A1.L) = ADDRESS OF FIRST COLUMN
  11   11   8000               |*          (A2.L) = ADDRESS OF SECOND COLUMN
  12   12   8000               |*
  13   13   8000               |* OUTPUT : (D3.W) = ABSOLUTE VALUE OF RESULT
  14   14   8000               |*          (D4.W) = 0 : ERROR OCCURRED
  15   15   8000               |*                   1 : RESULT IS POSITIVE
  16   16   8000               |*                  -1 : RESULT IS NEGATIVE
  17   17   8000               |*
  18   18   8000 48E7 6060     |SUMDIF MOVEM.L D1/D2/A1/A2,-(SP) ;SAVE REGISTERS
  19   19   8004 4243          |          CLR.W     D3            ;SUM = O
  20   20   8006 4244          |          CLR.W     D4            ;SET STATUS = ERROR
  21   21   8008 4A41          |          TST.W     D1            ;IF COLUMNS ARE EMPTY
  22   22   800A 6700 0026     |          BEQ       DONE          ; THEN RETURN ERROR
  23   23   800E               |*
  24   24   800E 3419          |LOOP   MOVE.W    (A1)+,D2      ;COMPUTE
  25   25   8010 945A          |          SUB.W     (A2)+,D2      ;((A1)) - ((A2))
  26   26   8012 6900 001E     |          BVS       DONE          ;EXIT IF OVERFLOW
  27   27   8016 D642          |          ADD.W     D2,D3         ;SUM THE DIFFERENCES
  28   28   8018 6900 0018     |          BVS       DONE          ;ON OVERFLOW,
  29   29   801C               |*                                ;  EXIT WITH ERROR
  30   30   801C 5341          |          SUBQ.W    #1,D1         ;DECREMENT COUNT
  31   31   801E 66EE          |          BNE       LOOP          ;LOOP UNTIL FINISHED
  32   32   8020               |*
  33   33   8020 4A43          |          TST       D3            ;IF RESULT IS POSITIVE
  34   34   8022 6C00 000C     |          BGE       POS           ; THEN PROCESS IT
  35   35   8026 383C FFFF     |          MOVE.W    #-1,D4        ;  ELSE STATUS =0 NEG
  36   36   802A 4443          |          NEG.W     D3            ;TAKE ABSOLUTE VALUE
  37   37   802C 6000 0004     |          BRA       DONE
  38   38   8030               |*
  39   39   8030 7801          |POS    MOVEQ     #1,D4         ;STATUS = POSITIVE
  40   40   8032 4CDF 0606     |DONE   MOVEM.L (SP)+,D1/D2/A1/A2 ;RESTORE REGISTERS
  41   41   8036 4E75          |          RTS
  42   42   8038               |          END
```

Figure 7.4   Routine to compute the sum of differences.

EXERCISES

7.2.1. Assume that (D1) = $0000 FFFF before each instruction below executes. Determine the results, including the setting of the condition codes after the following.

    (a) ADDI.B   #1,D1
    (b) ADDQ.L   #1,D1
    (c) SUBQ.B   #1,D1
    (d) NEG.W   D1
    (e) SUB.L   D1,D1

7.2.2. Show that using only one carry bit after an m-bit unsigned addition is sufficient to assure that no information is lost.

7.2.3. For an m-bit subtraction operation N3 = N2 − N1, show that the proper result is obtained when N1 is negative and N2 is positive, if the sum of the magnitudes of N1 and N2 is equal to or less than $2^{m-1} - 1$. Refer to Chapter 3 for the two's-complement formulas.

7.2.4. Write a routine to multiply two unsigned 16-bit integers by using repeated addition.

7.2.5. Modify the program in Example 7.2 to compute the sum of the absolute value of the differences between the two columns of numbers.

7.3 MULTIPLICATION AND DIVISION

Instructions for integer multiplication and division are included in the instruction set of most of the 16-bit microprocessors. The MC68000 provides separate instructions for multiplication and division of unsigned integers and for multiplication and division of two's-complement integers. This section presents these instructions and discusses the operations, numerical ranges, and possible errors involved in their use. Table 7.3 shows the syntax and the operations of the divide (DIV) instruction and the multiply (MUL) instruction. The suffix "U" for unsigned integers or "S" for signed integers must be specified each time they are used. The multiplicand for the multiplication operation and the divisor for the division operation are specified by a 16-bit source operand. Only address register direct addressing is prohibited for the source. The multiplier and dividend are always held in data registers.

Unsigned Multiplication. The MULU instruction multiplies two unsigned 16-bit operands to yield a 32-bit product. For example, the instruction

MULU $10,D2

Table 7.3 Multiply and Divide Instructions

Syntax	Operation

Multiplication
  MULU $\langle EA \rangle, \langle Dn \rangle$
  MULS $\langle EA \rangle, \langle Dn \rangle$
	$(Dn)[31:0] \leftarrow (Dn)[15:0] \times (EA)[15:0]$

Division
  DIVU $\langle EA \rangle, \langle Dn \rangle$
  DIVS $\langle EA \rangle, \langle Dn \rangle$
	$(Dn)[31:0]/(EA)[15:0];$
	$(Dn)[15:0] \leftarrow$ quotient
	$(Dn)[31:16] \leftarrow$ remainder

Notes:

1. Only data addressing modes are allowed for the source address $\langle EA \rangle$ (i.e., An is prohibited).
2. In division, a zero divisor causes a *trap;* an overflow is indicated by $V = \{1\}$.
3. In signed division, a remainder has the sign of the dividend.

with (D2)[15:0] = $0002 results in (D2) = $0000 0020 or 32 decimal. Because the 16-bit multiplicand and multiplier may each range from 0 to 65,535, the product cannot exceed 4,294,836,225, which is less than $2^{32} - 1$. Therefore, no overflow is possible and the condition code bits C and V are always cleared after the MULU instruction. N and Z are set according to the result. In the case of unsigned integers, N = {1} indicates that the product equals or exceeds 2^{31} in magnitude.

Signed Multiplication. When signed integers are multiplied, the result is positive or negative depending on the signs of the multiplicand and multiplier. The range of each is -2^{15} to $+2^{15} - 1$, or $-32,768$ to $+32,767$. If the two most negative values are multiplied, the result is 1,073,741,824, or 2^{30}. The largest possible negative result is

$$-2^{15} \times (2^{15} - 1) = -1,073,709,056.$$

Therefore, no out-of-range condition can occur for a 32-bit product, and both V and C are always cleared. For MULS, the N bit indicates a negative product when N = {1}, as expected. If the result is zero, then Z = {1}. The instruction

MULS   #−1,D2

with (D2)[15:0] = $0002 results in (D2) = $FFFF FFFE, or −2 in two's-complement notation. The N bit is set to indicate a negative result.

Example 7.3 _____

Although the use of a single multiply instruction cannot result in an overflow, the use of these instructions in an equation that requires several multiplies could produce a result that exceeds an allowable maximum magnitude. For example, to calculate the sum of squares as

$$Z = \sum_{i=1}^{N} (X_i^2 + Y_i^2)$$

the individual products cannot overflow in 32 bits, although the sum of several terms or the entire sum can. The program in Figure 7.5 computes the sum of squares of N pairs of signed 16-bit integers. The numbers are stored in the order

X(1), Y(1), X(2), Y(2), . . . , X(N), Y(N)

as a table or column of 16-bit words whose first address is given by (A1) when the program is entered. The length N is assumed to be in the low-order 16 bits of D1. The 32-bit result is accumulated in D3 unless an error occurs. After execution, an error is indicated by (D4)[15:0] = −1 and any program making use of the result should check the status in D4 before the result is accepted. The address register indirect with displacement mode is used to address the operands, so if an overflow occurs, (A1) points to the current pair of operands, if the program

```
abs. rel.    LC    obj. code    source line
---- ----    ----  ---------    -----------
   1    1    0000                        TTL      'FIGURE 7.5'
   2    2    0000                        LLEN     100
   3    3    8000                        ORG      $8000
   4    4    8000                        OPT      P=M68000
   5    5    8000              |*
   6    6    8000              |* SUM OF SQUARES
   7    7    8000              |*
   8    8    8000              |* INPUT :  (D1.W) = LENGTH OF COLUMN
   9    9    8000              |*          (A1.L) = ADDRESS OF COLUMN OF NUMBERS
  10   10    8000              |*                   STORED X1,Y1,X2,Y2,...,XN,YN
  11   11    8000              |*
  12   12    8000              |* OUTPUT:  (D3.L) = RESULT
  13   13    8000              |*          (D4.W) = 0 : SUCCESSFUL
  14   14    8000              |*                  -1 : ERROR
  15   15    8000              |*
  16   16    8000  48E7 6040   |SUMSQ  MOVEM.L  D1/D2/A1,-(SP) ;SAVE REGISTERS
  17   17    8004  4283        |       CLR.L    D3             ;SUM := 0
  18   18    8006  383C FFFF   |       MOVE.W   #-1,D4         ;SET DEFAULT TO ERROR
  19   19    800A  5341        |       SUBQ.W   #1,D1          ;IF LENGTH IS ZERO
  20   20    800C  6D00 0020   |       BLT      DONE           ; THEN EXIT WITH ERROR
  21   21    8010              |*
  22   22    8010  3411        |LOOP   MOVE.W   (0,A1),D2      ;COMPUTE
  23   23    8012  C5C2        |       MULS.W   D2,D2          ;XN**2
  24   24    8014  D682        |       ADD.L    D2,D3          ;ADD TO SUM
  25   25    8016  6900 0016   |       BVS      DONE           ;ON ERROR, EXIT
  26   26    801A  3429 0002   |       MOVE.W   (2,A1),D2      ;COMPUTE
  27   27    801E  C5C2        |       MULS.W   D2,D2          ;YN**2
  28   28    8020  D682        |       ADD.L    D2,D3          ;ADD TO SUM
  29   29    8022  6900 000A   |       BVS      DONE           ;ON ERROR, EXIT
  30   30    8026  5889        |       ADD.L    #4,A1          ;INCREMENT TO NEXT PAIR
  31   31    8028  51C9 FFE6   |       DBRA     D1,LOOP        ;DECREMENT COUNTER AND
  32   32    802C              |*                             ;  CONTINUE UNTIL -1
  33   33    802C              |*
  34   34    802C  4244        |       CLR.W    D4             ;SET STATUS TO SUCCESS
  35   35    802E  4CDF 0206   |DONE   MOVEM.L  (SP)+,D1/D2/A1 ;RESTORE REGISTERS
  36   36    8032  4E75        |       RTS
  37   37    8034              |       END
```

Figure 7.5   Sum-of-squares program.

is being traced for debugging. Otherwise, (A1) becomes its initial value after the last MOVEM instruction.

Unsigned Division.   The MC68000 instruction DIVU performs the division

$$Y/W = Q + R/W$$

where Y is a 32-bit unsigned integer, W is a 16-bit unsigned integer, Q is a 16-bit quotient, and R is a 16-bit remainder. For example, the instruction

DIVU   #2,D1

divides the 32-bit operand in D1 by 2. The result, as indicated in Table 7.3, is a quotient in the low-order word of D1 and the remainder, or zero, in the upper word of D1.

Thus, if D1 contained $0000 0005 before the instruction executed, the result is (D1) = $0001 0002 because $5/2 = 2 + 1/2$.

Two special conditions may arise when performing a division operation:

(a) division by zero, or
(b) an overflow of the quotient.

Both of these situations are indicated by error conditions. Exception processing in a trap routine occurs in the case of division by zero. An overflow can occur because the range of the dividend is 0 to $2^{32} - 1$ but the length of the quotient is only 16 bits. Obviously, dividing an integer greater than $2^{16} - 1$ by 1 would cause an overflow. Or, more generally, if the dividend exceeds the divisor in magnitude by 2^{16} or greater, an overflow will occur. The overflow is indicated by V = {1} even though unsigned arithmetic is being performed. If overflow occurs, the operands are not changed.

Signed Division. The instruction DIVS executes in the same manner as the DIVU instruction, but the operands are signed integers. Motorola's convention is that the sign of the remainder, if any, is the same as the sign of the dividend. Thus the instruction

DIVS   #3,D1

with (D1) = $FFFF FFF6 calculates $-10/3$ with the result

(D1) = $FFFF FFFD

or Q = -3 and R = -1. The condition code N is set to {1} to indicate that the quotient is negative.

In signed division, the quotient can range from -2^{15} to $+2^{15} - 1$. Therefore, overflow will occur unless the magnitude of the 32-bit dividend is less than 2^{15} times that of the divisor. The V bit is set to {1} if overflow occurs. A trap occurs if a divisor is zero.

Remainder in Division. Consider the unsigned division

$$Y/W = Q + R/W$$

where R/W is the remainder, which must be less than 1. Therefore, R/W has the representation

$$d_{-1} \times 10^{-1} + d_{-2} \times 10^{-2} + \cdots$$

and the positional number Y/X can be written

$$Q \cdot d_{-1}d_{-2} \cdots$$

as long as the division operation did not overflow. If only the fraction is considered, multiplying R/W by 10 yields d_{-1} as the first integer with a remainder of

$$d_{-2} \times 10_{-1} + \ldots .$$

Successive multiplications of R by 10 followed by a division by W yields the decimal digits as the quotient for as many places as desired. As an example, 22/7 is 3.142 . . . , which approximates π to three decimal places. The divisions and multiplications yield

$$22/7 = 3 + 1/7$$
$$10/7 = 1 + 3/7$$
$$30/7 = 4 + 2/7$$
$$20/7 = 2 + 6/7$$

and so on until the result 3.142 . . . is computed. Of course, the operations will be done in binary in the computer, but each digit can be converted to binary-coded decimal or ASCII for output if desired. The examples in Section 7.6 consider such conversions.

```
abs. rel.   LC   obj. code    source line
----  ----  ----  ---------   -----------
   1    1   0000               |      TTL      'FIGURE 7.6'
   2    2   0000               |      LLEN     100
   3    3   8000               |      ORG      $8000
   4    4   8000               |      OPT      P=M68000
   5    5   8000               |*
   6    6   8000               |* COMPUTATION OF AVERAGE
   7    7   8000               |*
   8    8   8000               |* INPUT  :  (A1.L) = ADDRESS OF COLUMN OF 16-BIT
   9    9   8000               |*                    NUMBERS
  10   10   8000               |*          (D1.W) = LENGTH OF COLUMN
  11   11   8000               |*
  12   12   8000               |* OUTPUT :  (D3)[15:0] = AVERAGE
  13   13   8000               |*          (D3)[31:16] = REMAINDER OF SUM/LENGTH
  14   14   8000               |*
  15   15   8000 48E7 6840     |AVG   MOVEM.L D1-D2/D4/A1,-(SP) ;SAVE REGISTERS
  16   16   8004               |*
  17   17   8004 4A41          |      TST.W   D1              ;IF LENGTH = 0
  18   18   8006 6700 0014     |      BEQ     DONE            ; THEN FINISHED
  19   19   800A               |*
  20   20   800A 4282          |      CLR.L   D2
  21   21   800C 4283          |      CLR.L   D3              ;SUM=0
  22   22   800E 3801          |      MOVE.W  D1,D4           ;SET COUNTER
  23   23   8010 5344          |      SUBQ    #1,D4           ; TO LENGTH - 1
  24   24   8012               |*
  25   25   8012 3419          |LOOP  MOVE.W  (A1)+,D2        ;LOOP TO SUM
  26   26   8014 D682          |      ADD.L   D2,D3           ;  NUMBERS
  27   27   8016 51CC FFFA     |      DBRA    D4,LOOP
  28   28   801A               |*
  29   29   801A 87C1          |      DIVS.W  D1,D3           ;SUM/LENGTH
  30   30   801C               |*
  31   31   801C 4CDF 0216     |DONE  MOVEM.L (SP)+,D1-D2/D4/A1 ;RESTORE REGISTERS
  32   32   8020 4E75          |      RTS
  33   33   8022               |      END
```

Figure 7.6 Program for averaging values.

Example 7.4 _____

The subroutine shown in Figure 7.6 computes the average of a series of numbers stored in vector or column form and addressed by A1. If the number of values, which is in D1, is not zero, the 32-bit sum is formed in D3. The sum is then divided by the number of values and the low-order word of D3 contains the quotient and any remainder is in the high-order word.

_____

EXERCISES

7.3.1. Determine the quotient and the remainder in the following divisions when the instruction listed is executed with the dividend and divisor as shown.

   (a)   10/5; DIVU

   (b) −10/5; DIVU

   (c) −10/5; DIVS

   (d) −5/2; DIVS

   The negative values should be written in two's-complement notation to perform the divisions.

7.3.2. Suppose two signed integers are multiplied by the MULU instruction. Show that unsigned binary multiplication will cause an error if one or both of the numbers are negative. Test this by multiplying $(-1) \times (-1)$ in two's-complement notation but with unsigned multiplication. How can the result be corrected?

7.3.3. If $N_2 < 2^m \times N_1$ in the unsigned binary division N_2/N_1, prove that overflow cannot occur if the dividend has $2m$ bits and the quotient is m bits.

7.3.4. Write a routine to compute a 32-bit quotient and a 32-bit remainder when overflow occurs with the DIVU instruction. The result can be obtained by writing Y/W as

$$(Y2 \times 2^{16} + Y1)/W$$

where Y2 is the upper 16 bits of the dividend and Y1 represents the lower 16 bits.

7.4 MULTIPLE-PRECISION ARITHMETIC

In scientific measurements, the term *accuracy* refers to the correctness of a measurement, that is, to its freedom from mistake or error. *Precision* refers to the amount of detail used to represent a measurement. For numerical values, the amount of precision is usually expressed by giving the number of significant digits in the numerical value. If a quantity is judged to have insufficient precision for a given application, additional significant digits may be used to produce a more precise result.

Arithmetic units in microprocessors operate on a maximum of m digits when performing arithmetic operations. We shall call this maximum length the *single-precision* length. The MC68000 maximum single-precision length is 32 bits, but 8-bit or 16-bit quantities can also be handled. Sequences of greater length cannot be handled as a single arithmetic operand by the processor. Therefore, to extend the precision, several m-digit operands can be considered mathematically as a single value. If k

Table 7.4 Extended Arithmetic Instructions

	Addressing Modes	
Syntax	**Source**	**Destination**
Add or Subtract Extended		
ADDX.⟨l⟩ ⟨Dm⟩,⟨Dn⟩	⟨Dm⟩	⟨Dn⟩
SUBX.⟨l⟩ ⟨Dm⟩,⟨Dn⟩		
ADDX.⟨l⟩ −(Am),−(An)	Predecrement	Predecrement
SUBX.⟨l⟩ −(Am),−(An)		
Negate with Extend		
NEGX.⟨l⟩ ⟨EA⟩	—	Data alterable

Note: ⟨l⟩ denotes B, W, or L.

operands were combined, the value would be $k \times m$ digits long. Double-precision values, for example, have $k = 2$. Thus the MC68000 double-precision length would be 2×32, or 64 bits.

Arithmetic operations with multiple-precision operands are performed by using the processor instructions on each m-digit portion of the values, and then combining the results. This procedure yields the correct answer when mathematical details such as carries or borrows between the intermediate results are treated properly.

The MC68000 provides special instructions to facilitate addition, subtraction, and negation of double-precision integers. This section is concerned primarily with the use of two 32-bit values to yield 64-bit double-precision numbers.

The instructions Add with Extend (ADDX), Subtract with Extend (SUBX), and Negate with Extend (NEGX) are defined in Table 7.4. The difference between these extended instructions and the instructions for addition, subtraction, and negation discussed previously is the use of the condition code bits X and Z by the extended operations.

As shown in Table 7.5, the extended instructions utilize the Extend (X) bit in their operation. If the X bit was set by a previous operation, the instructions ADDX,

Table 7.5 Operation of Extended Instructions

Instruction	Operation
ADDX.⟨l⟩ ⟨src⟩,⟨dst⟩	(dst)[l] ← (src)[l] + (dst)[l] + X
SUBX.⟨l⟩ ⟨src⟩,⟨dst⟩	(dst)[l] ← (dst)[l] − (src)[l] − X
NEGX.⟨l⟩ ⟨EA⟩	(EA)[l] ← 0 − (EA)[l] − X

Notes:
1. C, N, and V condition code bits set as for any arithmetic operation.
2. Z is cleared if the result is nonzero; otherwise, it is unchanged.
3. X is set the same as the C bit.
4. ⟨l⟩ denotes B, W, or L.
5. [l] indicates corresponding bits in the operation.

SUBX, and NEGX take this setting into account when they are executed. The primary use of the extend bit is to add a carry (ADDX) or subtract a borrow (SUBX) when the upper m bits of a double-precision value are being manipulated. The carry or borrow would have resulted from the single-precision operation on the lower m bits. For example, the sequence

ADD.L   D1,D3

ADDX.L   D2,D4

adds the double-precision value in D2/D1 to the 64-bit value in D4/D3. The X bit is set to {1} if the addition of the low order portions (D1 and D3) caused a carry. The second instruction adds the carry value to the sum.

In the case of arithmetic instructions, the X bit is set to the same value as the C bit. In general, most MC68000 instructions that are not used for arithmetic operations do not affect the X bit, so the C bit and the X bit should not be considered the same. For example, if D4 in the example just given was tested for zero by the instruction

TST.L   D4

the carry bit would be cleared but the X bit (set by the ADDX instruction) would not be changed.

The zero condition code bit, Z, is also treated in a special way by the extended instructions. The setting of Z, after an extended instruction is executed, is based on both the previous setting of the Z bit and the value of the current operand. Consider the 64-bit integer

$$0000\ 0000\ 0000\ 0001_{16}$$

in which the lower 32-bit value is nonzero. If each 32-bit half is tested for zero separately, Z would be set to {0} for the low-order portion. However, Z would become {1} when the high-order portion is tested. If a conditional test is subsequently made, the results would be based on a zero value!

To obtain the correct results for double-precision conditions, the instructions ADDX, SUBX, and NEGX set the Z bit according to the logical equation

$$Z = Z_2\ \text{AND}\ Z_1$$

in which *both* Z_1 and Z_2 must be {1} to set $Z = \{1\}$. Here Z_1 was the setting before the extended instruction was executed and Z_2 is the result from the extended operation. This is assumed to involve the high-order portion of a double-precision operand, as in the instruction sequence just given for addition. Thus if $Z_1 = \{0\}$, then $Z = \{0\}$ regardless of the setting of Z_2. Only if both portions of the double-precision value are zero will Z be set to {1}. In the example of the 64-bit number, $Z = \{0\}$ indicates that the result is nonzero when the value is computed using the extended instructions.

A double-precision integer can be written in positional notation as

$$(b_{2m-1}b_{2m-2}\ .\ .\ .\ b_m b_{m-1}\ .\ .\ .\ b_0)$$

where the digits $b_0 b_1 \ldots b_{m-1}$ represent the single-precision length. Two double-precision operands N_1 and N_2 can be written as

$$N_1 = N_{1U} + N_{1L}$$

and

$$N_2 = N_{2U} + N_{2L}$$

where N_{iL} refers to digits 0 through $m - 1$ and N_{iU} refers to the digits m through $2m - 1$, with $i = 1$ or 2. This notation will be used, when it is necessary, to distinguish the lower-precision from the upper-precision portions of a value.

7.4.1 Addition and Subtraction

The sequences of instructions to perform double-precision addition or subtraction require that the operation ADD be followed by ADDX or SUB be followed by SUBX. For addition, any carry generated from the ADD of the lower portion is indicated by both the C and the X condition code bits. The upper sum is then computed by ADDX, which adds the X bit into the result. A carry generated when the ADDX instruction executes indicates an unsigned result that is too large for the double-precision representation. If signed integers are being represented, an overflow condition is indicated by the condition code bit $V = \{1\}$.

When double-precision integers are subtracted, any borrow required by the low-order subtraction is indicated by the X bit. This is subtracted from the difference of the high-order values. A high-order out-of-range condition is indicated after SUBX executes by $C = \{1\}$ for unsigned integers or $V = \{1\}$ if signed integers were subtracted.

The sequence

NEG
NEGX

performs negation of a double-precision integer when NEG operates on the lower-precision portion and NEGX on the upper-precision portion. An overflow indication ($V = \{1\}$) occurs if the most negative integer is negated.

Example 7.5

Examples of multiple-precision operations are shown in Figure 7.7. For simplicity the single-precision lengths are eight binary digits. The states of the relevant condition codes are also shown after each portion of the multiple-precision operation. In each case, the lower 8 bits of the operands are treated first, and this is followed by the extended instruction operating on the upper 8 bits.

Example 7.6

The subroutine shown in Figure 7.8 adds two columns or vectors of N unsigned integers element by element. If X[i] represents the locations of the ith element in the first array and Y[i] is the corresponding element in the second, the operation is

$$(Y[i]) \leftarrow (X[i]) + (Y[i])$$

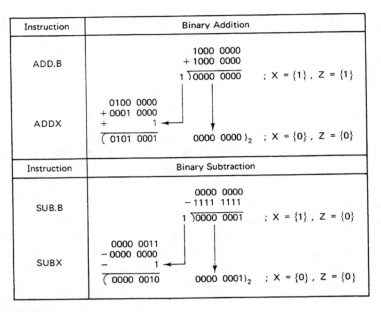

Instruction	Binary Addition
ADD.B	1000 0000 + 1000 0000 1)0000 0000 ; X = {1}, Z = {1}
ADDX	0100 0000 + 0001 0000 + 1 (0101 0001 0000 0000)₂ ; X = {0}, Z = {0}

Instruction	Binary Subtraction
SUB.B	0000 0000 − 1111 1111 1)0000 0001 ; X = {1}, Z = {0}
SUBX	0000 0011 − 0000 0000 − 1 (0000 0010 0000 0001)₂ ; X = {0}, Z = {0}

Figure 7.7   Multiple-precision operations.

with $i = 1, 2, \ldots, N$. In memory, each 64-bit integer is stored with the least significant 32 bits at the *higher* address of two longword locations. The values are stored with the last element, X[N] or Y[N], at the lowest memory address and each array requires $2N$ longword locations or $8N$ bytes. This storage scheme takes advantage of the predecrement addressing capability of the MC68000 using extended instructions.

When the subroutine is entered, A1 and A2 should point to the next longword location following the first and second array, respectively. Address register A3 must contain the address of the last element in the second array (i.e., it should point to Y[N]).

If no overflow occurs, the additions continue until (A2) = (A3) to indicate the location of the last value to be added. The Compare Address (CMPA) instruction might change the C condition code but leaves X unaffected. When the two addresses are equal, the branch test is FALSE and the loop is terminated. In programs where the state of the X bit must be preserved but the C bit is used for conditional tests, having separate C and X bits is an advantage because the X bit does not have to be saved before the compare operations.

7.4.2 Multiplication

The MC68000 instructions MULU and MULS form a 32-bit product when two 16-bit numbers are multiplied. In order to multiply 32-bit operands to yield a 64-bit product, the multiply instruction can be used repeatedly to form partial products. These partial products are added together to produce the result. For example, consider the multiplication

$$(x + y) \times (w + z) = x \times w + y \times w + x \times z + y \times z$$

```
abs. rel.   LC   obj. code   source line
---- ----   ----  ---------   -----------
   1    1   0000                       |         TTL       'FIGURE 7.8'
   2    2   0000                       |         LLEN      100
   3    3   8000                       |         ORG       $8000
   4    4   8000                       |         OPT       P=M68000
   5    5   8000               *
   6    6   8000               * ADD TWO VECTORS OF 64-BIT UNSIGNED INTEGERS
   7    7   8000               *    Y(I) <-- Y(I) + X(I)     FOR I = 1,N
   8    8   8000               *
   9    9   8000               * INPUTS :  (A1.L) = LAST ADDRESS OF FIRST VECTOR+4
  10   10   8000               *           (A2.L) = LAST ADDRESS OF SECOND VECTOR+4
  11   11   8000               *           (A3.L) = ADDRESS OF LAST ELEMENT IN
  12   12   8000               *                      SECOND VECTOR (ADDRESS OF Y(N))
  13   13   8000               *
  14   14   8000               * OUTPUTS:  (A2.L) = ADDRESS OF SUMS
  15   15   8000               *           (D1.B) = 0 : ERROR DETECTED
  16   16   8000               *                    NOT 0 : SUCCESSFUL
  17   17   8000               *
  18   18   8000               * NOTES:
  19   19   8000               *   1.   64-BIT NUMBERS ARE STORED THIS WAY:
  20   20   8000               *
  21   21   8000               *        XN [63:32]   LOW MEMORY
  22   22   8000               *        XN [31:0]      FIRST LONGWORD ADDRESS
  23   23   8000               *        .
  24   24   8000               *        .
  25   25   8000               *        .
  26   26   8000               *        X1 [63:32]          HIGHER MEMORY
  27   27   8000               *        X1 [31:0]      LAST LONGWORD ADDRESS
  28   28   8000               *
  29   29   8000               *   2.   A1 AND A2 POINT TO X1+4
  30   30   8000               *        A3 POINTS TO XN [63:32]
  31   31   8000               *
  32   32   8000 48E7 8060     |SERIES MOVEM.L D0/A1-A2,-(SP) ;SAVE REGISTERS
  33   33   8004 4201          |        CLR.B     D1              ;SET STATUS = FAIL
  34   34   8006 2021          |LOOP    MOVE.L    -(A1),D0
  35   35   8008 D1A2          |        ADD.L     D0,-(A2)
  36   36   800A D589          |        ADDX.L    -(A1),-(A2)
  37   37   800C 6500 000A     |        BCS       ERROR           ;OVERFLOW
  38   38   8010 B7CA          |        CMPA.L    A2,A3           ;IF NOT LAST NUMBER
  39   39   8012 65F2          |        BCS       LOOP            ;THEN CONTINUE
  40   40   8014               *                                 ; ELSE FINISHED
  41   41   8014 123C 00FF     |        MOVE.B    #-1,D1          ;SET STATUS = SUCCESS
  42   42   8018 4CDF 0601     |ERROR   MOVEM.L   (SP)+,D0/A1-A2  ;RESTORE REGISTERS
  43   43   801C 4E75          |        RTS
  44   44   801E               |        END
```

Figure 7.8   Multiple-precision addition program.

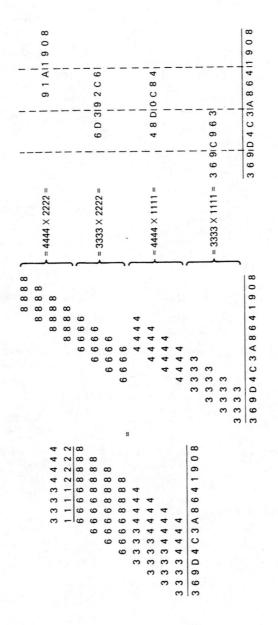

Note: All values are hexadecimal.

Figure 7.9 32-bit multiply example.

232

```
abs. rel.   LC    obj. code    source line
---- ----   ----  ----------   -----------
   1    1   0000                |              TTL      'Figure 7.10'
   2    2   0000                |              LLEN     120
   3    3   8000                |              ORG      $8000
   4    4   8000                |              OPT      P=M68000
   5    5   8000                |*
   6    6   8000                |* 32 X 32 BIT MULTIPLY
   7    7   8000                |*
   8    8   8000                |* INPUT : (D1.L) = FIRST VALUE
   9    9   8000                |*          (D2.L) = SECOND VALUE
  10   10   8000                |*
  11   11   8000                |* OUTPUT : (d2/D1)[63:0] = RESULT OF D1 * D2
  12   12   8000                |*
  13   13   8000  48E7 1C00     |MULR     MOVEM.L D3/D4/D5,-(SP)   ;SAVE REGISTERS
  14   14   8004                |*
  15   15   8004  2601          |         MOVE.L  D1,D3                ;COPY MULTIPLICAND
  16   16   8006  2801          |         MOVE.L  D1,D4
  17   17   8008  2A02          |         MOVE.L  D2,D5                ;COPY MULTIPLIER
  18   18   800A  4844          |         SWAP    D4
  19   19   800C  4845          |         SWAP    D5
  20   20   800E                |*
  21   21   800E  C2C2          |         MULU    D2,D1   ;PARTIAL 1 : NUM1[15:0] * NUM2[15:0]
  22   22   8010  C4C4          |         MULU    D4,D2   ;PARTIAL 2 : NUM1[31:16] * NUM2[15:0]
  23   23   8012  C6C5          |         MULU    D5,D3   ;PARTIAL 3 : NUM2[31:16] * NUM1[15:0]
  24   24   8014  C8C5          |         MULU    D5,D4   ;PARTIAL 4 : NUM2[31:16] * NUM1[31:16]
  25   25   8016                |*                        ;    (HIGH-ORDER)
  26   26   8016  4841          |         SWAP    D1
  27   27   8018  D242          |         ADD.W   D2,D1                    ;
  28   28   801A  4285          |         CLR.L   D5
  29   29   801C  D985          |         ADDX.L  D5,D4                ;CARRY 1
  30   30   801E  D243          |         ADD.W   D3,D1
  31   31   8020  D985          |         ADDX.L  D5,D4                ;CARRY 2
  32   32   8022  4841          |         SWAP    D1
  33   33   8024                |*
  34   34   8024  4242          |         CLR.W   D2
  35   35   8026  4243          |         CLR.W   D3
  36   36   8028  4842          |         SWAP    D2
  37   37   802A  4843          |         SWAP    D3
  38   38   802C  D483          |         ADD.L   D3,D2
  39   39   802E  D484          |         ADD.L   D4,D2
  40   40   8030                |*
  41   41   8030  4CDF 0038     |         MOVEM.L (SP)+,D3/D4/D5
  42   42   8034  4E75          |         RTS
  43   43   8036                |         END
```

Figure 7.10 Double-precision multiply routine.

233

which requires four multiplications and four additions. Double-precision multiplication is similar in theory if y and z represent the lower-precision values of the operands and x and w the upper. In machine computation, however, the magnitude ranges of the different partial products is not the same and this must be taken into account. The appropriate calculation can be determined for unsigned numbers by writing the double-precision operand in the form

$$N = N_U \times 2^m + N_L.$$

This is a $2m$ digit number in which N_U and N_L are the m-digit integers formed by the upper and lower portions, respectively. The product of two double-precision integers N_1 and N_2 becomes

$$N_1 \times N_2 = 2^{2m} \times (N_{2U} \times N_{1U}) + 2^m \times (N_{2L} \times N_{1U} + N_{2U} \times N_{1L}) + N_{2L}N_{1L}.$$

The total length of the product is $4m$ digits and each partial product has $2m$ digits. The machine algorithm that performs the multiplication operation must align the partial products properly before adding them in the same manner that multiplication by hand is achieved. For example, multiplying 1201 times 1501 involves four partial products. Each product after the first must be shifted left by one decimal place. The MC68000 instructions for shifting (discussed in Chapter 8) can be used to align the partial results, or the SWAP instruction can be used, as shown in Example 7.8.

When multiple-precision signed integers are multiplied, the scheme just described for unsigned integers fails if one or both of the operands to be multiplied is negative. A mathematical investigation using the two's-complement representation as described in Chapter 3 yields an algorithm that is suitable for this case. Several references in the Further Reading section at the end of this chapter discuss the approach. An alternative approach is to change the sign of (negate) any negative operands and correct the sign after performing unsigned multiplication. The program to accomplish this is left as an exercise.

Example 7.7 ——————————————————————————————

Figure 7.9 shows the multiplication of two 32-bit integers in hexadecimal, which produces a 64-bit result. The partial sums are shown aligned as they must be for computer implementation of the algorithm.

Example 7.8 ——————————————————————————————

The subroutine shown in Figure 7.10 multiplies

$(D1) \times (D2)$

as 32-bit unsigned integers and returns a 64-bit product in (D2)/(D1). Here D2 holds the upper 32 bits and D1 the lower 32 bits. Two observations must be noted to understand the program fully. First, the unsigned multiply instruction MULU and the ADD.W instruction operate only on the low-order 16 bits of the registers used as operands. Thus the SWAP instruction is required to move the high-order 16 bits into the low-order portion of a register, where they can be multiplied or added. Second, a diagram showing the contents of the registers involved during execution of the program is almost mandatory to create such a program. The version shown here is from Motorola; however, comments were added.

EXERCISES

7.4.1. Determine the range of unsigned integers, fractions, and signed integers for a 64-bit representation. Express the answers as powers of 10.

7.4.2. Modify the double-precision multiplication routine to multiply two's-complement integers. (Determine the sign of the result based on the signs of the factors; then perform the multiplication on the absolute values.)

7.5 DECIMAL ARITHMETIC

The MC68000 provides instructions for addition, subtraction, and negation of decimal values represented in Binary-Coded Decimal (BCD). This code was defined in Chapter 3 and this section applies many of the mathematical principles presented here. The three instructions for BCD arithmetic are defined in Table 7.6. The instructions allow addition, subtraction, and negation of BCD values. For each instruction the operand length if 8 bits, which represents two BCD digits. For example, the Add Decimal with Extend (ABCD) instruction

ABCD   D1,D2

performs decimal addition between byte-length operands. The operation is

$$(D2)[7:0] \leftarrow (D1)[7:0] + (D2)[7:0] + X$$

Table 7.6   Decimal Arithmetic Instructions

Syntax		Operation
Addition		
ABCD	⟨Dm⟩,⟨Dn⟩	$(Dn)[7:0] \leftarrow (Dn)[7:0] + (Dm)[7:0] + X$
ABCD	−(Am),−(An)	$(dest) \leftarrow (dest) + (src) + X$
Subtraction		
SBCD	⟨Dm⟩,⟨Dn⟩	$(Dn)[7:0] \leftarrow (Dn)[7:0] - (Dm)[7:0] - X$
SBCD	−(Am),−(An)	$(dest) \leftarrow (dest) \leftarrow (src) - X$
Negation		
NBCD	⟨EA⟩	$(EA) \leftarrow 0 - (EA) - X$

Notes:
1. All operations perform decimal arithmetic on two BCD digits.
2. N and V condition code bits are undefined.
3. C is set if a decimal carry (or borrow) occurs.
4. Z is cleared if the result is nonzero; otherwise it is unchanged.

Notice that this instruction adds the value of the X condition code bit into the sum to facilitate multiple-precision additions. However, the X bit must be cleared before the first ABCD is executed. After the addition operation, X = {1} indicates that a decimal carry occurred because the sum was greater than 99. The Z bit is cleared if the sum is not zero. Otherwise, it is unchanged to allow tests for zero to be performed after multiple-precision operations. The Subtract Decimal with Extend (SBCD) operates similarly, but the source operand and the value of the X bit are both subtracted from the destination value.

The operations of the BCD instructions are similar to those for extended-precision arithmetic as far as the use of condition code bits is concerned. The exception lies in the fact that the N and V bits are not defined for BCD operations. Moreover, the BCD instructions restrict the operand length to 8 bits and the addressing modes to data register direct or predecrement for BCD addition and subtraction. The instruction Negate Decimal with Extend (NBCD) forms the ten's complement of a two-digit operand when X = {0} before the operation. If X = {1}, NBCD forms the nine's complement. It allows any data-alterable addressing mode for the effective address in the symbolic form

NBCD $\langle$EA$\rangle$

This excludes an operand in an address register or one addressed relative to the program counter.

Table 7.7   Example of BCD Operations

(a) Addition: ABCD (src), (dest)

Before Execution				After Execution		
(src)	(dest)	X	Z	(dest)	X	Z
65	17	0	1	82	0	0
65	17	1	1	83	0	0
42	77	0	0	19	1	0
0	0	0	0	00	0	0

(b) Subtraction: SBCD (src), (dest)

Before Execution			After Execution	
(src)	(dest)	X	(dest)	X
32	77	0	45	0
65	17	0	52	1
65	17	1	51	1
35	35	1	99	1

Note: The contents of the source and destination locations are decimal values.

Example 7.9

Table 7.7 shows the effect of decimal addition and subtraction for various operands and condition code settings. Addition of the values 65 and 17 yields 82 if the X bit is cleared or 83 if it is set. Adding 42 and 77 yields a result of 19 with an indication of a carry. The proper value of 119 would require an additional BCD digit. The incorrect nonzero indication that results for the addition of 0 and 0 without the Z bit set is also shown.

Subtraction of two BCD digits yields a correct result for unsigned numbers in the range 0 to 99 or signed numbers between -10 and $+9$. The subtraction of the unsigned numbers 77 minus 32 yields 45 as expected. But 17 minus 65 leaves the ten's-complement result of -48 (52) with a borrow indication. If the X bit is set before the operation, the nine's complement of -48 (51) results when 65 is subtracted from 17. When two equal values are subtracted with the X bit set, the result is 99 or the nine's complement of 0 with a borrow indication.

Multiple-Precision Decimal Arithmetic. Operations on BCD numbers with more than two digits are normally performed on operands held in memory rather than in a register. This is because the ABCD and SBCD instructions can operate only on the low-order byte of a data register. If a decimal string of digits is held in a data register, the rotate instructions introduced in Chapter 8 would be needed to shift the digits being manipulated to the low-order byte. This is avoided by performing memory-to-memory operations using predecrement addressing. For example, the instruction

ABCD $-(A1),-(A2)$

first decrements (A1) by 1 and then (A2) by 1. Then the two digits in the addressed byte locations plus the X bit value are added into the destination location addressed by A2. To perform operations on more-than-two-digit numbers, a decimal string is stored in memory with the least significant two digits at the highest byte address. Thus the decimal number 123456 at location $1000 would be stored as follows:

(1000) = 12
(1001) = 34
(1002) = 56

An addition or subtraction of this value should start with the beginning address initialized at $1003 when the predecrement modes are used.

Example 7.10

The program in Figure 7.11 adds two six-digit BCD integers. Initially, the addresses of the operands as just described are stored in A1 and A2. The sum is left in the location addressed by A2. The X bit must be cleared and the Z bit must be set before the addition begins. If the result is nonzero, the Z bit will be cleared by the addition. If the integers are restricted to positive values, the C bit indicates an overflow condition after the additions.

```
abs. rel.    LC   obj. code    source line
---- ----    ---- ---------    -----------
   1    1    0000                    TTL      'FIGURE 7.11'
   2    2    0000                    LLEN     100
   3    3    8000                    ORG      $8000
   4    4    8000                    OPT      P=M68000
   5    5    8000             *
   6    6    8000             * BCD ADDITION
   7    7    8000             *
   8    8    8000             * INPUTS :   (A1.L) = ADDRESS OF THE BYTE FOLLOWING
   9    9    8000             *                      FIRST 6-DIGIT BCD NUMBER
  10   10    8000             *            (A2.L) = ADDRESS OF THE BYTE FOLLOWING
  11   11    8000             *                      SECOND 6-DIGIT BCD NUMBER
  12   12    8000             *                      (ADDRESS OF NUMBER + 3)
  13   13    8000             * OUTPUTS :  (A2.L) = ADDRESS OF HIGH ORDER BYTE OF
  14   14    8000             *                      6-DIGIT BCD RESULT
  15   15    8000             *
  16   16    8000             * NOTES:
  17   17    8000             *   1.  BCD NUMBERS ARE STORED 2 DIGITS/BYTE
  18   18    8000             *     BCD [6:5]
  19   19    8000             *     BCD [4:3]
  20   20    8000             *     BCD [2:1]
  21   21    8000             *
  22   22    8000             *   2.  ADDRESS REGISTERS POINT TO BCD [2:1] + 1
  23   23    8000             *       SO THAT PREDECREMENT ADDRESSING CAN BE
  24   24    8000             *       USED
  25   25    8000             *
  26   26    8000             *   3.  NO TEST FOR OVERFLOW
  27   27    8000             *
  28   28    8000             |
  29   29    8000 48E7 0060   |ADDBCD MOVEM.L A1/A2,-(SP) ;SAVE REGISTERS
  30   30    8004 44FC 0004   |       MOVE.W  #4,CCR       ;CLEAR X BIT
  31   31    8008             |*                          ; SET Z BIT
  32   32    8008 C509        |       ABCD    -(A1),-(A2) ;ADD THE BCD NUMBERS
  33   33    800A C509        |       ABCD    -(A1),-(A2)
  34   34    800C C509        |       ABCD    -(A1),-(A2)
  35   35    800E 4CDF 0600   |       MOVEM.L (SP)+,A1/A2 ;RESTORE REGISTERS
  36   36    8012 4E75        |       RTS                 ; AND RETURN
  37   37    8014             |       END
```

Figure 7.11   Six-digit BCD addition.

EXERCISES

7.5.1. Express the following BCD numbers in binary using four digits. Show the machine representation using ten's-complement notation.

    (a) +37

    (b) −37

    (c) −1319

7.5.2. Using the MC68000 formats, perform the following operations on BCD integers.

    (a) 1754 − 1319

    (b) 9375 + 3470

How are the results in part (b) interpreted when only four-digit unsigned integers are allowed?

7.5.3. Perform the addition 127 plus 299 using binary arithmetic but with the values in BCD notation. Adjust the binary result by adding 6 to any digit greater than 9 and add the carry to the next-higher digit. (The Motorola MC68000 BCD instructions perform this decimal adjustment automatically.)

7.5.4. Write a subroutine to add or subtract two BCD integers with up to eight digits each. Signed integers are represented in ten's-complement notation. What is the decimal range of the valid integers? What is the decimal range of their sum or difference?

7.5.5. Write a subroutine to multiply two four-digit positive BCD numbers in MC68000 format. The routine might perform multiplication by repeated addition, or an algorithm can be devised to perform decimal multiplication.

7.6 I/O AND CONVERSIONS

This section is concerned primarily with techniques that enable a programmer to transfer data between the memory of a computer system and the operator's terminal. Fortunately, many of the details of these transfers are handled by routines that are part of the operating system or monitor. These details are introduced in Subsection 7.6.1 but are not discussed completely until Chapter 12. After introduction of the physical or hardware aspects of I/O transfers, a number of macroinstructions are defined. The macroinstructions call routines in the monitor (via the trap mechanism) to complete the transfers. Finally, conversions between ASCII, binary, and BCD values are presented. For reference, the ASCII character set is presented in Appendix I.

Figure 7.12 shows the simplified structure of the computer system with emphasis on the I/O transfer hardware and software. This system is similar to that presented in Chapter 2 but with the interface to the terminal shown as a peripheral chip. The chip performs serial-to-parallel conversion for input from the keyboard and parallel-to-serial conversion for output to the display screen. It is programmed for input or output by the I/O chip-control routines that are part of the monitor in this discussion. In an MC68000-based system these I/O transfers could be made by the serial peripheral chip, such as the MC68681 described in Chapter 12.

The I/O macros discussed are written to call I/O routines. Conversion routines must be written to perform ASCII-to-binary or BCD conversion for data input. After conversion, the numbers may be manipulated as required by the applications program. Conversely, results from mathematical operations held in memory must be converted to ASCII characters for display at the terminal. An I/O macro library could be created for convenience to hold the I/O and conversion routines. This software module would be linked to the application program when I/O transfers are required.

The Arnewsh single-board computer with a TUTOR monitor is used for the examples in this section.[1] Other details concerning the operation of the monitor were

---

[1] Arnewsh Inc., Fort Collins, CO 80527.

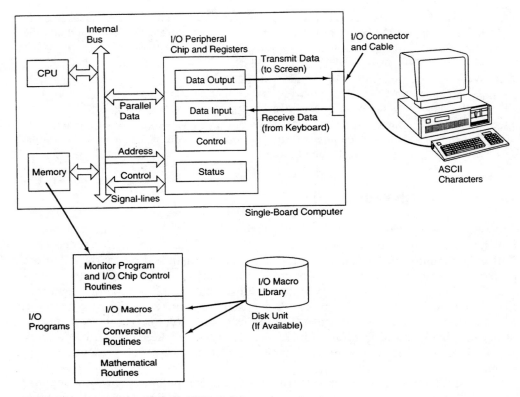

Figure 7.12 Simplified diagram of I/O operations.

presented in Chapter 5. In that discussion, the monitor served primarily to load and execute programs. Here the capability of the monitor to facilitate I/O transfers is most important.

7.6.1 I/O Transfer

To allow the operator to input a character from the keyboard, the I/O chip in Figure 7.12 must be programmed to accept the character. The simplified procedure for the I/O chip-control routine is as follows:

(a) Read the status of the I/O chip from its status register.
(b) If the chip is not busy, program its control register to accept a character. If busy, wait and try again.
(c) Once the character is received, transfer the character from the data input register to memory.

This procedure is termed *conditional* I/O transfer because the I/O routine must wait until the chip is ready for transfer. Other types of transfers including those controlled by interrupts are described in Chapter 12.

If the I/O chip is ready for input when the operator strikes a key, the character is received as a serial train of pulses and converted to an 8-bit value for storage in memory. In the present discussion, the 8-bit character is an ASCII character. Similarly, the chip is programmed for output to transmit an ASCII character when the terminal is ready to receive it. The peripheral chip in both cases provides electrical and timing compatibility between the computer and the terminal. In MC68000 systems, the registers of the I/O chip are addressed in the same manner as memory locations.

As you will see in Chapter 12, programming the peripheral chip is tedious. It is, in fact, unnecessary when this type of standard I/O transfer is required. Obviously, the monitor must input characters from the keyboard and respond to commands by displaying requested information. To utilize this capability of the TUTOR, it is only necessary to execute the TRAP #14 instruction in a program. Other operating systems and monitors may require a different mechanism to invoke I/O transfers, but all modern computer systems will have an equivalent feature. In most cases, a user-mode program cannot program the I/O chips directly. When an operating system is present, most computer systems require that calls to the operating system be made if a user–mode program requires I/O transfers. The TUTOR monitor in these examples serves as a rudimentary operating system for the Arnewsh single-board computer.

7.6.2 I/O System Calls and Macros

The TUTOR monitor provides various routines that are invoked by the TRAP #14 instruction. These routines are typically used by a programmer to facilitate I/O and other operations called *functions*. The code for a specific function must be defined in the least significant byte of D7 prior to the TRAP #14 instruction. The codes are decimal numbers in the range 0 to 253. Only a few of the I/O functions are considered here.

One TUTOR function was encountered in previous examples where the instruction

```
MOVE.B      #228,D7
TRAP        #14
```

was used to return control to the monitor after a program executed to completion. The use of the TRAP #14 instruction in this manner is sometimes termed a *system call*. Table 7.8 lists several of these system calls with their TRAP #14 codes and describes their function.

The input character function (.INCHE) can receive one character at a time from the terminal keyboard and the output character function (.OUTCH) sends one character to the terminal screen. Other I/O functions input or output a *string* of characters. The string is a series of ASCII characters considered as a unit. Usually, an input string is ended with a carriage return (CR). An output string may end with a carriage return and a line feed (LF). In this case, the output string creates one line of text on the display unit.

The input string (.PORTIN1) function can receive a string of characters terminated by a carriage return (CR). For this input function, the memory locations for the characters called the *buffer* area must be defined before the call is made. The method for defining the buffer address will be explained later.

Table 7.8 Tutor System Calls

Function	TRAP #14 Code	Description
.INCHE	247	Read one character into (D0)[B]
.PORTIN1	241	Read a string of characters up to 127 characters in length followed by a (CR); input buffer address is defined before call. The character string is echoed to the display unit.
.OUTCH	248	Output a character from (D0)[B]
.OUTPUT	243	Output a character string without a (CR)(LF)
.OUT1CR	227	Output a character string (line) followed by (CR)(LF)
.RETURN	228	Return to monitor

Notes:

1. (CR) = carriage return: ASCII ($0D)
   (LF) = line feed: ASCII ($0A)

2. A *string* is limited in length to 127 characters for the TUTOR monitor. A *line* is a string of characters followed by (CR) which typically corresponds to one line of characters on a CRT screen.

Output routines (.OUTCH, .OUTPUT, .OUT1CR) are used to display messages or data to the operator. Several forms in Table 7.8 are available to allow the programmer to format the terminal screen as desired. The .RETURN call is an example of a useful system call not related directly to I/O transfers. It causes the TRAP #14 operation using code 228. The TUTOR monitor also has a number of other functions not shown in Table 7.8.

Invoking System Calls. For each system call for I/O of strings, the programmer must define the locations to be used for input or output. The definitions must conform to the calling conventions defined in Table 7.9. For the transfer of one character, the low order byte of D0 is used.

Table 7.9 Calling Conventions for System Calls

Macro System Call (SYSCALL)	Calling Convention
.INCHE	(D7)[B]=247; read character from (D0)[B] when complete. (A0, D1 modified)
.PORTIN1	Define input buffer address in A5 and A6; (D7)[B]=241.
.OUTCH	Put character in (D0)[B];(D7)[B]=248. (A1, D1 modified)
.OUTPUT	Define beginning output buffer address in A5 and last address plus 1 in A6;(D7)[B]=243.
.OUT1CR	Define beginning output buffer address in A5 and last address plus 1 in A6;(D7)[B]=227. (A0, D0, D1 modified)

Notes:

1. The "calling convention" means that the programmer must perform the operations defined here before the system call is issued.

2. The system stack is used with these SYSCALLS.

The .INCHE function, for example, returns a single character to (D0)[B]. Thus, the program segment

```
MOVE.B    #241,D7    DEFINE FUNCTION
TRAP      #14        CALL .INCHE
```

(CHARACTER IN D0 AFTER KEYSTROKE)

waits for a character to be input, and the TRAP routine transfers the character to register D0. The .PORTIN1 function requires that the address of the input area or buffer in memory be placed in address register A5 before the call. Register A6 must contain the address of the byte one past the location of the last byte. Usually, the two addresses are the same before calling the function unless several strings are to be held in the input buffer. The maximum number of characters (bytes) is 127 for the string. After a string is received, A6 points to the last character location plus one. Similar conventions are required for each routine as defined exactly for use of the system calls. Examples of each call will be given subsequently.

Caution. Table 7.9 also lists the registers whose contents are modified by the TUTOR functions. If the contents are to be preserved, the values should be saved on the system stack before any TUTOR function is called. When the I/O function is complete, the register values can be restored. Of course, (D0) should not be saved or restored when used with the .INCHE or OUTCH function. Refer to the *User's Manual* for the single-board computer you are using for more details about the TUTOR functions. If another monitor is being used, the functions and calling conventions may differ substantially from those listed in Table 7.8 and Table 7.9.

Macrodirectives for System Calls. One way to simplify the programming of I/O functions and simultaneously improve the readability of a program is to employ *macro instructions,* usually called *macros* for short. The I/O macros considered here simply invoke system calls by name as required in a program. The basic macro structure for this purpose is

```
SYSCALL   MACRO
          MOVE.B    #\1,D7
          TRAP      #14
          ENDM
```

where the \1 indicates a value to be passed to the macro instruction when it is assembled. Each time the instruction SYSCALL is used, the assembly instructions are assembled into the program. Thus, the instruction

```
SYSCALL   .INCHE
```

assembles as

```
MOVE.B    #247,D7
TRAP      #14
```

if .INCHE is equated to the value 247. This is done by the directive

.INCHE   EQU   247

in the calling program.

Example 7.11 _____

The program in Figure 7.13 shows the use of system calls to perform various I/O operations. MACRO directives are used to define the calling structure followed by the parameters for the particular function. When executed, the program first displays the ASCII text at label LINE 1 on the operator's terminal. Before the output of characters defined by LINE 1, LINE 2, LINE

```
abs. rel.   LC   obj. code    source line
---- ----   ----  ---------    -----------
   1    1   0000                           |                 TTL      'FIGURE 7.13'
   2    2   8000                           |                 ORG      $8000
   3    3   8000                           |                 OPT      P=M68000
   4    4   8000               |*
   5    5   8000               |*    I/O SYSCALLS
   6    6   8000               |*
   7    7   8000               |SYSCALL          MACRO                 ;MACRO= SYSCALL
   8    8   8000               |                 MOVE.B   #\1,D7
   9    9   8000               |                 TRAP     #14
  10   10   8000               |                 ENDM
  11   11   8000               |*
  12   12   8000               |*    I/O SYSCALL PARAMETERS
  13   13   8000               |*
  14   14        0000 00F7     |.INCHE           EQU      247          ;INPUT CHARACTER
  15   15        0000 00F1     |.PORTIN1         EQU      241          ;INPUT LINE
  16   16   8000               |*
  17   17        0000 00F8     |.OUTCH           EQU      248          ;OUTPUT CHARACTER
  18   18        0000 00F3     |.OUTPUT          EQU      243          ;OUTPUT STRING
  19   19        0000 00E3     |.OUT1CR          EQU      227          ;OUTPUT STRING WITH (CR),(LF)
  20   20        0000 00E4     |.RETURN          EQU      228          ;RETURN TO MONITOR
  21   21   8000               |*
  22   22   8000               |*    OUTPUT "TEST OF I/O SYSCALLS"
  23   23   8000               |*
  24   24   8000 4BF9 0000     |STARTIO          LEA      LINE1,A5     ;ADDRESS OF MESSAGE-
  24        8004 8096          |
  25   25   8006 4DF9 0000     |                 LEA      ELIN1,A6     ;LINE 1
  25        800A 80AB          |
  26   26   800C               |                 SYSCALL .OUT1CR       ;WRITE TO SCREEN
  27   1m   800C 1E3C 00E3     +                 MOVE.B   #.OUT1CR,D7
  28   2m   8010 4E4E          +                 TRAP     #14
  29   27   8012               |*
  30   28   8012               |*    INPUT AND ECHO A CHARACTER
  31   29   8012               |*
  32   30   8012 4BF9 0000     |                 LEA      LINE2,A5     ;PROMPT FOR CHARACTER
  32        8016 80AC          |
  33   31   8018 4DF9 0000     |                 LEA      ELIN2,A6
  33        801C 80BE          |
  34   32   801E               |                 SYSCALL .OUT1CR
  35   1m   801E 1E3C 00E3     +                 MOVE.B   #.OUT1CR,D7
  36   2m   8022 4E4E          +                 TRAP     #14
```

(continued)

Figure 7.13   Program examples of I/O system calls.

```
37   33   8024                    |*
38   34   8024                    |                SYSCALL  .INCHE              ; INPUT
39   1m   8024  1E3C 00F7  +                       MOVE.B   #.INCHE,D7
40   2m   8028  4E4E      +                        TRAP     #14
41   35   802A                    |                                            ;CHARACTER IN (D0)
42   36   802A  4BF9 0000 |                        LEA      LINE3,A5           ;ECHO CHARACTER
42        802E  80BF      |
43   37   8030  4DF9 0000 |                        LEA      ELIN3,A6
43        8034  80CE      |
44   38   8036            |                        SYSCALL  .OUTPUT
45   1m   8036  1E3C 00F3 +                        MOVE.B   #.OUTPUT,D7
46   2m   803A  4E4E      +                        TRAP     #14
47   39   803C            |                        SYSCALL  .OUTCH
48   1m   803C  1E3C 00F8 +                        MOVE.B   #.OUTCH,D7
49   2m   8040  4E4E      +                        TRAP     #14
50   40   8042  4BF9 0000 |                        LEA      PCRLF,A5           ;NEXT LINE
50        8046  80ED      |
51   41   8048  4DF9 0000 |                        LEA      ECRLF,A6
51        804C  80EF      |
52   42   804E            |                        SYSCALL  .OUTPUT

53   1m   804E  1E3C 00F3 +                        MOVE.B   #.OUTPUT,D7
54   2m   8052  4E4E      +                        TRAP     #14
55   43   8054            |*
56   44   8054            |*
57   45   8054            |*       INPUT AND ECHO A STRING OF CHARACTERS
58   46   8054            |*
59   47   8054  4BF9 0000 |                        LEA      PCRLF,A5           ;SKIP A LINE
59        8058  80ED      |
60   48   805A  4DF9 0000 |                        LEA      ECRLF,A6
60        805E  80EF      |
61   49   8060            |                        SYSCALL  .OUTPUT
62   1m   8060  1E3C 00F3 +                        MOVE.B   #.OUTPUT,D7
63   2m   8064  4E4E      +                        TRAP     #14
64   50   8066  4BF9 0000 |                        LEA      LINE4,A5           ;PROMPT FOR STRING
64        806A  80CF      |
65   51   806C  4DF9 0000 |                        LEA      ELIN4,A6
65        8070  80DF      |
66   52   8072            |                        SYSCALL  .OUT1CR
67   1m   8072  1E3C 00E3 +                        MOVE.B   #.OUT1CR,D7
68   2m   8076  4E4E      +                        TRAP     #14
69   53   8078            |*
70   54   8078  4BF9 0000 |                        LEA      INBUF,A5           ;ADDRESS OF BUFFER
70        807C  80F0      |
71   55   807E  4DF9 0000 |                        LEA      INBUF,A6
71        8082  80F0      |
72   56   8084            |
73   57   8084            |                        SYSCALL  .PORTIN1           ;READ THE STRING
74   1m   8084  1E3C 00F1 +                        MOVE.B   #.PORTIN1,D7
75   2m   8088  4E4E      +                        TRAP     #14
76   58   808A            |                        SYSCALL  .OUTPUT            ;ECHO THE STRING
77   1m   808A  1E3C 00F3 +                        MOVE.B   #.OUTPUT,D7
78   2m   808E  4E4E      +                        TRAP     #14
79   59   8090            |*
80   60   8090            |                        SYSCALL  .RETURN            ;RETURN
81   1m   8090  1E3C 00E4 +                        MOVE.B   #.RETURN,D7
82   2m   8094  4E4E      +                        TRAP     #14
83   61   8096            |*
```

(continued)

Figure 7.13   *(continued)*

```
84   62   8096                    |  *        MESSAGES AND DATA
85   63   8096                    |  *
86   64   8096  2054 4553         | LINE1            DC.B     ' TEST OF I/O SYSCALLS'
86   64   809A  5420 4F46         |
86   64   809E  2049 2F4F         |
86   64   80A2  2053 5953         |
86   64   80A6  4341 4C4C         |
86   64   80AA  53                |
87   65   80AB  00                | ELIN1            DC.B     $00
88   66   80AC  2049 4E50         | LINE2            DC.B     ' INPUT A CHARACTER'
88   66   80B0  5554 2041         |
88   66   80B4  2043 4841         |
88   66   80B8  5241 4354         |
88   66   80BC  4552             |
89   67   80BE  00                | ELIN2            DC.B     $00
90   68   80BF  2043 4841         | LINE3            DC.B     ' CHARACTER IS  '
90   68   80C3  5241 4354         |
90   68   80C7  4552 2049         |
90   68   80CB  5320 20           |
91   69   80CE  00                | ELIN3            DC.B     $00
92   70   80CF  2049 4E50         | LINE4            DC.B     ' INPUT A STRING '
92   70   80D3  5554 2041         |
92   70   80D7  2053 5452         |
92   70   80DB  494E 4720         |
93   71   80DF  00                | ELIN4            DC.B     $00
94   72   80E0  2053 5452         | LINE5            DC.B     ' STRING IS',$0D,$0A
94   72   80E4  494E 4720         |
94   72   80E8  4953 0D0A         |
95   73   80EC  00                | ELIN5            DC.B     $00
96   74   80ED  0D0A              | PCRLF            DC.B     $0D,$0A
97   75   80EF  00                | ECRLF            DC.B     $00
98   76   80F0                    | *
99   77   80F0                    | INBUF            DS.B     127        ;INPUT BUFFER
100  78   8170                    |                  END
```

Figure 7.13   *(continued)*

3, LINE 4, and LINE 5 are displayed, each address is defined in A5 at the appropriate place in the program with the Load Effective Address (LEA) instruction.[2]

Figure 7.14 shows the text on the display screen as seen by the operator. After the text corresponding to LINE 2 is displayed, the operator types a character followed by (CR). This character is not displayed on the screen as it is typed, but it is stored in register D0. It is redisplayed or "echoed" on the screen after the text LINE 3 by the syscall .OUTCH.

Next a string of characters is received and echoed after the text of LINE 4 is displayed. A 128 byte area is reserved for the input string as INBUF in the data section of the program. The syscall .PORTIN1 reads and echoes the string of characters after the operator types it followed by a carriage return. The call leaves A6 pointing to the next byte after the last character. This buffer also serves as the output buffer for the next system call to .OUT1CR to echo the string again.

---

[2] The LEA instruction is studied in Chapter 9.

```
TUTOR MONITOR/DEBUG V1.32
(C) ARNEWSH, INC.
P.O. BOX 270352
FORT COLLINS, CO 80527-0352

TUTOR  1.32> DF
PC=00008000 SR=2700=.S7..... US=0000FC00 SS=00010000
D0=00000000 D1=00000000 D2=00000000 D3=00000000
D4=00000000 D5=00000000 D6=00000000 D7=00000000
A0=00000000 A1=00000000 A2=00000000 A3=00000000
A4=00000000 A5=00000000 A6=00000000 A7=00010000
--------------------008000    4BF900008090       LEA.L    $00008090,A5

TUTOR  1.32> GO
PHYSICAL ADDRESS=00008000
 TEST OF I/O SYSCALLS
 INPUT A CHARACTER
 CHARACTER IS  A

 INPUT A STRING
THIS IS A GOOD STRING

TUTOR  1.32> GO
PHYSICAL ADDRESS=00008000
 TEST OF I/O SYSCALLS
 INPUT A CHARACTER
 CHARACTER IS  H

 INPUT A STRING
THIS STRING IS SIX WORDS LONG.
THIS STRING IS SIX WORDS LONG.
TUTOR  1.32>
```

Figure 7.14   Trace of the program in Figure 7.13.

Calls to .OUT1CR display a line or string of characters followed by a (CR) and (LF). In contrast, a syscall to .OUTPUT would leave the cursor of the CRT unit in a position immediately following the displayed characters unless carriage return and linefeed characters were part of the string as they are in the text of LINE 5. The programmer can thus format the display as desired by transmitting format control characters to the terminal. When the text of LINE 5 is output for example, the (CR) character ($0D) and the (LF) character ($0A) are also sent. By selecting the appropriate format control characters, the programmer can cause actions such as clearing of the screen or tabulating any displayed results. The exact ASCII code for such formatting depends on the format required by the operator's terminal.

7.6.3 Conversions Between ASCII, Binary, and BCD

The standard representations for the MC68000 include binary and BCD for integers and ASCII for characters. For input and output ASCII is generally the code used for data being transferred between the computer system and peripheral devices, such as line printers or CRT terminals. Conversions between these representations are therefore frequently required because the arithmetic processing requires binary or BCD values in memory.

Figure 7.15(a) shows the typical steps to convert a decimal number in ASCII to a binary representation. The ASCII characters for the decimal digits are first converted to binary numbers in the range 0–9. The 4 bits for each digit are also the BCD value in memory. Then the string of BCD digits is converted to a binary number. This conversion takes into account the positional value of each BCD digit. For example, the ASCII string '123' as an input value is stored in memory as $31, $32, and $33. This is converted to three BCD digits, 1, 2, and 3, and then to the binary value 0111 1011.

The output of binary values that are to be printed as decimal numbers requires the opposite conversion from binary to BCD and then to ASCII. Of course, the binary

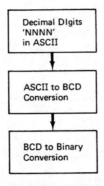

(a) Conversion of Input Data

Meaning	Binary Representation in Memory	To Convert to ASCII	ASCII in Memory
Decimal Digits 0–9 (BCD)	0000 0000 0000 0001 ⋮ 0000 1001	Add 0011 0000$_2$ ($30)	$30–$39
Hexadecimal Digits 0–9	0000 0000 0000 0001 ⋮ 0000 1001	Add $30	$30–$39
Hexadecimal Digits A–F	0000 1010 0000 1011 ⋮ 0000 1111	Add $37	$41–$46

(b) Decimal or Hexadecimal to ASCII for Output

Figure 7.15   Conversions of data values.

value in memory could be printed in binary, hexadecimal, or another code. The binary number in the input example just given has the hexadecimal value 7B. This could be output in ASCII as $37, $42.

The examples in this subsection use a table-lookup scheme to convert a hexadecimal digit to ASCII or BCD and a subroutine to convert a string of BCD digits to binary. The table in Figure 7.15(b) shows the ASCII equivalents for decimal and hexadecimal digits.

Example 7.12 _____

Figure 7.16 shows a subroutine to determine the ASCII or BCD value of a hexadecimal digit stored in the low-order 4 bits of D1. If (D2) [7:0] contains a 0 upon entry, the ASCII code is placed in the low-order byte of D2. The value is found by addressing the table ASCTAB and

```
abs. rel.   LC   obj. code   source line
----  ----  ----  ---------   -----------
  1    1    0000                        TTL     'FIGURE 7.16'
  2    2    0000                        LLEN    100
  3    3    8000                        ORG     $8000
  4    4    8000                        OPT     P=M68000
  5    5    8000              |*
  6    6    8000              |*    HEXADECIMAL TO BCD/ASCII
  7    7    8000              |*
  8    8    8000              |*    INPUT  :  (D2.B) = 0 : ASCII REQUESTED
  9    9    8000              |*                    NOT 0 :   BCD REQUESTED
 10   10    8000              |*              (D1.W) = HEXADECIMAL DIGIT TO CONVERT
 11   11    8000              |*                       MUST BE VALID = 0-F
 12   12    8000              |*
 13   13    8000              |*    OUTPUT :  (D2.B) = BCD OR ASCII
 14   14    8000              |*
 15   15    8000              |*
 16   16    8000              |*
 17   17    8000  2F09        |CONHAS  MOVE.L  A1,-(SP)          ;SAVE REGISTER
 18   18    8002  227C 0000   |        MOVE.L  #ASCTAB,A1        ;ASSUME ASCII WAS
 18        8006  801C        |
 19   19    8008              |*                                ; REQUESTED
 20   20    8008  4A02        |        TST.B   D2                ;CHECK REQUEST
 21   21    800A  6700 0008   |        BEQ     INDEX             ;IF BCD REQUESTED
 22   22    800E  227C 0000   |        MOVE.L  #BCDTAB,A1        ; THEN CHANGE TO BCD TABLE
 22        8012  802C        |
 23   23    8014              |*
 24   24    8014  1431 1000   |INDEX   MOVE.B  0(A1,D1.W),D2     ;LOOK UP VALUE
 25   25    8018  225F        |        MOVE.L  (SP)+,A1          ;RESTORE REGISTERS
 26   26    801A  4E75        |        RTS
 27   27    801C  3031 3233   |ASCTAB  DC.B    '0123456789ABCDEF'
 27   27    8020  3435 3637   |
 27   27    8024  3839 4142   |
 27   27    8028  4344 4546   |
 28   28    802C  0001 0203   |BCDTAB  DC.B    0,1,2,3,4,5,6
 28   28    8030  0405 06     |
 29   29    8033  0708 0910   |        DC.B    7,8,9,$10,$11
 29   29    8037  11          |
 30   30    8038  1213 1415   |        DC.B    $12,$13,$14,$15
 31   31    803C              |        END
```

Figure 7.16 Table lookup for hexadecimal conversion.

```
abs. rel.    LC    obj. code    source line
---- ----    ----  ---------    -----------
   1    1    0000                |                TTL     'FIGURE 7.17'
   2    2    0000                |                LLEN    100
   3    3    8000                |                ORG     $8000
   4    4    8000                |                OPT     P=M68000
   5    5    8000                |*
   6    6    8000                |*      BCD TO BINARY
   7    7    8000                |*
   8    8    8000                |*      INPUTS  :  (D2.W) = NUMBER OF DIGITS IN BCD NUMBER
   9    9    8000                |*                 (A1.L) = ADDRESS OF THE MOST SIGNIFICANT
  10   10    8000                |*                                 BCD DIGIT
  11   11    8000                |*
  12   12    8000                |*      OUTPUTS :  (D1.L) = BINARY VALUE
  13   13    8000                |*                 (D4.W) = 0 :   ERROR DETECTED
  14   14    8000                |*                       NOT 0 :   SUCCESSFUL
  15   15    8000                |*
  16   16    8000                |*      NOTES:
  17   17    8000                |*      1.   BCD DIGITS ARE STORED ONE/BYTE AND MUST BE
  18   18    8000                |*                   VALID. LIMIT IS 8 DIGITS.
  19   19    8000                |*      2.   ONLY POSITIVE BCD NUMBERS ARE ALLOWED
  20   20    8000                |*
  21   21    8000                |*
  22   22    8000                |*
  23   23    8000 48E7 3440      |BCDBN    MOVEM.L D2/D3/D5/A1,-(SP) ;SAVE REGISTERS
  24   24    8004 4281           |         CLR.L   D1              ;VALUE      := 0
  25   25    8006 4244           |         CLR.W   D4              ;SET DEFAULT TO ERROR
  26   26    8008 4285           |         CLR.L   D5              ;CLEAR ACCUMULATOR
  27   27    800A 4A42           |         TST.W   D2              ;IF LENGTH IS ZERO
  28   28    800C 6700 002E      |         BEQ     DONE            ; THEN EXIT WITH ERROR
  29   29    8010 0C42 0008      |         CMPI.W  #8,D2           ;IF LENGTH >8
  30   30    8014 6E00 0026      |         BGT     DONE            ; THEN EXIT WITH ERROR
  31   31    8018                |
  32   32    8018                |*
  33   33    8018 1A19           |LOOP     MOVE.B  (A1)+,D5        ;DIGIT IN D5
  34   34    801A D285           |         ADD.L   D5,D1           ;ADD TO SERIES
  35   35    801C 5342           |         SUBQ.W  #1,D2           ;DECREMENT COUNTER
  36   36    801E 6700 0018      |         BEQ     SUCCESS         ; IF FINISHED (LAST DIGIT)
  37   37    8022                |*                               ;      EXIT
  38   38    8022                |*
  39   39    8022                |*      MULTIPLY BY 10 AS  10X = (2X)*4+2X
  40   40    8022                |*
  41   41    8022 E389           |         LSL.L   #1,D1           ;MULTIPLY BY TWO
  42   42    8024 6B00 0016      |         BMI     DONE            ; ON OVERFLOW , EXIT
  43   43    8028                |*
  44   44    8028 2601           |         MOVE.L  D1,D3           ;SAVE THE RESULT OF MULTIPLY
  45   45    802A E589           |         LSL.L   #2,D1           ;MULTIPLY BY FOUR (VALUE*8)
  46   46    802C 6B00 000E      |         BMI     DONE            ; ON OVERFLOW , EXIT
  47   47    8030                |*
  48   48    8030 D283           |         ADD.L   D3,D1           ;VALUE = VALUE * 10
  49   49    8032 6B00 0008      |         BMI     DONE            ; ON OVERFLOW , EXIT
  50   50    8036                |*
  51   51    8036 60E0           |         BRA     LOOP            ;PROCESS NEXT DIGIT
  52   52    8038 383C 0001      |SUCCESS  MOVE.W  #1,D4           ;SET STATUS TO SUCCESS
  53   53    803C 4CDF 022C      |DONE     MOVEM.L (SP)+,D2/D3/D5/A1 ;RESTORE REGISTERS
  54   54    8040 4E75           |         RTS
  55   55    8042                |         END
```

Figure 7.17   BCD-to-binary conversion.

```
abs. rel.   LC   obj. code    source line
----  ----  ----  ---------    -----------
  1    1   0000                        |          TTL      'FIGURE 7.18'
  2    2   0000                        |          LLEN     100
  3    3   0000                        |          OPT      P=M68000
  4    4   0000                        |*
  5    5   0000                        |* BINARY TO DECIMAL CONVERSION AND DISPLAY
  6    6   0000                        |*
  7    7   0000                        |* THIS PROGRAM CONVERTS A TABLE OF UNSIGNED BINARY NUMBERS
  8    8   0000                        |* INTO DECIMAL VALUES AND CREATES ASCII STRINGS. THE RESULTS
  9    9   0000                        |* ARE DISPLAYED ON THE TERMINAL.
 10   10   0000                        |*
 11   11   0000                        |*    INPUTS:  BINARY NUMBERS IN TABLE BINNUM
 12   12   0000                        |*    OUTPUTS: DECIMAL EQUIVALENT DISPLAYED ON TERMINAL
 13   13   0000                        |*
 14   14   0000                        |* THE PROGRAM BEGINS AT LABEL START.
 15   15   0000                        |*
 16   16   0000                        |
|***************************************************************************
 17   17   0000                        |* MACROS AND TUTOR CALLS
 18   18   0000                        |*
 19   19   0000                        |SYSCALL MACRO
 20   20   0000                        |          MOVE.B   #\1,D7
 21   21   0000                        |          TRAP     #14
 22   22   0000                        |          ENDM
 23   23   0000                        |
 24   24         0000 00E3             |.OUT1CR EQU       227              ;OUTPUT WITH (CR),(LF)
 25   25         0000 00E4             |.RETURN EQU       228              ;RETURN TO MONITOR
 26   26   0000                        |*
 27   27   0000                        |
 28   28   0000                        |* DATA AREA
 29   29   7F00                        |          ORG      $7F00           ;WORD LENGTH BIN NUMBERS
 30   30   7F00 3039 007B              |BINNUM    DC.W     12345,123,0,9,98
 30   30   7F04 0000 0009              |
 30   30   7F08 0062                   |
 31   31         0000 7F0A             |EBINNUM EQU       *
 32   32   7F0A                        |DECSTR  DS.B      5                ;FIVE DIGITS
 33   33         0000 7F10             |E_DECSTR EQU      *
 34   34   7F10                        |*
 35   35   7F10                        |* INITIALIZATION
 36   36   8000                        |          ORG      $8000
 37   37   8000 247C 0000              |START     MOVE.L   #BINNUM,A2       ;GET ADDR OF INPUT TABLE
 37        8004 7F00                   |
 38   38   8006 363C 0004              |          MOVE.W   #(EBINNUM-BINNUM)/2-1,D3
 39   39   800A                        |                                   ;COUNT OF BINARY WORDS
 40   40   800A 323C 000A              |          MOVE.W   #10,D1           ;SET DIVISOR FOR CONVERSION
 41   41   800E                        |*
 42   42   800E                        |* OUTER LOOP -- GET NEXT BINARY NUMBER
 43   43   800E                        |*
 44   44   800E 301A                   |LOOP1     MOVE.W   (A2)+,D0         ;GET BIN NUM TO CONVERT
 45   45   8010 207C 0000              |          MOVE.L   #DECSTR,A0       ;GET ADDR OF OUTPUT STRING
 45        8014 7F0A                   |
 46   46   8016 343C 0004              |          MOVE.W   #E_DECSTR-DECSTR-1,D2
 47   47   801A                        |*                                  ;INITIALIZE OUTPUT POSITION
 48   48   801A                        |*                                  ; IN DECIMAL STRING
```

(continued)

Figure 7.18 Program for binary-to-decimal conversion and display.

```
50   50   801A                    |* INNER LOOP -- CONVERT USING REPEATED DIVISION BY 10
51   51   801A                    |*
52   52   801A 80C1              |LOOP2   DIVU    D1,D0          ;REMAINDER HAS NEXT DIGIT
53   53   801C 4840              |        SWAP    D0             ;MOVE REMAINDER TO LOW WORD
54   54   801E 0000 0030         |        ORI.B   #$30,D0        ;CONVERT DECIMAL TO ASCII
55   55   8022 1180 2000         |        MOVE.B  D0,(A0,D2.W)   ;STORE DIGIT FOR OUTPUT
56   56   8026 4240              |        CLR.W   D0             ;PREPARE FOR NEXT DIVISION
57   57   8028 4840              |        SWAP    D0             ;  AND RESTORE QUOTIENT
58   58   802A 51CA FFEE         |        DBF     D2,LOOP2       ;ADJUST OUTPUT POSITION AND
59   59   802E                   |*                             ; GET NEXT DIGIT UNTIL LAST.
60   60   802E                   |*
61   61   802E                   |* DISPLAY THE ASCII RESULT
62   62   802E                   |*
63   63   802E 2A7C 0000         |        MOVE.L  #DECSTR,A5     ;SETUP THE CALL TO OUTPUT THE
63        8032 7F0A              |
64   64   8034 2C7C 0000         |        MOVE.L  #E DECSTR,A6   ;  ASCII RESULT.
64        8038 7F0F              |
65   65   803A                   |        SYSCALL .OUT1CR        ;OUTPUT THE CONVERTED STRING
66   1m   803A 1E3C 00E3         +        MOVE.B  #.OUT1CR,D7
67   2m   803E 4E4E              +        TRAP    #14
68   66   8040                   |*
69   67   8040                   |* KEEP GOING UNTIL LAST BINARY NUMBER
70   68   8040                   |*
71   69   8040 51CB FFCC         |        DBF     D3,LOOP1       ;GET NEXT NUMBER UNTIL EOT
72   70   8044                   |        SYSCALL .RETURN        ;NO MORE NUMBERS TO CONVERT,
73   1m   8044 1E3C 00E4         +        MOVE.B  #.RETURN,D7
74   2m   8048 4E4E              +        TRAP    #14
75   71   804A                   |*                             ;  SO RETURN TO TUTOR
76   72   804A                   |        END
```

Figure 7.18   (continued)

indexing based on the hexadecimal digit in D1. If (D2) $[7:0]$ is 1 upon entry, the BCD digit corresponding to the hexadecimal digit is placed in D2. The value is obtained from the table BCDTAB.

Example 7.13 _____

If a string of decimal digits is stored as $D_{n-1}D_{n-2} \cdots D_0$ in separate bytes in memory, the conversion to binary is easily accomplished because the numerical value is

$$(\cdots (D_{n-1} \times 10 + D_{n-2}) \times 10 + \cdots + D_1 \times 10) + D_0$$

as explained in Chapter 3. The binary sum is formed by computing the terms in parentheses using binary arithmetic and adding each term into the total. Figure 7.17 shows a subroutine to accomplish this conversion. On entry, the decimal digits are assumed to be stored right-justified in the byte locations addressed by A1. If no out-of-range condition occurs, the binary value is returned in D1.

Example 7.14 _____

·The program in Figure 7.18 converts 16-bit unsigned binary numbers in the table defined by label BINNUM and displays the decimal values using the TUTOR I/O macros previously discussed. Before the program executes, the data area at label BINNUM must be loaded with the numbers to be converted. The label EBINNUM defines the end of the table of binary numbers. Several (five) sample values are shown in the program.

The binary-to-decimal conversion algorithm is discussed in Example 3.19 of Chapter 3. As each decimal digit is produced it is converted to ASCII format. After the conversion is complete, the ASCII string of decimal values to be displayed is held in the data area defined by label DECSTR. The string is limited to 5 digits because the maximum binary value in a word is 65535 in decimal.

EXERCISES

7.6.1. Write a subroutine to convert an ASCII string to signed binary representation when the range of the input can be up to eight decimal digits plus a sign.

7.6.2. Write a subroutine to convert a string of binary digits to ASCII.

7.6.3. To convert from binary to decimal, it is possible to follow the procedure shown in this section using the decimal expansion for the number to be converted. Repeatedly dividing the number by 10 will yield the decimal digits, as remainders, in ascending order. Write a subroutine to convert a 16-bit unsigned integer to the equivalent BCD value.

7.6.4. Write a routine to convert a string of ASCII digits to BCD when the input string contains a decimal point ($2E). Determine the number of digits and leave the scale factor in a register; that is, determine the number of decimal places and store the BCD number as an integer.

REVIEW QUESTIONS AND PROBLEMS

Multiple Choice

7.1. After the execution of an ADD or SUB instruction with unsigned operands, the condition code bit that determines if the answer is correct is
  (a) the V bit
  (b) the C bit
  (c) the Z bit

7.2. The branch tests that are valid after ADD.W X,Y when (X) and (Y) are unsigned operands are
  (a) BEQ, BNE, BPL
  (b) BEQ, BMI, BNE
  (c) BEQ, BNE

7.3. The instruction ADDI.B #20,(A1) causes
  (a) adds $20 to the byte addressed by A1
  (b) adds the contents of location 20 to the byte addressed by A1
  (c) adds 20 to the byte addressed by A1

7.4. Let (1000)=$01. Execution of the instruction NEG.B $1000 causes
  (a) (1000)=$FF
  (b) (1000)=$F1
  (c) (1000)=00

7.5. After adding unsigned numbers with the instruction ADD.W (A1)+,D2, which branch instruction branches to ERROR with the sum is out of range for 16 bits?
   (a) BCC ERROR
   (b) BCS ERROR
   (c) BVS ERROR

7.6. Let (D2)[W]=$0002. The result after MULU.W #$10,D2 executes is
   (a) (D2)[W]=20
   (b) (D2)[W]=200
   (c) (D2)[W]=32

7.7. To multiply to 16-bit signed numbers to form a 32-bit product, the instruction used is
   (a) MULU.W
   (b) MULS.W
   (c) MULS.L

7.8. The errors that can occur when a DIV instruction is executed are
   (a) divide by zero
   (b) overflow
   (c) (a) and (b) above

7.9. The instruction MULS.L D0,D6:D7 computes
   (a) the 64-bit product in D6 and D7
   (b) the 32-bit product in D7
   (c) the 32-bit product in D0

7.10. The instruction ABCD D1,D2 computes
   (a) the 16-bit BCD value of the sum in D2
   (b) the 2 digit BCD sum in D2
   (c) the single-digit BCD sum in D2

7.11. If (D1)[B] holds the BCD value 65 and (D2)[B] contains 17 and X=0, the BCD value in (D2)[B] after the instruction ABCD D1,D2 executes read in binary is
   (a) 0101 0010
   (b) 1000 0010
   (c) 1000 0011

7.12. The conversions necessary to display a binary number in memory as a decimal value on a terminal screen are
   (a) binary to BCD and send to terminal
   (b) send binary to terminal
   (c) binary to BCD to ASCII and send to terminal

7.13. The ASCII value $31 represents
   (a) the number 1
   (b) the ASCII value for 1
   (c) the value 31

7.14. If the ASCII E is $45, the ASCII string $45 $4E $44 represents
   (a) END
   (b) ENM
   (c) EOT

7.15. The instruction to multiply a BCD digit in D1 by 10 is
   (a) MULU.W #10,D1
   (b) MULU.W #$10,D1
   (c) MULU.W #1010,D1

Programs

7.16. Write subroutines to perform the basic arithmetic functions on pairs of binary numbers. The input numbers should be 32-bit signed integers and the result should be a 32-bit signed integer. Also return an indication of success or failure (due to overflow or division by zero).
(a) Write a subroutine to add two numbers.
(b) Write a subroutine to subtract two numbers.
(c) Write a subroutine to multiply two numbers.
(d) Write a subroutine to divide one number by another.

7.17. Perform some basic statistical calculations on a table of 100 random integers. Consider using the subroutines developed in Problem 7.16. Use the following data table to test your program.

```
TABLE    DC.W 92,14,27,81,53,87,27,6,68,91
         DC.W 50,40,97,49,42,76,51,50,69,56
         DC.W 10,41,88,67,17,51,58,25,7,86
         DC.W 70,31,53,4,32,72,33,80,35,1
         DC.W 41,31,19,44,44,16,71,16,80,96
         DC.W 93,40,26,1,67,36,86,33,29,94
         DC.W 42,5,32,94,46,82,19,28,4,94
         DC.W 31,33,54,6,20,63,53,38,88,39
         DC.W 32,81,36,29,67,93,68,81,56,72
         DC.W 7,33,5,33,91,28,66,52,37,4
```

Write a program that
(a) finds the sum of the values
(b) finds the average, or mean, of the values
(c) finds the variance of the values with respect to the mean

(*Note:* The variance is the sum of the squares of the deviation from the mean.)

7.18. Consider a 32-bit number of the form [int].[frac] where [int] is a 16-bit signed binary integer and [frac] is an unsigned 16-bit binary fraction (where the most significant bit is the 2^{-1} place). Develop the following routines that input and output numbers of this form:
(a) add two numbers
(b) subtract two numbers
(c) convert from ASCII to the number
(d) convert from a number to ASCII

7.19. Imagine a digital tachometer that outputs a six-digit packed-BCD value representing revolutions per minute (RPM). You are asked to produce a moving average using five consecutive measurements from the tachometer.
(a) Write a program to produce the moving average using a table of 20 tachometer inputs in the range 0 to 30,000. The results should be stored sequentially in a 16-entry table.
(b) Modify the program to accept BCD input in the range −30,000 to 30,000.
(c) Modify the program to accept BCD input in the range −50,000 to 50,000.
Create test data that covers the full range of expected input. Use BCD arithmetic for calculating the sum of five consecutive values. The initial average is calculated using the sum of the first five values, but subsequent averages are calculated by subtracting the oldest value from the sum of the previous five and then adding the new value to that result.

7.20. Using a monitor program such as TUTOR, modify the programs in Problem 7.19 to accept the tachometer measurements as input from the keyboard, and display the moving average on the console.

7.21. Using TUTOR or a similar monitor, create your own version of the classic Lunar Lander game. In this game you control the main engine of a lunar lander falling toward the moon's surface. Prompt for an initial altitude and an initial velocity. At some regular time interval, display the elapsed time, current velocity (negative for descent, positive for ascent), and current altitude, remaining fuel, engine status, and then prompt for thrust duration in seconds (a zero indicates no thrust).

Assume the module holds 10,000 lbs of fuel, the lunar gravity is 5 ft/sec^2, the main engine produces an upward acceleration of 20 ft/sec^2, and that the main engine burns 200 lbs of fuel per second. Continue displaying status until landing module impacts the surface. Here are the equations for velocity and position:

$$v = v_0 + at$$
$$d = d_0 + v_0t + (at^2)/2$$
$$\text{where} \quad v_0 = \text{the initial velocity}$$
$$d_0 = \text{the initial altitude.}$$

FURTHER READING

Knuth's volume on seminumerical algorithms contains a number of useful algorithms and other information for those doing sophisticated mathematical programming. Grappel lists a 32-bit divide routine for the MC68000 in his article. A number of such routines are available, and Motorola representatives usually have a repertoire for customers. The textbook by Stein and Monro presents the rigorous basis for machine arithmetic.

Grappel, Robert D. "68000 Routine Divides 32-Bit Numbers." *EDN* 26, no. 5 (March 4, 1981), 161–162.

Knuth, Donald E. *The Art of Computer Programming,* Vol. 2: *Seminumerical Algorithms.* Reading, Mass.: Addison-Wesley, 1968).

Stein, Marvin L., and William D. Monro. *Introduction to Machine Arithmetic.* (Reading, Mass.: Addison-Wesley, 1971).

CHAPTER 8

Logical and Bit Operations

This chapter introduces three new categories of MC68000 instructions: logical instructions, shift and rotate instructions, and bit-manipulation instructions. The *logical operations* treat an operand as a collection of separate logical variables. This category includes the instructions AND, OR, EOR (Exclusive OR), and NOT. The second category includes the instructions ASL, ASR, LSL, and LSR to *shift* the bits within an operand. Both arithmetic shifts and logical shifts are provided. The instructions ROL, ROR, ROXL, and ROXR rotate the bits of an operand in a cyclic fashion.

Instructions for *bit manipulation* form a separate category of instructions for the MC68000. Separate instructions are provided to test, set to {1}, clear, and change an individual bit within an operand. In order, they have the mnemonics BTST, BSET, BCLR, and BCHG. Two other instructions show the result of a conditional test by modifying an indicator variable called a flag. They are the Set According to Condition (SCC) and the Test and Set (TAS) instructions.

8.1 LOGICAL OPERATIONS

In some applications, it is convenient to treat each bit in an operand as an individual logical variable. The condition code register, for example, contains five independent bits, and the bits may be tested singly. Each logical variable has only two possible states, which are defined variously according to the application as TRUE or FALSE, ON or OFF, or {1} or {0}, among other possibilities. Therefore, the m-bit computer word holds m logical variables. In the MC68000, logical instructions may operate on 8, 16, or 32 such variables simultaneously.

If x and y are considered to be logical variables, the truth tables of Table 8.1 define the operations that correspond to MC68000 logical arithmetic. A collection of m logical variables is written in positional notation

$$(x_{m-1}x_{m-2} \cdot \cdot \cdot x_0)$$

as it would be stored in an m-bit word. This is called an m-tuple of variables. For example, the MC68000 instruction

Table 8.1   Results of Logical Operations

x	y	NOT x	x AND y	x OR y	x EOR y
0	0	1	0	0	0
0	1	1	0	1	1
1	0	0	0	1	1
1	1	0	1	1	0

Note: x and *y* are logical variables. The results for each operation are defined by the "truth" table for the operation. For example, (*x* OR *y*) is true or {1} if either *x* or *y* or both is {1}.

    NOT.W   X

will complement each bit of an operand containing 16 logical variables, in the memory location addressed by X. The other MC68000 instructions for logical operations perform their operation between the logical variables in the source location and those in the destination. The result is stored in the destination location. Thus the operation

    AND.W   D1,D2

leaves the results of the operation between 16 variables in D1 and 16 variables in D2 in the low-order word of D2. The operation performed is

(D2)[15:0]←(D2)[15:0] AND (D1)[15:0]

The logical instructions are listed in Table 8.2, which shows the assembler syntax and addressing modes for each instruction. The instructions with suffix "I" allow only an immediate value for the source operand. None of the instructions operate on address registers. Also, no memory-to-memory operations are possible with the AND, EOR, and OR instructions. The condition code bits C and V are always cleared after any logical operation and the N and Z bits are set according to the result. The Z bit would be set to {1} if the result is all zeros. The N bit is set to {1} if the most significant bit of the result is a {1}.

The AND and OR instructions allow the same addressing modes for their operands. For either instruction, if the destination is a data register, the source operand is addressed by any data mode. This allows all the source addressing modes except address register direct addressing. Alternatively, if Dn contains the source operand, the destination must be addressed by a memory-alterable mode. Thus only register direct and program counter relative addressing are prohibited for the destination.

The ANDI and ORI instructions use an immediate value as the source operand and any data-alterable location for the destination. For example, the instruction

    ANDI.W   #$000F,D1

clears all but the low-order 4 bits of D1. This instruction might be used to isolate a single BCD digit for subsequent mathematical calculations, for example. If the operand is held in memory, it may be addressed by all the addressing modes except PC relative.

Table 8.2 Instructions for Logical Operations

| Syntax | Addressing Modes | |
	Source	Destination
Logical AND		
AND.⟨l⟩ ⟨EA⟩,⟨Dn⟩	Data	⟨Dn⟩
AND.⟨l⟩ ⟨Dn⟩,⟨EA⟩	⟨Dn⟩	Memory alterable
ANDI.⟨l⟩ #⟨d⟩,⟨EA⟩	⟨d⟩	Data alterable
Logical OR		
OR.⟨l⟩ ⟨EA⟩,⟨Dn⟩	Data	⟨Dn⟩
OR.⟨l⟩ ⟨Dn⟩,⟨EA⟩	⟨Dn⟩	Memory alterable
ORI.⟨l⟩ #⟨d⟩,⟨EA⟩	⟨d⟩	Data alterable
Exclusive OR		
EOR.⟨l⟩ ⟨Dn⟩,⟨EA⟩	⟨Dn⟩	Data alterable
EORI.⟨l⟩ #⟨d⟩,⟨EA⟩	⟨d⟩	Data alterable
NOT		
NOT.⟨l⟩ ⟨EA⟩	—	Data alterable

Notes:
1. ⟨l⟩ denotes B, W, or L.
2. ⟨d⟩ is an 8-, 16-, or 32-bit logical variable as an immediate value.
3. The condition code bits C and V are always closed; N and Z are set according to the result.
4. The destination location is modified according to the result.

The Exclusive OR (EOR) instruction has slightly different addressing restrictions not conforming to the requirements for AND and OR. It requires a data register for the source location. Also, the destination must be a data-alterable location, as was the case for both ANDI and ORI. The immediate form, EORI, has the same addressing requirements for the destination operand. It is used to perform EOR between an immediate value and an operand in either a data register or in memory. Clearly, the regularity in addressing of most MC68000 instructions is missing among its logical instructions to some extent.

The NOT instruction complements each bit of an operand in a data register or in memory using any type of memory addressing except PC relative. The NOT can be interpreted as either forming the logical NOT, if the operand is composed of m logical variables, or the one's complement, if the operand is considered to be a number. The mathematical description of the one's-complement operation was given in Chapter 3.

System Control Using Logical Instructions. The immediate forms of ANDI, EORI, and ORI can reference the status register or the condition code register[1] as the destination operand. These forms are usually used for system control and will be discussed in Chapter 10. The reader is referred to that chapter for a complete discussion.

---

[1] Condition code register (CCR) refers to the low-order byte of the status register (SR[7:0]).

Example 8.1

Table 8.3 shows several examples of logical instructions operating on various operands. For simplicity, the examples limit the operand length to 8 bits and show only immediate-to-register or register-to-register operations. Before each instruction executes, the low-order byte of D1 contains the value {1101 0001}, which represents eight logical variables. Similarly, (D2) [7:0] = {1101 0101} initially.

The ANDI instruction, as used in the example, serves to "mask" or set to zero the low-order 4 bits of D1. The ORI instruction does the opposite by setting the designated bits (bits 0 and 1) to {1} and leaving the other bits in the destination unchanged. After either of these instructions execute, condition code bit N would be set to {1} in the examples shown.

The NOT instruction inverts the low-order 8 bits of D1. If the original value in D1 is interpreted as the decimal number −46 in 8-bit, one's-complement notation, the inversion produces +46, as expected.

The EOR instruction causes a logical variable in the result to be set to {1} when the two variables in the corresponding bit positions of the operands are different. For example, if D1 contains the first reading of 8 status bits taken from an external device and D2 contains a second reading taken later, any change in status would be indicated by a nonzero result (Z = {0}) after the EOR operation. If the readings had been the same, the result would have been all zeros with Z = {1}. A conditional branch instruction such as BEQ or BNE could be used after the EOR instruction to determine the subsequent program path.

Example 8.2

To illustrate the use of logical instructions with various addressing modes, this example implements the equations for a "two-line to four-line decoder" using MC68000 instructions. One of four possible output values is determined by the value of two input variables. The input variables are two logical variables, designated as A and B in the example. They are stored in memory in two consecutive bytes with B first and are addressed by (A1). The output is to be four variables designated Y0, Y1, Y2, and Y3 stored in four consecutive bytes addressed by (A2). All the logical variables are right-justified in their locations. That is, the variable {A} has the memory representation {0000 000A}. The truth table and the equations are shown in Table 8.4, which shows that only one output can be {0} for each pair of inputs. If A and B together are interpreted as a two-digit binary number, {BA}, then Yn = {0} when the value of the input is n, for n = 0, 1, 2, or 3. All the other output variables are {1}. Hence the output line corresponding

Table 8.3 Examples of Logical Operations

Instruction	Operands	Result
ANDI.B #$F0,D1	{1111 0000} AND {1101 0001}	(D1)[7:0] = {1101 0000}
ORI.B #03,D1	{0000 0011} OR {1101 0001}	(D1)[7:0] = {1101 0011}
NOT.B D1	NOT {1101 0001}	(D1)[7:0] = {0010 1110}
EOR.B D1,D2	{1101 0001} EOR {1101 0101}	(D2)[7:0] = {0000 0100}

Note: (D1)[7:0] = {1101 0001} and (D2)[7:0] = {1101 0101} before each instruction executes.

Table 8.4   Truth Table for Decoder

Input		Output			
B	**A**	**Y0**	**Y1**	**Y2**	**Y3**
0	0	0	1	1	1
0	1	1	0	1	1
1	0	1	1	0	1
1	1	1	1	1	0

Notes:
Y0 = NOT (NOT A AND NOT B) = A OR B
Y1 = NOT (A AND NOT B) = NOT A OR B
Y2 = NOT (NOT A AND B) = A OR NOT B
Y3 = NOT (A AND B) = NOT A OR NOT B

to the value of the inputs is selected. In this case, the selected line is considered "TRUE" or "ON" when its value is {0}.

The subroutine in Figure 8.1 first transfers A and B to the low-order bytes of D0 and D1, respectively. The complement of each input is then formed in the low-order bit of another register. Each equation for Yn is coded in a straightforward way. The computed result is stored in memory using the appropriate displacement from the base address held in A2.

EXERCISES

8.1.1. Write a simple subroutine to convert ASCII values to BCD, and vice versa, using logical instructions. The conversion was discussed in Chapter 7.

8.1.2. Improve the decoder program used as an example in this section. Make the program more general to allow 4-line to 16-line decoding.

8.1.3. Write a subroutine to exchange the contents of two *m*-bit words in memory using only the EOR instruction and appropriate data transfer instructions.

8.2   SHIFT AND ROTATE INSTRUCTIONS

The shift and rotate instructions of the MC68000 move the bits in an operand to the right or left a designated number of places. The three different possibilities provided include:

(a) arithmetic shifts;
(b) logical shifts;
(c) rotates.

Arithmetic shifts to the left in effect multiply a signed binary integer by a power of 2. Arithmetic right shifts accomplish division by powers of 2. Of course, the shifted number must be within a valid range or the result is in error. Logical shifts are used to shift an *m*-tuple of logical variables right or left. The rotate instructions cause bits shifted off one end of the *m*-tuple to reappear at the other end in a *cyclic* shift.

```
abs.  rel.    LC    obj. code     source line
----  ----   ----  ----------    -----------
  1     1    0000                        |          TTL      'FIGURE 8.1'
  2     2    0000                        |          LLEN     100
  3     3    8000                        |          ORG      $8000
  4     4    8000                        |          OPT      P=M68000
  5     5    8000                        |*
  6     6    8000                        |*  2 TO 4 LINE DECODER
  7     7    8000                        |*  INPUT : (A1.L) = ADDRESS OF 2 BYTES CONTAINING LOGICAL
  8     8    8000                        |*                                 VARIABLES
  9     9    8000                        |*          (A2.L) = ADDRESS OF 4 BYTES FOR DECODED VALUES
 10    10    8000                        |*  OUTPUT : 4 BYTES DECODED INTO LOCATION  POINTED TO
 11    11    8000                        |*                      BY (A2)
 12    12    8000                        |*
 13    13    8000                        |*  NOTE :   LOGICAL VARIABLES ARE STORED IN THE LS BIT
 14    14    8000                        |*                 OF THE BYTE
 15    15    8000                        |*
 16    16          0000 0000     |A        EQU      0
 17    17          0000 0001     |B        EQU      1
 18    18          0000 0000     |Y0       EQU      0
 19    19          0000 0001     |Y1       EQU      1
 20    20          0000 0002     |Y2       EQU      2
 21    21          0000 0003     |Y3       EQU      3
 22    22    8000                        |*
 23    23    8000 48E7 F800     |DECODER MOVEM.L D0-D4,-(SP)   ;SAVE REGISTERS
 24    24    8004                        |*
 25    25    8004 1011                   |          MOVE.B  A(A1),D0      ;GET A
 26    26    8006 1229 0001              |          MOVE.B  B(A1),D1      ;GET B
 27    27    800A                        |*
 28    28    800A 1400                   |          MOVE.B  D0,D2
 29    29    800C 4602                   |          NOT.B   D2
 30    30    800E 0202 0001              |          ANDI.B  #01,D2        ;COMPLEMENT OF A
 31    31    8012                        |*
 32    32    8012 1601                   |          MOVE.B  D1,D3
 33    33    8014 4603                   |          NOT.B   D3
 34    34    8016 0203 0001              |          ANDI.B  #01,D3        ;COMPLEMENT OF B
 35    35    801A                        |*
 36    36    801A 1800                   |          MOVE.B  D0,D4
 37    37    801C 8801                   |          OR.B    D1,D4
 38    38    801E 1484                   |          MOVE.B  D4,Y0(A2)     ;A .OR. B
 39    39    8020                        |*
 40    40    8020 1802                   |          MOVE.B  D2,D4
 41    41    8022 8801                   |          OR.B    D1,D4
 42    42    8024 1544 0001              |          MOVE.B  D4,Y1(A2)     ;NOT A .OR. B
 43    43    8028                        |*
 44    44    8028 1800                   |          MOVE.B  D0,D4
 45    45    802A 8803                   |          OR.B    D3,D4
 46    46    802C 1544 0002              |          MOVE.B  D4,Y2(A2)     ;A .OR. NOT B
 47    47    8030                        |*
 48    48    8030 8602                   |          OR.B    D2,D3
 49    49    8032 1543 0003              |          MOVE.B  D3,Y3(A2)     ;NOT A .OR. NOT B
 50    50    8036                        |*
 51    51    8036 4CDF 001F              |          MOVEM.L (SP)+,D0-D4   ;RESTORE REGISTERS
 52    52    803A 4E75                   |          RTS
 53    53    803C                        |          END
```

Figure 8.1 Program for a two-line to four-line decoder.

Figure 8.2 shows the operation of the arithmetic shifts (ASL, ASR), the logical shifts (LSL, LSR), and the rotate instructions (ROL, ROR). The rotate instructions have extended variations for shifting multiple-precision operands. The instructions ROXL and ROXR include the X bit in the cyclic shift. The assembler language syntax for the instructions is presented in Table 8.5.

Three different formats are available to designate the shift count and the operand to be shifted. The count can be held in a register or specified as an immediate value for operands in a data register. A one-place shift of a word-length operand in memory is allowed. These are shown in Table 8.6. For example, the instruction

LSR.W   D3,D2

shifts the 16-bit operand in the low-order word of D2 to the right by the number of places designated in D3. The number in D3 is treated as modulo 64, so shifts from 0 to 63 places are possible. Of course, after 16 logical shifts left or right in a word-length operand, the value left in the low-order word contains all zero bits. The instruction

LSL.W   #5,D3

performs a left logical shift of the low-order word of D3 by five places. An immediate shift length can range from 1 to 8. The third form of specifying operands allows a word-length operand in memory to be shifted or rotated one place at a time. Any operand referenced by a memory-alterable addressing mode can be manipulated. For example, the instruction

ASR   (A2)

performs a single-place, right arithmetic shift on the 16-bit operand addressed by A2.

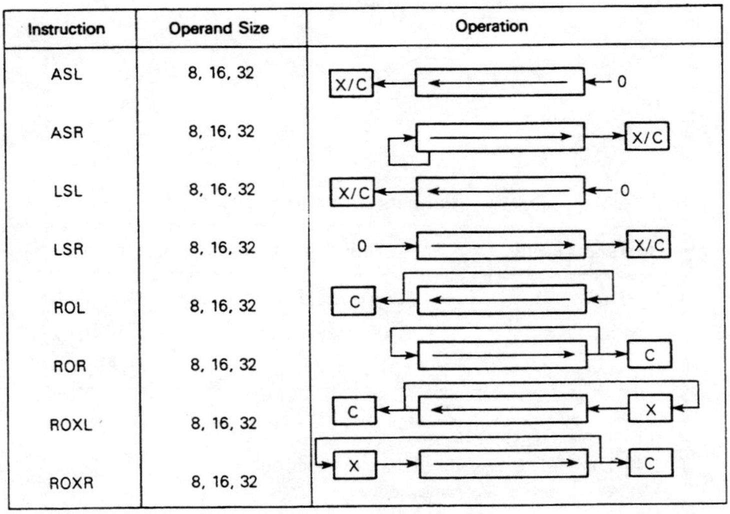

Instruction	Operand Size	Operation
ASL	8, 16, 32	
ASR	8, 16, 32	
LSL	8, 16, 32	
LSR	8, 16, 32	
ROL	8, 16, 32	
ROR	8, 16, 32	
ROXL	8, 16, 32	
ROXR	8, 16, 32	

Figure 8.2   Shift and rotate instructions. (Courtesy of Motorola, Inc.)

Table 8.5 Assembly Language Syntax for Shift and Rotate Instructions

Arithmetic Shift Left		Arithmetic Shift Right	
ASL.⟨l⟩	⟨Dm⟩,⟨Dn⟩	ASR.⟨l⟩	⟨Dm⟩,⟨Dn⟩
ASL.⟨l⟩	#⟨d⟩,⟨Dn⟩	ASR.⟨l⟩	#⟨d⟩,⟨Dn⟩
	ASL ⟨EA⟩		ASR ⟨EA⟩

Logical Shift Left		Logical Shift Right	
LSL.⟨l⟩	⟨Dm⟩,⟨Dn⟩	LSR.⟨l⟩	⟨Dm⟩,⟨Dn⟩
LSL.⟨l⟩	#⟨d⟩,⟨Dn⟩	LSR.⟨l⟩	#⟨d⟩,⟨Dn⟩
	LSL ⟨EA⟩		LSR ⟨EA⟩

Rotate Left		Rotate Right	
ROL.⟨l⟩	⟨Dm⟩,⟨Dn⟩	ROR.⟨l⟩	⟨Dm⟩,⟨Dn⟩
ROL.⟨l⟩	#⟨d⟩,⟨Dn⟩	ROR.⟨l⟩	#⟨d⟩,⟨Dn⟩
	ROL ⟨EA⟩		ROR ⟨EA⟩

Notes:

1. ⟨l⟩ denotes B, W, or L when ⟨Dn⟩ is the destination; ⟨Dm⟩ or #⟨d⟩ specifies the shift count.
2. When the destination is a memory location, only word-length operands are allowed for ⟨EA⟩.
3. Only memory-alterable addressing modes are allowed for ⟨EA⟩, excluding register direct and PC-relative addresses.
4. ROXL and ROXR have the same syntax as the rotate instructions.
5. Condition code bits N and Z are set according to the result; V is cleared except by ASL.

Table 8.6 Operand Formats for Shifts and Rotate

Operand Format	Shift Count	Destination
⟨Dm⟩,⟨Dn⟩	(Dm); range 0–63	(Dn)[l]
#⟨d⟩,⟨Dn⟩	#⟨d⟩; range 1–8	(Dn)[l]
⟨EA⟩	1	(EA)[15:0]

Notes:

1. ⟨l⟩ in B, W, or L for register operands.
2. [l] indicates corresponding bits in the operation.
3. ⟨EA⟩ is a memory-alterable address.
4. A shift count of zero in Dm has a special meaning.

A shift instruction using the shift count held in the data register performs *dynamic shifting* because the shift count can be changed under program control. A shift count of zero will affect only the condition codes, as shown in Appendix IV. For a nonzero shift count, the arithmetic and logical shifts preserve the last bit shifted off in the C and X bits. The rotate instructions affect only the C bit. Rotate with extend operations shift the previous value of the X bit into one end of the operand while saving the latest value rotated out the other end in the C and X bits.

After any shift or rotate operation, the N and Z condition codes are set according to the result, just as they were set for the arithmetic operations. The overflow condition code V is set to {0} after every operation except an arithmetic left shift. The ASL instruction multiplies the operand by 2^r if it is shifted r bits left. If the result exceeds the numerical range of a signed m-bit operand, V set to {1} indicates an out-of-range condition. For unsigned numbers, the carry bit C set to {1} indicates an overflow after an ASL instruction.

Example 8.3

Several shift and rotate instructions are shown in Table 8.7. The results shown in the table assume that the low-order byte of D1 contained the binary value {1110 0101} and that the X bit was {0} before each instruction was executed.

In the case of the ASR instruction, the sign bit is extended to the right at each shift. The original number was −27 in two's-complement notation. It becomes −14 and then −7 after successive shifts, thus simulating integer division by 2 each time. Note that −14, however, is not the expected result of the integer division of −27 by 2. Use of the DIVS instructions would result in a quotient of −13 and a remainder of −1. This truncation in the wrong direction will occur whenever an odd, negative integer is shifted to the right.

The LSL instruction, with a shift count of 4, shifts the low-order 4 bits of the byte to the upper 4 bits. The N condition code bit would be set to {0} to indicate that the most significant bit is zero. Rotating a byte four places to the right with the ROR instruction swaps 4 bits and leaves each unchanged. Rotating the original value with X = {0} one place to the right with the ROXR instruction causes a zero to be shifted into the most significant bit. The bit shifted off the right end is saved in both C and X. Notice from Figure 8.2 that if X = {1} before the ROXR executed, the most significant bit of the operand would become {1}.

Table 8.7 Examples of Shift and Rotate Operations

Instruction	Before (D1)[7:0]	After (D1)[7:0]	C	X
ASR.B #2,D1	{1110 0101}	{1111 1001}	0	0
LSL.B #4,D1	{1110 0101}	{0101 0000}	0	0
ROR.B #4,D1	{1110 0101}	{0101 1110}	0	—
ROXR.B #1,D1	{1110 0101}	{0111 0010}	1	1

Note: X = {0} before each instruction executes.

```
abs. rel.   LC  obj. code   source line
---- ----   ---- ---------   -----------
   1    1   0000                |           TTL      'FIGURE 8.3'
   2    2   0000                |           LLEN     100
   3    3   8000                |           ORG      $8000
   4    4   8000                |           OPT      P=M68000
   5    5   8000                |*
   6    6   8000                |*   POWER OF TWO MULTIPLY
   7    7   8000                |*      INPUT : (D1.B) = POWER OF TWO MULTIPLIER
   8    8   8000                |*              (D4.L) = INTEGER TO BE MULTIPLIED
   9    9   8000                |*                         (MUST BE POSITIVE)
  10   10   8000                |*
  11   11   8000                |*    OUTPUT : (D5/D4)[63:0] = RESULT
  12   12   8000                |*               (D2.B)  = 0 : SUCCESS
  13   13   8000                |*                        -1 : ERROR DETECTED
  14   14   8000                |*
  15   15   8000 2F01           |POWER2  MOVE.L   D1,-(SP)   ;SAVE REGISTER
  16   16   8002 4285           |        CLR.L    D5         ;ZERO RESULT
  17   17   8004 143C 00FF      |        MOVE.B   #-1,D2     ;SET DEFAULT TO ERROR
  18   18   8008                |*
  19   19   8008 E38C           |LOOP    LSL.L    #1,D4      ;SHIFT LEFT (MULTIPLY BY 2)
  20   20   800A E395           |        ROXL.L   #1,D5      ;SHIFT CARRY OUT OF D4 INTO D5
  21   21   800C 6500 0008      |        BCS      DONE       ;IF CARRY OUT OF D5,
  22   22   8010                |*                                EXIT WITH ERROR
  23   23   8010 5301           |        SUBI.B   #1,D1      ;DECREMENT COUNT
  24   24   8012 66F4           |        BNE      LOOP       ;   UNTIL (D1) = 0
  25   25   8014                |*
  26   26   8014 4202           |        CLR.B    D2         ;SET STATUS TO SUCCESS
  27   27   8016                |*
  28   28   8016 221F           |DONE    MOVE.L   (SP)+,D1   ;RESTORE REGISTER
  29   29   8018 4E75           |        RTS
  30   30   801A                |*
  31   31   801A                |        END
```

Figure 8.3   Program to multiply a number by 2^N.

Example 8.4

The subroutine shown in Figure 8.3 multiplies the 32-bit positive integer in D4 by a positive power of 2 specified in data register D1. The number is passed to the subroutine in D4 and the 64-bit result is held in registers D5 and D4. If the result causes a carry out of D5, the low-order byte of D2 is set to −1 to indicate the overflow. Otherwise, the error flag in D2 is zero. The loop is an example of dynamic shifting because the power of 2 in D1 is used as a counter and is decremented after each shift.

EXERCISES

8.2.1. Why do the MC68000 instructions allow shifts of up to 63 places when the longest operand is only 32 bits?

8.2.2. State in words and use an equation to define the conditions for an error when a *signed* integer is shifted left or right by the ASL or ASR instructions. Consider both even and odd integers.

8.2.3. Determine the numerical value when the binary number

{1110 0101}

is shifted left three places.

8.2.4. Show that the correct rule for doubling a one's-complement number is a left cyclic shift.

8.2.5. How can the two upper bytes in a 32-bit register be accessed for use by a MC68000 instruction operating on a byte-length operand?

8.2.6. Write subroutines that pack and unpack 4-bit operands into byte-length values. Assume that the operands are held in memory.

8.2.7. Modify the subroutine shown in Example 8.4 to multiply signed integers by 2^N.

8.2.8. Write a subroutine to multiply a signed integer by 2^r, where r can be positive or negative. If a negative value is truncated by right shifts, correct the result for proper integer division if necessary.

8.3   BIT-MANIPULATION AND FLAG-SETTING INSTRUCTIONS

It is convenient in many applications to employ a logical variable to indicate one of two possible results of an operation. A logical variable used in this way is called a *flag*. The flag variable indicates that a condition has occurred and is used to communicate this fact to a routine that must test the flag variable. For example, the flags that indicate the results of arithmetic operations for the MC68000 collectively form the condition code register. In this case, a conditional branch instruction may test one or more condition codes to determine subsequent action. Another common application of a flag is to indicate the status of a peripheral device. The status would determine if the device were "busy" or ready to accept data transfers. Usually, a single bit is set by the device when it is ready and this information is used by the processor to begin the transfer of data. In such an application, the logical variable can be termed an *event flag* and it serves to synchronize the operation of the processor with the external device. The MC68000 instructions to test, set, clear, or change a single bit within an operand are useful for manipulating such flag bits as well as for other operations.

Another important use of logical variables is to communicate information between independent programs or between routines of an operating system or other sophisticated software systems. The conditions for which a flag would be set or cleared can be very complicated and involve a number of conditional tests. The MC68000 provides a Set According to Condition (Scc) instruction, which allows a variable to be set to indicate a TRUE or FALSE result based on the condition code values.

When operating system routines or several processors share an area of memory, a synchronization problem can occur if a routine or processor can access a memory area or location during the time another routine or processor is using this same memory area or location. For example, this can occur in a real–time system where the program execution sequence is controlled in part by external events, signaled to the processor as an interrupt. In a multiprocessor system, the synchronization problem requires a partial hardware solution if the processors act independently. The TAS instruction is useful for such situations.

8.3.1 Bit-Manipulation Instructions

The MC68000 bit-manipulation instructions operate on a single bit in a Data Register or in memory. Each instruction tests a specified bit and sets only the Z condition code bit based on the value. As a logical variable, the Z bit indicates the *complement* of the designated bit. The Bit Test (BTST) instruction sets the Z bit according to the state of the tested bit. The instructions BCHG, BCLR, and BSET set the Z bit accordingly and then cause the designated bit to be changed, cleared, and set to {1}, respectively. The bit-manipulation instructions are shown in Table 8.8.

The number of the bit to examine is either contained in a data register for "dynamic" specification or it is specified as an immediate value in the instruction. For example, if

(D1) = $0000 0001
$\qquad$ and
(D2) = $0000 88FF

to specify bit 1 of register D2, the instruction

BCHG$\qquad$D1,D2

first causes the Z condition code to be set to {0}, the complement of (D2)[1]. Then the operand in D2 becomes $0000 88FD because bit 1 of D2 is complemented by the

Table 8.8 Bit-Manipulation Instructions

Syntax	Operation
Bit Change	
BCHG ⟨Dn⟩,⟨EA⟩	Z = NOT(bn)
BCHG #⟨bn⟩,⟨EA⟩	THEN bn ← NOT(bn)
Bit Clear	
BCLR ⟨Dn⟩,⟨EA⟩	Z = NOT(bn)
BCLR #⟨bn⟩,⟨EA⟩	THEN bn ← {0}
Bit Set	
BSET ⟨Dn⟩,⟨EA⟩	Z = NOT(bn)
BSET #⟨bn⟩,⟨EA⟩	THEN bn ← {1}
Bit Test	
BTST ⟨Dn⟩,⟨EA⟩	Z = NOT(bn)
BTST #⟨bn⟩,⟨EA⟩	

Notes:

1. If ⟨Dn⟩ is the destination, the length is 32 bits; otherwise, the length of the destination operand is 1 byte.
2. ⟨bn⟩ is the bit number of the operand.
3. BTST allows all addressing modes for the destination except address register direct (data modes only).
4. BCHG, BCLR, and BSET allow only data-alterable addressing modes for the destination.

operation. The value in D1 should be a number between 0 and 31 because the destination is another data register. The instruction

BCLR   #1,D2

would have the same effect on D2.

When the operand is held in memory, only a byte-length operand is allowed. The BCHG, BCLR, and BSET instructions allow only data-alterable addressing modes. Therefore, address register direct and program counter relative addressing is prohibited. The BTST instruction, however, allows all addressing modes except address register direct.

Example 8.5

The subroutine shown in Figure 8.4 generates odd parity for a single ASCII character held in a location addressed by A1 and leaves the result in the same location. The odd-parity bit (bit 7) is defined to be {0} when the number of {1} bits in the character is odd. Otherwise, the parity bit is set to {1} so that the number of nonzero bits in the byte is always an odd number.

The routine begins by setting (D0) = $FF, indicating that the parity bit should be {1}. In the loop, encountering a {0} bit has no effect and the next bit in sequence is tested. The bits

```
abs. rel.    LC   obj. code    source line
---- ----    ----  ---------    -----------
   1    1    0000                |         TTL    'FIGURE 8.4'
   2    2    0000                |         LLEN   100
   3    3    8000                |         ORG    $8000
   4    4    8000                |         OPT    P=M68000
   5    5    8000                |*
   6    6    8000                |*  PARITY GENERATOR
   7    7    8000                |*  INPUT : (A1.L) = ADDRESS OF CHARACTER
   8    8    8000                |*
   9    9    8000                |*  OUTPUT : PARITY BIT (BIT 7) ADDED TO CHARACTER AT (A1)
  10   10    8000                |*
  11   11    8000 48E7 C000      |PARITY   MOVEM.L  D0/D1,-(SP)    ;SAVE REGISTERS
  12   12    8004 123C 0006      |         MOVE.B   #6,D1          ;SET UP COUNTER
  13   13    8008 103C 00FF      |         MOVE.B   #$FF,D0        ;PARITY INDICATOR = 1'S
  14   14    800C                |*
  15   15    800C 0311           |LOOP     BTST     D1,(A1)        ;CHECK BIT
  16   16    800E 6700 0004      |         BEQ      NEXT           ;IF NOT ZERO
  17   17    8012 4600           |         NOT.B    D0             ;   THEN COMPLEMENT PARITY
  18   18    8014                |*                                         INDICATOR
  19   19    8014 51C9 FFF6      |NEXT     DBRA     D1,LOOP        ;CONTINUE, UNTIL (D1) = -1
  20   20    8018                |*
  21   21    8018 4A00           |         TST.B    D0             ;IF PARITY SHOULD BE 0
  22   22    801A 6700 0006      |         BEQ      DONE           ;   THEN SKIP (DO NOT CHG BIT)
  23   23    801E 08D1 0007      |         BSET     #7,(A1)        ;   ELSE SET PARITY BIT
  24   24    8022                |*
  25   25    8022 4CDF 0003      |DONE     MOVEM.L  (SP)+,D0/D1    ;RESTORE REGISTERS
  26   26    8026 4E75           |         RTS
  27   27    8028                |*
  28   28    8028                |         END
```

Figure 8.4   Subroutine to generate an odd-parity bit.

are tested in order 6, 5, 4, 3, 2, 1, and 0. When a {1} bit is found, (D0) is complemented. Because (D0) was initialized to indicate that a parity bit was needed, finding 1, 3, or 5 bits with value {1} will cause the routine to complement the initial value in D0. Otherwise, if 0, 2, 4, or 6 bits have the value {1}, (D0) will indicate that the parity bit must be set to {1}. The BSET instruction accomplishes this.

8.3.2 Set According to Condition Instruction

The instruction Scc sets all 8 bits of a destination location to {1}s if the condition "cc" is true. Otherwise, the byte is set to all {0}s when the condition is false. The conditions (carry clear, carry set, etc.) are the same as those for the DBcc instruction discussed in Chapter 6. For example, if the Z condition code bit is {1}, the instruction Set if Equal (to zero)

SEQ   D1

writes $FF into the low-order byte of D1, indicating a TRUE condition to any program testing the flag in D1.

The destination ⟨EA⟩ for the instruction must be a data-alterable location. Thus all addressing modes except address register direct and program counter relative are allowed.

After the operand is set to all {1}s or all {0}s, its condition might be tested using the TST.B instruction, which sets Z = {1} when the flag has the value {0000 0000}. In fact, all of the logical instructions operating on a byte-length operand can be used to manipulate the logical variable created by the Scc instruction. For example, if D1 contains the value $FF from the SEQ instruction just shown, the instruction

EORI.B   #$FF,D1

causes Z = {1} and reverses the flag.

8.3.3 Test and Set Instruction

The TAS instruction is used to test and modify a byte–length operand held either in a data register or in memory. Its operation is defined in Figure 8.5(a) and the instruction has the symbolic form

TAS   ⟨EA⟩

If the operand is zero, the condition code bit Z is set to {1}. Otherwise, the Z bit is cleared. If bit 7 of the operand is {1}, N is set to {1}. In this regard, TAS operates much like the Test (TST) instruction operating on a byte value. However, after the operand is examined by the TAS instruction and N and Z are set accordingly, the most significant bit of the operand is set to {1}.

For example, if the byte used as a flag and addressed by A1 has the initial value of $00, the instruction

TAS   (A1)

causes Z = {1} and changes the operand to $80 after the instruction is executed. If the initial zero value indicated that a memory area or other resource was free for use, the Z condition code bit indicates this after the TAS instruction has executed. However, now the flag has been altered. A subsequent test of the flag would indicate that the resource is in use. This subsequent test might be made by another program running concurrently (perhaps activated by an interrupt) or by another processor in a multiprocessor system.

Notice that if a flag variable set to $80 is tested in a loop such as

```
LOOP   TAS     (A1)        ;test flag
       BNE     LOOP        ;branch back if not clear
```

the next instruction in sequence cannot be executed until some other program or processor clears the flag. When the flag is already set as in this example, the TAS instruction does not change it.

Figure 8.5   TAS instruction.

If (EA) = 0
    THEN set Z = {1}
    ELSE set Z = {0}

If (EA)[7] = {1}
    THEN set N = {1}
    ELSE set N = {0}

Set (EA)[7] = {1}

Notes:
(1) (EA) is a byte-length operand addressed by data alterable addressing modes.
(2) The read-modify-write cycle is indivisible.

(a) TAS Operation   (TAS < EA >)

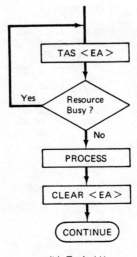

(b) Typical Usage

As noted in Figure 8.5(a), the operand being tested by TAS cannot be held in an address register or be referenced using program counter relative addressing. Also, the C and V bits are always cleared by the TAS operation. The flowchart in Figure 8.5(b) shows a typical use of the TAS instruction, followed by a conditional test. Here a program is testing to determine if some resource such as a line printer or memory area is available to it. The conditional test is most likely implemented by a conditional branch instruction that causes looping until the flag is clear. With such usage, the processor executing the test is busy executing the loop until either an interrupt occurs that allows the flag to be cleared or until another processor clears it.

If the resource being shared is critical to the continued execution of the program making the test, a wait loop is necessary with the TAS instruction. After the resource is made available, processing can continue utilizing the resource. The flag should be cleared after processing is completed to release the resource for other users. Thus the TAS instruction may be used as a means to set flags used to communicate between processes, such as real-time routines, or interrupt routines of an operating system. It

```
abs. rel.     LC    obj. code    source line
----  ----    ----  ---------    -----------
   1     1    0000                      |           TTL      'FIGURE 8.6'
   2     2    0000                      |           LLEN     100
   3     3    8000                      |           ORG      $8000
   4     4    8000                      |           OPT      P=68000
   5     5    8000                      |*
   6     6    8000                      |*   RESOURCE SHARING USING THE TAS INSTRUCTION
   7     7    8000                      |*   INPUT  : FREEMEM = ADDRESS OF FREE SPACE AREA
   8     8    8000                      |*
   9     9    8000                      |*   OUTPUT : (D1.B) = 0 : FOUND A BLOCK
  10    10    8000                      |*                  NOT 0 : NO BLOCK IS FREE
  11    11    8000                      |*           (A1.L) = ADDRESS OF THE BLOCK
  12    12    8000                      |*
  13    13          0000 9000   |FREEMEM EQU       $9000           ;SET UP FREE AREA
  14    14    8000  2F03         |SHARE  MOVE.L    D3,-(SP)        ;SAVE REGISTER
  15    15    8002  4201         |       CLR.B     D1              ; SET DEFAULT TO FOUND
  16    16    8004  227C 0000    |       MOVEA.L   #FREEMEM,A1     ;STARTING ADDRESS OF MEMORY
  16          8008  9000         |
  17    17    800A  7607         |       MOVEQ     #7,D3           ;SET UP COUNTER
  18    18    800C               |*
  19    19    800C  4AD1         |LOOP   TAS       (A1)            ;IF BLOCK IS FREE
  20    20    800E  6700 000E    |       BEQ       FOUND           ;   THEN EXIT WITH ADDRESS
  21    21    8012  D3FC 0000    |       ADDA.L    #256,A1         ;   ELSE ADVANCE TO NEXT BLOCK
  21          8016  0100         |
  22    22    8018               |*                                ;       AND CHECK FLAG
  23    23    8018  51CB FFF2    |       DBRA      D3,LOOP         ;CONTINUE UNTIL (D3) = -1
  24    24    801C               |*
  25    25    801C  1203         |       MOVE.B    D3,D1           ;SET STATUS TO NOT FOUND
  26    26    801E               |*
  27    27    801E  261F         |FOUND  MOVE.L    (SP)+,D3        ;RESTORE REGISTER
  28    28    8020  4E75         |       RTS
  29    29    8022               |*
  30    30    8022               |       END
```

Figure 8.6 Memory allocation subroutine.

has further use, however, to *synchronize* accesses to a shared resource when timing is a consideration. It is the synchronization aspect of the TAS instruction that is critically important in many multiprocessor systems in which several processors share resources such as memory areas.

Synchronization with the TAS Instruction. The TAS instruction is used to prevent access by other programs to a shared resource when one program has control of the resource. This is sometimes termed *lockout*. When hardware access to the shared resource is possible, as in a multiprocessing system, a problem could arise if two processors examine the flag simultaneously and both consider the resource to be available. To prevent this, the TAS instruction has an indivisible, read-modify-write cycle. Once the operand is addressed by the MC68000 executing the TAS instruction, the system bus is not available to any other device, including another processor, until the instruction completes. The first processor executing the TAS instruction controls the shared resource until the flag is cleared. Thus accesses are synchronized at the hardware level. Details of the hardware operation of the MC68000 are discussed in Chapter 13.

Example 8.6

The subroutine shown in Figure 8.6 illustrates the use of the TAS instruction to allocate and lock a block of memory locations for the calling program. The memory is segmented into eight blocks of 256 bytes with the blocks numbered 0, 1, 2, . . . , 7. The first byte of each block contains a flag or "lock" used by the TAS instruction.

On entry to the subroutine, the label FREEMEM must define the start of the free-space area. Up to eight 256-byte memory blocks are examined to determine whether one is available for use. If a free block is located, its lock byte is set by TAS, its address is returned in A1 and D1 is set to {0} to indicate a free block was found. If no free block is located, D1 is returned with a nonzero value.

EXERCISES

8.3.1. Give the bit test needed to determine the following:
   (a) An integer is even.
   (b) A signed integer m bits long is negative.
   (c) A character is an uppercase or lowercase ASCII alphabetic character.

8.3.2. Compare the following instructions with their equivalent bit manipulation instructions:
   (a) ANDI.W   #$7FFF,D2
   (b) ORI.W   #$8000,D2
   (c) EORI.W   #$8000,D2
   (d) TST.W   D2

8.3.3. Give the steps needed to generate the arithmetic (or logical) values {1} and {0} from the TRUE and FALSE conditions set by Scc instructions.

8.3.4. Write a subroutine to change a string of alphabetic characters (ASCII) from uppercase to lowercase, or vice versa. See Appendix I, which contains the character set.

8.3.5. Compare the hardware and software operation of the

TAS   ⟨EA⟩

instruction with the

BSET   #7,⟨EA⟩

instruction. Determine under what conditions a busy-wait loop using the TST and BNE instructions followed by the BSET instruction rather than TAS might not be sufficient to provide a lock of a shared resource. Consider both single- and multiple-processor systems.

8.3.6. Let a "bit map" contain N bits, indicating the availability of N 256-byte blocks of memory. If a bit is {0}, the block is free. Assume that the bit map is held in memory at a fixed location and the first address of the contiguous memory area is known. Write a subroutine to find the first collection of k contiguous blocks. Return the starting address of these blocks and set the proper bits in the bit map to {1} to indicate the memory blocks are in use. If fewer than k blocks are found or if k is not an integer between 1 and N, return an error indication to the calling program.

REVIEW QUESTIONS AND PROBLEMS

Multiple Choice

8.1. The instructions NOT, AND, OR, and EOR are used for
  (a) Boolean algebra operations
  (b) system control
  (c) both (a) and (b)

8.2. The instruction ANDI.W #$000F,D1
  (a) clears all but the low-order 4 bits of D1
  (b) clears the low-order word of D1
  (c) adds 15 to the contents of D1

8.3. Let (D1)[7:0]={1101 0001}. The instruction ORI.B #03,D1 yields D1[7:0]=
  (a) {0000 0011}
  (b) {1101 0011}
  (c) {1101 0000}

8.4. The expression NOT(NOT A AND NOT B) is equivalent to
  (a) NOT(A OR B)
  (b) A OR B
  (c) NOT A OR NOT B

8.5. The EOR instruction yields a TRUE value when the corresponding logical variables are
  (a) the same
  (b) ONE's
  (c) different

8.6. If (D1)[7:0]={1110 0101} before, the instruction ASR.B #2,D1 yields in D1[7:0]
  (a) {0011 1001}
  (b) {1111 1001}
  (c) {1001 0100}

8.7. Let (D1)=$0000 0001 and (D2)=$0000 88FF. The instruction BCHG.L D1,D2 yields
- **(a)** $0000 88FD
- **(b)** $0000 88FC
- **(c)** $0000 88FE

8.8. Suppose that the instruction SEQ D1 sets the flag in D1 to TRUE. The result in D1 is
- **(a)** {1}
- **(b)** {1000 0000}
- **(c)** {1111}

8.9. The instruction TAS (A1)
- **(a)** sets the operand addressed by A1 to $80
- **(b)** creates an indivisible, read-modify-write cycle
- **(c)** does both (a) and (b)

8.10. Which instruction is equivalent to ANDI.W #$7FFF,D2?
- **(a)** BSET #15,D2
- **(b)** BCLR #15,D2
- **(c)** BCHG #15,D2

Programs

8.11. Given a table of 100 random integers in the range 1–100, count the number of entries found in the first quartile (1–25) and in the last quartile (76–100). Use the data table from Problem 7.17.

8.12. Write a subroutine that compares two strings, A and B, and returns a value −1 if A < B, 0 if A = B, and 1 if A > B. Use "C-style" strings that are terminated by a null character ($00). Pass the address of each string to the subroutine and return the result in a data register.

8.13. Consider a printer interface that has a status byte containing flag bits indicating the state of the device. The meaning of each flag bit in the status byte is as follows:
- **(a)** Bit 7 is set if the printer is busy processing the last input.
- **(b)** Bit 6 is set if there was an error processing the last input.
- **(c)** Bit 5 is unused.
- **(d)** Bit 4 is unused.
- **(e)** Bit 3 is set if the printer is in compressed-print mode.
- **(f)** Bit 2 is set if the printer is in graphics mode.
- **(g)** Bit 1 is set if the printer is out of paper.
- **(h)** Bit 0 is set if the printer is offline.

Write a subroutine to display the meaning of the status byte. Call the subroutine for each value in the following table.

```
VALUES   DC.B   $00, $80, $80, $00, $04
         DC.B   $0C, $08, $88, $C8, $09
         DC.B   $00, $04, $74, $06, $07
```

FURTHER READING

A number of the textbooks referenced in previous chapters deal with logical and bit operations (Eckhouse and Morris, Tannenbaum, and Wakerly, for example). The mathematics of shifting is discussed by Stein and Munro, as referenced in Chapter 3.

Discussions of the concept of interlocking shared resources are generally found in texts dealing with operating systems. The article by Denning, listed here, defines in greater detail some of the concepts introduced in this chapter.

Denning, Peter J. "Third Generation Computer Systems." *Computing Surveys* 3, no. 4 (December 1971), 175–216.

CHAPTER 9

Programming Techniques

In the preceding chapters, the MC68000 instruction set was introduced by separating the instructions into categories. These categories discussed instructions that are used to create programs to move data, perform calculations, or provide simple functions. For the most part, these programs were designed to satisfy a particular application, such as ASCII-to-binary conversion. In contrast, this chapter explores various programming techniques that are useful in creating more sophisticated programs. The emphasis is on the powerful addressing capability of the MC68000.

Some MC68000 instructions, such as DIVU and MULU, prohibit operations on address registers directly. This restriction is very slight because address registers and data registers can be easily exchanged. In addition, however, the MC68000 provides special instructions to manipulate addresses. These instructions are introduced in the first section because they provide the flexibility necessary for advanced programming techniques.

One special technique involves the creation of *position-independent code*. Program counter relative addressing and base register addressing can be used in programs so that they will execute independent of their starting address in memory.

Manipulation of typical *data structures* such as arrays and lists involves the use of advanced programming methods. Although only a few of the many topics concerning the creation and manipulation of data structures is discussed in this chapter, the power and flexibility of the MC68000 instructions are revealed.

Subroutines were introduced in Chapter 6 as a common technique to aid the programmer in creating modular programs. The linkage between a calling program and a subroutine via the return address on the system stack was discussed there in some detail. Passing data values or addresses between program modules was accomplished using processor registers to hold the values. More sophisticated methods of transferring data values are discussed in this chapter. The techniques to create a stack frame (LINK and UNLK instructions) and to write reentrant subroutines are also presented.

9.1 INSTRUCTIONS THAT MANIPULATE ADDRESSES

Except for values in registers or values specified by the immediate mode of addressing, operands of MC68000 instructions are referenced by their addresses. A distinction is also made between data and addresses because they have separate output signal lines and are held in data registers and address registers, respectively. The address signal lines specify a location in memory (or in the I/O space of the system) and the separate data signal lines are used to transfer values. Internally, both data values and addresses can be manipulated with various processor instructions.

The MC68000 instruction set provides instructions to operate specifically on addresses. These include instructions to perform a comparison of two addresses (CMPA), to perform arithmetic operations (ADDA, SUBA), and to transfer addresses (MOVEA, LEA, PEA). Except for the CMPA instruction, these instructions do not alter the condition code register.

For all operations involving addresses, the valid address range for the MC68000 is 0 to $FFFFFF (hexadecimal) due to the limit of 24 address lines. For instructions that allow word-length operands, the 16-bit addresses are sign-extended to 32 bits before being used by the instructions. Therefore, short addresses (16 bits) have a valid range of either 0 to $7FFF or $FF8000 to $FFFFFF.

9.1.1 Arithmetic Address Manipulation

The instructions Add Address (ADDA), Subtract Address (SUBA), and Compare Address (CMPA) operate on a source operand that can be addressed by any mode. However, the destination operand must be held in an address register. The syntax and operation for these instructions are shown in Table 9.1. These instructions are similar to the ADD, SUB, and CMP operations on data values.

The ADDA instruction adds a value in a register or memory to the value in the destination address register. A common use of the ADDA instruction is to add a constant, which is specified by the immediate addressing mode, to the value in the address register involved. For example, the instruction

ADDA.L   #20,A1

Table 9.1 Operation of ADDA, SUBA, and CMPA

Instruction	Syntax	Operation	Condition Codes Affected
Add Address	ADDA.⟨1⟩ ⟨EA⟩,⟨An⟩	(An) ← (An) + (EA)	None
Subtract Address	SUBA.⟨1⟩ ⟨EA⟩,⟨An⟩	(An) ← (An) − (EA)	None
Compare Address	CMPA.⟨1⟩ ⟨EA⟩,⟨An⟩	(An) − (EA)	N, Z, V, C

Notes:

1. ⟨1⟩ denotes W or L only.
2. If a word operand (W) is specified, it is sign-extended to 32 bits.
3. All addressing modes are allowed for ⟨EA⟩.

increments (A1) by 20 when it is executed. In a loop, the instruction allows A1 to address every twentieth element in a data structure. The instructions that operate on addresses provide a much greater flexibility, however, because the source operand can be specified by any addressing mode. For example, an instruction in the form

ADDA.L   A2,A2

doubles the value in A2. This could be used to convert an entry number held in A2 to an index that references a table of 16-bit words. For example, if (A2) = \$100 indicates the 256th word in a memory block, then $2 \times \$100$, or \$200, is the address of that word as an offset (in bytes) from the starting location of the block.

    The SUBA instruction forms the difference between a destination address register and the source operand. The instruction

SUBA.L   D1,A1

leaves the 32-bit value of

    (A1) − (D1)

in A1. Like the ADDA instruction, SUBA does not modify the condition codes.

    The CMPA instruction is used to determine the order of two addresses (lower, equal, or higher). The condition codes are set in the same manner as for the CMP instruction discussed in Chapter 6. Addresses should be considered as unsigned integers and tests on the C bit and the Z bit have the interpretation as explained in Chapter 6. The tests BHI (higher), BLS (lower or same), BNE (not equal), or BEQ (equal) are also useful to implement tests comparing addresses. For example, the instructions

CMPA.L   A2,A1   ;form (A1) − (A2)
BHI   LOOP   ;branch if (A1) > (A2)

cause a branch if (A1) is larger than (A2). Otherwise, a branch is not taken.

    The CMPA instruction calculates

(destination) − (source)

and sets the condition codes based on the result. Table 9.2 summarizes the conditional branch instructions that are valid after a CMPA instruction.

Table 9.2   Comparison of Addresses

Instruction	Branch	True Condition
CMPA.L A1,A2	BHI (higher)	(A2) > (A1)
	BLS (lower or same)	(A2) ≤ (A1)
	BCC (high or same)	(A2) ≥ (A1)
	BCS (low)	(A2) < (A1)
	BNE (not equal)	(A2) ≠ (A1)
	BEQ (equal)	(A2) = (A1)

The instruction variations Add Quick (ADDQ) and Subtract Quick (SUBQ) can also be used to add and subtract a constant in the range 1 to 8 from the contents of an address register. The source operand is the immediate value of the constant as discussed in Chapter 7. These instructions represent one-word instructions that are used to efficiently increment or decrement an address.

Example 9.1 _____

The subroutine of Figure 9.1 counts the number of negative 8-bit integers in a block or table of consecutive memory bytes. The starting address of the block is passed to the subroutine in A0, and A1 must contain the address of the last byte + 1 upon entry. Each byte is then tested and the count accumulated in the low-order word of D0.

In each loop, A0 is incremented by 1 with the ADDA instruction. The testing terminates when the value in the addressing register A0 becomes equal to the final address in A1.

_____

9.1.2 Transfer of Addresses

The MOVEA, LEA, and PEA instructions are used to transfer addresses. The MOVEA (Move Address) instruction loads an address register with an operand that can be held in a register or memory location or be specified as an immediate value. The LEA (Load Effective Address) instruction computes the effective address of a

```
abs. rel.   LC   obj. code    source line
---- ----   ---  ---------    -----------
   1    1   0000                     TTL      'FIGURE 9.1'
   2    2   0000                     LLEN     100
   3    3   8000                     ORG      $8000
   4    4   8000                     OPT      P=M68000
   5    5   8000              *
   6    6   8000              *  COUNT THE NUMBER OF NEGATIVE FIRST BYTES IN TABLES
   7    7   8000              *
   8    8   8000              *  INPUT :  (A0.L) = STARTING ADDRESS OF FIRST TABLE
   9    9   8000              *           (A1.L) = ENDING ADDRESS OF LAST TABLE
  10   10   8000              *
  11   11   8000              *  OUTPUT : (D0.W) = NUMBER OF NEGATIVE BYTES
  12   12   8000              *
  13   13   8000 2F08         CNTNEG   MOVE.L   A0,-(SP)      ;SAVE REGISTER
  14   14   8002 4240                  CLR.W    D0            ;CLEAR COUNTER
  15   15   8004              *
  16   16   8004 4A10         LOOP     TST.B    (A0)          ;IF BYTE IS NOT NEGATIVE,
  17   17   8006 6A02                  BPL.S    NEXT          ;   THEN PROCESS NEXT TABLE
  18   18   8008 5240                  ADDQ.W   #1,D0         ; ELSE COUNT IT
  19   19   800A D1FC 0000    NEXT     ADDA.L   #12,A0        ;INCREMENT TO NEXT TABLE
  19        800E 000C
  20   20   8010 B3C8                  CMPA.L   A0,A1         ;IF NOT FINISHED,
  21   21   8012 62F0                  BHI      LOOP          ;   THEN CONTINUE
  22   22   8014 205F                  MOVE.L   (SP)+,A0      ;RESTORE REGISTER
  23   23   8016 4E75                  RTS
  24   24   8018                       END
```

Figure 9.1 Examples of address manipulation.

location in memory and transfers the address value to an address register. The PEA (Push Effective Address) instruction calculates an effective address and pushes it onto the system stack. LEA and PEA differ from most instructions because the effective address calculated for an operand is transferred rather than the contents of the addressed location. The syntax and addressing modes for these three instructions that transfer addresses are shown in Table 9.3.

Move Address Instruction.   The MOVEA instruction has the symbolic form

MOVEA.⟨l⟩   ⟨EA⟩,⟨An⟩

where ⟨l⟩ = W or L and ⟨EA⟩ is designated by any addressing mode. The operation results in the transfer

$$(An)[31:0] \leftarrow (EA)$$

with any 16-bit references being sign-extended to 32-bit quantities before the transfer. When a data register, an address register, or a memory location is the source, the contents of it are transferred. The instruction

MOVEA.L   TABLE,A1

moves the 32-bit word at location TABLE into A1. The value in this location is considered an address, and TABLE in the discussion might refer to the first address of a table of addresses. Register A1 can then be used to reference the value pointed to by the address at location TABLE. For example, if the MOVEA instruction given above is followed with the instruction

MOVE.W   (A1),D1

the 16-bit word referenced by the operand (address) at location TABLE is transferred to the low-order word of D1. The two operations thus perform the transfer

$$(D1)[15:0] \leftarrow ((TABLE)[31:0])[15:0]$$

Table 9.3   Instructions to Transfer Addresses

Instruction	Syntax	Operand Length (bytes)	Addressing Modes Source	Addressing Modes Destination
Move Address	MOVEA.⟨l⟩ ⟨EA⟩,⟨An⟩	16 or 32	All	An
Load Effective Address	LEA ⟨EA⟩,⟨An⟩	32	Control modes	An
Push Effective Address	PEA ⟨EA⟩	32	Control modes	−(SP)

Notes:

1. Condition codes are not affected.

2. ⟨l⟩ denotes W or L only.

3. For word-length operands, the source operand is sign-extended to 32 bits, and all 32 bits are loaded into the address register.

In contrast, when the immediate mode is used with a label, the address is moved. In the instruction

MOVEA.L   #TABLE,A1

the address of the location TABLE, which is the source operand, is transferred to A1. Thus the immediate form loads the address of the table itself. Without the immediate symbol, the MOVEA instruction would be used to load the first address within a table of addresses starting at location TABLE into A1.

Load Effective Address.   The LEA instruction is used to calculate an effective address based on the source addressing mode and transfer it to an address register. The symbolic form is

LEA   ⟨EA⟩,An

where ⟨EA⟩ is specified by a control addressing mode. Thus register direct, postincrement, and predecrement addressing is not allowed. The instruction

LEA   TABLE,A1

results in the operation

$$(A1)[31:0] \leftarrow TABLE$$

where TABLE is an address in an assembly language program. This instruction is equivalent to the MOVEA instruction with the immediate form of the source operand address.

When other addressing modes are used for the source operand in the LEA instruction, the instruction performs a function that is not possible with other data transfer instructions. The LEA instruction allows an address to be calculated during program execution and transferred to an address register.

Consider the instruction sequence

```
LEA       2(A1,D1.W),A0    ;CALCULATE ADDRESS
MOVE.W    (A0),D2          ;PUT VALUE IN D2
MULU      #4,D2            ;4 × VALUE
MOVE.W    D2,(A0)          ;SAVE IT
```

This sequence first calculates an address based on the indirect with indexing addressing mode as

$$(A1) + (D1)[15:0] + 2$$

and transfers it into A0. The operand addressed by (A0) is moved into D2, modified, and saved. The indirect reference (A0) is more efficient in the use of memory than the indirect with indexing reference 2(A1,D1.W), which requires an extension word for the displacement to its instruction. Also, fewer machine cycles are required to calculate the indirect address than to calculate the indirect with index address. Without the LEA instruction, both move instructions would need the indexed addressing to calculate the operand location.

Push Effective Address. The PEA instruction calculates an effective address and uses the system stack as the destination. The symbolic form

PEA   ⟨EA⟩

causes a 32-bit effective address to be calculated for an operand specified by one of the control addressing modes. The value calculated is pushed onto the stack by the CPU using the sequence

$$(SP) \leftarrow (SP) - 4$$
$$((SP)) \leftarrow \langle EA \rangle$$

where the system stack pointer is first decremented by 4 and then used to point to the longword location for ⟨EA⟩.

Example 9.2

Table 9.4 shows the results of executing the PEA, LEA, and MOVEA instructions. For the PEA instruction, the stack pointer is initialized to $0000 7FFE. The instruction pushes (A0) as shown. Because the system stack is used, the more significant bytes of A0 are stored at lower memory addresses. LEA loads A0 with the source operand itself. To accomplish a similiar result, MOVEA with an immediate value would be used. In the next example, if the source operand were specified for MOVEA as an absolute address rather than immediate, the contents of the word location at $8000 would be transferred to A0. As it is, the immediate value $8000 is sign-extended to $FFFF 8000 before it is used.

Example 9.3

Figure 9.2 lists the instruction sequences that are equivalent to LEA and PEA. In each case, A0 holds the first address of a table representing blocks of byte-length operands in memory. The low-order word of D0 is an index to select the starting address of a particular block when added to (A0). The OFFSET selects the address of a particular byte when used to calculate the effective address.

Table 9.4   Address Manipulation Examples

Memory Contents (hexadecimal)	Instruction	Results
4850	PEA (A0)	($7FFA) = $00
		($7FFB) = $00
		($7FFC) = $10
		($7FFD) = $20
41F9	LEA $00012345,A0	(A0) = $0001 2345
0001		
2345		
307C	MOVEA.W #$8000,A0	(A0) = $FFFF 8000
8000		

Note: Initially, (A0) = $0000 1020 and (SP) = $0000 7FFE.

```
abs. rel.    LC   obj. code    source line
----  ----   ----  ----------  -----------
  1    1    0000                |        TTL     'FIGURE 9.2'
  2    2    0000                |        LLEN    100
  3    3    8000                |        ORG     $8000
  4    4    8000                |        OPT     P=M68000
  5    5    8000                |*
  6    6          0000 000A     |OFFSET  EQU     10
  7    7    8000                |*
  8    8    8000                |* COMPARISON OF LEA AND PEA
  9    9    8000                |*
 10   10    8000                |*  INPUT  : (A0.L) = ADDRESS OF TABLE
 11   11    8000                |*           (D0.W) = INDEX INTO TABLE
 12   12    8000                |*            OFFSET = BYTE WITHIN ELEMENT OF PREDEFINED TABLE
 13   13    8000                |*
 14   14    8000                |*  OUTPUT : (A1.L) = EFFECTIVE ADDRESS (LEA)
 15   15    8000                |*
 16   16    8000                |*  OPERATION OF THE LEA TO LOAD THE ADDRESS OF AN ELEMENT
 17   17    8000                |*
 18   18    8000 43F0 000A      |        LEA     (OFFSET,A0,D0.W),A1
 19   19    8004                |*
 20   20    8004                |*  LOAD THE ADDRESS OF AN ELEMENT WITHOUT USING LEA
 21   21    8004                |*
 22   22    8004 2248           |        MOVEA.L A0,A1
 23   23    8006 D2C0           |        ADDA.W  D0,A1
 24   24    8008 D2FC 000A      |        ADDA.W  #OFFSET,A1
 25   25    800C                |*
 26   26    800C                |*   OPERATION OF THE PEA TO SAVE AN ADDRESS
 27   27    800C                |*
 28   28    800C 4870 000A      |        PEA     (OFFSET,A0,D0.W)
 29   29    8010                |*
 30   30    8010                |*  SAVE AN ADDRESS WITHOUT USING PEA
 31   31    8010                |*
 32   32    8010 2248           |        MOVEA.L A0,A1
 33   33    8012 D2C0           |        ADDA.W  D0,A1
 34   34    8014 D2FC 000A      |        ADDA.W  #OFFSET,A1
 35   35    8018 2F09           |        MOVE.L  A1,-(SP)
 36   36    801A                |*
 37   37    801A                |        END
```

Figure 9.2 Comparison of LEA and PEA.

EXERCISES

9.1.1. Let (A1) = $0000 1000 before each of the following instructions is executed. Compute the address in A1 after each instruction executes.

   (a) ADDA.W   #$2000,A1

   (b) ADDA.L   #$9000,A1

   (c) SUBA.W   #$2000,A1

9.1.2. Let the hexadecimal values in A0 and A1 be

    (A0) = $0000 1000

    (A1) = $0001 1F00

before the operation

CMPA.W   A0,A1

Which conditional branch statements following the compare will cause a branch?

9.1.3. If the instruction

LEA   1(A1,D1.W),A2

is executed with

(D1) = $0000 8000
(A1) = $0000 1FFF

what is the address loaded into A2?

9.1.4. How can LEA or PEA be used to debug a program by verifying the addresses from which data are fetched?

9.1.5. The MC68000 allows indirect addressing only using an address register. Suppose that it is desired to hold an indirect address in a given memory location that itself is indirectly addressed. Show how the LEA instruction is used to access the indirect address if the address is at location ADDR1. Then generalize the concept to allow two levels of indirect addressing with the addresses in memory.

9.2   POSITION-INDEPENDENT CODE AND BASE ADDRESSING

In most of the previous programming examples, the program occupied fixed locations in memory and the starting address was defined by the origin (ORG) directive of the assembler. If these programs were to be moved to another area in memory, reassembly with a new origin would be required. The fact that the programs cannot be moved or relocated in memory without reassembly is inconvenient in some cases. It is not acceptable at all for ROM-based programs. In these cases, the programs must be able to be relocated after they are assembled.[1] Some operating systems may need to relocate an applications program even after its execution has already started.

A program is said to be *statically position independent* if it can be loaded and executed from any starting address in memory. Most programs in ROM are statically position independent because the starting address of the ROM program is defined by the system designer based on the requirements of the system. A floating-point routine in ROM, for example, may have a starting address of $2000 in one system and start at location $10000 in another. Perhaps the second system requires a much greater contiguous RAM area. Writing *position-independent code* is a technique of coding a routine so that the starting address is arbitrary. A major difference in this type of program is that it does not contain any absolute addresses except those dictated by hardware definitions such as I/O device addresses.

---

[1] Relocation using a linkage editor is discussed in a number of the references in the Further Reading section for this chapter.

Dynamic position independence allows a program to be moved after it has begun execution. This is required in systems with *virtual* memory. Stated simply, the virtual memory system permits programs to be written without regard to the limitations of the physical memory. If relocation is required, the operating system and memory management circuitry control the actual (physical) addressing automatically.

Position independence is provided in MC68000 programs by several addressing modes. All program counter relative addresses are position independent. Also, address register indirect with index or displacement addressing can be used to create position-independent code in a scheme called *base register* addressing.

9.2.1 Position-Independent Code with (PC)

When a location referenced in a program is at a fixed distance from the instruction making the reference, the program counter relative addressing mode can be used to create position-independent code. As long as the relative displacement is not changed, the program will execute correctly anywhere in memory. If all memory references use PC relative addressing, the program will be *dynamically* position independent because the effective addresses will be calculated as each instruction executes. Moving the program and restarting it at the point where it was temporarily suspended will not cause a problem as long as the program and data are moved together as a block. Unfortunately, it is difficult to write MC68000 programs that use only the PC relative addressing mode because destination operands in memory cannot be referenced this way by most instructions.

Table 9.5 shows the type of instructions and memory references that are inherently position independent. Any of the conditional branch instructions or BRA use program counter with displacement addressing, allowing up to a 16-bit signed displacement. Immediate values are obviously unaffected by moving a program. Fixed addresses have system-defined values that are not changed. These are typically defined in the program using the Equate (EQU) directive. In addition, relative memory addresses using (PC) can be used by the programmer to ensure that the code is position independent.

As shown in Table 9.6, most of the MC68000 instructions that specify two operands allow only the source operand to be designated by the PC relative addressing mode. Thus the instruction

MOVE.W   X,Y

could have only X specified as PC relative. The destination location for BRA, Bcc, DBcc, BTST, JMP, or JSR, however, can be specified by a PC relative mode. Basically, the MC68000 does not allow an operand that can be altered to be referenced by PC relative addressing. The MC68000 designers consider a reference using the PC to be a reference to a program instruction, not a data reference.[2] The only exception is the Bit Test (BTST) instruction.

---

[2] This distinction is explained when the function code lines of the MC68000 are discussed in Chapter 12.

Table 9.5   Position-Independent References

Category	Addressing Mode

Branch instructions
  BRA
  Bcc
  DBcc
	PC relative with displacement

Immediate instructions
  Logical (ORI, ANDI, EORI)
  Arithmetic (ADDI, SUBI, CMPI, ADDQ, SUBQ)
  Transfers (MOVEI)
	Immediate

Absolute reference to fixed locations	Absolute long, Absolute short
Relative memory references	PC relative with displacement, PC relative with index

Assemblers may differ in the manner in which relative addressing is specified by the programmer. The Motorola Cross-Assembler, for example, requires that the directive Relative Origin (RORG) be used in place of the ORG directive. Following the RORG directive, instructions that reference labels defined in the program will be automatically assembled using PC relative addressing instead of absolute addressing. Motorola's Resident Structured Assembler requires that the suffix "(PC)" be appended to a label to force PC relative addressing or that the directive OPT PCS (PC relative option) be used. The manual for the system being used will define the specific procedure. Our examples will use various methods to create position-independent code.

Table 9.6   Relative Addressing for Instructions

Source	Destination
ADD, ADDA	BRA, Bcc, DBcc
AND	BSR
CHK	BTST
CMP, CMPA	JMP, JSR
DIV	
LEA	
MOVE, MOVEA	
MOVEM	
MUL	
OR	
PEA	
SUB, SUBA	

```
abs.  rel.   LC   obj. code   source line
----  ----   ----  ---------   -----------
   1    1   0000                  |           TTL     'FIGURE 9.3'
   2    2   0000                  |           LLEN    100
   3    3   0000                  |           OPT     P=M68000
   4    4   0000                  |*
   5    5   0000                  |*  ABSOLUTE VS. RELATIVE ASSEMBLY
   6    6   0000                  |*
   7    7   0000                  |*  MOVE A BLOCK OF DATA
   8    8   0000                  |*
   9    9   0000                  |*  INPUT : (D1.W) = NUMBER OF BYTES TO MOVE
  10   10   0000                  |*          (A0.L) = ADDRESS OF BLOCK
  11   11   0000                  |*
  12   12   0000                  |*  OUTPUT : BUFFER IS LOADED WITH BYTES
  13   13   0000                  |*
  14   14   0000                  |*
  15   15   0000                  |*
  16   16   0000                  |*          RELOCATABLE SECTION OF CODE
  17   17   0000                  |*
  18   18   0000                  |
  19   19   0000 48E7 40C0  |ABSMOVE MOVEM.L D1/A0-A1,-(SP)  ;SAVE REGISTERS
  20   20   0004 227C 0000  |         MOVEA.L #BUFABS,A1     ;GET ADDRES OF BUFFER
  20        0008 0016       |
  21   21   000A 12D8       |LOOP1    MOVE.B  (A0)+,(A1)+    ;MOVE A BYTE INTO BUFFER
  22   22   000C 5341       |         SUBQ.W  #1,D1          ;DECREMENT COUNTER
  23   23   000E 66FA       |         BNE     LOOP1          ;CONTINUE UNTIL
  24   24   0010            |*                               ;    (D1)=0
  25   25   0010 4CDF 0302  |         MOVEM.L (SP)+,D1/A0-A1 ;RESTORE REGISTERS
  26   26   0014 4E75       |         RTS
  27   27   0016            |BUFABS   DS.B    100            ;SET UP BUFFER (ABSOLUTE)
  28   28   0016            |*
  29   29   007A            |*          POSITION INDEPENDENT SECTION OF CODE
  30   30   007A            |*
  31   31   007A            |
  32   32   0000            |           SECTION 1
  33   33   007A            |*
  34   34   0000 48E7 40C0  |RELMOVE MOVEM.L D1/A0-A1,-(SP)  ;SAVE REGISTERS
  35   35   0004 43FA 002B  |         LEA     (BUFREL,PC),A1 ;PC RELATIVE
  36   36   0008 12D8       |LOOP2    MOVE.B  (A0)+,(A1)+    ;MOVE A BYTE; RELATIVE
  37   37   000A            |*                               ; REFERENCE TO DESTINATION
  38   38   000A 5341       |         SUBQ.W  #1,D1          ;DECREMENT COUNTER
  39   39   000C 66FA       |         BNE     LOOP2          ;CONTINUE UNTIL (D1)=0
  40   40   000E 4CDF 0302  |         MOVEM.L (SP)+,D1/A0-A1 ;RESTORE REGISTERS
  41   41   0012 4E75       |         RTS
  42   42   0014            |*
  43   43   0014            |BUFREL   DS.B    100            ;SET UP BUFFER (RELATIVE)
  44   44   0014            |*
  45   45   0078            |         END
```

Figure 9.3   Absolute and relative assembly.

Example 9.4 _____

Figure 9.3 shows two subroutines that compare the use of absolute and relative addresses. In the first routine starting at ABSMOVE, the starting address of the block or buffer of bytes is defined by BUFABS. The absolute location is $1016 and it is part of the machine language instruction at $1004 to load the address into A1. If the program with the buffer is moved in memory, the address of the buffer must be changed.

In contrast, when the buffer location is referenced relatively as shown in the subroutine RELMOVE, only the offset is specified in the LEA instruction at $1004. Because (PC) = $1006 when the address is calculated and the offset is $0E, a little hand calculation will show that (A1) = $1014, as it should before the loop is entered.

9.2.2 Base Register Addressing

In the discussion of position-independent code, the program counter was used as a *base address* to which a displacement and possibly an index value was added to locate an operand in memory. The address register indirect with displacement and the address register indirect with index addressing modes of the MC68000 can be used to accomplish *base register* addressing using any of the address registers. As the instruction using base register addressing executes, the operand location is calculated by adding an offset to the base address.

In MC68000 programs, the base register addressing is typically used to access data in an array or similar structure or to pass the base address of a data area to a subroutine. This addressing method is particularly useful when the relative position of a data item can be located by a displacement or an index value but the starting address of the structure is not known at the time of assembly.

Example 9.5

Figure 9.4 illustrates the possibility of segmenting memory into blocks using base register addressing. Registers A1, A2, and A3 address different segments. Memory references in a program using A1 reference the first segment, and an index value could be added to access a specific location. Similarly, the other address registers could be used to locate data or instructions in other segments. If a segment were moved, the operating system would reload the proper address register with the new starting address.

Figure 9.4 Base register addressing.

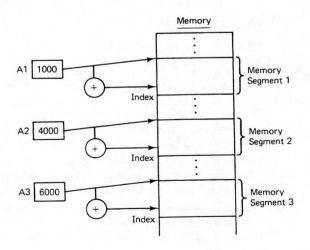

EXERCISES

9.2.1. Consider the simple program segment

```
              RORG      0               ;SYMBOLS ARE RELATIVE
START         MOVE.W    FIRST,D1
              ADD.W     SECOND,D1
              MOVE.W    D1,RESULT
              RTS
FIRST         DC.W      1
SECOND        DS.W      1
RESULT        DS.W      1
              END
```

The program attempts to calculate

$$(RESULT) = (FIRST) + (SECOND)$$

but will not work. Also, the program cannot be moved in memory from location 0000. Correct the program so that it will work anywhere in memory.

9.2.2. Show that the following program is statically position independent:

```
              ORG       0
START         LEA       *+0,A0              ;GET PC
              ADDA.L    #(DATA-START),A0    ;RELOCATE POINTER
              MOVE.W    #19,D1              ;COUNTER
LOOP          ADD.W     (A0)+,D2            ;SUM ARRAY
              DBF       D1,LOOP
              RTS
DATA          DS.W      20
              END
```

Show that the pointer to the data array (A0) contains the proper address of DATA after the program is moved. DBRA is also used for DBF in many assemblers.

9.2.3. Suppose that the instructions

```
LEA       Y(PC),A1
ADD.W     X(PC),(A1)
```

are used to create position-independent code. Relative addresses are specified for both the source X and destination Y. After execution of the LEA instruction, A1 contains the address of Y calculated from the (PC) value plus displacement. Show that the program is not dynamically position independent. (*Hint:* Consider the case when the program is moved after the LEA executes but before the ADD instruction executes.)

9.2.4. Compare indexed addressing with base register addressing. In the MC68000, the two addressing modes are identical as far as machine language code is concerned but have different purposes. Discuss the use of each.

9.2.5. Write a position-independent program to sum the values in five locations reserved by a DS directive in the program. Test it by writing another program that moves the program to sum the values and executes it after it is moved.

9.3 DATA STRUCTURES

The fundamental data types for the MC68000 are signed or unsigned integers, BCD integers, and Boolean variables. They are considered to be fundamental or primitive data types because MC68000 instructions are available to manipulate them directly. In contrast, strings of characters must be created and manipulated in algorithms that are devised by the programmer. These represent a data type not available at the assembly language level. The definition of new data types and the logical relationship defining their organization leads to the study of *data structures*.

An *array* is an example of a data structure. Arrays consist of a set of items of a single data type stored in contiguous locations in memory. The terms "array" and "table" are usually used synonymously. Arrays of numbers and tables of addresses have been used in examples in previous chapters. In this section, the concepts are generalized for both single- and multidimensional arrays.

A *string* is typically a sequence of ASCII characters treated as a single unit for input or output. The MC68000 instruction set allows efficient string manipulation routines to be created. The *queue* is another data structure that is convenient for storing data in certain applications. Programming techniques for string and queue manipulations are introduced in this section.

The *linked list* is another type of data structure useful in many applications. This structure allows data items to be stored in noncontiguous storage locations using pointers to indicate the location of the next item in the list. The MC68000, with its extensive address manipulation capability, is well suited for programs using linked lists. Review of the MC68000 addressing modes described in Chapter 5 may be helpful.

9.3.1 One-Dimensional Arrays

The one-dimensional array is a structure consisting of a collection of items in which each element is identified uniquely by an index value corresponding to its position in the array. Because each item in the array is of the same data type, the structure is homogeneous. In mathematics, the one-dimensional array of numbers is called a *vector*, with the position of each element specified by a subscript. If the first subscript is arbitrarily chosen as 1, the vector elements are

$$V_1, V_2, \ldots, V_N$$

for an N-element vector. The FORTRAN convention begins the vector with subscript 1 and has elements

$$V(1), V(2), \ldots, V(N)$$

where $V(i)$ indicates the address of element V_i. Languages such as Pascal or ALGOL allow arbitrary specification of the first index. In assembly language, complete flexibility is available. For this discussion, V_i will refer to the ith element itself and $V(i)$ will be the address of the ith element.

Address Calculation. Sequential storage of the elements of a one-dimensional array allows each element to be easily referenced according to its address. If an *N*-element array X starts at location X(1) and each element occupies *C* bytes, then the *j*th element has the address

$$X(j) = X(1) + C * (j - 1) \qquad 1 \leq j \leq N$$

where C is a constant.[3] The array length in bytes is

$$\text{length} = (X(N) - X(1)) + C$$

when X(N) and X(1) represent the last and first addresses of the elements, respectively. Table 9.7 shows the address calculations for various arrays, and Figure 9.5 illustrates the general relationship in memory.[4]

Array Addressing with the MC68000. Addressing elements of an array is accomplished using the indirect or relative addressing modes of the MC68000. If the starting or base address of an array is held in an address register, the indirect modes can be used to locate elements in the array as shown in Table 9.8. When the program counter is used to perform relative addressing, a displacement value or an index plus displacement value is added to the (PC) to reference an element of the array. The fixed displacement is calculated by the assembler when relative addressing is specified in an instruction and the array storage is allocated with a Define Storage (DS) directive at a labeled statement.

The postincrement addressing mode can be used to reference array elements in sequence. For byte, word, or longword elements, the postincrement addressing mode allows an element to be examined and manipulated, after which the address register contains the address of the next element in the array. The instruction

MOVE.W   (A1)+,D1

Table 9.7   One-Dimensional Array Addressing

Size of Element in Table	C	*j*th Location	Length of Array (bytes)
Byte	1	$X(j) = X(1) + (j - 1)$	N
Word	2	$X(j) = X(1) + 2 * (j - 1)$	$2N$
Longword	4	$X(j) = X(1) + 4 * (j - 1)$	$4N$
Strings (fixed length)	k	$X(j) = X(1) + k * (j - 1)$	$k \times N$

Notes:
1. N is the number of elements each of length C bytes.
2. The index range is $1 \leq j \leq N$. The first address is $X(1)$, containing the value X_1.

---

[3] If $X(0)$ is the first address, then the jth element has address $X(j) = X(0) + c \times j$; $j = 0, 1, 2, \ldots, N - 1$.

[4] There is no reason that the array could not have elements with higher indices occupying lower memory addresses and thus have the physical ordering be opposite the logical ordering. In this case, the equations given in this section would require modification.

Figure 9.5 Array storage in memory.

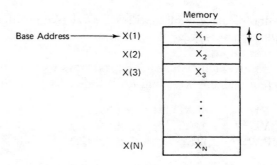

Notes:
(1) C is the length of one element in bytes.
(2) X(k) is the address of element X_k.

transfers the 16-bit value addressed by A1 to (D1) [15:0]. Register A1 is then incremented by 2 to point to the next value in the array. For an array whose elements are stored at decreasing memory locations from the base address, the predecrement mode permits the array to be scanned in the reverse order.[5] Without additional programming, only byte, word, or longword elements can be addressed with these modes in a sequential fashion.

The address register indirect with displacement mode can be used to address a specific element in an array. The address register holds the base address of the array

Table 9.8 MC68000 Array Addressing

MC68000 Addressing Mode	Typical Use
Predecrement	To address byte, word, or longword elements in descending sequence
Postincrement	To address byte, word, or longword elements in ascending sequence
(PC) with displacement or (An) with displacement	To locate an element at a fixed position from the base address
(PC) with index or (An) with index	To locate an element at an arbitrary position using an index register

---

[5] The predecrement and postincrement modes are useful for manipulating stacks and queues of byte, word, or longword entries. These are *dynamic* arrays because the length varies with program execution. Several of the references in the Further Reading section at the end of this chapter discuss these structures in detail.

and the displacement specifies the relative position of the element as an offset. The displacement is a 16-bit signed integer. This mode is often used to compare elements in separate arrays.

As an example, consider two arrays with word-length elements. If (A1) points to the first element X(1) and (A2) points to Y(1), the instruction sequence

```
MOVE.W      4(A1),D1      ;(X(3))
MOVE.W      4(A2),D2      ;(Y(3))
CMP.W       D1,D2         ;COMPARE: (Y(3)) − (X(3))
```

compares Y_3 and X_3 located at addresses Y(3) and X(3), respectively. Notice that the third element is located by a 4-byte offset from the base address according to the addressing equation for X(j) previously given. The offset in the indirect mode and in the program counter relative with displacement mode cannot be modified after the program is assembled, so the use of these modes does not allow indexing through an array.

Flexibility is provided by using either the address register indirect with index or PC relative with index addressing mode. In these modes, the base address consists of a register value (address register or PC) and possibly a sign-extended 8-bit displacement value. The index into the array may be calculated before it is used by evaluation of an index *expression* of any complexity. For example, in the instruction

```
MOVE.W      0(A1,D1.W),D1
```

the base address is held in A1 and the low-order 16-bit value in D1 contains the index. The value in D1 could be calculated by any mathematical expression before it is used as an index. If the 16-bit index is not sufficient, a long index of 32 bits may be specified according to the discussion of these addressing modes given in Chapter 5.

Example 9.6 _____

The subroutine ERRMSG in Figure 9.6 calculates the address of a specified element in an array consisting of five elements, each 13 bytes in length. The array of error messages begins at location TABLE and an offset from that address is calculated according to the value in the low-order byte of D0. The address of the element is calculated by multiplying the error number by 13 and adding this value to the base address in A0. The addressing equation in this example would be

$$TABLE(k) = TABLE + 13 \times (k)$$

where k = 0, 1, 2, 3, or 4 is the error number.

Example 9.7 _____

The subroutine of Figure 9.7 performs a "bubble" sort on an array of 8-bit values. When completed, the routine leaves the largest value in the first location with the following values in descending numerical order. The starting address, holding the first value of the array, is supplied in A0 and D0 should contain N, the number of elements in the array. The array is sorted by comparing elements starting with the last value in the array for each pass.

```
abs. rel.   LC    obj. code    source line
---- ----   ----  ---------    -----------
  1    1    0000                        |        TTL      'FIGURE 9.6'
  2    2    0000                        |        LLEN     100
  3    3    8000                        |        ORG      $8000
  4    4    8000                        |        OPT      P=M68000
  5    5    8000                        |*
  6    6    8000                        |*    ERROR MESSAGES
  7    7    8000                        |*
  8    8    8000                        |*    INPUT : (D0.B) = ERROR NUMBER BETWEEN 0 AND 4
  9    9    8000                        |*
 10   10    8000                        |*    OUTPUT : (A0.L) = ADDRESS OF THE CORRESPONDING ERROR MESSAGE
 11   11    8000                        |*
 12   12    8000  2F00                  |ERRMSG   MOVE.L   D0,-(SP)      ;SAVE REGISTER
 13   13    8002  41F9 0000             |         LEA      TABLE,A0      ;GET ADDRESS OF START OF TABLE
 13         8006  8012                  |
 14   14    8008  C0FC 000D             |         MULU.W   #13,D0        ;13 BYTES PER STRING
 15   15    800C  D0C0                  |         ADDA.W   D0,A0         ;COMPUTE INDEX
 16   16    800E  201F                  |         MOVE.L   (SP)+,D0      ;RESTORE REGISTER
 17   17    8010  4E75                  |         RTS
 18   18    8012                        |*
 19   19    8012                        |*    DEFINE TABLE OF ERROR MESSAGES
 20   20    8012                        |*
 21   21    8012  4F56 4552             |TABLE    DC.B     'OVERFLOW     '
 21   21    8016  464C 4F57             |
 21   21    801A  2020 2020             |
 21   21    801E  20                    |
 22   22    801F  554E 4445             |         DC.B     'UNDERFLOW    '
 22   22    8023  5246 4C4F             |
 22   22    8027  5720 2020             |
 22   22    802B  20                    |
 23   23    802C  5355 4253             |         DC.B     'SUBSCRIPT    '
 23   23    8030  4352 4950             |
 23   23    8034  5420 2020             |
 23   23    8038  20                    |
 24   24    8039  5A45 524F             |         DC.B     'ZERO DIVIDE  '
 24   24    803D  2044 4956             |
 24   24    8041  4944 4520             |
 24   24    8045  20                    |
 25   25    8046  554E 494D             |         DC.B     'UNIMPLEMENTED'
 25   25    804A  504C 454D             |
 25   25    804E  454E 5445             |
 25   25    8052  44                    |
 26   26    8053                        |*
 27   27    8053                        |         END
```

Figure 9.6   Addressing of an array element.

This routine uses (D0) to set up two counters in D3 and D4. The inner loop at SORT20 uses the counter in D4 to index into the array and retrieve the elements to sort. The inner loop exchanges elements until the branch condition is true. Then if any elements were exchanged, it is indicated by the flag in D1 set to $FF. The outer loop is then used to check for elements to exchange again. If the operation of the subroutine is not clear from the program comments, more detailed explanations on bubble sorts are given in the references in the Further Reading section for the chapter.

```
os. rel.    LC   obj. code    source line
--- ----    ---- ----------   -----------
  1    1   0000              |          TTL      'FIGURE 9.7'
  2    2   0000              |          LLEN     100
  3    3   8000              |          ORG      $8000
  4    4   8000              |          OPT      P=M68000
  5    5   8000              |*
  6    6   8000              |*  SORT A TABLE OF 8-BIT NUMBERS IN DESENDING ORDER
  7    7   8000              |*
  8    8   8000              |*  INPUT : (A0.L) = ADDRESS OF TABLE TO SORT
  9    9   8000              |*                  (D0.L) = NUMBER OF ENTRIES IN TABLE (N)
 10   10   8000              |*
 11   11   8000 48E7 F882    |SORT      MOVEM.L  D0-D4/A0/A6,-(SP)   ;SAVE REGISTERS
 12   12   8004 5380         |          SUBQ.L   #1,D0
 13   13   8006 2400         |          MOVE.L   D0,D2               ;COUNTER 1 = N - 1
 14   14   8008 4201         |SORT10    CLR.B    D1                  ;FLAG = 0; ASSUME SORTED
 15   15   800A 2600         |          MOVE.L   D0,D3               ;COUNTER 2 = N - 1
 16   16   800C              |*
 17   17   800C 4DF0 3800    |SORT20    LEA      (0,A0,D3.L),A6      ;START AT BOTTOM OF ARRAY
 18   18   8010 182E FFFF    |          MOVE.B   (-1,A6),D4          ;GET VALUE (CNT2 + 1)
 19   19   8014 B816         |          CMP.B    (0,A6),D4           ;IF VALUE(CNT2+1) >= VALUE(CNT2)
 20   20   8016 6C00 000C    |          BGE      SORT30              ;   THEN SKIP
 21   21   801A 123C 00FF    |          MOVE.B   #$FF,D1             ;ELSE SET FLAG AND
 22   22   801E 1D56 FFFF    |          MOVE.B   (0,A6),(-1,A6)      ; SWAP VALUES
 23   23   8022 1C84         |          MOVE.B   D4,(0,A6)
 24   24   8024              |*
 25   25   8024 5383         |SORT30    SUBQ.L   #1,D3               ;DECREMENT CNT2
 26   26   8026 66E4         |          BNE      SORT20              ;   AND LOOP
 27   27   8028              |*
 28   28   8028 4A01         |          TST.B    D1                  ;IF FLAG = 0, SORTED
 29   29   802A 6700 0006    |          BEQ      DONE                ; THEN EXIT
 30   30   802E 5382         |          SUBQ.L   #1,D2               ;ELSE DECREMENT CNT1
 31   31   8030 66D6         |          BNE      SORT10              ; AND LOOP
 32   32   8032              |*
 33   33   8032 4CDF 411F    |DONE      MOVEM.L  (SP)+,D0-D4/A0/A6   ;RESTORE REGISTERS
 34   34   8036 4E75         |          RTS
 35   35   8038              |          END
```

Figure 9.7   Sorting example.

9.3.2 Two-Dimensional Arrays

A generalization of the one-dimensional array is a higher-dimensional array of elements. The elements in the multidimensional array are specified by more than one subscript, as is shown in Table 9.9(a) for the two-dimensional $M \times N$ array. Table 9.9(b) shows the 3×3 array as a specific example. In this notation, M is the number of horizontal rows and N is the number of vertical columns. The element in the ith row and jth column is designated X_{ij}, where $1 \leq i \leq M$ and $1 \leq j \leq N$. X(i,j) represents the address in memory of the element X_{ij} in the form X (row, column). An $M \times N$ array has $M \times N$ elements. The row index has a range of M values and the column index has a range of N values regardless of the starting indices. Questions pertinent to structuring the data in memory concern the method of storage by row and by column as well as the techniques necessary to compute the address of an element.

Table 9.9   Multidimensional Arrays

(a) General form

$$
\begin{array}{cccc}
X_{11} & X_{12} & \cdots & X_{1N} \\
X_{21} & X_{22} & \cdots & X_{2N} \\
\cdot & \cdot & & \cdot \\
\cdot & \cdot & & \cdot \\
\cdot & \cdot & & \cdot \\
X_{M1} & X_{M2} & \cdots & X_{MN}
\end{array}
$$

(b) 3×3 Example

$$
\begin{array}{ccc}
X_{11} & X_{12} & X_{13} \\
X_{21} & X_{22} & X_{23} \\
X_{31} & X_{32} & X_{33}
\end{array}
$$

(c) Column-major storage of a 3×3 array starting at location $1000

	Address (hexadecimal)	Array Element
X(1,1)	$1000	X_{11}
X(2,1)	$1002	X_{21}
X(3,1)	$1004	X_{31}
X(1,2)	$1006	X_{12}
X(2,2)	$1008	X_{22}
X(3,2)	$100A	X_{32}
X(1,3)	$100C	X_{13}
X(2,3)	$100E	X_{23}
X(3,3)	$1010	X_{33}

Array Storage.   If the $M \times N$ array X is stored sequentially by rows in addresses beginning with X(1,1) as

X(1,1), X(1,2), . . . , X(1,N), X(2,1), . . . , X(2,N), . . . , X(M,N)

the storage is said to be in *row-major* form. An alternative scheme is *column-major* form with successive element addresses

X(1,1), X(2,1), . . . , X(M,1), X(1,2), . . . , X(M,2), . . . , X(M,N)

as shown in Table 9.9(c). The figure shows the storage for a 3×3 array of words in the MC68000 memory starting at location $1000. This column-major form is required in standard FORTRAN and will be used for our examples in this subsection.

Once the storage form is chosen, the indices for a specific element can be computed in several ways. The *address polynomial* for an $M \times N$ array has the form

$$X(i,j) = \text{base address} + C_1 \times (j - i) + C_2 \times (i - 1)$$

for a two-dimensional array with the first address at location X(1,1). When the constants C_1 and C_2 are properly chosen, address calculation is straightforward on a processor with a multiply instruction.[6]

---

[6] In some cases, to save the time required by multiplication or to allow dynamic (during execution) allocation of array storage, special addressing methods are used. These include the use of a "dope vector" describing the array characteristics or addressing by indirection in which the row (or column) addresses are held in a table. This subject is developed further in several of the references in the Further Reading section at the end of this chapter.

Table 9.10 Multidimensional Array Addressing

Array Storage	Address X(i,j)
Column-major storage X(1,1), X(2,1), . . .	$Bo + C * [(i - 1) + M \times (j - 1)]$
Row-major storage X(1,1), X(1,2), . . .	$Bo + C * [(j - 1) + N \times (i - 1)]$

Notes:
1. Bo is the base address of array X(i,j) with elements C bytes in length.
2. The indices range as follows:
   rows: $1 \leq i \leq M$
   columns: $1 \leq j \leq N$

Table 9.10 shows the address polynomials for arrays with elements of length C bytes. The address of X_{ij} in an array stored by column-major form is

$$X(i,j) = Bo + C \times [(i - 1) + M \times (j - 1)]$$

in which

$$1 \leq i \leq M \quad \text{and} \quad 1 \leq j \leq N$$

and Bo is the base address. This addressing method is easily implemented with the MC68000 using indirect addressing with indexing if elements in a fixed column (or row) are being selected. If both indices are varied, one address must be calculated separately and added to the value of the effective address computed with indexed addressing.

Example 9.8

When the address of an element in a two-dimensional array is calculated, two indices must be added to the starting or base address. The address register indirect with index and the PC relative with index addressing modes of the MC68000 allow two separate offsets and calculate the effective address as

$$\langle EA \rangle = (R) + (Rn) + \langle d_8 \rangle$$

The register R here could be either an address register in the indirect mode or the program counter for the relative mode. The index register Rn is either an address register or data register. The displacement value $\langle d_8 \rangle$ is limited to an 8-bit signed integer. Figure 9.8 shows the use of this addressing method for an array stored in column-major form. If the number of bytes per element is 1, the address calculation becomes

$$X(i,j) = Bo + [(i - 1) + M \times (j - 1)]$$

The base address and one index can be stored in registers. The second index can be specified as the 8-bit offset in the address register indirect with index addressing mode. In the figure, the offset selects the row and remains fixed. The 3×3 array begins at hexadecimal location

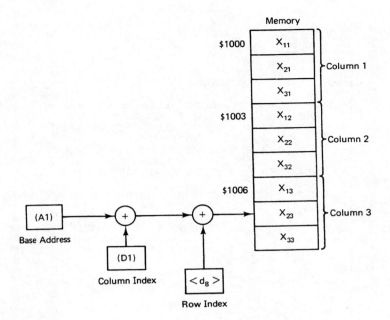

Example: Load X_{23} into D2
MOVE.B 1(A1, D1.L),D2   ;(D2)[7:0] ← ((A1) + (D1) + 1)

(A1) = $0000 1000 (Base)
(D1) = $0000 0006 (Column index = (j − 1)*M)
$< d_8 >$ =   .1   (Row index = i − 1)

Figure 9.8   Fixed-element addressing.

$1000, which is stored in A1, and the register D1 contains the column index $(j − 1) \times M$, which is $0006. The instruction

MOVE.B   1(A1,D1.L),D2

transfers element X_{23} to the low-order byte of D2. To select A_{2j} from another column, (D1) must be changed to indicate the column offset, which is computed as $3 \times (j − 1)$ with j = 1, 2, or 3.

Example 9.9 ──

The subroutine in Figure 9.9 performs a binary search of a table or array for a word-length bit pattern called a key value. The table is composed of M entries that are each N bytes long. The starting address of the table is supplied in A0 and the key to locate is in D0[7:0]. The low-order word of D1 contains the length (N) of each entry and D2[15:0] contains the number of entries (M) in the table.

```
abs. rel.    LC   obj. code    source line
---- ----    ----  ---------    -----------
   1    1   0000                |         TTL     'FIGURE 9.9'
   2    2   0000                |         LLEN    100
   3    3   8000                |         ORG     $8000
   4    4   8000                |         OPT     P=M68000
   5    5   8000                |*
   6    6   8000                |* SEARCH SUBROUTINE DOES A BINARY SEARCH OF A TABLE
   7    7   8000                |*  INPUT : (A0.L) = TABLE TO SEARCH
   8    8   8000                |*          (D0.B) = KEY TO SEARCH FOR IN TABLE
   9    9   8000                |*                     BYTE LENGTH
  10   10   8000                |*          (D1.W) = LENGTH IN BYTES OF EACH ENTRY
  11   11   8000                |*                     IN TABLE
  12   12   8000                |*          (D2.W) = NUMBER OF ENTRIES IN TABLE   (END)
  13   13   8000                |*
  14   14   8000                |* OUTPUT : (A6.L) = ADDRESS OF ITEM IN TABLE WITH VALUE OF
  15   15   8000                |*                     KEY (OR ZERO)
  16   16   8000                |*
  17   17   8000 48E7 FC00      |SEARCH   MOVEM.L  D0-D5,-(SP)       ;SAVE REGISTERS ON STACK
  18   18   8004 5342           |         SUBQ.W   #1,D2             ;INIT :  END (D2)
  19   19   8006 4283           |         CLR.L    D3                ;         BEGIN (D3)
  20   20   8008 2C43           |         MOVEA.L  D3,A6             ;         OUTPUT VALUE
  21   21   800A                |*
  22   22   800A B642           |SER10    CMP.W    D2,D3             ;IF BEGIN >= END
  23   23   800C 6E00 0028      |         BGT      EXIT              ;   THEN EXIT
  24   24   8010 3803           |         MOVE.W   D3,D4             ;   ELSE COMPUTE
  25   25   8012 D842           |         ADD.W    D2,D4             ;          INDEX = (BEGIN
  26   26   8014 E24C           |         LSR.W    #1,D4             ;               + END)/2
  27   27   8016                |*                                  ;ENDIF
  28   28   8016                |*
  29   29   8016 3A04           |         MOVE.W   D4,D5             ;COMPUTE ADDRESS
  30   30   8018 CAC1           |         MULU.W   D1,D5             ;INDEX IN TABLE OF KEY
  31   31   801A                |*                                  ;   TO TEST
  32   32   801A B030 5000      |         CMP.B    (0,A0,D5),D0      ;IF KEY >= ENTRY
  33   33   801E 6C00 0008      |         BGE      SER20             ;   THEN BRANCH TO MODIFY
  34   34   8022                |*                                  ;          BEGIN
  35   35   8022 5344           |         SUBQ.W   #1,D4             ;   ELSE SET
  36   36   8024 3404           |         MOVE.W   D4,D2             ;          END = INDEX - 1
  37   37   8026 60E2           |         BRA      SER10             ;   TRY AGAIN
  38   38   8028                |*                                  ;ENDIF
  39   39   8028                |*
  40   40   8028 6700 0008      |SER20    BEQ      SUCCESS           ;IF KEY = TABLE ENTRY
  41   41   802C                |*                                  ;   THEN BRANCH TO UPDATE
  42   42   802C                |*                                  ;          OUTPUT
  43   43   802C 5244           |         ADDQ.W   #1,D4             ;   ELSE
  44   44   802E 3604           |         MOVE.W   D4,D3             ;          BEGIN = INDEX + 1
  45   45   8030 60D8           |         BRA      SER10             ;   TRY AGAIN
  46   46   8032                |*                                  ;ENDIF
  47   47   8032                |*                                  ;
  48   48   8032 4DF0 5000      |SUCCESS  LEA      (0,A0,D5.W),A6    ;SAVE OUTPUT ADDRESS
  49   49   8036 4CDF 003F      |EXIT     MOVEM.L  (SP)+,D0-D5       ;RESTORE REGISTERS
  50   50   803A 4E75           |         RTS
  51   51   803C                |         END
```

Figure 9.9   Search routine.

If the search is successful, the address of the key value is returned in A6. Otherwise, A6 contains a zero. Because a binary search is being performed, the data are assumed to be sorted numerically in the table being searched.

9.3.3 Strings

For the examples in this textbook, a *string* is a sequence of ASCII characters considered as a unit. The string is usually stored as an array of bytes. This storage method allows an instruction to use the predecrement or postincrement addressing mode to access a particular character in a string and leave the pointer at the next character address. The string is defined in memory by its starting address and either its length in bytes or the location of the last byte.

For output, the string can be defined by its starting address and length as in the examples in Section 7.6. This defines a fixed length string. Alternatively, the string can be ended by a special character such as the ASCII NUL character. Chapter 12 presents examples of this method of terminating a string of variable length used for output. For input, a variable length string is usually terminated by a carriage return or other special ASCII character.

To perform more sophisticated operations on strings, programs must be created because the MC68000 CPU has no string manipulation instructions. A common operation is the comparison of two strings for equality. The example of Subsection 6.2.3 performs string comparison.

9.3.4 Queues

The *queue* is a data structure in which all insertions are made at one end and all deletions are made at the other end. This differs from the stack structure, which is accessed at only one end. The queue has an advantage in applications that require items to be accessed in the queue in the same order that they were inserted. In fact, a queue is sometimes called a FIFO (first-in-first-out) although hardware designers usually reserve this term for FIFO memory chips. Also, most software queues are *circular* queues in which the addressing pointer wraps around to the beginning when the last location assigned to the queue is encountered. Otherwise, the size of the queue could grow without limit if more and more items are inserted without the corresponding deletion of items.[7]

Figure 9.10 shows the various operations on a queue. In memory, the queue is defined to occupy locations between location New Queue Pointer (NEWQP) and End Queue Pointer (ENDQP). Two other address pointers indicate the active queue locations being discussed. QUEOUT points to the first item inserted and hence the next item to be accessed when an instruction uses QUEOUT to point to an item. In other terminology, QUEOUT points to the *head* or the *front* of the queue. QUEIN

---

[7] Knuth's textbook referenced in the Further Reading section of this chapter gives a more thorough discussion of queues.

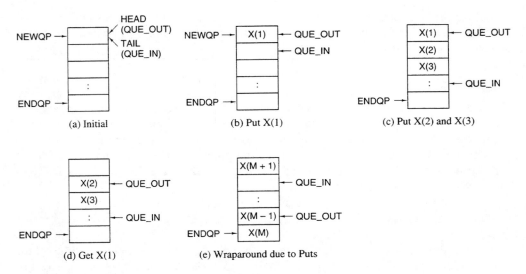

Figure 9.10 Queue structure.

points to a queue location that is empty. The next item to be inserted into the queue is placed in location QUEIN. The location indicated by QUEIN is often called the *tail* or the *rear* of the queue. As shown in Figure 9.10(a), the queue pointers QUEIN and QUEOUT both point to the head of the queues (NEWQP) when initialized. In fact, any time that the address pointers read

QUEIN = QUEOUT

the queue is considered empty.

The operation of adding an item to the queue is variously called an *enqueue* operation or a *put* operation. Figure 9.10(b) shows the queue after item X(1) is added. QUEIN now points to the next empty location. The queue has the form in Figure 9.10(c) after three items are added. After each put operation, QUEIN is incremented by the number of bytes for each item if the queue grows from low to high memory locations. For byte, word or longword operands, the postincrement addressing mode of the MC68000 could be used to put an item on the queue. Of course, the predecrement addressing mode would be used for a put operation if the queue grows from high to low memory.

After an item is accessed as in Figure 9.10(d), QUEOUT points to the next item to be used. The operation of accessing a queue item and changing QUEOUT is called a *dequeue* operation or a *get* operation. It is clear from Figure 9.10 that if the queue pointers QUEIN and QUEOUT become equal, the queue is empty. An *underflow* condition occurs when a program attempts to get an item from an empty queue. This could mean an error condition or simply that the program has accessed all the items in the queue. The underflow condition in the latter case indicates that all data in the queue have been read and the program should begin processing the data items as required.

Because the queue holds a finite number of data items, the size of the queue and its circular structure must be considered. For example, assume that the queue in Figure 9.10 holds M items. The first M-1 puts will leave the condition

QUEIN = ENDQP

When this occurs, the pointer QUEIN must be reinitialized with the value of NEWQP. This condition is shown in Figure 9.10(e) after QUEIN is reinitialized and one more item, X(M + 1), is put on the queue. As long as QUEOUT continues to "chase" QUEIN but does not become equal to it, the queue has space for additional items.

An *overflow* occurs when a put operation is attempted that causes

QUEIN = QUEOUT

This equality condition should only occur when the queue is empty and indicates an error at any other time.

In summary, a queue manipulation program should perform the following operations:

(a) initialize QUEIN = QUEOUT to the first location assigned to the queue;
(b) test for overflow before a put operation;
(c) test for underflow before a get operation;
(d) reinitialize QUEIN and QUEOUT if either pointer becomes equal to the last location assigned to the queue.

In an I/O application, the queue can serve as an input or output buffer area for characters or strings to be transferred. Because queues preserve the order of the data (FIFO) and can grow and shrink dynamically, queues are often used as buffers for I/O whenever the speed of data transfer does not match the speed of data processing. This is discussed further in Chapter 12.

Example 9.10 _____

The program in Figure 9.11 uses an input queue to store each character read from keyboard input and echoes the typed character to the screen using an output queue. After a line of characters, terminated by a carriage return, has been received, the entire line is displayed on the screen. The program segments are as follows:

(a) MAIN initializes the queues, receives and echoes characters, and outputs a line of characters;
(b) QINIT creates an empty queue;
(c) ENQUEUE adds a character to the queue;
(d) DEQUEUE removes a character from the queue.

```
abs.  rel.   LC    obj. code    source line
----  ----   ----  ---------    -----------
  1     1    0000                      |         TTL      'Figure 9.11'
  2     2    0000                      |         LLEN     100
  3     3    8000                      |         ORG      $8000
  4     4    8000                      |         OPT      P=M68000
  5     5    8000                      |*
  6     6    8000                      |* INPUT AND OUTPUT QUEUES
  7     7    8000                      |*
  8     8    8000                      |* THIS PROGRAM RECEIVES CHARACTERS FROM THE KEYBOARD AND
  9     9    8000                      |* ECHOES THE CHARACTERS ON THE SCREEN. AFTER ANY INPUT LINE
 10    10    8000                      |* IS TERMINATED * BY A CARRIAGE RETURN, THE ENTIRE LINE IS
 11    11    8000                      |* RE-DISPLAYED ON THE SCREEN.
 12    12    8000                      |*
 13    13    8000                      |* AN INPUT QUEUE IS USED FOR STORING THE CHARACTERS AND AN
 14    14    8000                      |* OUTPUT QUEUE IS USED FOR STORING DATA TO BE DISPLAYED
 15    15    8000                      |* ON THE SCREEN.
 16    16    8000                      |* THE PROGRAM BEGINS AT LABEL MAIN.
 17    17    8000                      |*
 18    18    8000                      |***********************************************************
 19    19    8000                      |* MACROS AND TUTOR CALLS
 20    20    8000                      |*
 21    21    8000                      |SYSCALL MACRO
 22    22    8000                      |         MOVE.B   #\1,D7
 23    23    8000                      |         TRAP     #14
 24    24    8000                      |         ENDM
 25    25    8000                      |
 26    26          0000 00F7           |.INCHE   EQU      247       ;INPUT A CHARACTER
 27    27          0000 00F8           |.OUTCH   EQU      248       ;OUTPUT A CHARACTER
 28    28    8000                      |*
 29    29    8000                      |* THESE ARE THE OFFSETS INTO THE QUEUE POINTER TABLES
 30    30          0000 0000           |NEWQP    EQU      0
 31    31          0000 0004           |ENDQP    EQU      4
 32    32          0000 0008           |QUEOUT   EQU      8
 33    33          0000 000C           |QUEIN    EQU      12
 34    34    8000                      |*
 35    35    8000                      |*
 36    36    8000 287C 0000            |MAIN     MOVE.L   #IN_QPTRS,A4    ;ADDR OF POINTER TABLE FOR
 36          8004 831E                 |
 37    37    8006                      |*                               ; INPUT QUEUE
 38    38    8006 2A7C 0000            |         MOVE.L   #INQUEUE,A5     ;ADDR OF INPUT QUEUE
 38          800A 811E                 |
 39    39    800C 2C7C 0000            |         MOVE.L   #E_INQUEUE,A6   ;ADDR OF END OF INPUT QUEUE
 39          8010 821E                 |
 40    40    8012 4EB9 0000            |         JSR      QINIT           ;CALL INITIALIZATION ROUTINE
 40          8016 80AA                 |
 41    41    8018 287C 0000            |         MOVE.L   #OUT_QPTRS,A4   ;ADDR OF POINTER TABLE FOR
 41          801C 832E                 |
 42    42    801E                      |*                               ; OUTPUT QUEUE
 43    43    801E 2A7C 0000            |         MOVE.L   #OUTQUEUE,A5    ;ADDR OF OUTPUT QUEUE,
 43          8022 821E                 |
 44    44    8024 2C7C 0000            |         MOVE.L   #E_OUTQUEUE,A6  ;ADDR OF END OF OUTPUT QUEUE
 44          8028 831E                 |
 45    45    802A 4EB9 0000            |         JSR      QINIT           ;CALL INITIALIZATION ROUTINE
 45          802E 80AA                 |
 46    46    8030                      |*
 47    47    8030                      |* WAIT FOR INPUT
 48    48    8030                      |KEYIN                            ;DEFINE LABEL FOR MACROCALL
 49    49    8030                      |         SYSCALL .INCHE          ;GET NEXT CHARACTER
 50    1m    8030 1E3C 00F7           +         MOVE.B   #.INCHE,D7
 51    2m    8034 4E4E               +         TRAP     #14
 52    50    8036                      |*
```

Figure 9.11   Queue program example.

```
53   51   8036                |* CHARACTER WAS RECEIVED, ADD TO QUEUE AND ECHO
54   52   8036 287C 0000      |         MOVE.L    #IN_QPTRS,A4    ;PASS ADDR OF POINTER TABLE
54        803A 831E           |
55   53   803C                |*                                 ;  TO INPUT QUEUE
56   54   803C 4EB9 0000      |         JSR       ENQUEUE         ;ADD CHARACTER TO INPUT QUEUE
56        8040 80BA           |
57   55   8042 287C 0000      |         MOVE.L    #OUT_QPTRS,A4   ;ECHO THE CHARACTER
57        8046 832E           |
58   56   8048 4EB9 0000      |         JSR       ENQUEUE         ;  BY ADDING TO OUTPUT QUEUE
58        804C 80BA           |
59   57   804E 0C00 000D      |         CMPI.B    #KEY_CR,D0      ;WAS IT A CARRIAGE RETURN?
60   58   8052 6600 003E      |         BNE       DISPLAY         ;NO, DISPLAY EVERYTHING IN
61   59   8056                |                                  ;  THE OUTPUT QUEUE
62   60   8056                |*
63   61   8056                |* END OF LINE, ADD LINE FEED TO INPUT AND OUTPUT QUEUE
64   62   8056                |*  PASS ENTIRE INPUT QUEUE TO OUTPUT QUEUE
65   63   8056                |*
66   64   8056 103C 000A      |         MOVE.B    #KEY_LF,D0      ;WRITE A LINE FEED CHARACTER
67   65   805A 287C 0000      |         MOVE.L    #IN_QPTRS,A4    ;  TO INPUT QUEUE,
67        805E 831E           |
68   66   8060 4EB9 0000      |         JSR       ENQUEUE         ;  FOR REDISPLAY WITH LINE.
68        8064 80BA           |
69   67   8066 287C 0000      |         MOVE.L    #OUT_QPTRS,A4   ;ADD LF TO OUTPUT QUEUE TO
69        806A 832E           |
70   68   806C 4EB9 0000      |         JSR       ENQUEUE         ;  MOVE DISPLAY TO NEXT LINE
70        8070 80BA           |
71   69   8072                |*
72   70   8072 287C 0000      |INOUT    MOVE.L    #IN_QPTRS,A4    ;PASS ADDR OF POINTER TABLE
72        8076 831E           |
73   71   8078                |*                                 ;  FOR INPUT QUEUE
74   72   8078 4EB9 0000      |         JSR       DEQUEUE         ;GET NEXT CHARACTER
74        807C 80F0           |
75   73   807E 4A46           |         TST.W     D6              ;IS THE QUEUE EMPTY?
76   74   8080 6600 0010      |         BNE       DISPLAY         ;YES, DISPLAY EVERYTHING IN
77   75   8084                |                                  ;  THE OUTPUT QUEUE
78   76   8084 287C 0000      |         MOVE.L    #OUT_QPTRS,A4   ;PASS ADDR OF POINTER TABLE
78        8088 832E           |
79   77   808A                |*                                 ;  AND ADD THE CHARACTER
80   78   808A 4EB9 0000      |         JSR       ENQUEUE         ;  TO OUTPUT QUEUE
80        808E 80BA           |
81   79   8090 60E0           |         BRA       INOUT           ;GET THE NEXT ITEM FROM
82   80   8092                |*                                 ;  INPUT QUEUE
83   81   8092                |*
84   82   8092                |* DISPLAY CONTENTS OF THE OUTPUT QUEUE
85   83   8092 287C 0000      |DISPLAY  MOVE.L    #OUT_QPTRS,A4   ;PASS ADDR OF POINTER TABLE
85        8096 832E           |
86   84   8098                |                                  ;  FOR OUTPUT QUEUE
87   85   8098 4EB9 0000      |         JSR       DEQUEUE         ;GET THE NEXT CHARACTER
87        809C 80F0           |
88   86   809E 4A46           |         TST.W     D6              ;IS THE QUEUE EMPTY?
89   87   80A0 668E           |         BNE       KEYIN           ;YES, GO BACK AND WAIT FOR
90   88   80A2                |                                  ;  NEXT CHARACTER
91   89   80A2                |         SYSCALL   .OUTCH          ;NO, OUTPUT THE CHARACTER
92   1m   80A2 1E3C 00F8      +         MOVE.B    #.OUTCH,D7
93   2m   80A6 4E4E           +         TRAP      #14
94   90   80A8 60E8           |         BRA       DISPLAY         ;  AND GET NEXT CHARACTER
95   91   80AA                |*                                 ;  FROM THE QUEUE.
96   92   80AA                |
97   93   80AA                |
98   94   80AA                |
99   95   80AA                |****************************************************************
```

Figure 9.11   *(continued)*

```
100    96   80AA            | * QINIT:            INITIALIZES A QUEUE AND ITS POINTER TABLE
101    97   80AA            | * INPUT:            A4.L   THE ADDR OF THE QUEUE POINTER TABLE
102    98   80AA            | *                   A5.L   THE ADDR OF THE START OF THE QUEUE
103    99   80AA            | *                   A6.L   THE ADDR OF THE END OF THE QUEUE
104   100   80AA            | * OUTPUT:           THE POINTERS IN THE QUEUE TABLE ARE
105   101   80AA            | *                       INITIALIZED TO CREATE AN EMPTY QUEUE.
106   102   80AA            | *
107   103   80AA 288D       | QINIT    MOVE.L   A5,NEWQP(A4)      ;SET ADDR OF START OF QUEUE
108   104   80AC 294E 0004  |          MOVE.L   A6,ENDQP(A4)      ;SET ADDR OF END OF QUEUE
109   105   80B0 294D 0008  |          MOVE.L   A5,QUEOUT(A4)     ;SET THE HEAD OF THE QUEUE
110   106   80B4 294D 000C  |          MOVE.L   A5,QUEIN(A4)      ;SET THE TAIL OF THE QUEUE
111   107   80B8 4E75       |          RTS
112   108   80BA            |
113   109   80BA            | ****************************************************************
114   110   80BA            | * ENQUEUE:   ADD A CHARACTER TO THE QUEUE
115   111   80BA            | * INPUT:          D0.B   THE CHARACTER TO BE ADDED
116   112   80BA            | *                 A4.L   THE ADDR OF THE QUEUE POINTER TABLE
117   113   80BA            | * OUTPUT:         D6.W   ZERO IF CHARACTER WAS ADDED;
118   114   80BA            | *                        -1 IF QUEUE WAS FULL
119   115   80BA 48E7 80A0  | ENQUEUE  MOVEM.L  A0/A2/D0,-(SP)    ;SAVE REGISTERS
120   116   80BE            | *
121   117   80BE 3C3C 0000  |          MOVE.W   #0,D6             ;SET GOOD RETURN CODE
122   118   80C2 206C 000C  |          MOVE.L   QUEIN(A4),A0      ;GET THE ADDR OF NEXT EMPTY LOC.
123   119   80C6 2448       |          MOVE.L   A0,A2             ;MAKE SURE THE TAIL DOES NOT
124   120   80C8 528A       |          ADDA.L   #1,A2             ;  WRAP AROUND TO THE HEAD.
125   121   80CA B5EC 0004  |          CMPA.L   ENDQP(A4),A2      ;  IF END OF THE QUEUE SPACE
126   122   80CE 6500 0004  |          BCS      ENQ_A             ;  IS REACHED, THEN SET  TAIL
127   123   80D2 2454       |          MOVE.L   NEWQP(A4),A2      ;  TO THE BEGINNING OF QUEUE.
128   124   80D4            | *
129   125   80D4 B5EC 0008  | ENQ_A    CMPA.L   QUEOUT(A4),A2     ;IS THE QUEUE FULL?
130   126   80D8 6600 000A  |          BNE      ENQ_OK            ;NO, SO ADD ITEM
131   127   80DC 3C3C FFFF  |          MOVE.W   #-1,D6            ;YES, SET THE RETURN CODE
132   128   80E0 6000 0008  |          BRA      ENQ_X             ;  AND EXIT THE ROUTINE
133   129   80E4            |
134   130   80E4 1080       | ENQ_OK   MOVE.B   D0,(A0)           ;STORE THE NEW INPUT
135   131   80E6 294A 000C  |          MOVE.L   A2,QUEIN(A4)      ;SAVE THE NEW TAIL POINTER
136   132   80EA            | *
137   133   80EA 4CDF 0501  | ENQ_X    MOVEM.L  (SP)+,D0/A0/A2    ;RESTORE REGISTERS
138   134   80EE 4E75       |          RTS
139   135   80F0            |
140   136   80F0            | ****************************************************************
141   137   80F0            | * DEQUEUE:   REMOVE A CHARACTER TO THE QUEUE
142   138   80F0            | * INPUT:          A4.L   THE ADDR OF THE QUEUE POINTER TABLE
143   139   80F0            | * OUTPUT:         D6.W   ZERO IF CHARACTER WAS REMOVED;
144   140   80F0            | *                        -1 IF QUEUE WAS EMPTY
145   141   80F0            | *                 D0.B   THE CHARACTER RETURNED
146   142   80F0            | *                        (UNDEFINED IF QUEUE EMPTY)
147   143   80F0            | *
148   144   80F0 2F08       | DEQUEUE  MOVE.L   A0,-(SP)          ;SAVE THE REGISTERS
149   145   80F2 3C3C 0000  |          MOVE.W   #0,D6             ;SET TO GOOD RETURN CODE
150   146   80F6 206C 0008  |          MOVE.L   QUEOUT(A4),A0     ;GET ADDR OF NEXT OUTPUT ENTRY
```

Figure 9.11   *(continued)*

```
151  147  80FA B1EC 000C  |           CMPA.L  QUEIN(A4),A0    ;IS THE QUEUE EMPTY?
152  148  80FE 6600 000A  |           BNE     DEQ_OK          ;NO, GET THE ITEM
153  149  8102 3C3C FFFF  |           MOVE.W  #-1,D6          ;YES, SET THE RETURN CODE
154  150  8106 6000 0012  |           BRA     DEQ_X           ;   AND EXIT THE ROUTINE
155  151  810A 1018       |DEQ_OK     MOVE.B  (A0)+,D0        ;GET THE ITEM
156  152  810C B1EC 0004  |           CMPA.L  ENDQP(A4),A0    ;IF END OF QUEUE SPACE IS
157  153  8110 6500 0004  |           BCS     DEQ_A           ; REACHED, SET NEW POINTER
158  154  8114 2054       |           MOVE.L  NEWQP(A4),A0    ; TO THE BEGINNING OF QUEUE.
159  155  8116 2948 0008  |DEQ_A      MOVE.L  A0,QUEOUT(A4)   ;SAVE NEW HEAD POINTER
160  156  811A            |*
161  157  811A 205F       |DEQ_X      MOVE.L  (SP)+,A0        ;RESTORE REGISTERS
162  158  811C 4E75       |           RTS
163  159  811E            |
164  160  811E            |****************************************************************
165  161  811E            |* DATA
166  162  811E            |*
167  163  811E            |* THESE ARE THE QUEUES
168  164  811E            |INQUEUE    DS.B 256      ;INPUT QUEUE
169  165       0000 821E  |E_INQUEUE  EQU  *        ;END OF INPUT QUEUE
170  166  821E            |OUTQUEUE   DS.B 256      ;OUTPUT QUEUE
171  167       0000 831E  |E_OUTQUEUE EQU  *        ;END OF OUTPUT QUEUE
172  168  831E            |*
173  169  831E            |* THESE ARE THE QUEUE POINTER TABLES
174  170  831E            |           EVEN          ; ALIGN ON WORD BOUNDARY
175  171       0000 831E  |IN_QPTRS   EQU  *
176  172  831E            |           DS.L 1        ;POINTS TO BEGINNING OF QUEUE (NEWQP)
177  173  8322            |           DS.L 1        ;POINTS TO END OF QUEUE (ENDQP)
178  174  8326            |           DS.L 1        ;HEAD OF QUEUE (QUEOUT)
179  175  832A            |           DS.L 1        ;TAIL OF QUEUE (QUEIN)
180  176       0000 832E  |OUT_QPTRS  EQU  *
181  177  832E            |           DS.L 1        ;POINTS TO BEGINNING OF QUEUE (NEWQP)
182  178  8332            |           DS.L 1        ;POINTS TO END OF QUEUE (ENDQP)
183  179  8336            |           DS.L 1        ;HEAD OF QUEUE (QUEOUT)
184  180  833A            |           DS.L 1        ;TAIL OF QUEUE (QUEIN)
185  181  833A            |*
186  182  833E            |* THESE ARE SOME CHARACTER DEFINITIONS
187  183       0000 000D  |KEY_CR     EQU  $0D      ;CARRIAGE RETURN
188  184       0000 000A  |KEY_LF     EQU  $0A      ;LINE FEED
189  185  833E            |           END
```

Figure 9.11 *(continued)*

Each character to be echoed and the entire line are output to the display screen using the
TUTOR TRAP #14 I/O command.

The MAIN program first passes the addresses for the input queue pointers to subroutine
QINIT. The subroutine QINIT is called to establish the initial values of the pointers in the
queue pointer table. Thus, the routine initializes the input queue to be empty. Next, the addresses
for the output queue are passed to QINIT and the subroutine initializes the output queue to
be empty.

Two queues are used. INQUEUE holds the input characters and OUTQUEUE holds the
characters to be displayed. Each queue is described by a table of four pointers as follows:

(a) the address of the start of the queue (NEWQP);
(b) the address of the last byte beyond the end of the queue (ENDQP);

(c) the position of the next entry to be dequeued (OUTQUE);

(d) the position of the next entry to be enqueued (INQUE).

After the queues are initialized, the MAIN program issues a TRAP #14 command to TUTOR and waits until a character is input. The subroutine ENQUEUE is called to add the character to INQUEUE. In order to echo the typed characters to the screen, ENQUEUE is called again to add the character to OUTQUEUE.

If the input character was not a carriage return ($0D), then every character in OUTQUEUE is dequeued, using the subroutine DEQUEUE, and displayed on the screen. If the captured character is a carriage return ($0D), then every character in INQUEUE is dequeued and added to OUTQUEUE. Then, every character in OUTQUEUE is dequeued and displayed on the screen. Finally, the MAIN program returns to label KEYIN to receive a new character. Notice that if either queue is full, any input characters are discarded.

9.3.5 Linked Lists

When dealing with arrays, the successor of an item being addressed is located by adding a constant to the address of the present item. For example, in a one-dimensional array

$$X(j + 1) = X(j) + C$$

where C is the number of bytes occupied by each element. The elements, ordered successively, occupied contiguous blocks of memory, as shown in Figure 9.12. In comparison, the *linked list* is a data structure that does not require contiguous storage of its elements. The linked list is treated in an introductory manner in this subsection. Advanced operations on such lists, including management of the memory space occupied by the list and ordering of list elements, is discussed in several of the references in the Further Reading section at the end of the chapter.

The sketch in Figure 9.13 shows a sample linked list of five items. Each element in the list contains both a pointer (address) or a *link* to the next element in the list and the data item. The list shown has a *one-way* link because only an item's successor can be found. Also, the sample list is unordered because the data items do not follow

Figure 9.12   Sequential array storage.

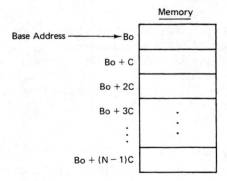

Notes:
(1) Bo is the base address.
(2) Each element is C bytes long.

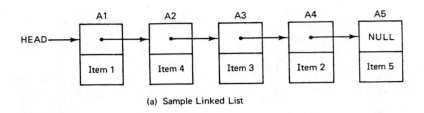

(a) Sample Linked List

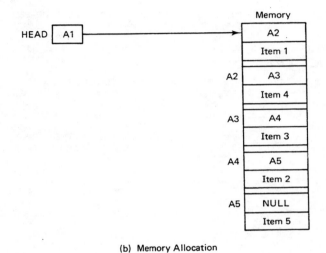

(b) Memory Allocation

Figure 9.13   Allocation for an unordered linked list.

one another in numerical order. The pointer to the list is stored at address HEAD. The last element, item 5 in the figure, contains a special symbol. It is designated NULL to indicate the end of the list. A NULL value of zero, for example, could be used in MC68000 programs because no data item would be stored at location 0000. If location HEAD contains the value NULL, the list is empty. Figure 9.13(b) illustrates how the sample list could be stored in memory.

Example 9.11

Figure 9.14 shows a subroutine that creates a linked list in a free area of memory with beginning address AVAIL. The nodes or elements in the list are each 8 bytes in length and the list will be initialized to contain 10 entries. The loop starting at label LINK computes the address of the next node in the list and then stores the address as the link to that node. After the loop terminates, the last link that was stored is rewritten with a NULL value.

EXERCISES

9.3.1. Write a subroutine to return the sine of an angle in degrees when an angle from 0 to 360 degrees is specified. Use a lookup table with a 1-degree resolution in angle. Assume that the table with starting address SINE is already provided and contains 16-bit values

```
abs.   LC    obj. code   source line
----   ----  ---------   -----------
  1   0000                 |        TTL     'FIGURE 9.14'
  2   0000                 |        LLEN    100
  3   8000                 |        ORG     $8000
  4   8000                 |        OPT     P=M68000
  5   8000                 |*
  6         0000 0000      |NULL    EQU     0            ;SET UP VALUE FOR NULL
  7   8000                 |*
  8   8000                 |*   CREATE A LINKED LIST
  9   8000                 |*
 10   8000                 |*   INPUT : AVAIL IS THE ADDRESS OF A FREE BLOCK OF MEMORY
 11   8000                 |*           HEAD IS THE POINTER TO THE TOP OF THE LIST
 12   8000                 |*
 13   8000                 |* OUTPUT : LINKS ARE STORED IN THE BLOCK AT AVAIL AND
 14   8000                 |*           HEAD POINTS TO THE FIRST NODE IN THE LIST
 15   8000                 |*
 16   8000 48E7 C0C0       |LNKLST  MOVEM.L D0-D1/A0-A1,-(SP) ;SAVE REGISTERS
 17   8004 41F9 0000       |        LEA     AVAIL,A0
      8008 802E            |
 18   800A 23C8 0000       |        MOVE.L  A0,HEAD          ;SET HEAD TO POINT TO AVAIL
      800E 802A            |
 19   8010 7008            |        MOVE.L  #8,D0            ;SET UP BYTES/NODE
 20   8012 720A            |        MOVE.L  #10,D1           ;SET UP NUMBER OF NODES
 21   8014                 |*
 22   8014 2248            |LINK    MOVE.L  A0,A1            ;NODE ADDRESS INTO A1
 23   8016 D1C0            |        ADD.L   D0,A0            ;COMPUTE NEXT NODE
 24   8018 2288            |        MOVE.L  A0,(A1)          ;STORE LINK TO NEXT NODE
 25   801A 5381            |        SUBQ.L  #1,D1            ;DECREMENT NUMBER OF NODES
 26   801C 66F6            |        BNE     LINK             ;UNTIL COUNT REACHES ZERO
 27   801E                 |*
 28   801E 22BC 0000       |        MOVE.L  #NULL,(A1)       ;REWRITE LAST LINK
      8022 0000            |
 29   8024                 |*
 30   8024 4CDF 0303       |        MOVEM.L (SP)+,D0-D1/A0-A1 ;RESTORE REGISTERS
 31   8028 4E75            |        RTS
 32   802A                 |*
 33   802A                 |HEAD    DS.L    1                ;POINTER TO TOP OF LINKED LIST
 34   802E                 |AVAIL   DS.L    20               ;MEMORY BLOCK FOR LIST
 35   802E                 |*
 36   807E                 |        END
```

Figure 9.14   Subroutine to create a linked list.

for the sines of angles from 0 to 89 degrees. If the angle is greater than 90 degrees, compute the sine of the angle using trigonometric identities. (*Note:* The value could be calculated using the Taylor series expansion for SIN(X):

$$SIN(X) = X - X^3/3!) + (X^5/5!) - \cdots$$

The series method will yield the sine value to any accuracy but is far slower than the table lookup if the table has sufficient resolution. Create an abbreviated table of sine values and test your routine.

9.3.2. Write a subroutine to clear a three-dimensional array *without* calculating the three-dimensional address polynomial. Assume that the array is stored in column-major form and the subroutine is passed the starting address, the size of the array ($M \times N \times O$), and the number of bytes in each element.

9.3.3. Write a routine to multiply two 2×2 matrices.

9.3.4. Write the address polynomial for a *k*-dimensional array if $I_1, I_2, I_3, \ldots, I_k$ are the indices and $L_1, L_2, L_3, \ldots, L_k$ are the lengths.

9.3.5. Write a subroutine to copy a string from one buffer to another.

9.3.6. Write a subroutine to count the number of characters in a string.

9.3.7. Write a subroutine to compare two strings that allows different length strings.

9.3.8. Write a program that gets input from a temperature probe and stores those values in a queue along with a time stamp (hours, minutes, seconds). Every tenth reading a subroutine is called that empties the queue and displays the average of all the values found in the queue, as well as the time of the first entry and the last entry. Let the queue itself contain nothing but pointers (addresses) to an entry in a table holding a measurement and the time stamp. Create test data of at least 30 entries to test your program.

9.3.9. Modify the program of Example 9.10 to ring the bell when INQUEUE becomes full. You should bypass OUTQUEUE and directly write the ASCII Bell character to the screen.

9.3.10. Modify the program of Example 9.10 to display the line with the order of the characters reversed. Do this by pushing each character of the input line onto a stack and then, once the entire line is written to the stack, popping them off the stack and placing them in the output queue.

9.3.11. Write a subroutine to remove an entry from the top of the linked list created in Example 9.11. What happens if the list is empty [i.e., (HEAD) = NULL]?

9.4 SUBROUTINE USAGE AND ARGUMENT PASSING

The use of *subroutines* or procedures is an important programming technique to create modular programs in which each subroutine performs a specific task within the overall program. The method of transfer of control between the calling program and the subroutine is called *subroutine linkage*. In the MC68000, the call to the subroutine is performed by the instruction

JSR ⟨SUBR⟩

which first causes the return address within the calling program to be pushed onto the system stack. Then control is transferred to the subroutine at address ⟨SUBR⟩. The address may be specified by any of the control addressing modes of the MC68000. Thus transfer of control is accomplished very simply in the MC68000. When data must be passed between the calling program and the subroutine, a number of methods are available to transfer the information. The method is selected when the program is designed, and this choice constitutes an important part of the program design.

The information needed by the subroutine is defined in terms of *parameters* that allow the subroutine to handle general cases rather than operate on specific values. Each call to the subroutine allows different values, called *arguments,* to be supplied for the parameters. Some typical FORTRAN subroutine references are shown in

Figure 9.15. The subroutine is named SUBR and has the parameters A, B, and C. It may be called with various arguments as long as the arguments are the same data type (integer, floating–point, etc.) as the parameters in the subroutine definition. The names or values of the arguments are arbitrary. The symbolic names for the arguments in the example are actually addresses assigned by the compiler. The specific values 1.0 and 3.0 in the second call can be substituted to take advantage of the flexibility of the FORTRAN language. In assembly language, the distinction between actual values and the addresses of arguments is important.

The mechanics of defining parameters and transmitting arguments to subroutines are more complex in assembly language. Processor registers, the system stack, or fixed locations in memory can hold arguments. Also, the LINK and UNLK instructions of the MC68000 can be used to create *stack frames* as a subroutine is called. This frame is a block of memory reserved on the stack that holds the return address, arguments, and local variables, if any. Recursive subroutines can be implemented using this method of data handling in the subroutine.

```
!-------------------------------------------------------------------!
!                                   !                               !
!  Calling Program                  !  Subroutine                   !
!                                   !                               !
!-----------------------------------+-------------------------------!
!                                   !                               !
!                                   !  SUBROUTINE SUBR (A, B, C)     !
!              .                    !   .                           !
!              .                    !   .                           !
!              .                    !   .                           !
!  CALL SUBR (X, Y, Z)              !   .                           !
!              .                    !   .                           !
!              .                    !  RETURN                       !
!              .                    !  END                          !
!  CALL SUBR (1.0, 3.0, RESULT)     !                               !
!              .                    !                               !
!              .                    !                               !
!              .                    !                               !
!  CALL SUBR (A(1), W, ANS)         !                               !
!              .                    !                               !
!              .                    !                               !
!              .                    !                               !
!                                   !                               !
!  END                              !                               !
!                                   !                               !
!-----------------------------------+-------------------------------!
```

Figure 9.15   FORTRAN subroutine usage.

9.4.1 Passing Arguments to Subroutines

The parameters, which define the arguments to be transferred between a subroutine and the calling program, can be data values, addresses, or combinations of both. When only a small number of arguments are to be transferred, they are passed directly between the programs in processor registers. When many variables are passed or when a data structure such as an array is being referenced, the address of the group of variables or data structure is transferred. The distinctions here have broad significance in many high-level language programs when the operation of the compiler is considered. In assembly language, the distinction between values and addresses is important because the method of passing the parameters determines how the arguments are accessed.

Several techniques used to pass values or addresses between programs are listed in Table 9.11. The calling program sets up the calling sequence, including the definition of the arguments to be transferred to the subroutine. The subroutine then accesses the arguments for processing and possibly returns values or addresses to the calling program. The arguments passed to the subroutine are defined as the *input parameters*. The results are values or addresses corresponding to the *output parameters* for the subroutine. Of course, a combination of the techniques listed in Table 9.11 could be used when a complicated set of input and output parameters are defined.

Register Transfer. The simplest method of passing arguments is using the MC68000 register set. Data values can be passed in any of the eight data registers. Similarly, the address registers can be used to pass addresses that may point to data values or may contain the starting addresses of data structures. The register passing scheme has the advantages of simplicity, small memory requirements, and minimum execution time. The number of arguments that can be passed is limited to the number of registers available: 15 for the MC68000. The designer of the subroutine and the calling routine

Table 9.11 Methods to Pass Arguments

Type	Description	Comments
Register	The calling routine loads predefined registers with values or addresses.	Number of parameters is limited Dynamic
Stack	The calling routine pushes values or addresses on the stack.	Return address must be saved during processing and restored before return
Parameter areas	Memory areas are defined that contain the values or addresses.	Static, if areas are defined during assembly Dynamic, if base address of area is passed in a register
In-line	Values or addresses are stored following the call. The subroutine computes the location of the parameters.	Static

need only agree on which registers are used to pass the arguments. For example, the instruction sequence

```
MOVE.W      VALUE,D1      ;DATA
MOVEA.L     ADDTAB,A1     ;POINTER
LEA         HEAD,A2       ;ADDRESS OF HEAD
JSR         SUBR
```

sets up a 16-bit value in D1, the address pointer in location ADDTAB in A1 and the address HEAD in A2. The subroutine SUBR can then access the values in the registers directly to perform its function.

Stack Transfer. A stack can be used to pass arguments by having the calling routine push values or addresses on the stack before the call. A private stack, which can be defined in MC68000 programs using address register A0, A1, . . . , or A6 to represent the stack pointer, could be used. The values are pushed using the predecrement or postincrement mode of addressing in the calling program. Popping the arguments in the subroutine allows access to the values or addresses. For a private stack, the modification of the stack pointer during execution does not affect system operation and is handled at the discretion of the programmer. The sequence to set up A1 as the stack pointer and pass two values might be

```
LEA         STACKP,A1     ;ADDRESS OF STACK IN A1
MOVE.W      VAL1,(A1)+    ;PUSH FIRST VALUE
MOVE.W      VAL2,(A1)+    ;PUSH SECOND VALUE
JSR         SUBR
```

where STACKP is the bottom of the stack in memory and the stack increases into higher memory locations. The subroutine accesses the values by the sequence

```
MOVE.W      -(A1),D2      ;SECOND VALUE
MOVE.W      -(A1),D1      ;FIRST VALUE
```

if the object is to load the values into data registers. The stack pointer A1 now contains its original value STACKP. In this example, the argument in register A1 contained the address of a data structure (the stack) in memory.

When the system stack is used to pass arguments, the return address is at top of the stack when the subroutine begins execution. This value must be removed before the subroutine can extract the arguments from the stack. When the processing of the subroutine is complete, the return address must be placed back on the top of the stack before the RTS instruction can be executed. The calling sequence for this method of parameter passing could be

```
PEA         ADDR          ;PUSH ADDRESS
MOVE.W      VAL1,-(SP)    ;PUSH VALUE
JSR         SUBR
```

The address ADDR is placed on the stack first, then the value in location VAL1, and finally the return address. Becaue (PC) is at the top of the stack, it can be saved and then restored before the return. The sequence to accomplish this might be

```
MOVE.L      (SP)+,A1      ;SAVE (PC) TEMPORARILY
MOVE.W      (SP)+,D1      ;GET DATA
MOVEA.L     (SP)+,A2      ;GET ADDRESS
            .
            .             ;PROCESS
            .
MOVE.L      A1,-(SP)      ;RESTORE (PC)
RTS
```

Because this method is typically used by a program in the user mode, the active stack pointer is the USP. Modifying the stack pointers does not interfere with interrupt processing and similar system operations that use the supervisor stack pointer (SSP).

Memory Locations for Arguments. When large numbers of parameters are to be passed, a *parameter area* can be set up in memory. This area contains, in a predetermined sequence, the values or addresses that are accessed by the subroutine after it has been passed the starting address of the area. The same area could be used by several subroutines requiring different parameters, as long as the area is large enough to hold the maximum number of arguments.

Another use of this method is common in systems that have subroutines in ROM such as the TUTOR monitor described in Chapter 5. The parameter area in RAM memory is defined according to the system requirements and the address is passed to the subroutines to use as a base address in accessing the arguments.[8]

The calling program could set up a parameter area in the following way:

```
         MOVE.W    VAL1,PRMAREA         ;STORE DATA
         MOVE.W    VAL2,PRMAREA+2       ;SECOND WORD
                   .
                   .

         MOVE.W    VAL5, PRMAREA+8      ;FIFTH WORD
         LEA       PRMAREA,A1           ;PUSH ADDRESS
         JSR       SUBR
                   .
                   .
                   .
PRMAREA  DS.W      5                    ;RESERVED AREA
         END
```

---

[8] Another version, used when FORTRAN "common" areas are specified, defines the address of the parameter area to both the calling program and the subroutine during compilation.

in which five words are defined as parameters. The subroutine could access the values using indirect addressing with displacement. For example, the instruction

MOVE.W 6(A1),D1

transfers the fourth value to D1. Many variations are possible to define the parameter areas in memory.

In-Line Coding. Another method of passing values to a subroutine is to code the values following the call to the subroutine. This method, called *in-line coding,* defines argument values that are constant and will not change after assembly. These values can be defined by DC directives following the call. Consider the instruction sequence

```
JSR    SUBR
DC.W   1            ;IN-LINE ARGUMENT
```

The 32 bits of (PC) are pushed on the system stack by the subroutine call. This address points to the location of the *argument* in the instruction sequence. The following sequence can be executed by the subroutine to load the argument into the low-order word of D1 and point the return address on the stack to the word beyond the value:

```
MOVEA.L    (A7),A0        ;GET PC VALUE
MOVE.W     (A0)+,D1       ;GET ARGUMENT, INCREMENT
                          ;PUSH NEW RETURN ADDRESS
MOVE.L     . A0,(A7)
       .                  ;PROCESS
       .
       .
RTS
```

The first instruction loads the (PC) into A0 from the stack. Register A0 then addresses the argument and is incremented by 2 after D1 is loaded. After A0 is incremented, it points to the next instruction in the calling program following the in-line argument. The next instruction in the subroutine pushes the correct return address on the stack, overwriting the value saved by the JSR instruction. The RTS instruction is used here to restore (PC) and return control to the calling program. The reference to A7 indicates the system stack pointer; either USP or SSP, depending on whether the program mode is user or supervisor, respectively.

9.4.2 Stack Frames

One of the principal issues in the design of subroutines involves the concept of *transparency.* Simply stated, when a subroutine finishes executing, it should have no "visible" effect except as defined by its linkage to the calling program. For example, a subroutine should not change the values in any registers, unless a register is used to return a result. In the previous example programs, this was accomplished by pushing the contents of the registers used by the subroutine on the stack upon entry to the subroutine. The values were restored before returning to the calling program. The return address was automatically saved and restored by the JSR and RTS instructions.

The use of the system stack to save and restore the return address and the contents of registers used within the subroutine assured that the details of the subroutine operation were transparent to the calling program. If a subroutine itself made a subroutine call, the use of the stack for temporary storage of register contents by each subroutine and for each return address allowed such nesting of subroutine calls without difficulty. This concept of using the stack to store data temporarily during subroutine execution can be extended by defining a *stack frame*.

The stack frame is a block of memory in the stack that is used for return addresses, input parameters, output parameters, and local variables. It is the area of the stack accessed by a subroutine during its execution. Local variables are those values used during the subroutine execution that are not transferred back to the calling routine. A loop counter, for example, which changes as the subroutine performs each iteration, might be defined as a local variable. On each call to the subroutine, a new set of parameters, local variables, and return addresses can be accessed by a subroutine using the stack frame technique. If the subroutine is called before it is completely finished, the values in the stack frame will not be destroyed.

MC68000 Stack Frames. In multiprogramming systems, several independent tasks may use the same subroutine. For example, two CRT terminals may be connected to the system but share the same I/O routine. As the operating system switches control between the terminals, it is possible that the I/O routine of one is interrupted and control passed to the other terminal temporarily. All the data associated with the first terminal used by the I/O routine must be saved so that when the first terminal regains control, the I/O routine begins where it left off. Such usage requires *reentrant* routines in which no data in the program memory itself changes value during execution. Any values that change are placed on the stack for storage. Thus the program or code is separated completely from the data on which it operates. A special case is the *recursive* routine, which calls itself and so is self-reentrant. The stack frame allows reentrant and recursive routines to be created easily.

The stack frame is created by the calling program and the subroutine using the MC68000 instructions LINK and UNLK. The syntax and operation of these instructions are shown in Table 9.12. Access to variables on the stack by the subroutine is accomplished by using offsets (or indexing) from a base register called a *frame pointer*. Although the stack pointer may change value as items are pushed or popped, the

Table 9.12   Operation of LINK and UNLK

Instruction	Syntax	Operation
Link	LINK ⟨An⟩,#⟨disp⟩	1. (SP) ← (SP) − 4; ((SP)) ← (An) 2. (An) ← (SP) 3. (SP) ← (SP) + ⟨disp⟩
Unlink	UNLK ⟨An⟩	1. (SP) ← (An) 2. (An) ← ((SP)); (SP) ← (SP) + 4

Note: ⟨disp⟩ is a 16-bit signed integer. A negative displacement is specified to allocate stack area.

frame pointer does not change during the subroutine's execution. Figure 9.16 illustrates a possible sequence in the calling program and the operation of the subroutine. Figure 9.17 shows the stack contents for this example. Many variations are possible according to the application. In the case shown, the calling routine first reserves N bytes on the stack for arguments to be returned by the subroutine. Then an input value and an address are pushed on the stack. The JSR instruction pushes the return address and transfers control to the subroutine. At this point, the stack contains N bytes of space for the result, the 32-bit contents of location ARG, the address X, and the return address.

The subroutine first executes the LINK instruction to create a stack frame and define the frame pointer. This instruction saves the value of $\langle An \rangle$ on the stack and replaces $\langle An \rangle$ with the value of the stack pointer using A1 in our example. The frame pointer thus points to the *bottom* of the local area for the subroutine. Then the displacement is added to the stack pointer so that (SP) points M bytes farther down in memory. Local variables are stored in this area and accessed by displacements from the value in the frame pointer. Once the input arguments are processed and the outputs are stored on the stack, the UNLK instruction is executed. This instruction as defined in Table 9.12 releases the local area and restores the stack pointer contents so that it points to the return address.

Specifically, the UNLK instruction first loads (SP) with the value in the frame pointer A1, which points to the old value of A1 saved on the stack by the LINK instruction. Then A1 is restored to its previous value using the autoincrement mode of addressing so that (SP) now points to the return address. The RTS instruction returns control to the calling program, with (SP) indicating the location of the top of the parameter area set up by this program as indicated in Figure 9.17. The calling routine then adds 8 to (SP) by the instruction

ADD.L   #8,SP

which skips over the input parameter area and leaves (SP) pointing to the output arguments. These arguments now represent input values to the calling program. After these are popped, the stack pointer has its original contents.

EXERCISES

9.4.1. Compare the passing of addresses rather than data values as arguments when a subroutine processes an array.

9.4.2. Discuss the advantages and disadvantages of in-line parameter passing.

9.4.3. The dynamic nature of the stack used to hold arguments can result in a considerable savings of memory space when compared to the assignment of individual parameter areas for each subroutine. How does one compute the required maximum size of the stack to hold parameters?

9.4.4. Write a subroutine to compare two multiple-precision integers (64 bits) and place the largest value in a given location. Pass the addresses on the stack with the first integer in location N1, the second in N2, and the result to be placed in location MAX. Be sure to

```
abs.   LC   obj. code    source line
----   ----  ---------    -----------
  1   0000                        |            TTL     'FIGURE 9.16'
  2   0000                        |            LLEN    100
  3   8000                        |            ORG     $8000
  4   8000                        |            OPT     P=M68000
  5   8000                        |*
  6   8000                        |*  CALLING PROGRAM
  7   8000                        |*
  8         0000 0008      |N          EQU 8 ; 8 BYTES FOR OUTPUTS
  9         0000 0008      |M          EQU 8 ; 8 BYTES FOR LOCAL VARIABLES
 10   8000                        |*          .
 11   8000                        |*          .
 12   8000 DFFC FFFF       |            ADD.L   #-N,SP              ; OUTPUT AREA
      8004 FFF8            |
 13   8006 2F39 0000       |            MOVE.L  ARG,-(SP)           ; INPUT ARGUMENT FOR SUBROUTINE
      800A 801E            |
 14   800C 4879 0000       |            PEA     X                   ; INPUT ADDRESS X
      8010 8022            |
 15   8012 4EB9 0000       |            JSR     SUBR
      8016 80EA            |
 16   8018 508F            |            ADDQ.L  #8,SP               ; SKIP OVER INPUTS
 17   801A 221F            |            MOVE.L  (SP)+,D1            ; READ OUTPUTS FROM SUBROUTINE
 18   801C 241F            |            MOVE.L  (SP)+,D2
 19   801E                        |*          .
 20   801E                        |*  CONTINUE PROCESSING AS REQUIRED
 21   801E                        |*          .
 22   801E                        |*          .
 23   801E                        |*
 24   801E                        |*          (END OF MAIN PROGRAM PROCESSING)
 25   801E                        |*
 26   801E 0123 4567       |ARG        DC.L    $01234567           ; ARGUMENT TO PASS
 27   8022                        |X          DS.B    200                 ; TABLE WHOSE ADDRESS IS PASSED
 28   8022                        |*
 29   80EA                        |*  SUBROUTINE
 30   80EA                        |*
 31   80EA 4E51 FFF8       |SUBR       LINK    A1,#-M              ; SAVE OLD FP
 32   80EE                        |*          .
 33   80EE                        |*          .
 34   80EE 2379 0000       |            MOVE.L  LOCAL1,(-4,A1)      ; SAVE LOCAL VARIABLES
      80F2 8112 FFFC       |
 35   80F6 2379 0000       |            MOVE.L  LOCAL2,(-8,A1)
      80FA 8116 FFF8       |
 36   80FE                        |*          .
 37   80FE                        |*          .
 38   80FE 52A9 FFFC       |            ADD.L   #1,(-4,A1)          ; CHANGE LOCAL VARIABLE
 39   8102 2469 0008       |            MOVEA.L (8,A1),A2           ; GET X
 40   8106                        |*          .
 41   8106                        |*          .
 42   8106 2379 0000       |            MOVE.L  OUTPUT1,(16,A1)     ; PUSH AN OUTPUT
      810A 811C 0010       |
 43   810E                        |*          .
 44   810E                        |*          .
 45   810E 4E59            |            UNLK    A1                  ; RESTORE SP AND RETURN
 46   8110 4E75            |            RTS
 47   8112                        |*
 48   8112 9876 5432       |LOCAL1     DC.L    $98765432           ; LOCAL VARIABLES
 49   8116 8765 4321       |LOCAL2     DC.L    $87654321
 50   811C 4142 4344       |OUTPUT1    DC.L    'ABCD'              ; OUTPUT VALUE
 51   8120                        |            END
```

Figure 9.16   Program operation to create a stack frame.

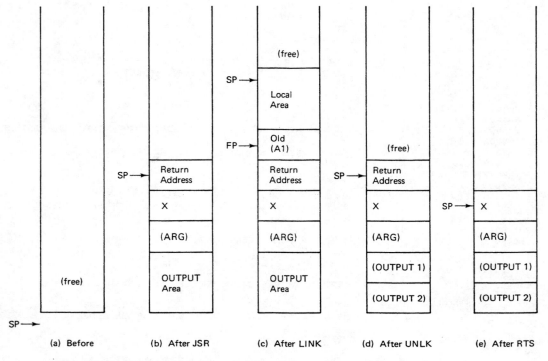

	(free)		
SP →			
	Local Area		
FP →	Old (A1)	(free)	
SP → Return Address	Return Address	SP → Return Address	
X	X	X	SP → X
(ARG)	(ARG)	(ARG)	(ARG)
OUTPUT Area	OUTPUT Area	(OUTPUT 1)	(OUTPUT 1)
		(OUTPUT 2)	(OUTPUT 2)

(a) Before (b) After JSR (c) After LINK (d) After UNLK (e) After RTS

Figure 9.17 Stack contents using stack frame.

correct the stack pointer value to "collapse" the stack before the subroutine returns control to the calling program.

9.4.5. Compare the instruction sequence

```
LEA     $2000,A3
LEA     $1FF0,SP
```

with the instruction

```
LINK    A3,−#$10
```

if (SP) = $2000 when the LINK instruction is executed.

9.4.6. Write a program that produces the sum and the average value of N positive 16-bit integers stored in a fixed area of memory. Use a stack frame to pass all the parameters between the program segments.

REVIEW QUESTIONS AND PROBLEMS

Multiple Choice

9.1. The instruction ADDA.L #20,A1
  (a) adds 20 to the value in A1 and sets Z = {1}
  (b) adds 20 to the value in A1 and sets Z = {0}
  (c) adds 20 to the value in A1

9.2. True or False: The instruction SUBA can never result in a negative number.
   (a) True
   (b) False

9.3. The instruction sequence

    CMPA.L A1,A2
    BHI   LOOP

   (a) branches if $(A1) > (A2)$
   (b) branches if $(A2) > (A1)$
   (c) branches if $(A2) \geq (A1)$

9.4. The instructions LEA and PEA transfer
   (a) operands in memory
   (b) data being addressed
   (c) addresses

9.5. The instruction LEA TABLE,A1 is equivalent to
   (a) MOVEA.L TABLE,A1
   (b) MOVEA.L #TABLE,A1
   (c) MOVE.L TABLE,A1

9.6. When a program is written with Program Counter relative references only, it is called
   (a) reentrant
   (b) recursive
   (c) position independent

9.7. The instruction MOVE.W (A1)+,D1 could be used to sequential address
   (a) an array of word-length operands
   (b) an array of longword operands
   (c) a stack of word-length operands on a stack growing into higher memory

9.8. Assume that A1 points to the first element of a word length array. The instruction MOVE.W (4,A1),D1 points to
   (a) the second element
   (b) the third element
   (c) the fourth element

9.9. A sequence of ASCII characters input from a keyboard would most likely be stored as a
   (a) list
   (b) queue
   (c) string

9.10. A data structure that does not require contiguous storage of its elements is a
   (a) string
   (b) stack
   (c) linked list

9.11. To transfer data to a subroutine, the fastest method is
   (a) stack transfer
   (b) register transfer
   (c) in-line transfer

9.12. The instruction to push an address on the stack for subroutine usage is
   (a) PEA
   (b) MOVE.L
   (c) MOVEA.L

9.13. Use of subroutines saves
   (a) execution time
   (b) memory locations
   (c) register space

9.14. The instructions LINK and UNLK create
   (a) stack frames
   (b) parameter areas
   (c) subroutines

9.15. The instruction LINK A1,#−8 is used to
   (a) add 8 to the stack pointer
   (b) reserve 8 bytes on the stack
   (c) create a frame pointer of 8 bytes

Programs

9.16. Write a program to sort a set of variable-length strings in ascending order. Build a table of addresses pointing to the strings, and use a simple Bubble Sort, as in Example 9.7, to rearrange the addresses in the table. When the sort is complete, display the strings in sorted order with each string on a separate display line

Use the following 10 strings as a test case:

STRINGA	DC.B	'Saha, P.K.',0
STRINGB	DC.B	'United States of America',0
STRINGC	DC.B	'United Colonies of Luna',0
STRINGD	DC.B	'Willis, Connie',0
STRINGE	DC.B	'Adams, Douglas',0
STRINGF	DC.B	'Feynman, Richard',0
STRINGG	DC.B	'Le Guin, Ursula',0
STRINGH	DC.B	'Von Neumann, John',0
STRINGI	DC.B	'Knuth, Donald',0
STRINGJ	DC.B	'Tyler, Annie',0

9.17. Create a subroutine to search for a substring within a string and return to its relative position or −1 if not found.

9.18. Create a subroutine to create a new string from a substring of another string. Inputs would be the address of original string, the starting relative position of the substring, the length of the substring, and the address of the output buffer. Outputs would be the modified output buffer and a return code indicating success or failure (position not found in string, unexpected end of string, etc.).

9.19. Create a subroutine to insert a substring within another string. Inputs would be the address of original string, the starting relative position of the substring, the length of the substring, the address of the target string, and the starting relative position for the insertion. Outputs would be the modified output string and a return code indicating success or failure (position not found in string, unexpected end of string, etc.).

9.20. Create a linked list of the strings from Problem 9.16 in ascending sequence. Start with an empty list and insert the strings in order (this is called an insertion sort). Display the resulting list, with each string on a separate line, from the beginning of the linked list to the end.

FURTHER READING

Position-independent coding is discussed in most textbooks written about the PDP–11 computer. In particular, Tanenbaum explains this technique and also includes discussions of base register addressing and dynamic relocation of programs. Stritter and Gunter discuss a number of the MC68000 instructions used in this chapter. Knuth gives an excellent discussion of data structures, as well as subroutine usage, with emphasis on assembly language programming. The stack frame technique is well discussed in Wakerly's book, which includes MC68000 programming examples.

Knuth, Donald E. *The Art of Computer Programming* Vol. 1: *Fundamental Algorithms.* (Reading, Mass.: Addison–Wesley, 1969).

Stritter, Edward, and Tom Gunter. "A Microprocessor Architecture for a Changing World: The Motorola 68000." *IEEE Computer* 12, no. 2 (February 1979), 43–52.

Tanenbaum, Andrew S. *Structured Computer Organization,* 2nd ed. (Englewood Cliffs, N.J.: Prentice–Hall, 1984).

Wakerly, John F. *Microcomputer Architecture and Programming, The 68000 Family.* (New York, John Wiley and Sons, 1989).

CHAPTER 10

System Operation

The previous chapters have considered the MC68000 applications programs. The emphasis was, therefore, on the instruction set of the MC68000 and programming techniques. The next three chapters discuss the operation of MC68000 systems, emphasizing the interaction between the various components. These components include the operating system or supervisor program, the CPU, memory, and associated hardware.

The CPU can be running a program *normally,* processing an *exception,* or simply waiting in a *halted* state for a signal to reset the system. The reset sequence is associated with system initialization and is initiated by external circuitry. The operation of the system reset is shown in Figure 10.1(a). The initial values of certain registers and the vector addresses associated with system operation are initialized by the system hardware and the supervisor program. Then control is passed to an applications program, which operates as illustrated in Figure 10.1(b). Program execution continues until its task is complete or a supervisor service is required. The program may cause a trap to return control to the supervisor in order to perform a service, such as inputting or outputting data.

The trap is an example of a condition that causes exception processing in the supervisor mode, as shown in Figure 10.1(c). Each exception initiates execution of the appropriate exception routine. Upon completion of the routine, control may be returned to the program that was executing when the exception occurred. The exact sequence depends completely on the design of the system.

This chapter discusses MC68000 system operation in the various states. Instructions used for system control and initialization are included. Exception processing is covered in Chapter 11 using program examples. System design considerations and interfacing requirements are discussed in Chapter 13. A review of Chapters 2 and 4 is suggested, as those chapters contain preliminary discussions of many of the concepts presented in the following text.

10.1 PROCESSOR STATES AND MODES

The MC68000 operates in one of three processing states: normal, exception, or halted. When a program is executing, the processor operation is further described by its mode or privilege state. In normal operation, the mode is either supervisor or user, as

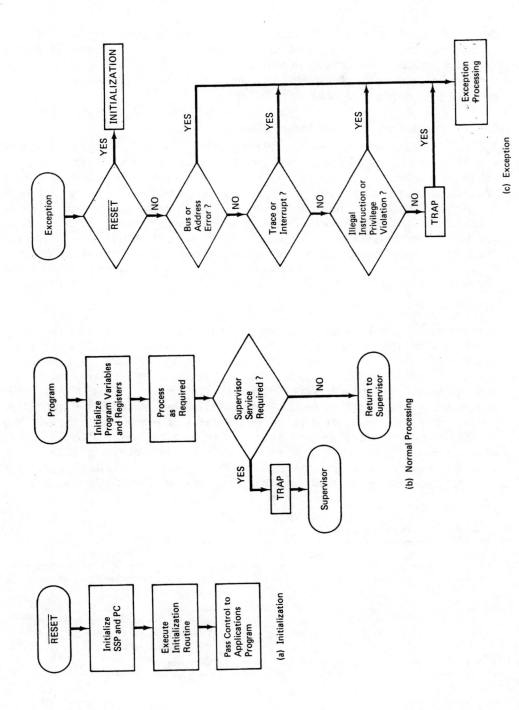

Figure 10.1 System operation.

(a) Initialization

(b) Normal Processing

(c) Exception

determined by the setting of the status bit ("S" bit) in the status register. When an exception occurs, the processor is automatically put in the supervisor mode. The states and modes affect both the programming and the hardware operation of the system. Programming aspects are considered here, and a discussion of the external electrical activity of the processor is presented in Chapter 13.

10.1.1 Normal, Exception, and Halted States

Table 10.1 summarizes the three states of the processor. The *normal* state is associated with program execution in either the supervisor or the user mode. In this state, the processor is fetching instructions and operands from memory to process. Programs not associated with exception processing always execute in the normal state. The state is typically entered after initialization of the system. Unless a hardware failure is detected, the state of the processor changes only when an exception occurs.

The only special case during operation in the normal state is the *stopped* condition. Here the processor no longer executes instructions but waits for an external event to initiate continued execution. A stopped condition in the normal state occurs when the instruction

STOP   #$\langle d_{16} \rangle$

is executed. The 16-bit immediate value $\langle d_{16} \rangle$ replaces the contents of the status register and the (PC) is advanced to point to the next instruction. Until an interrupt or other event is recognized, the processor stops fetching and executing instructions. The STOP instruction must be executed by a program in the supervisor mode or a privilege violation (trap) will occur.[1] Processor activity resumes when an interrupt is recognized or when a system reset is performed. In practice, the STOP instruction can be considered a "wait for interrupt" instruction used only in special applications.

The *exception* state is entered when a reset, hardware error condition, trace, interrupt, or trap is recognized. The processor is automatically placed in the supervisor mode and program execution begins at the location in memory indicated in the exception vector table for the particular exception that occurred. The addresses in the table

Table 10.1   Processor States and Modes

State	Condition	CPU Activity
Normal	Processing	Supervisor or user mode program executing
	Stopped	Waiting for interrupt
Exception	Reset	Initialization
	Interrupt	Interrupt acknowledge and processing
	Trap	Trap processing
	Trace	Single instruction trace
Halted	System error condition	No activity

---

[1] A trace exception will occur if the trace condition is indicated when the STOP instruction is executed.

are supplied when the system is initialized. When a program is executing in the user mode, only exception processing can change the mode to the supervisor mode.

The *halted* state provides system protection by causing the CPU to cease all external signaling activity. As explained in Chapter 13, the processor halts if certain types of errors are detected while the processor is already processing another error. These error conditions should occur only after a catastrophic hardware failure for which recovery is not possible, so a system reset is necessary to restart the halted processor. During the halted state, the CPU indicates its condition on the system bus via a signal line explicitly for this purpose. External circuitry or an operator must then determine whether to restart the halted processor.

A processor in the stopped condition or halted state cannot be restarted by a program because external signal lines control the CPU activity after one of these conditions has occurred. In contrast, during normal or exception processing, a program controls the operation of the processor. The distinction between supervisor mode and user mode programs is important when the processor operates in the normal state. During normal operations, the mode defines the privilege conditions for a program.

10.1.2 Supervisor and User Modes

Some of the differences between the supervisor mode and the user mode of the MC68000 were presented in earlier chapters. The use of a separate stack pointer for each mode and other important distinctions are listed in Table 10.2. Basically, the supervisor mode represents the level with more privilege. A program in this mode

Table 10.2   Distinctions Between Supervisor and User Modes

	Supervisor Mode	**User Mode**
Enter mode by	Recognition of a trap, reset, or interrupt	Clearing status bit "S"
System stack pointer	Supervisor stack pointer	User stack pointer
Other stack pointers	User stack pointer and registers A0–A6	Registers A0–A6
Status bits available		
Read	$C, V, Z, N, X, I_0–I_2, S, T$	$C, V, Z, N, X, I_0–I_2, S, T$
Write	$C, V, Z, N, X, I_0–I_2, S, T$	C, V, Z, N, X
Instructions available	All, including:	All except those listed
	STOP	under supervisor mode
	RESET	
	MOVE to SR	
	ANDI to SR	
	ORI to SR	
	EORI to SR	
	MOVE USP to ⟨An⟩	
	MOVE to USP	
	RTE	
Function code line FC2 =	1	0

can execute any MC68000 instruction and can change the status register and user stack pointer. User mode programs are allowed a restricted instruction set with no privilege to execute certain instructions that control the operation of the system.

As indicated in Table 10.2, the supervisor mode is entered when any exception is recognized by the CPU. A program operating in the supervisor mode can change to the user mode by modifying the status bit, $(SR)[13]$, in the status register. Whenever $(SR)[13] = \{1\}$, the processor is operating in the supervisor mode and the transition to user mode can be accomplished by setting $(SR)[13] = \{0\}$. Because only the supervisor program can change modes, the instructions listed in the table that indicate SR as destination are privileged instructions. In addition, the contents of the user stack pointer (USP) can only be transferred or changed by a program in the supervisor mode. Attempted execution of a privileged instruction by a user mode program causes a trap that indicates a privilege violation.

In the user mode, an MC68000 program may use the MOVE instruction to read the contents of the entire status register, but it may write only to the condition code register (CCR), or $(SR)[7:0]$. The other system control instructions, STOP, RESET, and RTE, are also privileged. The MC68010 processor differs from the MC68000 in that it allows no user mode access to the supervisor part of the status register $(SR)[15:8]$. It allows a user program to read or write using the condition code register only.

Finally, the function code signal lines indicate the processor mode to external devices. These signal lines are used as the basis for memory protection and for control of external interrupt circuitry.[2]

EXERCISES

10.1.1. Draw a diagram showing the possible states of the MC68000 and the transitions between them.

10.1.2. Compare the stopped condition with the halted state.

10.1.3. Discuss possible applications for the STOP instruction and the stopped condition of the processor.

10.1.4. List the protection mechanisms provided for an MC68000 system and define the purpose and possible application of each. Include both hardware and software considerations.

10.2 SYSTEM CONTROL INSTRUCTIONS

Special MC68000 instructions are provided to change the system operation dynamically in some way. The primary group of these instructions are privileged instructions that can be executed only in the supervisor mode. These include instructions to change

---

[2] These lines indicate the supervisor or user mode and program or data reference for an executing program. They also indicate the acknowledgment of an interrupt by the MC68000. The use of the function code signal lines for system protection is discussed in Chapter 13.

Table 10.3 Instructions to Modify
the Processor Status

Syntax		Operation
ANDI.W	#$\langle d_{16} \rangle$,SR	(SR) ← (SR) AND $\langle d_{16} \rangle$
EORI.W	#$\langle d_{16} \rangle$,SR	(SR) ← (SR) EOR $\langle d_{16} \rangle$
MOVE.W	$\langle EA \rangle$,SR	(SR) ← (EA)
ORI.W	#$\langle d_{16} \rangle$,SR	(SR) ← (SR) OR $\langle d_{16} \rangle$

Notes:
1. All instructions are privileged.
2. MOVE to SR requires a data addressing mode for $\langle EA \rangle$.

the contents of the status register (SR) or the user stack pointer (USP). Another group of system control instructions is available for user mode programs to modify the contents of the condition code register (CCR). In an MC68000 system, a user mode program can also examine the contents of the status register but cannot modify it. For completeness, the Return From Exception (RTE) and the RESET instructions are also considered in this section. However, applications of these two system control instructions are described more completely in the next two chapters.

10.2.1 Status Register Modification

Table 10.3 lists the instructions available to a supervisor mode program that can modify the contents of the status register. The logical instructions (ANDI, EORI, ORI) operate in the manner described in Chapter 8. However, in this application, the destination location is the status register, which makes them privileged instructions. Each of these instructions performs the logical operation designated using the 16-bit immediate value and the entire contents of the status register.

The MOVE instruction is used to transfer the source operand to the status register, thereby replacing the previous contents. In the form

MOVE $\langle EA \rangle$,SR

the effective address $\langle EA \rangle$ is specified by any data addressing mode, which includes all addressing modes except address register direct. To transfer the contents of the status register to a memory location or a data register, the instruction

MOVE.W   SR,$\langle EA \rangle$

is used. This instruction is available to both supervisor and user mode programs in MC68000 systems. The MC68010 restricts this move from SR instruction to supervisor mode programs. A new instruction, move from CCR, is provided for user mode programs.

Example 10.1

Figure 10.2 shows the MC68000 status register as it is divided into user byte SR[7:0] and system byte SR[15:8]. The values in the AND and OR columns are given in hexadecimal. The AND

Bit Number	15	14	13	12	11	10	9	8	7	6	5	4	3	2	1	0
	T		S			I_2	I_1	I_0				X	N	Z	V	C

←——————— System Byte ———————→←——————— User Byte ———————→

Condition	Status Bit	AND {MASK}	OR {ENABLE}
Trace Mode	(SR) [15] = T	{7FFF}	{8000}
Supervisor Mode	(SR) [13] = S	{DFFF}	{2000}
Interrupt Level	(SR) [10:8] = level	{F8FF}	{0700}
Extend	(SR) [4] = X	{FFEF}	{0010}
Negative	(SR) [3] = N	{FFF7}	{0008}
Zero	(SR) [2] = Z	{FFFB}	{0004}
Overflow	(SR) [1] = V	{FFFD}	{0002}
Carry	(SR) [0] = C	{FFFE}	{0001}

Notes:
  (1) SR [15:8] is System Byte; SR [7:0] is User Byte or CCR.
  (2) ((SR) AND {MASK}) sets bit to {0} ; ((SR) OR {ENABLE}) sets bit to {1} .
  (3) For the interrupt level, the value {$I_2 I_1 I_0$} is interpreted as a 3-bit code.

Figure 10.2   Status register operation.

mask clears any bit in the status register that corresponds to a {0} in the mask. The OR of (SR) with the bit pattern shown as {ENABLE} sets the corresponding bit to {1}. Thus the instruction

ANDI   #$7FFF,SR

sets T = {0} to disable the trace mode, but does not affect the other bits. The instruction

ORI   #$8000,SR

enables the trace mode. The use of the supervisor mode bit is similar. When bit 13 is {0}, the processor is operating in the user mode.

The interrupt level value {I_2, I_1, I_0} is not treated as a logical variable but as a 3-bit integer. A value of 0, {000}, indicates that all interrupt levels will be accepted with increasing priority from 1 to 7. During interrupt processing, the value indicates the current level as explained in Chapter 11. Interrupts at that level and below are ignored. The instruction

ANDI   #$F8FF,SR

enables all interrupt levels by setting the interrupt bits to {000}. The interrupt levels are disabled by the instruction

ORI   #$0700,SR

which disables all interrupts below level 7 because a level 7 interrupt cannot be masked (disabled).

Operations on the user byte or CCR are also shown in Figure 10.2. The immediate values listed do not affect the system byte when used with the ANDI or ORI instructions as shown. The instruction

ANDI   #$FF00,SR

would clear the CCR, for example. Thus use of ANDI with the {MASK} value shown clears the corresponding bit to {0} in the CCR. The bit is set to {1} when ORI is used with the immediate value given as {ENABLE} in Figure 10.2.

10.2.2 User Stack Pointer Manipulation

A program executing in the supervisor mode can save, restore, or change the contents of the user stack pointer. The privileged instruction

MOVE.L   USP,⟨An⟩

copies (USP) into address register An. The opposite transfer has the form

MOVE.L   ⟨An⟩,USP

and is used to initialize or modify (USP). In each case, a 32-bit transfer occurs.

As would be expected, only a program operating in the supervisor mode has control of the contents of the user stack pointer. Therefore, the address of the user stack must be loaded into USP by the supervisor program during initialization. The proper address could be transferred to ⟨An⟩ and then to the user stack pointer by the instructions

MOVEA.L   #USERSTK,⟨An⟩
MOVE.L   ⟨An⟩,USP

where USERSTK is the address of the bottom of the user's stack.

10.2.3 Status Register Access in User Mode

The instructions listed in Table 10.4 are available to user mode programs. The logical instructions allow the contents of the CCR to be modified using the 8-bit immediate value. For example, the instruction

ORI.B   #$01,CCR

sets the carry bit C = {1} and does not modify any other condition code bits. The entire contents of the CCR can be modified by the instruction

MOVE.W   ⟨EA⟩,CCR

in which (⟨EA⟩)[7 : 0] contains the new condition code bits for the CCR. The addressing

Table 10.4 User Mode Access to (SR)

Syntax		Operation
Modify CCR		
ANDI.B	#⟨d₈⟩,CCR	$(SR)[7:0] \leftarrow (SR)[7:0]$ AND $\langle d_8 \rangle$
EORI.B	#⟨d₈⟩,CCR	$(SR)[7:0] \leftarrow (SR)[7:0]$ EOR $\langle d_8 \rangle$
MOVE.W	⟨EAs⟩,CCR	$(SR)[7:0] \leftarrow (EAs)[7:0]$
ORI.B	#⟨d₈⟩,CCR	$(SR)[7:0] \leftarrow (SR)[7:0]$ OR $\langle d_8 \rangle$
Move SR		
MOVE.W	SR,⟨EAd⟩	$(EA) \leftarrow (SR)$

Notes:

1. MOVE to CCR requires a data addressing mode for ⟨EAs⟩, the source.
2. MOVE SR to ⟨EAd⟩ requires a data-alterable addressing mode for ⟨EAd⟩, the destination. This is not allowed in MC68010 systems.
3. CCR is (SR)[7:0].

for ⟨EA⟩ requires a data addressing mode, which allows all modes except address register direct. Note that the operation requires a word operand but only the low-order byte is used to update the condition codes. This means that the address of the operand ⟨EA⟩ must be an even address in memory.

An MC68000 user mode program can also transfer the contents of the status register by executing the instruction

MOVE.W   SR,⟨EA⟩

where ⟨EA⟩ is a data-alterable location that excludes address register direct and program counter relative addressing. As pointed out previously, this is a privileged operation in MC68010 systems.

10.2.4 RTE Instruction

The Return from Exception (RTE) instruction is a privileged instruction used to load the status register and the program counter with values stored on the supervisor stack. The operation of the RTE instruction for the MC68000 is

(a) Load (SR)

$(SR)[W] \leftarrow ((SSP))$;POP (SR)
$(SSP) \leftarrow (SSP) + 2$;POINT TO (PC)

(b) Load (PC)

$(PC)[L] \leftarrow ((SSP))$;POP (PC)
$(SSP) \leftarrow (SSP) + 4$

as shown in Figure 10.3. The RTE is most frequently used as the last instruction in an exception-handling routine because (PC) and then (SR) are pushed on the supervisor stack when an exception occurs. The RTE instruction is also used to pass control to a user mode program during the system initialization procedure, as discussed in the next section.

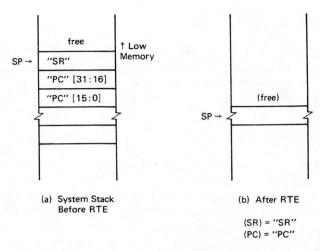

(a) System Stack
Before RTE

(b) After RTE

(SR) = "SR"
(PC) = "PC"

Note: "SR" and "PC" are values to replace the current (SR) and (PC), respectively.

Figure 10.3   Operation of RTE instruction.

As indicated in Section 4.2, the MC68010 stack contents during an exception can occupy either 4 or 29 words of the stack area in memory. Thus the RTE instruction of the MC68010 does not operate in the manner just described. The reader should consult the MC68010 manual from Motorola for specific details. The operation shown in Figure 10.3 applies then only to the MC68000 processor.

10.2.5 RESET Instruction

The RESET instruction is a privileged instruction used to reset external interfaces during system initialization. Its execution asserts a signal line used as an indication to external circuitry that the processor is requesting initialization of the appropriate interfaces. The exact function of the RESET instruction in terms of system operation is determined by the hardware design of the system. Most of the peripheral chips of the MC68000 family respond to this instruction (via a reset signal line) by initializing their internal circuitry.

EXERCISES

10.2.1. Determine the effect of the following instructions.
   (a) MOVE.W   #$0400,SR
   (b) ANDI.W   #$DFFD,SR
   (c) MOVE.W   #$2700,SR
   (d) EORI.W   #$2000,SR
   The instructions are executed by a program in the supervisor mode.

10.2.2. What is the effect of the following instructions executed by a program in the user mode?
   (a) MOVE.W   #$000C,CCR

    (b) ANDI.B   #$01,CCR
    (c) EORI.W   #$2700,SR

10.2.3. Write the sequence of instructions to initialize (USP) to hexadecimal value $FC00.

10.2.4. Why does a user mode program in an MC68000 system need to be able to transfer the contents of (SR)? How does a user mode program test the entire contents of the CCR? How does the MC68010 allow the user to check (CCR)?

10.3 SYSTEM INITIALIZATION

A *system initialization* sequence is used to place the system in a known state or condition before any application program is executed. The sequence normally begins with a reset signal supplied by external circuitry to cause the hardware to assume its initial state. Then an initialization routine is executed to initialize all constants, addresses, and other data values associated with the system operation. The initialization routine is usually held in read-only memory and can be executed automatically after a system reset. This routine not only initializes the data used for the system but can also cause the loading of the operating system from an external storage device such as a disk unit. The exact procedure, of course, depends on the hardware configuration and the application of the system. In this section the basic initialization procedure for MC68000-based systems is described according to the requirements of the processor.

    Initialization may produce the typical memory layout as shown in Figure 10.4. The vector table in MC68000 systems is lowest in memory and contains the addresses of exception routines. The SSP and USP point to the bottom of the supervisor stack area and the user stack area, respectively. The amount of space reserved for the program and stack areas is determined when the software system is designed. Enough space must be allocated so that the various areas do not overlap during operation. The stacks shown in Figure 10.4 are system stacks and grow toward lower memory locations as return addresses, register contents, and other data are pushed onto the stack.

    Once the system is initialized and the various programs to be executed are loaded into memory, the supervisor program passes control to an applications program. The techniques for transfer of control are also covered in this section.

    In MC68010 systems, the exception vector table may be relocated in memory. The processor contains a "vector base register" whose contents are added to the vector location (offset) in the table to compute the effective address of the vector in memory. During the initialization sequence, the instruction Move to/from Control Register (MOVEC) of the MC68010 is used to initialize to contents of the vector base register.

Example 10.2 _____

General-purpose computers with a disk unit typically have a simple initialization routine in ROM to load the operating system software into memory at power-up. The nonvolatile memory holding the loader routine is often called the *bootstrap* ROM. It contains only the routines to

Figure 10.4 Initial memory layout.

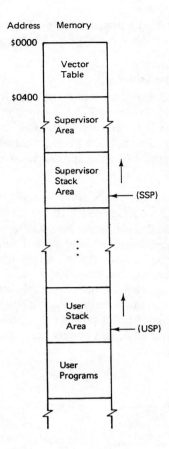

read the operating system from the disk unit into the supervisor area of RAM memory of the computer. The operating system in turn is used to load various application programs in response to operator commands. Having the disk-based operating system allows great flexibility for the programmer. Consequently, software development systems contain an associated disk unit. On the other hand, most of the MC68000-based products have no disk unit and thus their programs are held in ROM.

In a completed product used for a specific purpose, the software including system and applications programs are stored in ROM. The starting addresses of all the programs are set and are not intended to be altered unless the product and its software are redesigned. A certain amount of RAM is necessary for the system stacks and to store data that changes dynamically as the product operates. The initialization of products that store programs in ROM are considered in this section.

Once the product is initialized, the supervisor program passes control to an applications program. The techniques for transfer of control are covered in this section. Before the applications programs are executed however, the supervisor program initializes the values in various control registers associated with peripheral chips. The techniques to program peripheral chips for specific applications are not discussed until Chapter 12.

10.3.1 Initialization Procedure

For convenience of discussion, the initialization procedure is separated into two phases. In the first phase, the MC68000 CPU hardware is reset and causes execution of the routine at the location indicated by the initial value of the program counter. This portion of the initialization is fixed by the processor design and cannot be altered. The second phase begins when the initialization routine of the supervisor program executes.

Reset and Initialization of the Processor.   Initialization is accomplished by an external device asserting the $\overline{\text{RESET}}$ signal line to the CPU. The sequence of events is shown in Figure 10.5, in which the processor first sets the status register contents to indicate the supervisor mode and masks all interrupts. Then, in sequence, the supervisor stack pointer and the program counter are initialized from the first eight bytes in memory. If no errors occur, program execution begins at the location defined by the program counter contents. The entire procedure up to this point requires about 100 milliseconds.

In most systems, the lower memory locations that hold the vector addresses are in volatile memory. Therefore, if power is lost, the contents of such locations should be considered destroyed. During a system reset, the MC68000 or MC68010 processor

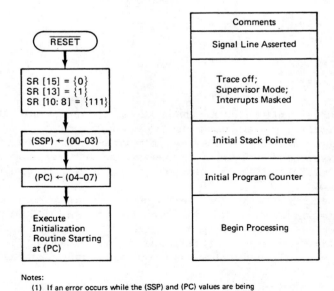

Notes:

(1) If an error occurs while the (SSP) and (PC) values are being fetched, the processor will enter the halted state.

(2) $\overline{\text{RESET}}$ is a signal line to the CPU.

Figure 10.5   Reset operation.

reads longword location (00–03) for the supervisor stack pointer value and location (04–07) for the program counter value. These eight bytes must be interpreted as permanent values. Therefore, external circuitry must be designed to accomplish this.[3] The initialization routine, which is then executed, can initialize the remaining values in the other vector locations and perform other necessary processing. In MC68010 systems, the vector table may be relocated in memory.

Initialization Routine. The initialization routine prepares the system for execution of the supervisor program. This preparation includes loading vector addresses into the locations shown in Figure 10.6 and initializing any external devices. When the routine completes, control may be passed to a supervisor routine, which determines which application program is to be executed. If there is no supervisor, control is passed directly to the application program immediately after initialization of the system is complete.

The initialization routine for the MC68000 usually performs the series of operations shown in Table 10.5, although many variations are possible for a specific application. After the system is properly initialized, the user stack pointer, status register, and program counter contents are set to pass control to the application program, which is assumed to operate in the user mode in the sequence shown in Table 10.5. As shown in that table, the supervisor program controls all systems operations and the initial setting of the stack pointer for the application program. If the memory utilization is also controlled by the supervisor, the application programs would be loaded into memory before control is passed to the user.

10.3.2 Initialization Example

Assume that a supervisor program is to pass control to a user mode program that starts at location $8000. Also, assume that the user stack is to begin at location $FC00. Figure 10.7 shows a brief initialization routine that could be used with the SBC68K single-board computer described in more detail in Chapter 12. The board has 64 KB of RAM from address $0000 to $FFFF. The peripheral chips include a programmable timer (MC68230), a serial communication chip (MC68681) and a disk controller chip. Details about programming these peripheral chips are presented in Chapter 12.

Example 10.3 _____

First in the program of Figure 10.7, the CPU is put in the supervisor mode with interrupts masked. Next, the SSP is initialized to location $10000 just above the last RAM address. This stack will use lower addresses to store data as values are pushed on the stack. The vectors for traps and interrupts are initialized next by the call to subroutine EXINIT (not shown). Chapter

---

[3] To initialize the values, the addresses could be translated by special circuits to an address in read-only memory, as is done with the Motorola Design Module. In any case, the first eight byte locations must not be volatile locations when accessed by the CPU for initialization. The reset sequence is initiated in the Design Module by a pushbutton that causes the $\overline{\text{RESET}}$ signal line of the CPU to be asserted.

Figure 10.6 Vector table for the MC68000 system. (Courtesy of Motorola, Inc.)

Vector Number(s)	Dec	Address Hex	Space	Assignment
0	0	000	SP	Reset: Initial SSP[2]
	4	004	SP	Reset: Initial PC[2]
2	8	008	SD	Bus Error
3	12	00C	SD/	Address Error
4	16	010	SD	Illegal Instruction
5	20	014	SD	Zero Divide
6	24	018	SD	CHK Instruction
7	28	01C	SD	TRAPV Instruction
8	32	020	SD	Privilege Violation
9	36	024	SD	Trace
10	40	028	SD	Line 1010 Emulator
11	44	02C	SD	Line 1111 Emulator
12[1]	48	030	SD	(Unassigned, Reserved)
13[1]	52	034	SD	(Unassigned, Reserved)
14[1]	56	038	SD	(Unassigned, Reserved)
15	60	03C	SD	Uninitialized Interrupt Vector
16-23[1]	64	040	SD	(Unassigned, Reserved)
	95	05F		–
24	96	060	SD	Spurious Interrupt[3]
25	100	064	SD	Level 1 Interrupt Autovector
26	104	068	SD	Level 2 Interrupt Autovector
27	108	06C	SD	Level 3 Interrupt Autovector
28	112	070	SD	Level 4 Interrupt Autovector
29	116	074	SD	Level 5 Interrupt Autovector
30	120	078	SD	Level 6 Interrupt Autovector
31	124	07C	SD	Level 7 Interrupt Autovector
32-47	128	080	SD	TRAP Instruction Vectors[4]
	191	0BF		
48-63[1]	192	0C0	SD	(Unassigned, Reserved)
	255	0FF		–
64-255	256	100	SD	User Interrupt Vectors
	1023	3FF		–

NOTES:
1. Vector numbers 12, 13, 14, 16 through 23, and 48 through 63 are reserved for future enhancements by Motorola. No user peripheral devices should be assigned these numbers.
2. Reset vector (0) requires four words, unlike the other vectors which only require two words, and is located in the supervisor program space.
3. The spurious interrupt vector is taken when there is a bus error indication during interrupt processing. Refer to Paragraph 5.5.2.
4. TRAP #n uses vector number 32 + n.

11 describes this initialization procedure in more detail, and an example in Chapter 12 presents the vector initialization routine. The calls to COMINIT and PITINIT initialize the communications chip and the timer chip on the SBC68K, respectively. Examples in Chapter 12 explain these routines.

The address USERSTK = $FC00 is used for the USP. Thus, the supervisor stack can use the memory locations from $FFFE down to this address. If the supervisor stack "overflows" beyond the lower address, the values in the user stack will be corrupted and errors will occur in the system.

Next, all of the register values except for the system stack pointer value (A7) are set to zero during initialization.

Table 10.5 Typical System Initialization Procedure

Operation	Comments
Initialize addresses for vectors as required	Locations $0008–$03FC (32-bit addresses)
Load any supervisor routines needed	Load from disk unit
Initialize contents of any memory locations needed by the supervisor	As needed
Initialize all peripheral devices for the system	As needed: RESET instruction for peripheral devices and other I/O control commands
Initialize USP	(USP) ← user stack address
Push starting address for user mode program	((SSP)) ← start of program
Push status for user mode program	((SSP)) ← user (SR)
Transfer control to user mode program	RTE

Notes:
1. Exception-handling routines have fixed starting addresses. These addresses are used to initialize the vectors.
2. After hardware initialization, the CPU is in the supervisor mode with all interrupts masked. Execution of the initialization begins at the address specified in location $0004.
3. Initialization of an MC68010 system is similar, although the vector table may be relocated.

To initialize the user mode program, the beginning address of the program (USERIN) is first pushed on the stack. Then, the user status register (SR) value of $0000 is pushed. When the RTE instruction is executed, the user mode program will begin execution at location USERIN ($8000) with all interrupts enabled.

EXERCISES

10.3.1. Assume that lower memory is volatile RAM. List (or diagram) the decision required by external circuitry to allow the CPU to address nonvolatile ROM locations during the reset sequence. Assume that the reset vectors are held in ROM beginning at hexadecimal location $20000. The initial (SSP) is $06B8 and the initial (PC) is $20008.

10.3.2. The instruction sequence

```
MOVE.W      #$0000,SR

JMP         USER
```

could be used to transfer control to a program at location USER. Why is the approach using RTE better? Consider the case in which the memory is segmented (and protected) into supervisor program space and user program space.

10.3.3. Design an interrupt-handling routine to respond to a level 7 interrupt. The routine should simply display "Level 7 Interrupt" on the operator's display screen when the interrupt is recognized. The vector address is $07C in a standard MC68000-based computer system. If possible, test the routine on the system you are using.

```
abs.   LC    obj. code    source line
----   ----  ---------    -----------
  1    0000                        |          TTL     'FIGURE 10.7'
  2    0000                        |          LLEN    100
  3    0000                        |          OPT     P=M68000
  4    0000                        |*
  5    0000                        |* SYSTEM INITIALIZATION EXAMPLE
  6    0000                        |*
  7    3000                        |          ORG     $3000
  8    0000                        |*
  9          0000 FC00            |USERSTK EQU       $0000FC00       ;INITIAL USER STACK POINTER
 10          0000 8000            |USERIN  EQU       $00008000       ;INITIAL USER PROGRAM COUNTER
 11          0000 3100            |EXINIT  EQU       $3100
 12          0000 3200            |COMINIT EQU       $3200
 13          0000 3300            |PITINIT EQU       $3300
 14    3000                        |***********************************************************
 15    3000                        |* SYS_INIT
 16    3000                        |*
 17    3000                        |* THIS PROGRAM INITIALIZES THE SYSTEM TO A KNOWN STATE.
 18    3000                        |*    PERIPHERAL DEVICES ARE INITIALIZED
 19    3000                        |*    DATA AND ADDRESS REGISTERS ARE CLEARED
 20    3000                        |*    THE USER STACK IS SET TO START AT $FC00
 21    3000                        |*    THE SUPERVISOR STACK IS SET TO START AT $10000;
 22    3000                        |*
 23    3000  46FC 2700            |SYSINIT MOVE.W   #$2700,SR       ;DISABLE INTERRUPTS
 24    3004  2E7C 0001            |        MOVE.L   #$10000,SP     ;SET SSP=10000
       3008  0000
 25    300A                        |*
 26    300A                        |* INITIALIZE EXCEPTION VECTORS
 27    300A  4EB8 3100            |        JSR      EXINIT         ;INITIALIZE EXCEPTION VECTORS
 28    300E                        |*
 29    300E                        |* INITIALIZE PERIPHERALS AND RELATED DATA AREAS
 30    300E  4EB8 3200            |        JSR      COMINIT        ;INITIALIZE SERIAL COMMUNICATION
 31    3012  4EB8 3300            |        JSR      PITINIT        ;INITIALIZE TIMER
 32    3016                        |*
 33    3016                        |* SET USP
 34    3016  207C 0000            |        MOVEA.L  #USERSTK,A0    ;SET USER STACK POINTER
       301A  FC00
 35    301C  4E60                 |        MOVE.L   A0,USP         ;
 36    301E                        |*
 37    301E                        |* CLEAR REGISTERS D0-D7, A0-A6
 38    301E  227C 0000            |        MOVEA.L  #RZ0,A1        ;ADDR OF FIRST LONGWORD
       3022  3046
 39    3024  303C 000E            |        MOVE.W   #14,D0         ;SET TO CLEAR 15 LOCATIONS
 40    3028  22FC 0000            |SYSLOOP MOVE.L   #0,(A1)+       ;MOVE THROUGH THE DATA AREA,
       302C  0000
 41    302E  51C8 FFF8            |        DBF      D0,SYSLOOP     ;  CLEARING EACH LONGWORD.
 42    3032                        |*
 43    3032  4CF9 7FFF            |        MOVEM.L  RZ0,D0-D7/A0-A6 ;CLEAR THE REGISTERS
       3036  0000 3046
 44    303A                        |*
 45    303A                        |* SET SR (USER MODE, INTERRUPTS ENABLED)
 46    303A  2F3C 0000            |        MOVE.L   #USERIN,-(SP)  ;SR VALUE WILL BE SET BY RTE
       303E  8000
 47    3040  3F3C 0000            |        MOVE.W   #0,-(SP)
 48    3044                        |*
 49    3044                        |*
 50    3044  4E73                 |GOTOUSR RTE                     ;PASS CONTROL TO PROGRAM
 51    3046                        |                               ;  IN USER MODE.
 52    3046                        |RZ0     DS.L     15             ;
 53    3082                        |        END
```

Figure 10.7 Simplified initialization routine.

10.3.4. Define several reasons for having the Vector Base Register (VBR) in the MC68010. Consider first the cases in which the VBR is used to relocate the exception vector table after system initialization. Next consider systems in which more than one exception vector table is needed.

REVIEW QUESTIONS AND PROBLEMS

Multiple Choice

10.1. When power is applied to an MC68000 system, the CPU fetches the initial values of
(a) the supervisor stack pointer and the PC
(b) the PC and status register
(c) the supervisor stack pointer and the user stack pointer

10.2. The "stopped" condition for the CPU is considered
(a) an exception state
(b) a normal state
(c) the halted state

10.3. An exception procession routine normally ends with the instruction
(a) RTS
(b) RTD
(c) RTE

10.4. How is the supervisor mode of the CPU entered?
(a) a user mode program makes a jump to the operating system
(b) an exception occurs
(c) the user mode program makes an error

10.5. The user stack pointer is initialized by
(a) a user mode program
(b) a reset of the CPU
(c) a supervisor mode program

10.6. In an MC68000 system, the first 256 longword addresses in memory are reserved for
(a) the vector table
(b) supervisor routines
(c) the supervisor stack

10.7. The *vectors* are 32-bit addresses that define
(a) the starting addresses of subroutines
(b) the starting addresses of exception handling routines
(c) the starting addresses of applications programs

10.8. After a reset, the order of initialization for an MC68000-based system is
(a) CPU, peripheral devices, modules, applications
(b) peripheral devices, CPU, modules
(c) CPU, modules, peripheral equipment

10.9. In a single-board computer, the Monitor resides in
(a) ROM
(b) RAM
(c) the disk

Programs

10.10. For the computer system you are using, print the exception vector table and label all the exceptions that are handled by the monitor program or the operating system. Refer to Figure 10.6 for the names of the specific exceptions.

FURTHER READING

The following article by Cates describes the circuitry that is required to translate the initial vector addresses from locations $0000–$0007 to an area of memory occupied by ROM. For a specific computer system, the user's manuals must be consulted for detailed information concerning system initialization and operation at the level discussed in this chapter.

Cates, Ron L. "Mapping an Alterable Reset Vector for the MC68000." *Electronics* 55, no. 15 (July 28, 1982).

CHAPTER 11

Exception Processing

The exception processing capability of the MC68000 is provided to assure an orderly transfer of control from an executing program to the supervisor program. Exceptions may be broadly divided into those caused by an instruction, including an unusual condition arising during its execution, and those caused by external events. Those in the first category are called *traps* and represent exceptional conditions caused by the program itself that are detected by the CPU. Hardware error exceptions and *interrupts* are caused by external events and these exceptions are initiated by circuitry outside the CPU. The hardware error is termed a *bus error* by Motorola.

Table 11.1 lists the MC68000 exceptions and their causes. The program exceptions are divided into those generated by the TRAP instruction, instructions used to test program conditions, unimplemented instructions, tracing, and various error conditions. Externally generated exceptions include bus errors and interrupts from external devices.

Figure 11.1 illustrates the processing sequence for most exceptions. Briefly stated, the processor enters the supervisor mode with tracing off and executes the exception routine at the address in the exception vector table. The previous values in the program counter and status register are saved on the supervisor stack so that a return from exception (RTE) can be performed when processing is complete. For certain exceptions, such as interrupts, the processing is slightly different and will be discussed in the appropriate section of this chapter. The vector number and its memory address for each exception are listed in Figure 11.2. These addresses are loaded during system initialization.

In MC68010 systems, at least four words are saved on the system stack when an exception is recognized as discussed in Section 4.2. The vector address in memory is calculated by adding the contents of the vector base register to the vector offset. In Figure 11.2 the vector offset is computed as four times the vector number. The vector base register of the MC68010 must first be initialized as described in Section 10.3.

The discussions of MC68000 exception processing in this chapter apply to the MC68010 in almost every detail. The MC68010 exception stack is four words long for

Table 11.1   MC68000 Exceptions

Type	Cause
Trap instruction TRAP #⟨N⟩ ;N = 0, 1, . . . , 15	Sixteen trap routines may be defined and used by the program
Program checks DIVS, DIVU	Trap when divisor is zero

Instructions
  TRAPV
  CHK   ⟨EA⟩,⟨Dn⟩
	
	Trap if overflow condition is detected
	Trap if register value is out of bounds

Unimplemented instruction	Trap if op code is {1010} or {1111}
Trace exception	Single instruction trace
Error conditions Privilege violation	 Trap if user mode program attempts to execute privileged instruction
Illegal instruction	Trap if op code is unrecognized
Address error	Trap if access of a word or longword location with odd address
Bus error	Externally generated exception request
Halt	Double error condition halts processor

Interrupt system
  Autovector
  Vectored
	
	Automatic vectoring for seven levels of priority
	192 user interrupt vectors; priority determined by hardware design

all exceptions but address errors and bus errors. In these two cases, 29 words of information are saved on the stack. Unless the exception stack is being manipulated during exception processing, the exact stack format is not of concern to the programmer. Also, after initialization, the vector address for the various exceptions is determined automatically by the MC68010. Therefore, programming routines to handle most of the exceptions is the same for both the MC68000 and the MC68010.

11.1   EXCEPTIONS CAUSED BY PROGRAM EXECUTION

As each instruction is executed, the processor tests the instruction to determine if an exception condition is generated. These exceptions are called *traps* and are caused by normal program execution. These exceptions are categorized as shown in Table 11.2 into those caused by either one of the following:

  (a) Program control instructions
  (b) Instructions to check program conditions

A TRAP instruction is used to return control to the supervisor program. The traps caused by division, TRAPV or CHK, indicate a program error. Two unimplemented instruction traps are available to allow the design of special routines.

Figure 11.1 Exception processing sequence.

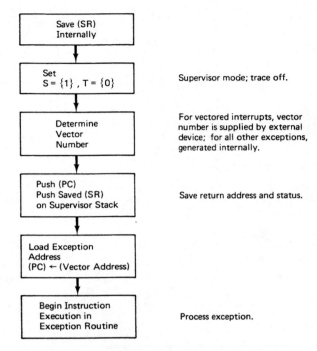

Save (SR) Internally

Set S = {1} , T = {0}   Supervisor mode; trace off.

Determine Vector Number   For vectored interrupts, vector number is supplied by external device; for all other exceptions, generated internally.

Push (PC) Push Saved (SR) on Supervisor Stack   Save return address and status.

Load Exception Address (PC) ← (Vector Address)

Begin Instruction Execution in Exception Routine   Process exception.

Note: Processing for reset, interrupts, address error, and bus error is slightly different from that shown.

The operation sequence of each type of trap is the same as far as the application program is concerned. Upon detection of a trap condition, the processor saves the contents of the status register (internally) and enters the supervisor mode with the trace condition off. Then the contents of the program counter and the saved value of the status register are pushed on the supervisor stack. After the vector address is calculated, program control is passed to the trap routine beginning at the address specified by the vector.

If control is to be returned to the program that caused the exception, the trap routine executes an RTE instruction, which restores the contents of the status register and program counter from the stack. For traps caused by unimplemented instructions, the contents of the program counter that is saved indicate the location of the instruction causing the trap. In order to return to the next instruction in sequence, the saved (PC) must be modified by the exception-handling routine for the unimplemented instructions.

11.1.1 The TRAP Instruction

Execution of the TRAP instruction with the format

TRAP   #⟨vector⟩

first causes the contents of the program counter and the status register, in that order, to be pushed on the supervisor stack. Then processing begins at the address specified

Figure 11.2 Vector allocation.

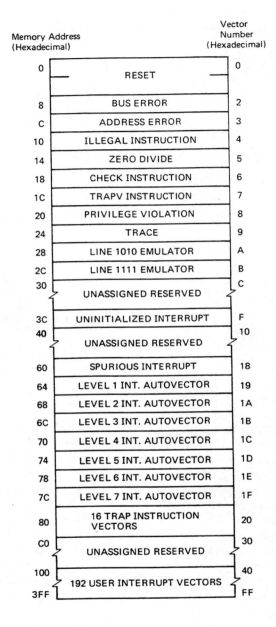

in the vector location. The value ⟨vector⟩ is an integer in the range 0–15 and is used to calculate the hexadecimal vector address as follows:

vector address = 80_{16} + 4*⟨vector⟩

The exception processing begins at the location loaded into the program counter with

(PC) ← (vector address)

Table 11.2   Program Traps

Type of Trap	Cause of Trap	Vector Address	Comments
Program control TRAP instruction	Normal execution	$080–$0BC	Normal operation to call executive program (PC) value is next instruction location
Unimplemented instruction	Normal execution	$028 or $02C	Op code {1010} or {1111} (PC) locates unimplemented instruction
Program condition CHK	Range checked is out of bounds	$018	The exception routine can attempt to recover from the condition and return control to the program or abort program
DIVS or DIVU	Divisor is zero	$014	
TRAPV	Execution with V = {1}	$01C	
			(PC) value is next instruction location

where each address is 32 bits in length. For convenience, the vector addresses are listed in Table 11.3. These locations must be initialized before the TRAP is executed.

The TRAP instruction has numerous uses. For a program operating in the user mode, its execution can return control to the supervisor program at the location of the designated trap routine. The 16 possible traps allow a user mode program to call the supervisor for processing, which must be executed at the supervisor level. For example, a call via the TRAP instruction might be used to input or output data using peripheral devices that are controlled by the supervisor. In effect, the trap is a software interrupt. This mechanism could be used in debugging operations to simulate interrupts. The TRAP instruction is also a means of returning control to the supervisor mode after the applications task is completed.

Example 11.1 _____

Figure 11.3(a) shows an application of the TRAP instruction. Before a program calls the TRAP routine, the address of the trap routine (vector) must be loaded into the CPU vector table and the trap routine must be stored in memory beginning at that address.

A CPU vector table in RAM can be initialized in several ways. For example, assume the trap–handling routine for a TRAP #0 instruction is to be loaded starting at location $5000. This TRAP instruction uses the vector at location $80 in the CPU vector table as shown in Figure 11.2.

The instruction

```
MOVE.L    $5000,$80    TRAP 0 VECTOR
```

Table 11.3 Trap Vector Addresses

TRAP #⟨N⟩ Instruction	Vector Address (hexadecimal)
TRAP #0	80
TRAP #1	84
TRAP #2	88
TRAP #3	8C
TRAP #4	90
TRAP #5	94
TRAP #6	98
TRAP #7	9C
TRAP #8	A0
TRAP #9	A4
TRAP #10	A8
TRAP #11	AC
TRAP #12	B0
TRAP #13	B4
TRAP #14	B8
TRAP #15	BC

Note:
Vector number is decimal.

```
abs.   LC    obj. code    source line
----   ----  ----------   -----------
  1    0000                         TTL     'FIGURE 11.3(a)'
  2    0000                         LLEN    100
  3    0000                         OPT     P=M68000
  4    0000              |*
  5    0000              |*          EXAMPLE TRAP #0 CALL TO INITIALIZE CPU REGISTERS.
  6    0000              |*
  7    8100                         ORG     $8100
  8    0000              |*
  9    8100  21FC 0000   |          MOVE.L  #$5000,$80      ;LOAD TRAP 0 VECTOR ADDRESS
       8104  5000 0080   |
 10    8108              |*
 11    8108  4E40        |T1         TRAP    #0             ;CALL TO TRAP ROUTINE
 12    810A              |*
 13    810A  4E71        |          NOP                     ;CONTINUE EXECUTION
 14    810C              |*          .
 15    810C              |*          .
 16    810C              |*          .
 17    810C  1E3C 00E4   |          MOVE.B  #228,D7
 18    8110  4E4E        |          TRAP    #14             ;RETURN TO MONITOR
 19    8112              |
 20    8112              |          END
```

Figure 11.3(a) Program example of TRAP call.

```
abs.   LC    obj. code    source line
----   ----  ---------    -----------
   1   0000              |        TTL       'FIGURE 11.3(b)'
   2   0000              |        LLEN      100
   3   0000              |        OPT       P=M68000
   4   0000              |*
   5   0000              |* REGISTER INITIALIZATION TRAP ROUTINE
   6   0000              |*
   7   5000              |        ORG       $5000
   8   0000              |*
   9   5000              |* CLEAR REGISTERS D0-D7, A0-A6
  10   5000              |*
  11   5000  227C 0000   |        MOVEA.L   #R2Z,A1         ;ADDR OF FIRST LONGWORD
       5004  501E        |
  12   5006  303C 000E   |        MOVE.W    #14,D0          ;NEED TO CLEAR 15 REGISTERS
  13   500A  22FC 0000   |REGLOOP MOVE.L    #0,(A1)+        ;MOVE THROUGH THE DATA AREA,
       500E  0000        |
  14   5010  51C8 FFF8   |        DBF       D0,REGLOOP      ;   CLEARING EACH LONGWORD.
  15   5014              |*
  16   5014  4CF9 7FFF   |        MOVEM.L   R2Z,D0-D7/A0-A6 ;CLEAR THE REGISTERS
       5018  0000 501E   |
  17   501C              |*
  18   501C  4E73        |RETURN  RTE                       ;RETURN TO CALLING PROGRAM
  19   501E              |
  20   501E              |R2Z     DS.L      15              ;DATA AREA
  21   505A              |        END
```

Figure 11.3(b)   TRAP routine.

initializes the TRAP #0 vector when the instruction executes. Another approach is to initialize the vector when the initialization program is loaded. For example, the instruction sequence

```
ORG    $80        TRAP 0 VECTOR
DC.L   $5000
```

causes $5000 to be written in location $80 after assembly and program loading.

In this example, the program calling the TRAP loads the address of the trap routine in the vector table before the trap routine is called. The TRAP #0 routine in Figure 11.3(b) is loaded at location $5000. When called, the routine clears all the register values except for the system stack pointer (A7). This is done by creating a table of zeros in memory and then using the MOVEM.L instruction to load the registers with zero values. After execution, the trap routine returns control to the program that executed the TRAP instruction.

11.1.2 Divide-by-Zero Trap and TRAPV

Certain arithmetic errors in an applications program can be detected and trapped by the CPU. In particular, execution of a signed divide (DIVS) or unsigned divide (DIVU) instruction with a divisor of zero automatically causes a trap through vector address $14. An overflow condition $(V = \{1\})$ will cause a trap if the instruction TRAPV is executed. Processing in this case begins at the address located by the vector at $1C. In the design of most systems, control is not returned to the program having an error when one of the arithmetic traps occurs.

Example 11.2 _____

Figure 11.4 shows an example of processing a "divide by zero" trap. The first section stores the address of the trap handler routine in the trap vector location at $14. Next, a short program segment causes a divide by zero to occur to test the trap routines. Finally, the trap handler routine begins at $5300 and sets the quotient to a particular value based on the value of the dividend, either the most positive or most negative integer allowed.

_____

11.1.3 CHK Instruction

The Check Register Against Bounds instruction (CHK) has the symbolic form

CHK   ⟨EA⟩,⟨Dn⟩

where ⟨EA⟩ is designated by a data addressing mode. This allows all addressing modes except address register direct. This instruction determines if the 16-bit contents of ⟨Dn⟩ is between 0 and the value contained in ⟨EA⟩ and causes a trap if the contents of ⟨Dn⟩ are not within this range. The upper bound held in (EA) is treated as a 16-bit, two's-complement integer. The operation is as follows:

IF $0 < (Dn)[15:0] \leq (EA)$, THEN continue
   ELSE *trap* and
      set $N = \{1\}$ if $(Dn)[15:0] < 0$ or
      set $N = \{0\}$ if $(Dn)[15:0] > (EA)$

The exception routine begins at the address in location $18 if the trap is taken. The CHK instruction is typically placed in a program after the calculation of an offset or an index value to assure that the limits of the value are not exceeded. This facilitates testing whether an array address fits within the dimensions of the array when the address register indirect with indexing addressing mode is used to locate array elements. For example, the sequence

CHK   #99,D1
MOVE.B   0(A1,D1.W),D2

would trap if $(D1)[15:0]$ exceeded 99 decimal. If A1 held the base address of the array with 100 byte-length elements, the addressing range of the MOVE instruction is limited to values within this array. The addressing of arrays in FORTRAN is not always protected in this manner, but array boundary checking is typically provided in languages such as Pascal.[1]

Other uses of the CHK instruction include the testing of the space used by a stack or keeping a program from accessing data outside its designated space. For these applications, a value in an address register must be transferred to a data register in order to perform the bounds check.

_____

[1] Whether or not array boundary checking is provided depends on the compiler and the system being used. In some cases, boundary checking is an optional feature of a compiler.

```
abs.    LC    obj. code    source line
----    ----  ---------    -----------
  1    0000                            TTL      'FIGURE 11.4(a)'
  2    0000                            LLEN     100
  3    8000                            ORG      $8000
  4    8000                            OPT      P=M68000
  5    8000            |*
  6    8000            |*    SET THE ZERO DIVIDE TRAP VECTOR
  7    8000            |*
  8    8000 21FC 0000  |      MOVE.L   #$5300,$14        ;ZERO-DIVIDE TRAP
       8004 5300 0014  |                                ; ROUTINE ADDR.
  9    8008            |*
 10    8008            |*
 11    8008            |*
 12    8008            |*
 13    8008            |*   EXECUTE A DIVIDE BY ZERO WITH
 14    8008            |*    INPUTS :     (D0.W) = DIVISOR (SET TO 0 FOR TEST)
 15    8008            |*                 (D1.L) = DIVIDEND.
 16    8008            |*    OUTPUT :     (D1.W) = LARGEST VALUE BASED ON SIGN
 17    8008            |*                              OF DIVIDEND
 18    8008            |*
 19    8008 4280       |      CLR.L    D0                ;SET DIVISOR TO ZERO
 20    800A 83C0       |      DIVS     D0,D1             ;DIVIDE D1[31:0] BY
 21    800C            |*                                ;   D0[15:0]=0
 22    800C            |*
 23    800C 1E3C 00E4  |      MOVE.B   #228,D7
 24    8010 4E4E       |      TRAP     #14               ;RETURN TO MONITOR
 25    8012            |*
 26    8012            |      END
```

```
abs.    LC    obj. code    source line
----    ----  ---------    -----------
  1    0000                            TTL      'FIGURE 11.4(b)'
  2    0000                            LLEN     100
  3    0000                            OPT      P=M68000
  4    0000            |*
  5    0000            |*      ZERO-DIVIDE TRAP HANDLER ROUTINE
  6    0000            |*
  7    5300                            ORG      $5300
  8    5300 4A81       |ZDIV   TST.L    D1
  9    5302 6B00 000C  |       BMI      ZDIV01            ;IF DIVIDEND IS POSITIVE,
 10    5306 223C 0000  |       MOVE.L   #$7FFF,D1         ; RETURN WITH
       530A 7FFF       |
 11    530C 6000 0008  |       BRA      RTN               ; (D1).W=LARGEST POSITIVE
 12    5310            |*                                 ;  VALUE
 13    5310            |*                                 ;
 14    5310 223C 0000  |ZDIV01 MOVE.L   #$8000,D1         ; ELSE RETURN LARGEST
       5314 8000       |                                  ;  NEGATIVE VALUE
 15    5316            |*
 16    5316 4E73       |RTN    RTE
 17    5318            |       END
```

Figure 11.4 Divide by zero trap.

351

11.1.4 Unimplemented Instruction Trap

The CPU recognizes instructions whose first bits are {1010} ($A) or {1111} ($F) as *unimplemented instructions*. These op codes are available to the programmer to cause execution of special routines that appear to be additional "macroinstructions" added to the MC68000 instruction set. Typical examples include routines for floating-point software, string manipulation, and fast Fourier transform algorithms. The other 12 bits in a word-length instruction with either of these op codes can be used to select the various options as interpreted by the special routines.

To cause the {1010} trap using vector location $28, the directive

DC.W $A000

could be inserted in the program where the trap is required. Similarly, the directive

DC.W $F000

would cause a trap using the vector at $2C. After the trap occurs, the value of the program counter saved on the supervisor stack points to the unimplemented instruction. The address can be obtained using the instruction

MOVEA.L   2(SP),A1

which skips over the contents of the status register saved on the stack and loads A1 with the value saved for the PC. This value can be used to locate the unimplemented instruction to decode any other bits used to select options for the trap routine.

EXERCISES

11.1.1. If the instruction

      TRAP   #1

is located at location $1004 and executes when (SR) = $A000 and (SP) = $7FFE, show the contents of the system stack before and after the TRAP instruction executes. Initialize the TRAP #1 vector so that the trap routine starts at location $1100.

11.1.2. Write a program that pushes word-length values on the user stack beginning at location $1000. If the stack length is 10 words maximum, use the CHK instruction to return control to the supervisor program when the stack overflows.

11.1.3. Write a signed division routine that will accept a divisor of zero. If the dividend is positive, load the quotient with the largest positive integer when the zero divide occurs. If the dividend is negative, set the quotient to the most negative number. Define the register used for the dividend in the user program. Why is it not possible to correct a division routine in which the location of the dividend is arbitrary?

11.1.4. Emulate a "macroinstruction" that moves a string of characters of arbitrary length from one area of memory to another. Assume that D1 holds the length, A1 the start of the string, and A2 the destination location for the first character to be moved.

Figure 11.5 Trace operation.

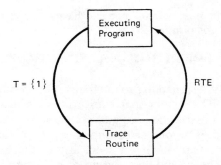

11.2 THE TRACE EXCEPTION

In testing a microcomputer system, it is sometimes desirable to execute instructions one at a time with the processor suspending execution between instructions. This single instruction mode of operation is provided by the trace feature of the MC68000.[2] Rather than suspending operation, the MC68000 causes execution of a trace routine after each instruction is executed whenever the trace status bit (SR)[15] or T = {1}. The vector holding the address of the first instruction of the trace routine is held in longword location $24. Figure 11.5 shows the general operation of the trace feature.

The trace operates for programs executing in either the supervisor mode or the user mode. However, the trace bit (T) can be changed only by an instruction executed in the supervisor mode. The return to the program being traced is accomplished by executing an RTE instruction, which should be the last instruction in the trace routine.

The trace routine is typically designed to display or record the contents of all the registers and memory locations that affect program execution or processing. These values can be output to a terminal or similar device to allow a programmer to follow changes in register contents or other values as the program executes each instruction in turn.

EXERCISES

11.2.1. Write a simple trace routine to print the contents of all the processor registers when the trace exception is taken. Test the tracing by writing a simple program in which each instruction is traced. (*Note:* The routine to print values is provided by the supervisor

---

[2] An alternative is *single-cycle* operation in which execution is suspended after each access to memory. The $\overline{\text{HALT}}$ signal line of the MC68000 can be used for this purpose if the proper circuitry is employed to implement it.

program in most systems. You may have to convert the hexadecimal values on the stack to ASCII for printing.)

11.3 ERROR CONDITIONS CAUSING TRAPS

The MC68000 is designed to protect the system from errors that could cause unpredictable behavior. The errors that cause traps are listed in Table 11.4. The privilege violation, illegal instruction, or address error traps usually occur when a program is being debugged. The bus error and the halted state typically indicate a hardware failure in the system. Recovery from any of these errors is usually not possible or desirable until the offending program or hardware is corrected.

In MC68010 systems, the bus error condition is used to implement virtual memory. If the processor were to attempt to access a location in the virtual memory space that is not residing in physical memory, a page fault would occur. The access to that location is temporarily suspended while the necessary program segments or data are fetched from secondary storage and placed in physical memory. Then the suspended access is completed. If the bus error is used to signal the page fault, the MC68010 can continue instruction execution of the suspended instruction after the physical memory has been updated.

11.3.1 Privilege Violations

If a program operating in the user mode attempts to execute a *privileged* instruction, a trap using the vector at location $20 is caused. The value of the PC saved on the stack is the address of the first word of the instruction causing the violation.

Table 11.4 Errors Causing Traps

Errors	Cause	Comments
Privilege violation	In user mode; attempt to execute privileged instruction	If S = {0}, attempt to execute: STOP, RESET, RTE, MOVE USP, ANDI to SR, EORI to SR, ORI to SR, or MOVE to SR
Illegal instruction	Bit pattern of op code not recognized	(PC) value on stack is address of illegal instruction
Address error	Attempted word access at odd address	Stack contains (PC), (SR), op word, access address, and status word (seven words)
Bus error	External request	Same stack contents as for address error
Halted state	Address or bus error during processing of bus error, address error, or reset	Reset required to restart processor

Note:
The address error or bus error exception in the MC68010 uses 29 words of the stack to allow instruction continuation after the exception is processed.

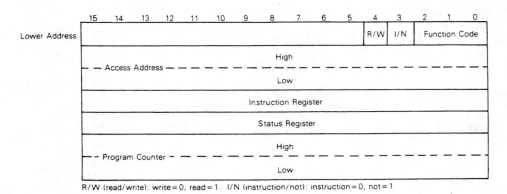

R/W (read/write): write = 0, read = 1. I/N (instruction/not): instruction = 0, not = 1

Figure 11.6   Stack contents after an address error. (Courtesy of Motorola, Inc.)

11.3.2 Illegal Instruction

The *illegal instruction* trap, with vector at location $10, is used to protect the system from the effects of incorrect machine code or a localized memory failure. The value of the PC saved on the stack points to the illegal instruction that caused the trap.

11.3.3 Address Error

If the processor attempts to access a word operand, longword operand, or an instruction at an odd address, an *address error* exception is generated using the vector at location $0C. Before the exception routine is executed, the following information is saved on the stack:

(a) the value of (PC), which may be advanced beyond the address of the first word of the instruction causing the error
(b) the value of (SR)
(c) the contents of the instruction register
(d) the address being accessed
(e) a status information word

The order and format of this information on the stack are shown in Figure 11.6. The information saved would not normally allow recovery from the error but is useful to aid in diagnosis of the problem.

Example 11.3

Figure 11.7 shows the printed output for a trace routine. The assembly language program shown causes an addressing error when the second instruction is executed. The first instruction, executed in the supervisor mode with trace yields the register contents shown. When the address error occurs, the supervisor stack pointer is used to address the stack and store the error information into lower memory locations.

```
abs.   LC    obj. code    source line
----   ----  ----------   -----------
  1    0000              |        TTL     'FIGURE 11.7(a)'
  2    0000              |        LLEN    100
  3    8000              |        ORG     $8000
  4    8000              |        OPT     P=M68000
  5    8000              | *
  6    8000              | *   CREATE AN ADDRESSING ERROR
  7    8000              | *
  8    8000  307C 3001   |        MOVE.W  #$3001,A0     ;ILLEGAL ADDRESS IN A0
  9    8004  3010        |        MOVE.W  (A0),D0       ; AND TRY TO ACCESS IT
 10    8006              |        END
```

```
TUTOR  1.32> T
PHYSICAL ADDRESS=00008000

3015 00003001 3010
ADDR TRAP ERROR
PC=00008006 SR=2700=.S7..... US=0000FC00 SS=00010000
D0=00000000 D1=00000000 D2=00000000 D3=00000000
D4=00000000 D5=00000000 D6=00000000 D7=00000000
A0=00003001 A1=00000000 A2=00000000 A3=00000000
A4=00000000 A5=00000000 A6=00000000 A7=00010000
--------------------008006    4E71              NOP
```

Figure 11.7   Address error example.

The trace exception was not taken because the address error exception has a higher priority. Instead, a TRAP ERROR is indicated by the monitor address error exception routine.

11.3.4 Bus Error

The bus error exception is generated when external circuitry reports the error via a processor signal line. The exact cause is determined by the hardware design of the system. Programmatically, the bus error is treated exactly as the address error just described but uses the vector at location $08. The stacked information is the same as for the address error exception. More details on the bus error and its system consequences are discussed in Chapter 13.

11.3.5 Halted State

If an address error or a bus error occurs during the processing of an address error, bus error, or system reset, the processor is halted. This response to double errors prevents the processor from unpredictable behavior when a catastrophic failure in the system is indicated. One of the processor signal lines is activated to indicate this condition to external devices.

EXERCISES

11.3.1. Write an address error routine to decode the information saved on the stack and output it to the user. (*Note:* The output routine to print the information is part of the supervisor program and depends on the system being used.)

11.3.2. Describe several ways an illegal instruction trap could occur in a system.

11.3.3. Assume that an illegal (odd) address is stored in the address error vector location. What would happen to the system if the processor detects an address error during execution of a program?

11.4 INTERRUPT PROCESSING BY THE MC68000

The interrupt system of the MC68000 allows an external device to interrupt the processor execution and causes program control to be passed to an interrupt–handling routine. This section is concerned with the programming aspects of the interrupt sequence for the MC68000, including the priority scheme for interrupts and the processor operation during interrupt processing. A discussion of the signal lines used for interrupt requests and acknowledgments is presented in Chapter 13.

An interrupt request from an external device can occur at one of seven levels of priority, which is determined by the value presented on the three interrupt signal lines of the processor. Priorities are assigned to interrupts so that level 1 is lowest and level 7 is highest. These priorities allow a routine that is processing a lower-level interrupt to be interrupted by a higher-level interrupt request. After the highest-level interrupt processing is completed, control returns to the next-lower-level interrupt routine that is waiting to complete execution. Once all of the interrupt requests are processed, control finally returns to any program that was interrupted.

From the processor's point of view, an interrupt is an externally generated request for exception processing. The interrupt request is considered active, pending, or disabled. An *active* request is processed immediately after the completion of any instruction currently executing provided that no higher priority exceptions take precedence. The request is *pending* if the processor is currently processing a higher-priority exception. Pending requests will be serviced when the higher-priority processing completes unless the processor enters the halted state. If an interrupt level is disabled, an interrupt request at that level is ignored until the level is enabled by changing the interrupt mask in the status register, (SR)[10:8].

The interrupt system is initialized by a supervisor program during system initialization by loading the starting addresses of each interrupt routine to be used into the appropriate location in the vector table. This initialization is performed with all interrupt levels disabled except level 7, which cannot be disabled. The interrupt levels are enabled just before control is passed from the supervisor program to the first application program to be executed.

The interrupt mask bits are shown in Figure 11.8, which includes the mask value for each level. In general, when the interrupt request is at the level masked or less, the interrupt request will not be accepted. The level 7 request is an exception to this rule and will be processed regardless of the setting of the interrupt mask. If the mask is set to {000}, all interrupt levels are enabled.

The vector table for interrupts is shown in Table 11.5. The spurious interrupt vector is used when the processor recognizes an interrupt request but some error condition exists externally. Rapid changes in the interrupt level signal lines (unstable

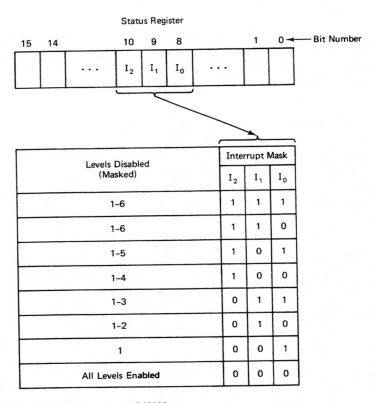

Figure 11.8   Interrupt mask for the MC68000.

value) or an indication of a bus error from an external device during an interrupt request will cause a spurious interrupt. Otherwise, the starting address of the interrupt routine for levels 1 through 7 will be taken from the appropriate vector location. The two classifications of interrupts are autovector and user interrupts. Selection between these modes of operation for the interrupt system is determined entirely by external circuitry. In either case, the address of the interrupt is calculated as four times the vector number. The difference in hardware operation has no effect on the design of interrupt routines.

11.4.1 Interrupt Processing

If the processor is executing instructions in the normal state, an interrupt request that is acknowledged and becomes active causes a sequence of events designed to pass control to a designated interrupt routine. This routine processes the interrupt as required, and then returns control to the interrupted program. This sequence of events is shown in Figure 11.9. The operations are performed by the hardware until control is passed to the interrupt routine. The processing required in the interrupt routine is entirely dependent on the application.

Table 11.5   Vector Table for Interrupt Routines

Vector Number (decimal)	Memory Location (hexadecimal)	Name
15	$003C	Uninitialized interrupt vector
24	$0060	Spurious
25	$0064	Level 1 autovector
26	$0068	Level 2 autovector
27	$006C	Level 3 autovector
28	$0070	Level 4 autovector
29	$0074	Level 5 autovector
30	$0078	Level 6 autovector
31	$007C	Level 7 autovector
•	•	•
•	•	•
•	•	•
64	$0100	User interrupt vector 1
65	$0104	User interrupt vector 2
•	•	•
•	•	•
•	•	•
255	$03FC	User interrupt vector 192

Notes:

1. Vector 15 should be provided by an uninitialized external device if the CPU requests a vector number.
2. A spurious interrupt occurs when the CPU detects an error during interrupt processing.

Use of interrupts for I/O programming is considered in Chapter 12. A program example in that chapter initializes and defines an interrupt routine for a programmable timer chip.

Example 11.4 _____

During the processing of a trap or interrupt, the contents of both the program counter (PC) and the status register (SR) are saved on the system stack. The supervisor stack pointer is the active stack pointer because the exception processing forces the system into the supervisor mode. The contents of the PC are pushed first, and then the 16-bit status register contents are pushed. Therefore, at least three words of the system stack are used for each response to an interrupt by the MC68000 processor. As discussed previously, the MC68010 uses at least four words of the stack.

As shown in Figure 11.10, an interrupt occurring when the initial contents of selected registers have the values

(SR) = $2008
(PC) = $00 164A
(SSP) = $0000 FFFE

causes a push of (PC) and then (SR) into the longword location $FFFA and the word location $FFF8, respectively. The interrupt routine may continue to use the system stack as required as long as (SSP) is restored to the value $0000 FFF8 before executing an RTE instruction to return control to the interrupted program. To return, (SR) are pulled from word location $FFF8 and

Figure 11.9 Interrupt processing.

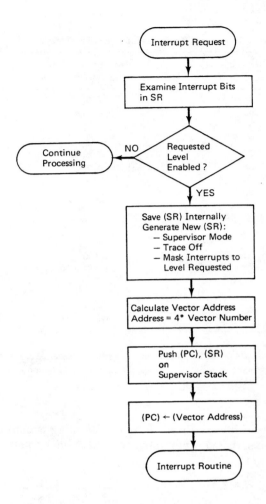

(PC) is restored from longword location $FFFA. Then processing continues with the instruction at location $00 164A. If the interrupt request is at autovectored level 5, the interrupt routine begins execution with

(PC) = ($0074)
(SR) = $2500

which allows execution at the vectored location with interrupts at level 5 and below disabled.

Example 11.5 _____

External signal lines determine whether the interrupt mode is *autovector* or *user interrupt*. The user interrupt is also called simply a vectored interrupt request. When the autovector mode is requested by an external device, the CPU automatically provides the vector location. Otherwise, in the user interrupt (vectored) mode, the external device must provide the vector number.

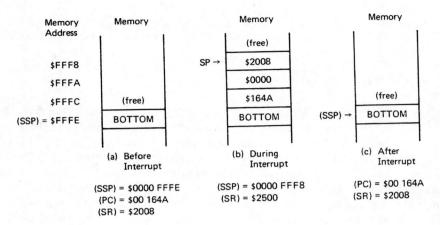

Memory Address	Memory	Memory	Memory
		(free)	
		SP → $2008	
$FFF8		$0000	
$FFFA		$164A	(free)
$FFFC	(free)		
(SSP) = $FFFE	BOTTOM	BOTTOM	(SSP) → BOTTOM

(a) Before Interrupt	(b) During Interrupt	(c) After Interrupt
(SSP) = $0000 FFFE	(SSP) = $0000 FFF8	(PC) = $00 164A
(PC) = $00 164A	(SR) = $2500	(SR) = $2008
(SR) = $2008		

Notes:
(1) The interrupt is a level 5 autovectored request.
(2) (SSP) indicates the Supervisor Stack Pointer.
(3) When the interrupt routine at level 5 begins execution, (PC) = ($0074).

Figure 11.10   System stack during interrupt processing.

This number is multiplied by 4 to give the address of the exception vector. The way that the vector number is supplied to the CPU from the external device is discussed in Chapter 13.

In either case, the programmer must know the following information to program an interrupt routine:

(a) the vector number (or address)
(b) the priority assigned by the hardware design of the system for masking or unmasking vectored interrupts (the priority for autovectors is fixed)
(c) details of the functional operation of the interrupt routine as determined by the hardware requirements

The location of the interrupt routine is typically decided during the software design phase and the priority for each interrupt is determined by the system requirements, primarily the timing constraints of external devices. The interrupt priority is important to the programmer only if the interrupt routine manipulates the interrupt mask.

EXERCISES

11.4.1. What are some system applications for the level 7, nonmaskable interrupt?

11.4.2. How many user interrupt vectored plus autovectored interrupt vectors are available according to the exception vector table? The vector number is an 8-bit integer allowing 256 entries to address the exception vector table, but not all entries can be associated with interrupts according to the table. Or can they? In other words, is a designer forbidden to use vectors 0 through 23 or vectors 32 through 47 as interrupt vectors?

11.4.3. Write an interrupt routine to update a real-time clock. The clock should give the time of day in hours, minutes, and seconds. Assume that a level 4 interrupt occurs once per second.

11.5 EXCEPTION PRIORITY

Exceptions can be categorized according to their priorities, which are fixed by the design of the processor and cannot be changed. Motorola divides exceptions into group 0, group 1, and group 2, in descending order of priority. Figure 11.11 lists the priority groups, the exceptions contained in each group, and the processing that occurs. The highest-priority exception is the reset exception, which causes system initialization. If this exception occurs, any other processing is immediately terminated. All the other exceptions are processed according to priority if two or more occur simultaneously.

If a group 0 exception is detected, the instruction being executed is aborted, possibly before it has completed execution. The instruction will complete its current hardware cycle, and then exception processing will begin. Within group 0, reset exceptions have the highest priority, followed by bus errors and, finally, address errors.

In group 1, the trace exception has the highest priority if the instruction being traced executes to completion. If execution is not completed because the instruction caused an illegal instruction trap or privilege violation trap, the trace exception will not occur. If an interrupt is pending following an instruction to be traced, the trace exception is processed first.

The exceptions in group 2 have no priorities within the group because only one instruction executes at a time. The execution of these instructions will always be completed unless a group 0 exception occurs during their execution. If the program and system are operating properly, only a reset exception would cause these instructions to be aborted.

EXERCISES

11.5.1. Describe the processing that occurs under the following conditions:
  (a) An illegal instruction is detected when the trace bit is {1}.
  (b) A bus error occurs during a TRAP instruction.
  (c) An interrupt occurs while an instruction is executing with $T = \{1\}$. Which routine gets final control?

Group	Exception	Processing
0	Reset Bus Error Address Error	Exception processing begins at the next minor cycle
1	Trace Interrupt Illegal Privilege	Exception processing begins before the next instruction
2	TRAP, TRAPV, CHK Zero Divide	Exception processing is started by normal instruction execution

Figure 11.11 Priority groups for exceptions. (Courtesy of Motorola, Inc.)

REVIEW QUESTIONS AND PROBLEMS

Multiple Choice

11.1. A CPU "trap" is recognized when
  (a) a hardware error occurs
  (b) a program instruction causes an exception
  (c) an interrupt occurs

11.2. An exception is always processed
  (a) in the supervisor mode
  (b) while a program is executing
  (c) when a program error occurs

11.3. The vector addresses for each exception handler are defined by the
  (a) hardware designer
  (b) applications programmer
  (c) system programmer

11.4. Which instruction initializes vector 32 (at address $80) with value $3600?
  (a) MOVE.L #$3600,$80
  (b) MOVE.L $3600,$80
  (c) MOVE.L 3600,80

11.5. Where does the PC point after the TRAP instruction executes in the sequence

  TRAP #15
  DC.W $0024

  (a) the next instruction
  (b) the constant
  (c) the TRAP instruction

11.6. There are program traps for which of the following errors?
  (a) divide by zero, overflow, syntax errors
  (b) out of memory, overflow, privilege violation
  (c) divide by zero, illegal instruction, divide by zero

11.7. What are typical examples of applications of the TRAP instruction?
  (a) I/O operations, instruction emulation, return to supervisor program
  (b) compile a program or execute a program
  (c) reset the processor

11.8. What is the use of the trace exception?
  (a) debugging the source program
  (b) debugging the object module
  (c) debugging the executing code

11.9. A privilege violation trap can only be caused by
  (a) a supervisor program
  (b) a user program
  (c) an exception handler

11.10. When a memory error occurs, error-correcting circuitry would cause
  (a) an addressing error
  (b) a bus error
  (c) an illegal instruction error

11.11. The MC68000 has how many priority levels of interrupts?
   (a) seven
   (b) six
   (c) eight

11.12. A level 5 interrupt occurs while the CPU is servicing a level 6 interrupt. The CPU
   (a) holds the level 5 request until the level 6 routine ends
   (b) ignores the level 5 request
   (c) services the level 5 request

11.13. What is the primary factor that determines the selection of priority for interrupts
   (a) the number of peripheral devices
   (b) the type of interrupt as vectored or autovectored
   (c) the timing requirements of external devices

11.14. When an autovectored interrupt occurs, the vector number is provided by the
   (a) CPU
   (b) interrupting device
   (c) interrupt routine

11.15. If an error occurs during exception processing, debugging includes analyzing the
   (a) status register
   (b) memory map
   (c) stack frame

Essay Questions

11.16. What is the disadvantage of a trace routine that only displays the values in every CPU register after the execution of each instruction?

11.17. Assume that the trace is enabled when the following conditions occur:
   (a) STOP #$2000 is executed.
   (b) An illegal instruction is encountered by the CPU in an instruction stream.
   What is the operation of the CPU in these cases?

11.18. Define the features of a general–purpose debugging routine that are desirable to aid a programmer in testing and debugging a program.

11.19. List the errors in hardware and software that could result in an illegal instruction being created.

11.20. Describe the cause of the possible errors or failures and the subsequent action of the bus error handling routine when a bus error exception occurs in the following cases:
   (a) during CPU access to external memory
   (b) during an interrupt acknowledgment cycle

11.21. Because there can be a number of sources of a bus error in an MC68000–based product, what methods might be used by the exception handler to determine the exact source? Assume that the information to determine the source of the error is not included in the exception stack frame.

11.22. What are some possible actions to be taken by the exception-handling routines when the CPU recognizes the following:
   (a) a spurious interrupt
   (b) an uninitialized interrupt

Programs

11.23. Create an illegal instruction exception handler to be located at location $5500. The handler should pass control to a supervisor program at location $4000 after an illegal instruction is recognized in a user mode program. Leave the faulted stack frame from the illegal instruction on the stack and create a new stack frame below it to pass control to location $4000.

11.24. Assume that an external watchdog timer causes a bus error when nonexistent memory is accessed. Write a routine to determine the amount of RAM in the system. Start the test of memory above the known locations of RAM that holds the monitor and the programs being developed. The address that caused the bus error is held in the stack frame at location (SP)+8.

FURTHER READING

The article by Hemenway and that by Starnes listed here discuss exception processing for the MC68000. Hemenway's article gives an example of an interrupt routine to service a terminal.

The article by Grappel gives an interesting application of the trace feature of the MC68000. His program calculates a histogram of memory usage as a program is traced.

The exceptions that are recognized by the Motorola 32–bit processors are described in detail in the user's manuals for those processors. Refer to these manuals for exact details concerning the various exceptions.

Grappel, Robert D. "MC68000 Charts Its Own Memory Usage." *EDN* 25, no. 21 (November 20, 1980), 115–117.

Hemenway, Jack, and Robert Grappel. "Use MC68000 Interrupts to Supervise a Console." *EDN* 25, no. 11 (June 5, 1980), 183–186.

Starnes, Thomas W. "Handling Exceptions Gracefully Enhances Software Reliability." *Electronics* 53, no. 20 (September 11, 1980), 153–157.

CHAPTER 12

Interfacing and
I/O Programming

This chapter discusses interfacing requirements and I/O programming for MC68000 systems. First, the general characteristics of interfaces are defined to describe typical MC68000 systems. Today, these systems are designed using peripheral chips to provide much of the interface circuitry. The programming of a few of these I/O chips is covered as their special requirements are defined. Basic techniques for programming the interfacing devices considered here apply to many of those available. Interfaces and I/O programming are presented in this chapter.

The third section presents programming examples using the SBC68K single-board computer. This unit contains peripheral chips for I/O functions and for timing applications. Motorola's MC68681 serial communications device is used to illustrate the techniques of I/O programming for communication. Another programming example for the MC68230 parallel interface and timer chip (PI/T) shows the initialization and interrupt handling for an interval timer.

12.1 INTERFACE DESIGN

Figure 12.1 shows the organization of a typical computer system, emphasizing the role played by the interfacing circuitry. This interface electrically connects the internal system bus with the device controller and resolves differences in timing or formats between the CPU and the external device. The CPU, through its logical circuitry and I/O routines, controls the operation of the system. The I/O routines prepare the interface for data transfers with the peripheral device, which itself is controlled by the device controller.

When the device is initialized and ready for transfer, it can perform its function, such as transmitting a character to the CPU or memory. Transfers to the CPU usually involve an interrupt request. High-speed transfers of blocks of characters are accomplished with Direct Memory Access (DMA) requests. If the device uses a DMA technique, for example, the transfer request passes through the interface to the bus control circuits and the memory.

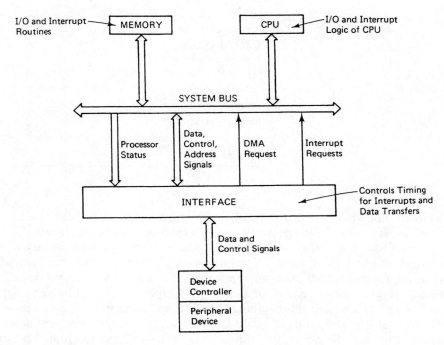

Figure 12.1   Typical interface organization.

The functional characteristics of an interface depend entirely on its application. For standard operations such as serial transfer of data to CRT terminals or parallel transfer to tape drives and disk units, the interface and perhaps part of the device controller are typically packaged together as a peripheral chip. When the interface is used to connect special-purpose devices, a custom interface may be required.

12.1.1 Functional Design of Interfaces

Most standard peripheral devices are connected to a system using peripheral chips to supply the interfaces. Therefore, it is important to understand the general principle of operation of these chips, which function as *programmable* interfaces. Such interfaces require initialization and act under the control of an I/O routine. The design, integration, and testing of these interfaces and the associated routines require cooperation between the hardware designer and the programmer.

Table 12.1 lists some of the items to be specified when describing the functional operation of an interface. This functional description is normally prepared by the hardware designer, who includes details of the programming required for the interface.

The functional description begins with a discussion of the purpose of the interface. This is governed by the requirements of the system design, particularly the type of peripheral device involved. Operational modes are described by defining the operation

Table 12.1 Functional Design of Interfaces

Purpose of the interface
Modes of operation
Initialization
Data transfers
Timing
Electrical characteristics
Physical characteristics
Programming: sequence of data and commands transferred
Test procedure

of the circuitry needed to perform initialization and data transfers. Timing considerations and similar hardware aspects of the interface must also be described.

Electrical and physical characteristics of the interface are also included in the functional description. The electrical details are dictated by the system bus and the electrical requirements of the device controller. Such details include specification of the voltage levels, rate of change of signals (rise times), and the like. The physical considerations include environmental requirements and space limitations on the circuit boards involved. The temperature and humidity ranges for the environment determine the type of chip and the packaging required for integrated circuit chips. The size of the circuit boards influences the number of chips and their placement on a board. Most designs involve standard-size circuit boards that plug into connectors to the system bus. The MC68000 itself is available in a number of different packages to meet various physical requirements.

Once the mode of operation of an interface has been specified, the programming requirements can be defined. An interface is controlled by sequences of logical variables or binary values written to the interfacing circuitry by an I/O routine. A sequence must be defined for each mode of operation. For example, the programming procedure to reset the interface and initialize it to some known state must be described. Programming a data transfer includes determining if the interface is ready for the transfer, performing the transfer, and checking for errors.

Finally, a test procedure for the interface is defined. This usually includes a hardware procedure to verify that the interface is functional, as well as the steps for testing of the interface under program control. The peripheral chips available as part of the MC68000 family of products serve to simplify the design and programming of interfaces.

12.1.2 Peripheral Chips as Interfaces

The MC68000 family includes a number of peripheral chips for interfacing. These integrated circuits provide the interface to particular peripheral devices and eliminate a great deal of the hardware design associated with interfacing. These chips are designed to connect to the MC68000 system bus directly or with a minimal amount of additional circuitry.

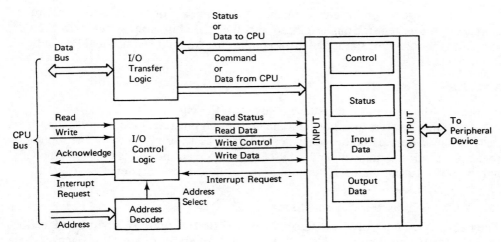

Figure 12.2 A typical peripheral chip.

Peripheral chips are programmed by writing into or reading from registers internal to the chips. The addresses of these registers are in the I/O space of the system. For MC68000 systems, these addresses appear to be memory locations because of the memory-mapped I/O scheme of the MC68000.

Figure 12.2 shows the structure of a typical peripheral chip. The output side of the peripheral chip has data and control lines designed to transfer control information and data to a peripheral device. The signal lines on the input side are used to select or "enable" the chip and transfer data between the CPU and internal registers of the chip. Each internal register has an I/O address defined during the system design. The address decoder shown enables the chip when an internal register address is placed on the address lines by the CPU.[1] The read/write control lines from the I/O control logic circuitry then select a particular register of the chip. After the peripheral chip acknowledges the request from the CPU, the I/O transfer logic acts as a buffer for data between the data bus and the peripheral chip.

The control register receives a sequence of bits from the CPU to control the chip's operation. For example, the MC68000 instruction

MOVE.B   #$\langle d_8 \rangle$,CREG

could be used to transfer 8 bits to the control register at address CREG. Various bits in the command byte $\langle d_8 \rangle$ typically determine whether a transfer operation is input or output and whether interrupts from the chip are enabled or not.

---

[1] In certain peripheral chips, the control and status registers have the same address. This is accomplished by using the "read" signal line to select a read-only status register. Similarly, the write signal line would be used to select the write-only control register. Many of the peripheral chips for the MC68000 family have this feature.

A status register on the chip contains information about the transfer or about the chip. Bits in this register might indicate whether the chip is ready for data transfer, as well as the interrupt status and error conditions. An 8-bit status could be read with the instruction

MOVE.B   SREG,D1

if SREG is the address of the status register on the chip.

Two data registers are shown on the chip in Figure 12.2. One receives data from the peripheral device and the other stores data to be transmitted to it. The input data register is read in a manner similar to that used to read the status register. The CPU writes to the output data register as it does to the control register.

12.1.3 MC68000 System Support Chips

The chips in the family of MC68000 products are listed in Figure 12.3. Their connections to the address and data bus of the CPU are illustrated in Figure 12.4. Each of these support chips fulfills a specific interfacing function and is programmed according to its unique requirements. Notice that when a memory management unit (MMU) is included in the system, it controls the addresses output on the address lines. In addition, a number of peripheral chips from the M6800 8-bit family that are directly compatible with the MC68000 are listed in Figure 12.5.

Product	Device No.	Title
MPU	MC68000	16 Bit MPU
	MC68008	16 Bit MPU with 8 Bit Data Bus
	MC68010	16 Bit Virtual Memory MPU
	MC68020	32 Bit MPU
MMU	MC68451	Memory Management Unit (MMU)
Math Processor	MC68881	Floating Point Co-Processor (FPCP)
Bus Controllers	MC68452	Bus Arbitration Module (BAM)
	MC68153	Bus Interrupter Module (BIM)
DMA Controllers	MC68440	DMA Controller (2-Channel) (DDMA)
	MC68450	DMA Controller (4-Channel) (DMAC)
General Purpose I/O	MC68230	Parallel Interface Timer (PI/T)
	MC68901	Multi-Function Peripheral (MFP)
Peripheral Controller	MC68120	Intelligent Peripheral Controller (IPC)
Data Communications	MC68561	Multi-Protocol Communications Controller II (MPCCII)
	MC68562	Dual Univeral Serial Communications Controllers
	MC68564	Serial Input/Output (SIO)
	MC68652	Multi-Protocol Communications Controller (MPCC)
	MC68653	Polynomial Generator Checker (PGC)
	MC68661	Enhanced Programmable Communications Interface (EPCI)
	MC68681	Dual Universal Asynchronous Receiver/Transmitter (DUART)
Disk Controller	MC68465	Floppy Disk Controller (FDC)

Figure 12.3   MC68000 family chips. (Courtesy of Motorola, Inc.)

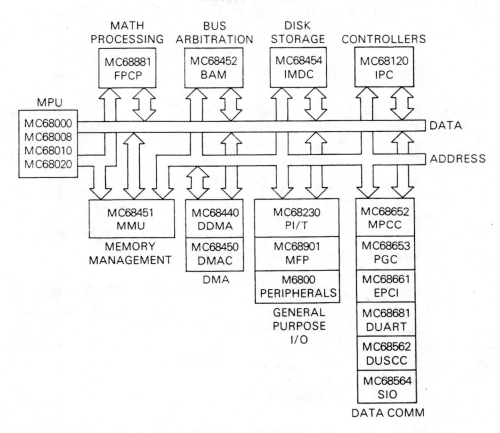

Figure 12.4 MC68000 system. (Courtesy of Motorola, Inc.)

EXERCISES

12.1.1. MC68000 systems use memory-mapped I/O for data transfer. Another scheme is called *isolated I/O*. With this technique, separate instructions for I/O are part of the CPU instruction set. Typically, there are several I/O instructions (IN or OUT) and 256 possible I/O locations or *ports*. These ports are accessed by separate signal lines that are not part of the system address bus. Describe the advantages and disadvantages of each scheme in terms of the system flexibility and the hardware requirements for interfacing.

12.1.2. Describe the steps in the program and the hardware sequence needed to initialize an interface to receive a byte of data from an external device. Define the operation of the interface and the use of its registers after the byte is received. What sequence is required of the processor itself to read the byte?

12.1.3. Refer to Motorola literature to describe the purpose and operation of the peripheral chips available for MC68000 systems.

M6800 FAMILY PERIPHERALS THAT ARE DIRECTLY COMPATIBLE WITH THE MC68000

MCM6810	128 X 8 Bit Static RAM
MC6821	Peripheral Interface Adapter
MC6840	Programmable Timer Module
MC6843	Floppy Disk Controller
MC6845	CRT Controller
MC6847	Video Display Generator
MC6850	Asynchronous Communication Interface Adapter
MC6852	Synchronous Serial Data Adapter
MC6854	Advanced Data Link Controller
MC6859	Data Security Device
MC6860	0 to 600 bps Digital Modem
MC6862	2400 bps Modulator
MC68488	General Purpose Interface Adapter

Figure 12.5   M6800 family chips. (Courtesy of Motorola, Inc.)

12.2 I/O PROGRAMMING

The use of memory-mapped I/O in MC68000 systems allows a great flexibility to the designer of I/O routines. Any instruction that references memory can be used to control or transfer data to and from a peripheral interface. The powerful instruction set and the various addressing modes of the MC68000 can be applied to I/O programming. In addition, two special instructions are available to control and access peripheral chips. These are introduced first in this section.

A general discussion of I/O transfers follows that differentiates between CPU initiated transfers and device initiated transfers. The SBC68K single-board computer is used to illustrate the I/O capability of various peripheral chips. The peripheral chips that are discussed are not described in complete detail; however, the manufacturer's literature about the chip can be referenced for further information.

12.2.1 RESET and MOVEP Instructions

The two instructions listed in Table 12.2 are used to control or access peripheral interfaces. The RESET instruction asserts a CPU signal line that is used to cause interfaces to assume their initial hardware state. This instruction can be executed only in the supervisor mode and does not affect the processor state. Execution continues with the next instruction.

The Move Peripheral Data (MOVEP) instruction transfers a word or longword value between a data register and *alternate* bytes of memory. This instruction is used to access peripheral chips whose register addresses are successive even or odd byte addresses in memory. For example, the instruction

MOVEP.L   D1,0(A1)

transfers four bytes from D1 to every other byte beginning at the first byte addressed by (A1). The high-upper byte, $(D1)[31:24]$, is transferred to location (A1); the middle-upper byte, $(D1)[23:16]$, to byte address (A1) + 2; and so on. This method to access interface registers is used for the M6800 family of peripheral chips and for some 16-bit peripheral chips from Motorola.

12.2.2 I/O Transfer Techniques

I/O transfers are divided into those initiated by the CPU and those initiated by the peripheral device and its controller. Table 12.3 lists transfer techniques in these categories and defines the required initialization and program operation. Before transfers begin, an I/O routine executed by the CPU performs the initialization for the interface.

The *unconditional* transfer requires that the peripheral device be ready for transfers at all times. A common example of this type of transfer is found in systems where

Table 12.2   RESET and MOVEP Instructions

Syntax	Operation
RESET	RESET signal line asserted
MOVEP.⟨l⟩ ⟨Dn⟩,⟨d₁₆⟩(An)	Transfer to (EA), (EA + 2), . . .
MOVEP.⟨l⟩ ⟨d₁₆⟩(An),⟨Dn⟩	Transfer from (EA), (EA + 2), . . .

Notes:
1. RESET is a privileged instruction.
2. ⟨l⟩ is W or L for MOVEP.

Table 12.3   I/O Transfer Techniques

Type of Transfer	Initialization by Program	Program Operation
CPU-initiated transfer		
Unconditional	None	Transfer data
Conditional	Set up device for direction of transfer	Test status of device and wait until ready; then transfer
Device-initiated transfer		
Interrupt transfer	1. Set up device for I/O transfer with interrupt 2. Enable interrupts	1. Transfer data when interrupt occurs 2. Clear interrupt request after transfer

DMA	1. Set up device for I/O transfer
	2. Load DMA registers
	  (a) Count
	  (b) Address
	3. Issue command to begin
		Process end-of-block interrupt

Note: Conditional I/O is sometimes called programmed I/O or polled I/O.

the "device" is a unit to display numbers or characters. The I/O routine simply transfers the data with a MOVE instruction to the proper address. Circuitry of the display unit is used to convert the binary word from the CPU to the proper display format. Another use of this method is to read a group of switches whose settings are coded into binary sequence.

Conditional transfers are sometimes called *programmed I/O* or *polled I/O*. The operation of these transfers is illustrated in Figure 12.6. Once the interface is initialized,

Figure 12.6   Conditional transfer.

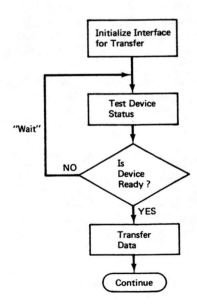

the I/O routine repeatedly checks the status register of the chip until the status indicates the device is ready. The routine then transfers the data to the peripheral chip. Because the CPU is in a wait loop until the device is ready, the usefulness of this transfer is limited. For example, a conditional transfer could be used to write into a control register during initialization of the interface.

For devices that transmit or receive data very slowly compared to the execution times of the I/O routines, interrupt-controlled transfer is preferred. This transfer method is shown in Figure 12.7. The interface is first initialized to transfer data. Then the CPU executes other programs until an interrupt occurs. When control is passed to the interrupt routine, a test of the peripheral chip status for errors is made. If an error is detected, the appropriate action is taken. Otherwise, the transfer occurs. Afterward, the interrupt request from the interface must be cleared before the next transfer can occur. This will usually be done automatically when the status register of the peripheral chip is read.

Figure 12.7   Interrupt-controlled transfer.

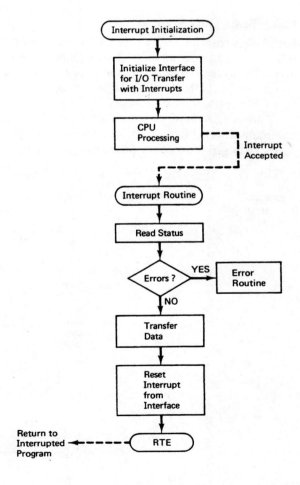

When a device is capable of transferring blocks of data at high speeds, the DMA method of transfer is frequently used. A DMA interface typically contains two programmable registers. A counter register contains the number of values to transfer and an address register contains the starting address of the block. The initialization program loads these registers and sends a command sequence to initiate the transfer. The CPU is then free to process other programs. The transfer of data between the interface and memory is entirely controlled by DMA circuitry without the intervention of the CPU.

For each DMA transfer of a data value, the DMA circuitry requests use of the system bus. The CPU relinquishes the bus for the time of transfer, which is usually one bus cycle. During the transfer, the DMA interface controls the memory just as the CPU normally does. Typically, less than one bus cycle out of five is used for DMA transfers. This "cycle stealing" has little effect on system performance in most cases. For comparison, the MC68000 requires at least four bus cycles to fetch an instruction. Upon completion of the transfer of an entire block of data, the DMA circuitry causes an interrupt. The CPU may now process the input data or initiate the next DMA output.

12.2.3 Serial Data Transfer

Serial data transfer refers to a communication method that transmits 1 data bit at a time via a single conductor. The stream of bits transmitted is encoded electrically according to the particular method used. A serial shift register connected to the signal conductor holds the data bits to be sent and those received. These data bits are stored in a CPU register or in memory in parallel form.

Figure 12.8 shows a simplified block diagram of the circuitry needed for serial communication. For output, circuits convert data from the parallel form stored in memory to serial data for transmission. Bits received in serial fashion are accumulated

Figure 12.8 Serial data communication circuitry.

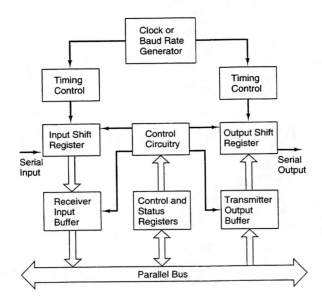

and stored in parallel form. The series of bits is sent to or received from an external device not shown in the figure. The precise characteristics of the serial bit stream being transferred depends on the particular serial communications scheme employed. The scheme is defined by the protocol used, the signaling characteristics, and the physical characteristics of the interfaces involved.

In serial transmission, the receiver must sample the signal at precise time intervals to determine whether a data bit is a {1} or a {0}. To do so, the receiver must know exactly when a transmission starts and stops. These and other issues are specified by a precisely defined set of rules for data transmission that are often called *protocols*. Standard protocols have been defined for both asynchronous and synchronous methods of transfer.

Asynchronous and Synchronous Communication. In *asynchronous communication,* information is transmitted as individual data items bracketed by a start bit and either one or two stop bits. A typical example is shown in Figure 12.9(a) for the transmission of 8 data bits bracketed by one start bit and one stop bit. This format could be used to transmit an ASCII character plus a parity bit. Chapter 8 presented a program example to add a parity bit to an ASCII character. The complete sequence of bits is called a *frame*.

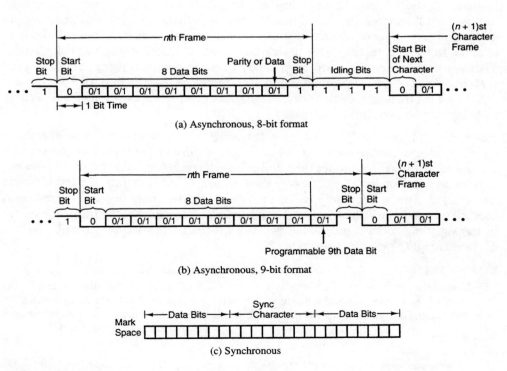

Figure 12.9 Asynchronous and synchronous frame formats.

Some interfaces allow a ninth bit to be added to the data frame. The format for each frame with the added bit is shown in Figure 12.9(b). The extra bit is used for communication between master and slave devices to address the slave devices.

The time between frames is called the *idle time* and it may vary from frame to frame. Figure 12.9(a) shows three idle bits after the nth frame. No idle bits between frames are shown in Figure 12.9(b). A frame is detected by the receiving device when the signal voltage drops from HIGH (logic {1}) to LOW (logic {0}) as the start bit is recognized. This synchronizes the frame between receiver and transmitter. However, the transmission method is called *asynchronous* because the receiver and transmitter have separate clocks that are not synchronized in time. The clocks determine the bit transmission rate (baud rate).

After the start bit is recognized, the receiver must determine the location of each bit boundary and sample the value. The receiver does this by measuring the voltage of the input signal a fixed number of times during the period of each bit. The bit is stored as a {0} or a {1} according to the results of the measurement.

The asynchronous method is often used for data transmission between a computer and a low-speed peripheral device such as an operator's terminal or printer. Because start and stop bits are needed in every frame, the method is not generally as efficient for the transfer of information as synchronous transfers.

In *synchronous communication*, information is transmitted in blocks. A possible format is shown in Figure 12.9(c). The receiver and transmitter clocks are synchronized in time because a clock signal is transmitted along with the data. The clock signal can be sent on a separate conductor or it may be encoded with the data in a self-clocking scheme. In either case, the start and stop bits of asynchronous transfer are eliminated. There is no idle time between blocks, but when no data are being sent, a "sync" (synchronizing) character is transmitted. The receiver's clock is thus always synchronized with the transmitter's clock. Very high transmission rates can be achieved reliably as compared to asynchronous transfer.

Baud Rate.   Once the type of format for serial data transmissions is selected, the signaling characteristics must be defined. One important parameter is the bit rate measured in bits per second technique. Because the bit is the smallest unit of information for serial communications, the signaling speed is defined in terms of bits per second transmitted. This is frequently called the *baud rate* when only two-level signaling is used.[2] The baud rate is the reciprocal of the length (in seconds) of the shortest element in the signaling code. If more than two voltage levels are used in transmission, the bit rate is higher than the baud rate.

Example 12.1 _____

The baud rate for two-level signaling is the reciprocal of the time duration of a bit. The frequency of the transmitter and receiver clocks determine the baud rate for an asynchronous communications channel. The clocks must not differ in frequency by more than several percent

---

[2] Baud is a contraction of the surname of J.M.F. Baudot according to the *Encyclopedia of Computer Science*, Van Nostrand Reinhold Company, 1976. Baudot invented a French telegraph code adopted in 1877.

or transmission errors may result. At 9600 baud for example, each bit has a duration of 1/9600 seconds, or about 0.104 milliseconds. A frame of 10 bits thus requires about a millisecond to be received or transmitted. This rate would allow the transmission of 1000 ASCII characters per second.

In most applications using interrupt controlled transfers, the CPU would be interrupted after each character is received. The interrupt handling routine would read the character from the receiver's input register and store it in memory. For transmission, the CPU is interrupted each time the transmitter is ready for a new character. The interrupt handling routine must transfer each character from one of the CPU registers or memory to the transmitter's output register for transmission.

EXERCISES

12.2.1. Compare interrupt-controlled and conditional transfer of data with respect to each of the following:
   (a) system speed of operation
   (b) programming complexity
   (c) interface complexity

12.2.2. Describe the program and system operation for a conditional I/O transfer routine when a string of characters is to be transferred. If a character is ready from an external unit every 0.1 second, compute the time required to input 10 characters if the CPU processing time is 10^{-5} seconds per character. What is the percentage of time the CPU is idle waiting for a character? This timing is typical for an operator's terminal transferring data at 110 baud.

12.2.3. Suppose an asynchronous transfer requires 11 bits per frame to transmit a seven-bit ASCII character. If the transfer rate is 1200 baud, calculate the following:
   (a) the overhead in percent used by the control bits for each character
   (b) the actual transmission rate in characters per second

12.2.4. Consider an asynchronous transmission in the full-duplex mode (simultaneous transmission in both directions) at 1200 baud. The format is seven bits plus parity per character with one start and one stop bit. If interrupt-controlled transfer is used, calculate the following:
   (a) the number of characters per second that are transmitted and received
   (b) the time the CPU has to respond to an interrupt indicating that the receiver buffer is full
   (c) the percentage of CPU time taken servicing the interrupts if each interrupt routine requires 10 microseconds to execute

12.2.5. Answer the questions in Exercise 12.2.4 if the baud rate is 500,000 baud for the system.

12.3 I/O FEATURES OF THE EVALUATION MODULES

The Arnewsh SBC68K is a single-board computer used to create and test prototype systems. Figure 12.10 is a block diagram of this module. The TUTOR monitor is contained in ROM and the RAM storage is 64 KB for user programs. The board

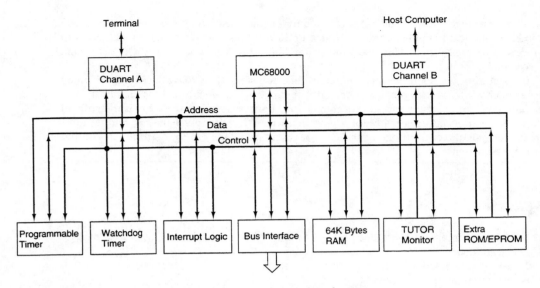

Figure 12.10   Block diagram of the SBC68K single-board computer.

contains peripheral chips for serial communications, timer applications, and disk control.

For serial communication, the SBC68K uses the Motorola MC68681 Dual Universal Asynchronous Receiver Transmitter (DUART) chip. The MC68230 Parallel Interface/Time (PI/T) chip provides timing functions as well as parallel I/O ports. The disk controller chip is not discussed in this section. Memory assignments for the module are listed in Table 12.4. These addresses will be used in the programming examples in this section.

Motorola MC68681 DUART.   The MC68681 is a dual-channel asynchronous communications chip typically used for communication between a computer and an operator's terminal. Figure 12.11 is a block diagram of the device. Channel A or Channel B is

Table 12.4   SBC68K Memory Assignment

Item	Address
RAM area (64 KB)	0-$FFFF
ROM (TUTOR)	$800000-$81FFFF
PI/T	$FE8000-$FEFFFF
DUART	$FF0000-$FF7FFF
Disk controller	$FF8000-$FFFFFF

Notes:
The TUTOR monitor uses the RAM area up to $08FF.
User programs should start above this address.

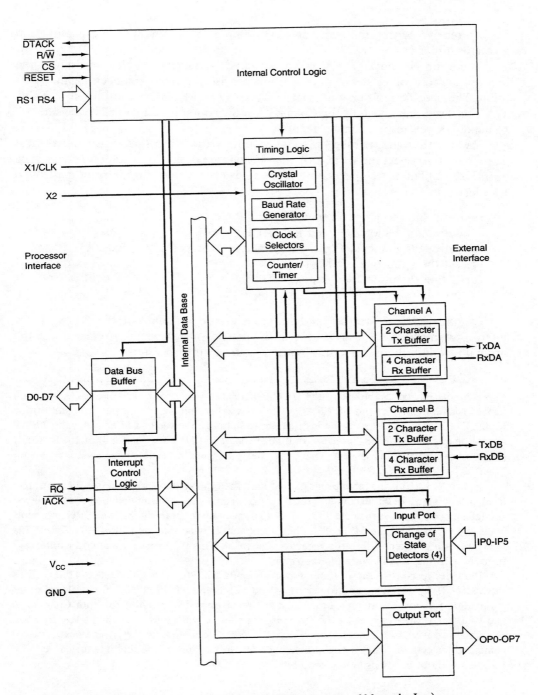

Figure 12.11   Block diagram of the MC68681 DUART. (Courtesy of Motorola, Inc.)

used to receive or transmit serial data. The chip also has parallel I/O capability and a counter/timer.[3]

Operation of the DUART is programmed by writing control words into the appropriate control registers. Status registers provide information concerning the operation. The specific registers and their addresses on the SBC68K are listed in Table 12.5. For our discussion, only the Channel A serial communications will be considered. Channel B operates in a similar manner.

The control registers used for serial communications are 8-bit registers that must be initialized to select the serial format and baud rate desired and other functions such as interrupts. The basic operation and the registers involved for interrupt controlled transfers is as follows:

(a) reset the receiver and transmitter (CRA)
(b) select the baud rate generator (ACR)
(c) select the number of bits, parity, and interrupts (MR1A, MR2A)
(d) set the baud rate (CSRA)
(e) define the interrupt vector (IVR)
(f) enable the receiver and transmitter (CRA)

Example 12.2 shows the initialization and programming of the DUART for interrupt-controlled serial I/O using an input and output queue. Comments in the program explain the initial values written to the control registers.

Example 12.2 _____

This example illustrates the use of the MC68681 for interrupt-driven serial I/O using separate queues for input and output. A TRAP #2 call is used to initiate the output. The subroutine EXINIT in Figure 12.12(a) has the instructions to initialize the vector addresses for the TRAP #2 routine and the MC68681 interrupt routine at vector 65 ($41). The MC68681 on the SBC68K board interrupts at CPU level 4. The CPU interrupts must be enabled for level 4 or lower. Thus, SR[10:8] = $3, $2, $1, or $0.

Figure 12.12(b) is the main program that initializes the system and processes the input line from the keyboard. In this example, the main program only causes the input line to be redisplayed and then loops to receive the next line. Notice that the main program disables the CPU interrupts while the vectors and the MC68681 is being initialized by subroutine COMINIT. Then CPU interrupts are enabled before the receiver is enabled. This is to prevent a spurious interrupt from the peripheral chip during initialization.

Routines to control the I/O are shown in Figure 12.12(c). Routine COMINIT initializes the MC68681 for receiving and transmitting and selects the baud rate (19200 baud) and other parameters for serial communication. First the receiver and transmitter for both Channel A and Channel B are reset, even though only Channel A will be used. Setting CRA[B] = $20 resets the receiver for Channel A and CRA[B] = $30 resets the transmitter. The next series of commands selects the clock source, interrupt-controlled transfer, and the data format as 8-bit transfers without parity using two stop bits.

---

[3] The parallel I/O ports and the counter/timer of the DUART are not discussed in this textbook.

Table 12.5 Register Addresses for the MC68681 DUART Chip

Register	Address	Meaning
MR1A	$FF0001	;MODE REGISTER A
MR2A	$FF0001	;MODE REGISTER A
SRA	$FF0003	;CHANNEL STATUS REGISTER A
CSRA	$FF0003	;CLOCK-SELECT REGISTER A
CRA	$FF0005	;COMMAND REGISTER A
RBA	$FF0007	;RECEIVER BUFFER A
TBA	$FF0007	;TRANSMITTER BUFFER A
IPCR	$FF0009	;INPUT PORT CHANGE REGISTER A
ACR	$FF0009	;AUXILIARY CONTROL REGISTER A
ISR	$FF000B	;INTERRUPT STATUS REGISTER
IMR	$FF000B	;INTERRUPT MASK REGISTER
CUR	$FF000D	;COUNTER MODE: CURRENT MSB OF CTR
CTUR	$FF000D	;COUNTER/TIMER UPPER REGISTER
CLR	$FF000F	;COUNTER MODE: CURRENT LSB OF CTR
CTLR	$FF000F	;COUNTER/TIMER LOWER REGISTER
MR1B	$FF0011	;MODE REGISTER B
MR2B	$FF0011	;MODE REGISTER B
SRB	$FF0013	;STATUS REGISTER B
CSRB	$FF0013	;CLOCK-SELECT REGISTER B
CRB	$FF0015	;COMMAND REGISTER B
RBB	$FF0017	;RECEIVER BUFFER B
TBB	$FF0017	;TRANSMITTER BUFFER B
IVR	$FF0019	;INTERRUPT VECTOR REGISTER
IPR	$FF001B	;UNLATCHED INPUT PORT
OPCR	$FF001B	;UNLATCHED OUTPUT PORT
SACC	$FF001D	;START COUNTER COMMAND
OPRBSC	$FF001D	;OUTPUT PORT REGISTER BIT SET
SOCC	$FF001F	;STOP COUNTER COMMAND
OPRBRC	$FF001F	;OUTPUT PORT REGISTER BIT RESET

The routine COMINIT also defines the interrupt vector ($41) used by the MC68681. When the routine completes, the chip is initialized for serial communication but the receiver and transmitter are not enabled. Because there are separate input and output queues, routine COMINIT initializes both queues. The COMINIT subroutine calls subroutine QINIT to create the initially empty queues as previously described in Section 9.3.

After the main program enables the CPU interrupts, the MC68681 receiver and its interrupts are enabled by a call to routine COMSTRT. Then the main program waits for an input character in the loop at label KEYIN. The character is held in the input queue after the receiver interrupt

```
abs. rel.    LC   obj. code    source line
---- ----    ----  ---------    -----------
  1    1    0000                |         TTL     'FIGURE 12.12(a)'
  2    2    0000                |         LLEN    100
  3    3    3100                |         ORG     $3100
  4    4    3100                |         OPT     P=M68000
  5    5    3100                |*
  6    6    3100                |* SUBROUTINE TO INITIALIZE EXCEPTION VECTORS
  7    7    3100                |*
  8    8    3100 21FC 0000      |EXINIT   MOVE.L  #$32AE,34*4      ;SET TRAP2 VECTOR (SENDCH)
  8         3104 32AE 0088      |
  9    9    3108                |*                                ; FOR OUTPUT
 10   10    3108 21FC 0000      |         MOVE.L  #$32DE,65*4      ;SERIAL COMMUNICATION
 10         310C 32DE 0104      |
 11   11    3110                |*                                ; INTERRUPT VECTOR (COMINT)
 12   12    3110                |*
 13   13    3110 4E75           |         RTS
 14   14    3112               |         END
```

Figure 12.12(a)   Program to initialize the exception vectors for serial data transfer.

routine stores the character. Subroutine DEQUEUE places a character from the queue into D0[B]. The main program causes the character to be echoed to the screen by a call to TRAP #2.

The TRAP #2 routine uses subroutine ENQUEUE to move the character from D0[B] to the output queue. Then, if the transmitter is not busy, the trap routine sets a flag (F—XMIT) and enables the transmitter. This causes an interrupt because the transmitter buffer register (TBA) is empty. The interrupt routine COMINT reads a character from the output queue and loads the character in TBA if the queue is not empty. After each character is transmitted, an interrupt occurs and another character is output to the screen. When the output queue is empty, the interrupt routine clears the flag F—XMIT and disables the transmitter.

Motorola MC68230 Parallel Interface/Timer (PI/T).   The PI/T is a multipurpose chip used for parallel data transfers and timing functions. Figure 12.13 shows the block diagram indicating the I/O ports and the timer. The timing function of the PI/T can be used to generate periodic interrupts or a single interrupt after a programmed time period. Thus, it can be used for elapsed time measurement or as a watchdog timer. This chip can also generate a square wave output with a programmed period.

As an interval timer, the time period between interrupts is programmable by writing a 24-bit number into three 8-bit registers called the *preload* registers. The value is decremented by one on every clock pulse to the timer. For continuous counting, a periodic clock signal is used. When the count reaches $000000, the preload register can be reset to the original value and the countdown begins again. The time in seconds corresponding to each decrement of the count is determined by the period of the clock signal that decrements the counter. The time interval for a given number of counts N is determined as

$$T_i = T_c \times N \text{ seconds}$$

where T_c is the clock period.

```
abs.  rel.   LC    obj. code     source line
----  ----   ----  ---------     -----------
  1     1    0000                        TTL     'FIGURE 12.12(b)'
  2     2    0000                        LLEN    100
  3     3    8000                        ORG     $8000
  4     4    8000                        OPT     P=M68000
  5     5    8000          *
  6     6    8000          * SERIAL I/O - INTERRUPT DRIVEN
  7     7    8000          *
  8     8    8000          * THIS PROGRAM RECEIVES AN INPUT LINE FROM THE KEYBOARD.
  9     9    8000          * AFTER AN INPUT LINE IS TERMINATED BY A CARRIAGE RETURN,
 10    10    8000          * THE ENTIRE LINE IS REDISPLAYED ON THE SCREEN.
 11    11    8000          *
 12    12    8000          * AN INPUT BUFFER (INBUF) IS USED FOR STORING THE CHARACTERS.
 13    13    8000          *
 14    14          0000 3100  EXINIT  EQU     $3100           ; INITIALIZE VECTORS ROUTINE
 15    15          0000 3200  COMINIT EQU     $3200           ; INITIALIZE MC68681 ROUTINE
 16    16          0000 329C  COMSTRT EQU     $329C           ; START RECEIVER
 17    17          0000 35B4  INQPT   EQU     $35B4           ; RECEIVER QUEUE POINTER TABLE
 18    18          0000 3386  DEQUEUE EQU     $3386           ; GET CHARACTER FROM QUEUE
 19    19          0000 000D  KEY_CR  EQU     $0D             ; CARRIAGE RETURN
 20    20          0000 000A  KEY_LF  EQU     $0A             ; LINE FEED
 21    21    8000          ********************************************************************
 22    22    8000 46FC 2700  MAIN    MOVE.W  #$2700,SR       ; MASK CPU INTERRUPTS
 23    23    8004 4EB8 3100          JSR     EXINIT          ; INITIALIZE VECTORS
 24    24    8008 4EB8 3200          JSR     COMINIT         ; INITIALIZE MC68681 FOR INPUT
 25    25    800C 46FC 0000          MOVE.W  #$0000,SR       ; ENABLE CPU INTERRUPTS; USER MODE
 26    26    8010 4EB8 329C          JSR     COMSTRT         ; START RECEIVER
 27    27    8014          *
 28    28    8014          * FIRST MOVE TO A NEW DISPLAY LINE
 29    29    8014 103C 000A          MOVE.B  #KEY_LF,D0      ; SEND A LINE FEED
 30    30    8018 4E42             TRAP    #2              ;
 31    31    801A 103C 000D          MOVE.B  #KEY_CR,D0      ; SEND A CARRIAGE RETURN
 32    32    801E 4E42             TRAP    #2              ;
 33    33    8020          |
 34    34    8020 43F9 0000          LEA     INBUF,A1        ; INPUT BUFFER
 34         8024 8068
 35    35    8026 323C FFFF  NEXT    MOVE.W  #-1,D1          ; BUFFER POINTER
 36    36    802A          * WAIT FOR INPUT
 37    37    802A 287C 0000  KEYIN   MOVEA.L #INQPT,A4       ; GET NEXT CHARACTER
 37         802E 35B4
 38    38    8030 4EB8 3386          JSR     DEQUEUE         ; D0.B = DATA; D6.W = RETURN CODE
 39    39    8034 4A46             TST.W   D6              ; WAS ANY DATA RECEIVED?
 40    40    8036 66F2             BNE     KEYIN           ; NO, SO KEEP CHECKING
 41    41    8038          *
 42    42    8038          * CHARACTER WAS RECEIVED, ECHO AND ADD TO BUFFER
 43    43    8038          *
 44    44    8038 4E42             TRAP    #2              ; ECHO CHARACTER
 45    45    803A 5241             ADDI.W  #1,D1           ; INCREMENT BUFFER POINTER
 46    46    803C 1380 1000          MOVE.B  D0,(A1,D1.W)    ; STORE CHARACTER
 47    47    8040 0C00 000D          CMPI.B  #KEY_CR,D0      ; WAS IT A CARRIAGE RETURN?
 48    48    8044 66E4             BNE     KEYIN           ; NO, GET THE NEXT CHARACTER
 49    49    8046          *
```

Figure 12.12(b)   Main program for MC68681 DUART.

```
50  50  8046            | * WHEN CR RECEIVED, INSERT LF TO BUFFER AND ECHO TO SCREEN
51  51  8046 103C 000A  |          MOVE.B  #KEY_LF,D0         ;CREATE LINE FEED CHARACTER
52  52  804A 1380 1000  |          MOVE.B  D0,(A1,D1.W)       ;REPLACE CR WITH LF IN BUFFER
53  53  804E 4E42       |          TRAP    #2                 ;ECHO LF TO THE SCREEN
54  54  8050 13BC 000D  |          MOVE.B  #KEY_CR,1(A1,D1.W)
54      8054 1001       |
55  55  8056            |                                     ;APPEND BUFFER WITH CR
56  56  8056            | *
57  57  8056            | * PROCESS DATA - PASS INPUT LINE TO THE OUTPUT QUEUE
58  58  8056 4241       |          CLR.W   D1                 ;CLEAR BUFFER POINTER
59  59  8058 1031 1000  |INOUT     MOVE.B  (A1,D1.W),D0       ;PASS CHARACTER TO OUTPUT QUEUE
60  60  805C 4E42       |          TRAP    #2                 ;DISPLAY CHARACTER
61  61  805E 0C00 000D  |          CMPI.B  #KEY_CR,D0         ;WAS IT A CARRIAGE RETURN?
62  62  8062 67C2       |          BEQ     NEXT               ;YES, START INPUT AGAIN
63  63  8064 5241       |          ADDI.W  #1,D1              ;INCREMENT BUFFER POINTER
64  64  8066 60F0       |          BRA     INOUT              ;GET NEXT ITEM FROM INPUT BUFFER
65  65  8068            | ****************************************************************************
66  66  8068            | * DATA BUFFER
67  67  8068            |INBUF     DS.B    256                ;INPUT LINE
68  68  8068            | *
69  69  8168            |          END
```

Figure 12.12(b)   *(continued)*

```
abs. rel.   LC    obj. code    source line
---- ----   ----  ---------    -----------
  1    1   0000                |          TTL     'FIGURE 12.12(c)'
  2    2   0000                |          LLEN    100
  3    3   3200                |          ORG     $3200
  4    4   3200                |          OPT     P=M68000
  5    5   3200                | *
  6    6   3200                | * SERIAL I/O - INTERRUPT DRIVEN
  7    7   3200                | *   INTERRUPT AND TRAP VECTORS MUST BE INITIALIZED BY EXINIT.
  8    8   3200                | * INITIALIZATION FOR COMMUNICATION, TRAP 2 AND INTERRUPT ROUTINES
  9    9   3200                | ****************************************************************************
 10   10   3200                | * REGISTER ADDRESSES FOR THE MC68681 DUART CHIP
 11   11        00FF 0001      |MR1A      EQU     $FF0001 ;MODE REGISTER A
 12   12        00FF 0001      |MR2A      EQU     $FF0001 ;MODE REGISTER A
 13   13        00FF 0003      |SRA       EQU     $FF0003 ;CHANNEL STATUS REGISTER A
 14   14        00FF 0003      |CSRA      EQU     $FF0003 ;CLOCK-SELECT REGISTER A
 15   15        00FF 0005      |CRA       EQU     $FF0005 ;COMMAND REGISTER A
 16   16        00FF 0007      |RBA       EQU     $FF0007 ;RECEIVER BUFFER A
 17   17        00FF 0007      |TBA       EQU     $FF0007 ;TRANSMITTER BUFFER A
 18   18        00FF 0009      |ACR       EQU     $FF0009 ;AUXILIARY CONTROL REGISTER A
 19   19        00FF 000B      |ISR       EQU     $FF000B ;INTERRUPT STATUS REGISTER
 20   20        00FF 000B      |IMR       EQU     $FF000B ;INTERRUPT MASK REGISTER
 21   21        00FF 0015      |CRB       EQU     $FF0015 ;COMMAND REGISTER B
 22   22        00FF 0019      |IVR       EQU     $FF0019 ;INTERRUPT VECTOR REGISTER
 23   23   3200                | ****************************************************************************
```

Figure 12.12(c)   Serial I/O routines for MC68681 DUART.

```
24   24   3200                    |* THESE ARE THE OFFSETS INTO THE QUEUE POINTER TABLES
25   25         0000 0000         |NEWQP   EQU     0
26   26         0000 0004         |ENDQP   EQU     4
27   27         0000 0008         |QUEOUT  EQU     8
28   28         0000 000C         |QUEIN   EQU     12
29   29   3200                    |*
30   30   3200                    |* INITIALIZE THE SERIAL COMMUNICATIONS PERIPHERAL.
31   31   3200                    |*
32   32   3200                    |* ONLY CHANNEL A IS BEING USED BY THIS APPLICATION, THEREFORE
33   33   3200                    |* CHANNEL B IS DISABLED.
34   34   3200                    |*
35   35   3200 13FC 000A          |COMINIT MOVE.B  #$0A,CRA        ;STOP THE RECV AND XMIT ON CHAN A
35        3204 00FF 0005          |
36   36   3208 1039 00FF          |DI_WAIT MOVE.B  SRA,D0          ;GET STATUS REGISTER FOR CHAN A
36        320C 0003               |
37   37   320E 0800 0003          |        BTST.L  #3,D0           ;IS IT DONE XMITTING?
38   38   3212 66F4               |        BNE     DI_WAIT         ;IF NOT, WAIT UNTIL DONE
39   39   3214 13FC 0020          |        MOVE.B  #$20,CRB        ;RESET THE RECEIVER B
39        3218 00FF 0015          |
40   40   321C 13FC 0030          |        MOVE.B  #$30,CRB        ;RESET THE TRANSMITTER B
40        3220 00FF 0015          |
41   41   3224 13FC 0020          |        MOVE.B  #$20,CRA        ;RESET THE RECEIVER A
41        3228 00FF 0005          |
42   42   322C 13FC 0030          |        MOVE.B  #$30,CRA        ;RESET THE TRANSMITTER A
42        3230 00FF 0005          |
43   43   3234 4239 0000          |        CLR.B   F_XMIT          ;NOT CURRENTLY TRANSMITTING
43        3238 35D4               |
44   44   323A                    |*
45   45   323A 13FC 0090          |        MOVE.B  #$90,ACR        ;SELECT BAUD RATE GENERATOR 2,
45        323E 00FF 0009          |
46   46   3242                    |                                ;CLOCK SOURCE = TRANSMITTER A,
47   47   3242                    |                                ;DISABLE INTERRUPTS
48   48   3242 13FC 0010          |        MOVE.B  #$10,CRA        ;RESET MODE REGISTER POINTER A
48        3246 00FF 0005          |
49   49   324A 13FC 0013          |        MOVE.B  #$13,MR1A       ;RTS DISABLED,
49        324E 00FF 0001          |
50   50   3252                    |                                ;RXRDY INTERRUPT,
51   51   3252                    |                                ;CHAR MODE ERROR REPORTING,
52   52   3252                    |                                ;NO PARITY, 8 BITS PER CHAR
53   53   3252 13FC 000F          |        MOVE.B  #$0F,MR2A       ;NORMAL MODE,
53        3256 00FF 0001          |
54   54   325A                    |                                ;RTS CONTROL DISABLED,
55   55   325A                    |                                ;CTS IGNORED,
56   56   325A                    |                                ;2 STOP BITS
57   57   325A 13FC 00CC          |        MOVE.B  #$CC,CSRA       ;SELECT 19.2 BAUD
57        325E 00FF 0003          |
58   58   3262 13FC 0041          |        MOVE.B  #$41,IVR        ;IDENTIFY INTERRUPT VECTOR
58        3266 00FF 0019          |
59   59   326A                    |
```

Figure 12.12(c)   *(continued)*

```
60    60    326A            |* INITIALIZE QUEUES
61    61    326A 287C 0000  |        MOVE.L   #INQPT,A4        ;PASS ADDR OF PTR TABLE FOR INPUT
61          326E 35B4       |
62    62    3270 2A7C 0000  |        MOVE.L   #INQUEUE,A5      ;  QUEUE, ADDR OF THE INPUT QUEUE,
62          3274 33B4       |
63    63    3276 2C7C 0000  |        MOVE.L   #E_INQUEUE,A6    ;  AND ADDR OF END OF THE QUEUE
63          327A 34B4       |
64    64    327C 4EB9 0000  |        JSR      QINIT            ;  TO INITIALIZATION ROUTINE
64          3280 3340       |
65    65    3282            |
66    66    3282 287C 0000  |        MOVE.L   #OUTQPT,A4       ;PASS ADDR OF PTR TABLE FOR OUTPUT
66          3286 35C4       |
67    67    3288 2A7C 0000  |        MOVE.L   #OUTQUEUE,A5     ;  QUEUE, ADDR OF OUTPUT QUEUE,
67          328C 34B4       |
68    68    328E 2C7C 0000  |        MOVE.L   #E_OUTQUEUE,A6   ;  AND ADDR OF END OF THE QUEUE
68          3292 35B4       |
69    69    3294 4EB9 0000  |        JSR      QINIT            ;  TO INITIALIZATION ROUTINE
69          3298 3340       |
70    70    329A 4E75       |        RTS                       ;COMMUNICATION INITIALIZED
71    71    329C            |*                                 ; BUT NOT ACTIVE
72    72    329C            |********************************************************************
73    73    329C            |* START RECEIVER
74    74    329C 13FC 0003  |COMSTRT MOVE.B   #$03,IMR          ;ENABLE INTERRUPTS
74          32A0 00FF 000B  |
75    75    32A4            |*                                 ; (RXRDYA AND TXRDYA)
76    76    32A4 13FC 0001  |        MOVE.B   #$01,CRA          ;START RECEIVER
76          32A8 00FF 0005  |
77    77    32AC 4E75       |        RTS
78    78    32AE            |********************************************************************
79    79    32AE            |* SENDCH TRAP 2 ROUTINE
80    80    32AE            |*
81    81    32AE            |* INPUT:       D0.B    BYTE TO BE TRANSMITTED
82    82    32AE            |* OUTPUT:      D6.W    -1 IF QUEUE IS FULL, 0 IF SUCCESS.
83    83    32AE            |*
84    84    32AE            |* QUEUES A BYTE INTO THE OUTPUT QUEUE AND STARTS THE TRANSMITTER
85    85    32AE            |*  IF NOT ALREADY TRANSMITTING.
86    86    32AE            |*
87    87    32AE 48E7 8008  |SENDCH  MOVEM.L  D0/A4,-(SP)       ;SAVE REGISTERS
88    88    32B2            |*
89    89    32B2 287C 0000  |        MOVE.L   #OUTQPT,A4        ;PASS ADDR OF PTR TABLE FOR OUTPUT
89          32B6 35C4       |
90    90    32B8 4EB9 0000  |        JSR      ENQUEUE           ;  QUEUE AND ENQUEUE THE CHARACTER
90          32BC 3350       |
91    91    32BE            |*
92    92    32BE 4A39 0000  |        TST.B    F_XMIT            ;ARE WE CURRENTLY XMITTING?
92          32C2 35D4       |
93    93    32C4 6600 0012  |        BNE      SB_EXIT           ;YES, GET OUT
94    94    32C8            |*
95    95    32C8 13FC 00FF  |        MOVE.B   #-1,F_XMIT        ;SET TRANSMITTING FLAG
95          32CC 0000 35D4  |
96    96    32D0 13FC 0004  |        MOVE.B   #$04,CRA          ;ENABLE THE TRANSMITTER
96          32D4 00FF 0005  |
97    97    32D8            |*                                 ; (WILL CAUSE AN INTERRUPT)
98    98    32D8 4CDF 1001  |SB_EXIT MOVEM.L  (SP)+,D0/A4
99    99    32DC 4E73       |        RTE
100   100   32DE            |
101   101   32DE            |********************************************************************
```

Figure 12.12(c)   *(continued)*

```
102  102   32DE              |* COMINT
103  103   32DE              |*
104  104   32DE              |* THIS INTERRUPT ROUTINE IS INVOKED WHEN A CHARACTER HAS BEEN
105  105   32DE              |* RECEIVED, TRANSMITTED, OR BOTH.
106  106   32DE              |*
107  107   32DE 48E7 9A08    |COMINT MOVEM.L D0/D3-D4/D6/A4,-(SP)        ;SAVE REGISTERS
108  108   32E2              |*
109  109   32E2 1639 00FF    |       MOVE.B  ISR,D3            ;GET INTERRUPT STATUS
109        32E6 000B         |
110  110   32E8 1839 00FF    |       MOVE.B  SRA,D4            ;GET CHANNEL STATUS A
110        32EC 0003         |
111  111   32EE 0804 0000    |       BTST.L  #0,D4             ;DID A BYTE GET RECEIVED?
112  112   32F2 6700 0014    |       BEQ     CWRITE            ;NO, CHECK THE XMIT STATUS
113  113   32F6 1039 00FF    |       MOVE.B  RBA,D0            ;YES, READ THE BYTE
113        32FA 0007         |
114  114   32FC 287C 0000    |       MOVEA.L #INQPT,A4         ;GET ADDR OF INPUT QUEUE PTRS
114        3300 35B4         |
115  115   3302 4EB9 0000    |       JSR     ENQUEUE           ;ADD THE BYTE -- IF QUEUE IS FULL
115        3306 3350         |
116  116   3308              |                                 ;  BYTE IS DISCARDED.
117  117   3308              |*
118  118   3308 0804 0002    |CWRITE BTST.L  #2,D4             ;DID A BYTE GET SENT?
119  119   330C 6700 002C    |       BEQ     CEXIT             ;NO, EXIT
120  120   3310              |*
121  121   3310 287C 0000    |       MOVEA.L #OUTQPT,A4        ;GET ADDR OF OUTPUT QUEUE PTRS
121        3314 35C4         |
122  122   3316 4EB9 0000    |       JSR     DEQUEUE           ;GET NEXT CHARACTER TO SEND
122        331A 3386         |
123  123   331C 4A46         |       TST.W   D6                ;IS QUEUE EMPTY?
124  124   331E 6700 0014    |       BEQ     CSEND             ;NO, SEND THE CHARACTER
125  125   3322              |*
126  126   3322 4239 0000    |       CLR.B   F_XMIT            ;YES, CLEAR XMIT FLAG
126        3326 35D4         |
127  127   3328 13FC 0008    |       MOVE.B  #08,CRA           ;DISABLE THE XMITTER
127        332C 00FF 0005    |
128  128   3330 6000 0008    |       BRA     CEXIT             ;  AND EXIT
129  129   3334              |*
130  130   3334 13C0 00FF    |CSEND  MOVE.B  D0,TBA            ;SEND THE BYTE
130        3338 0007         |
131  131   333A              |*
132  132   333A 4CDF 1059    |CEXIT  MOVEM.L (SP)+,D0/D3-D4/D6/A4   ;RESTORE REGISTERS
133  133   333E 4E73         |       RTE                       ;EXIT
134  134   3340              |***************************************************************
135  135   3340              |* QINIT:       INITIALIZES A QUEUE AND ITS POINTER TABLE
136  136   3340              |* INPUT:       A4.L  THE ADDR OF THE QUEUE POINTER TABLE
137  137   3340              |*              A5.L  THE ADDR OF THE START OF THE QUEUE
138  138   3340              |*              A6.L  THE ADDR OF THE END OF THE QUEUE
139  139   3340              |* OUTPUT:      THE POINTERS IN THE QUEUE TABLE ARE
140  140   3340              |*              INITIALIZED TO CREATE AN EMPTY QUEUE.
141  141   3340              |*
142  142   3340 288D         |QINIT  MOVE.L  A5,NEWQP(A4)      ;SET ADDR OF START OF QUEUE
143  143   3342 294E 0004    |       MOVE.L  A6,ENDQP(A4)      ;SET ADDR OF END OF QUEUE
144  144   3346 294D 0008    |       MOVE.L  A5,QUEOUT(A4)     ;SET THE HEAD OF THE QUEUE
145  145   334A 294D 000C    |       MOVE.L  A5,QUEIN(A4)      ;SET THE TAIL OF THE QUEUE
146  146   334E 4E75         |       RTS
147  147   3350              |
148  148   3350              |***************************************************************
```

Figure 12.12(c)   *(continued)*

```
149  149  3350                  | * ENQUEUE:  ADD A CHARACTER TO THE QUEUE
150  150  3350                  | * INPUT:            D0.B   THE CHARACTER TO BE ADDED
151  151  3350                  | *                   A4.L   THE ADDR OF THE QUEUE POINTER TABLE
152  152  3350                  | * OUTPUT:           D6.W   ZERO IF CHARACTER WAS ADDED;
153  153  3350                  | *                        -1 IF QUEUE WAS FULL
154  154  3350 48E7 80A0        |ENQUEUE MOVEM.L D0/A0/A2,-(SP)   ;SAVE REGISTERS
155  155  3354                  |*
156  156  3354 3C3C 0000        |        MOVE.W  #0,D6            ;SET GOOD RETURN CODE
157  157  3358 206C 000C        |        MOVE.L  QUEIN(A4),A0     ;GET THE ADDR OF NEXT EMPTY LOC.
158  158  335C 2448             |        MOVE.L  A0,A2            ;MAKE SURE THE TAIL DOES NOT
159  159  335E 528A             |        ADDA.L  #1,A2            ;   WRAP AROUND TO THE HEAD.
160  160  3360 B5EC 0004        |        CMPA.L  ENDQP(A4),A2     ;   IF END OF THE QUEUE SPACE
161  161  3364 6500 0004        |        BCS     ENQ_A            ;   IS REACHED, THEN SET  TAIL
162  162  3368 2454             |        MOVE.L  NEWQP(A4),A2     ;   TO THE BEGINNING OF QUEUE.
163  163  336A                  |*
164  164  336A B5EC 0008        |ENQ_A   CMPA.L  QUEOUT(A4),A2    ;IS THE QUEUE FULL?
165  165  336E 6600 000A        |        BNE     ENQ_OK           ;NO, SO ADD ITEM
166  166  3372 3C3C FFFF        |        MOVE.W  #-1,D6           ;YES, SET THE RETURN CODE
167  167  3376 6000 0008        |        BRA     ENQ_X            ;   AND EXIT THE ROUTINE
168  168  337A                  |
169  169  337A 1080             |ENQ_OK  MOVE.B  D0,(A0)          ;STORE THE NEW INPUT
170  170  337C 294A 000C        |        MOVE.L  A2,QUEIN(A4)     ;SAVE THE NEW TAIL POINTER
171  171  3380                  |*
172  172  3380 4CDF 0501        |ENQ_X   MOVEM.L (SP)+,D0/A0/A2  ;RESTORE REGISTERS
173  173  3384 4E75             |        RTS
174  174  3386                  |
175  175  3386                  |*********************************************************
176  176  3386                  | * DEQUEUE:  REMOVE A CHARACTER TO THE QUEUE
177  177  3386                  | * INPUT:            A4.L   THE ADDR OF THE QUEUE POINTER TABLE
178  178  3386                  | * OUTPUT:           D6.W   ZERO IF CHARACTER WAS REMOVED;
179  179  3386                  | *                        -1 IF QUEUE WAS EMPTY
180  180  3386                  | *                   D0.B   THE CHARACTER RETURNED
181  181  3386                  | *                        (UNDEFINED IF QUEUE EMPTY)
182  182  3386                  |*
183  183  3386 2F08             |DEQUEUE MOVE.L  A0,-(SP)         ;SAVE THE REGISTERS
184  184  3388 3C3C 0000        |        MOVE.W  #0,D6            ;SET TO GOOD RETURN CODE
185  185  338C 206C 0008        |        MOVE.L  QUEOUT(A4),A0    ;GET ADDR OF NEXT OUTPUT ENTRY
186  186  3390 B1EC 000C        |        CMPA.L  QUEIN(A4),A0     ;IS THE QUEUE EMPTY?
187  187  3394 6600 000A        |        BNE     DEQ_OK           ;NO, GET THE ITEM
188  188  3398 3C3C FFFF        |        MOVE.W  #-1,D6           ;YES, SET THE RETURN CODE
189  189  339C 6000 0012        |        BRA     DEQ_X            ;   AND EXIT THE ROUTINE
190  190  33A0 1018             |DEQ_OK  MOVE.B  (A0)+,D0         ;GET THE ITEM
191  191  33A2 B1EC 0004        |        CMPA.L  ENDQP(A4),A0     ;IF END OF QUEUE SPACE IS
192  192  33A6 6500 0004        |        BCS     DEQ_A            ;   REACHED, SET NEW POINTER
193  193  33AA 2054             |        MOVE.L  NEWQP(A4),A0     ;   TO THE BEGINNING OF QUEUE.
194  194  33AC 2948 0008        |DEQ_A   MOVE.L  A0,QUEOUT(A4)    ;SAVE NEW HEAD POINTER
195  195  33B0                  |*
196  196  33B0 205F             |DEQ_X   MOVE.L  (SP)+,A0         ;RESTORE REGISTERS
197  197  33B2 4E75             |        RTS
198  198  33B4                  |
199  199  33B4                  |*********************************************************
200  200  33B4                  | * DATA
201  201  33B4                  |*
202  202  33B4                  | * THESE ARE THE QUEUES
203  203  33B4                  |INQUEUE     DS.B 256   ;INPUT QUEUE
204  204       0000 34B4        |E_INQUEUE   EQU  *     ;END OF INPUT QUEUE
205  205  34B4                  |OUTQUEUE    DS.B 256   ;OUTPUT QUEUE
206  206       0000 35B4        |E_OUTQUEUE  EQU  *     ;END OF OUTPUT QUEUE
207  207  35B4                  |*
```

Figure 12.12(c)   *(continued)*

```
208  208  35B4                  |* THESE ARE THE QUEUE POINTER TABLES
209  209            0000 35B4    |INQPT    EQU    *
210  210  35B4                  |         DS.L   1      ;POINTS TO BEGINNING OF QUEUE (NEWQP)
211  211  35B8                  |         DS.L   1      ;POINTS TO END OF QUEUE (ENDQP)
212  212  35BC                  |         DS.L   1      ;HEAD OF QUEUE (QUEOUT)
213  213  35C0                  |         DS.L   1      ;TAIL OF QUEUE (QUEIN)
214  214            0000 35C4    |OUTQPT   EQU    *
215  215  35C4                  |         DS.L   1      ;POINTS TO BEGINNING OF QUEUE (NEWQP)
216  216  35C8                  |         DS.L   1      ;POINTS TO END OF QUEUE (ENDQP)
217  217  35CC                  |         DS.L   1      ;HEAD OF QUEUE (QUEOUT)
218  218  35D0                  |         DS.L   1      ;TAIL OF QUEUE (QUEIN)
219  219  35D0                  |*
220  220  35D4                  |* FLAGS
221  221  35D4                  |F_XMIT   DS.B   1      ;IF SET, CURRENTLY TRANSMITTING DATA
222  222  35D4                  |*
223  223  35D6                  |         END
```

Figure 12.12(c)   *(continued)*

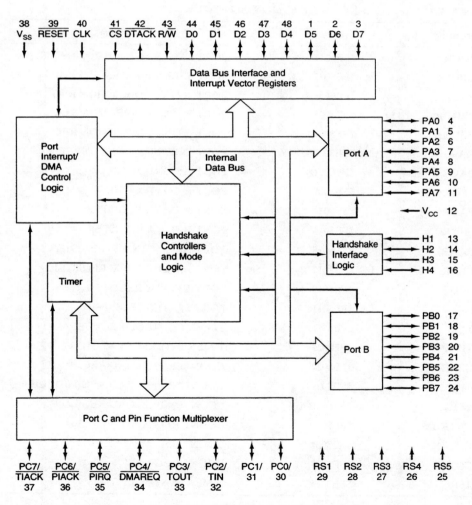

Figure 12.13   Block diagram of the MC68230 PI/T. (Courtesy of Motorola, Inc.)

An external clock signal can be used or the signal can be derived from the system clock. For the authors' SBC68K board the system clock frequency is 8 MHz. Thus, the period is $1/(8 \times 10^6) = 125$ nanoseconds (125×10^{-9} seconds). When the system clock is used by the PI/T, the time between counts is fixed at 32 times as long as the system clock period. In our case, this is $32 \times 125 \times 10^9$ seconds or 4 microseconds (4×10^{-6} seconds). Thus, for a one-second interval, an initial count of 250,000 must be loaded into the counter preload register as shown in Example 12.3.

Register addresses for the MC68230 are shown in Table 12.6. Figure 12.14 shows

Table 12.6   Register Addresses for the MC68230 PI/T Chip

Register	Address	Meaning
PGCR	$FE8001	;PORT GENERAL CONTROL REGISTER
PSRR	$FE8003	;PORT SERVICE REQUEST REGISTER
PADDR	$FE8005	;PORT A DATA DIRECTION REG
PBDDR	$FE8007	;PORT B DATA DIRECTION REG
PCDDR	$FE8009	;PORT C DATA DIRECTION REG
PIVR	$FE800B	;PORT INTERRUPT VECTOR REG
PACR	$FE800D	;PORT A CONTROL REG
PBCR	$FE800F	;PORT B CONTROL REG
PADR	$FE8011	;PORT A DATA REGISTER
PBDR	$FE8013	;PORT B DATA REGISTER
PAAR	$FE8015	;PORT A ALTERNATE REGISTER
PBAR	$FE8017	;PORT B ALTERNATE REGISTER
PCDR	$FE8019	;PORT C DATA REGISTER
PSR	$FE801B	;PORT STATUS REGISTER
TCR	$FE8021	;TIMER CONTROL REGISTER
TIVR	$FE8023	;TIMER INTERRUPT VECTOR REGISTER
CPRH	$FE8027	;COUNTER PRELOAD REG HIGH
CPRM	$FE8029	;COUNTER PRELOAD REG MIDDLE
CPRL	$FE802B	;COUNTER PRELOAD REG LOW
CNTRH	$FE802F	;COUNT REG HIGH
CNTRM	$FE8031	;COUNT REG MIDDLE
CNTRL	$FE8033	;COUNT REG LOW
TSR	$FE8035	;TIMER STATUS REG

the general initialization procedure for a timer. The specific procedure for interval timing uses the following sequence of steps and registers:

(a) select interrupts, continuous operation, and clock source (TCR)
(b) set initial count (CPRH, CPRM, CPRL)
(c) define interrupt vector (TIVR)
(d) enable interrupts for timer (TCR)

Example 12.3 presents a program for the MC68230 that creates an interval time of one second. Comments in the program explain the initial values written to the control registers.

Example 12.3

This example uses the MC68230 PI/T and MC68681 DUART to display elapsed time in minutes and seconds. The timer begins timing and display when the main program executes. Any keyboard input causes the elapsed time to be reset to 00:00.

Interrupt routines are used to receive input from the keyboard and display the elapsed time. An MC68230 timer interrupt is used to update the elapsed time each second. Trap routines are used to start the timer and reset it. Another trap routine is used to output characters to the screen as was previously described in Example 12.2. Figure 12.15(a) shows the subroutine EXINIT to initialize the interrupt and trap vectors. The MC68230 on the SBC68K board interrupts at CPU level 5. CPU interrupts must be enabled at level 5 or below to receive interrupts from the PI/T.

The main program is shown in Figure 12.15(b). It calls the subroutine COMINIT to initialize the MC68681. This subroutine was described in Example 12.2. The MC68230 is initialized by subroutine PITINIT shown in Figure 12.15(c). Options set by initialization cause the timer to decrement its counter until the counter reaches zero. Then, the timer automatically reloads the initial value and continues counting down to zero again. An interrupt occurs each time the counter reaches zero. The initial value in subroutine PITINIT was selected to cause the counter to reach zero every second by writing the value 250,000 ($03D090) into the counter preload register. The interrupt vector number (64) is also defined in this subroutine.

The main program next calls TRAP #2, presented in Example 12.2, to move the cursor to a new line on the screen. The TRAP #1 call (PITSTRT) starts the timer with interrupts enabled.

After initializing the peripheral chips, the main program goes into a loop to examine the input queue and the flag F_TICK. Whenever a character is pulled from the input queue, the main program calls TRAP #3 (PITRSET) to reset the elapsed timer. Whenever F_TICK is set, the main program clears F_TICK, then calls TIME2ASC to create the output string and display the new elapsed time.

When the timer counter is decremented to zero, the PI/T timer counter is reset to its initial value, an interrupt request is issued, and the timer resumes decrementing the counter. The timer interrupt routine INTPIT in Figure 12.15(c) increments the elapsed time by one second and sets the flag F_TICK to indicate that the time has been updated.

Figure 12.14   PI/T initialization for interrupt-controlled timing.

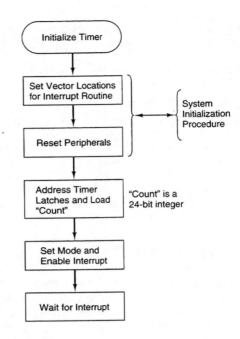

```
abs. rel.   LC   obj. code     source line
---- ----   ----  ---------    -----------
  1    1   0000                    |          TTL      'FIGURE 12.15(a)'
  2    2   0000                    |          LLEN     100
  3    3   3100                    |          ORG      $3100
  4    4   3100                    |          OPT      P=M68000
  5    5   3100                    |*
  6    6   3100                    |* SUBROUTINE TO INITIALIZE EXCEPTION VECTORS
  7    7   3100                    |*
  8    8   3100 21FC 0000  |EXINIT MOVE.L  #$3636,33*4      ;SET TRAP1 VECTOR (PITSTRT)
  8         3104 3636 0084  |
  9    9   3108 21FC 0000  |          MOVE.L  #$32AE,34*4      ;SET TRAP2 VECTOR (SENDCH)
  9         310C 32AE 0088  |
 10   10   3110 21FC 0000  |          MOVE.L  #$3648,35*4      ;SET TRAP3 VECTOR (PITRSET)
 10         3114 3648 008C  |
 11   11   3118                    |*
 12   12   3118 21FC 0000  |          MOVE.L  #$366E,64*4      ;TIMER INTERRUPT (PITINT)
 12         311C 366E 0100  |
 13   13   3120 21FC 0000  |          MOVE.L  #$32DE,65*4      ;SERIAL COMM. INTERRUPT (COMINT)
 13         3124 32DE 0104  |
 14   14   3128                    |*
 15   15   3128 4E75        |          RTS
 16   16   312A                    |          END
```

Figure 12.15(a)   Initialization routine for SBC68K exception vectors.

```
abs. rel.   LC    obj. code   source line
----  ----  ----  ---------   -----------
   1     1  0000                    |           TTL     'FIGURE 12.15(b)'
   2     2  0000                    |           LLEN    100
   3     3  8000                    |           ORG     $8000
   4     4  8000                    |           OPT     P=M68000
   5     5  8000                    |*
   6     6  8000                    |* PERIODIC INTERRUPT TIMER - WITH SERIAL I/O
   7     7  8000                    |*
   8     8  8000                    |* THE PROGRAM BEGINS AT LABEL MAIN.
   9     9  8000                    |*
  10    10        0000 3100         |EXINIT  EQU     $3100               ; INITIALIZE VECTORS ROUTINE
  11    11        0000 3200         |COMINIT EQU     $3200               ; INITIALIZE MC68681 ROUTINE
  12    12        0000 3600         |PITINIT EQU     $3600               ; INITIALIZE MC68230 ROUTINE
  13    13        0000 329C         |COMSTRT EQU     $329C               ; START RECEIVER
  14    14        0000 35B4         |INQPT   EQU     $35B4               ; RECEIVER QUEUE POINTER TABLE
  15    15        0000 3386         |DEQUEUE EQU     $3386               ; GET CHARACTER FROM QUEUE
  16    16        0000 36C8         |F_TICK  EQU     $36C8               ; NEW SECOND FLAG
  17    17        0000 36C9         |TIME_MM EQU     $36C9               ; CURRENT MINUTES
  18    18        0000 36CA         |TIME_SS EQU     $36CA               ; CURRENT SECONDS
  19    19  8000                    |
  20    20  8000                    |***********************************************************************
  21    21  8000 46FC 2700          |MAIN    MOVE.W  #$2700,SR           ; MASK CPU INTERRUPTS
  22    22  8004 4EB8 3100          |        JSR     EXINIT              ; INITIALIZE VECTORS
  23    23  8008 4EB8 3200          |        JSR     COMINIT             ; INITIALIZE MC68881 FOR INPUT
  24    24  800C 4EB8 3600          |        JSR     PITINIT             ; INITIALIZE MC68230 FOR TIMER
  25    25  8010 46FC 0000          |        MOVE.W  #$0000,SR           ; ENABLE CPU INTERRUPTS;USER MODE
  26    26  8014 4EB8 329C          |        JSR     COMSTRT             ; START RECEIVER
  27    27  8018                    |*
  28    28  8018                    |* MOVE TO NEW LINE ON SCREEN
  29    29  8018 103C 000A          |        MOVE.B  #$0A,D0             ;
  30    30  801C 4E42              |        TRAP    #2                  ; OUTPUT THE LINE FEED
  31    31  801E 103C 000D          |        MOVE.B  #$0D,D0             ;
  32    32  8022 4E42              |        TRAP    #2                  ; OUTPUT THE CARRIAGE RETURN
  33    33  8024                    |*
  34    34  8024                    |* START THE PIT
  35    35  8024 4E41              |        TRAP    #1                  ; START THE TIMER
  36    36  8026                    |*
  37    37  8026                    |* MAIN BODY OF THE ROUTINE
  38    38  8026 287C 0000          |KEYIN   MOVEA.L #INQPT,A4          ; GET NEXT KEYSTROKE
  38        802A 35B4               |
  39    39  802C 4EB8 3386          |        JSR     DEQUEUE             ; D0.B = DATA; D6.W = RETURN CODE
  40    40  8030 4A46              |        TST.W   D6                  ; WAS ANY DATA RECEIVED?
  41    41  8032 6600 0004          |        BNE     CHKSEC              ; NO, SO CHECK NEW SECOND FLAG
  42    42  8036                    |*
  43    43  8036 4E43              |        TRAP    #3                  ; RESET THE TIMER
  44    44  8038                    |*
  45    45  8038 4A38 36C8          |CHKSEC  TST.B   F_TICK              ; IS A NEW TIME AVAILABLE?
  46    46  803C 67E8              |        BEQ     KEYIN               ; NO, SO WAIT FOR KEYSTROKE
  47    47  803E                    |                                    ;   OR NEW TIME
  48    48  803E 4238 36C8          |        CLR.B   F_TICK              ; YES, RESET FLAG AND
  49    49  8042                    |*                                   ;   OUTPUT TIME
```

Figure 12.15(b)   Program for periodic timer and display.

```
50   50   8042              |* CONVERT BINARY TIME TO ASCII
51   51   8042 1038 36C9    |        MOVE.B   TIME_MM,D0        ;
52   52   8046 207C 0000    |        MOVEA.L  #TIMESTR,A0       ;
52        804A 8096         |
53   53   804C 4EB9 0000    |        JSR      TME2ASC          ;CONVERT MINUTES TO ASCII
53        8050 8070         |
54   54   8052 1038 36CA    |        MOVE.B   TIME_SS,D0        ;
55   55   8056 207C 0000    |        MOVEA.L  #TIMESTR+3,A0     ;
55        805A 8099         |
56   56   805C 4EB9 0000    |        JSR      TME2ASC          ;CONVERT SECONDS TO ASCII
56        8060 8070         |
57   57   8062              |*
58   58   8062              |* OUTPUT THE TIME STRING
59   59   8062 207C 0000    |        MOVEA.L  #TIMESTR,A0       ;GET ADDR OF OUTPUT STRING
59        8066 8096         |
60   60   8068 1018         |NEXTCH  MOVE.B   (A0)+,D0          ;GET NEXT CHAR OF STRING
61   61   806A 67BA         |        BEQ      KEYIN             ;IF END OF STR, WAIT FOR EVENT
62   62   806C 4E42         |        TRAP     #2                ;ELSE, OUTPUT THE CHARACTER
63   63   806E 60F8         |        BRA      NEXTCH            ;   AND GET NEXT CHARACTER
64   64   8070              |*****************************************************************
65   65   8070              |* TIME2ASC - CONVERT BINARY TIME COMPONENT TO 2-CHARACTER ASCII
66   66   8070              |*
67   67   8070              |*   INPUT:  D0.B    TIME COMPONENT (MINUTES, OR SECONDS);BINARY
68   68   8070              |*           A0.L    ADDRESS OF OUTPUT TWO-CHARACTER FIELD
69   69   8070              |*   OUTPUT: D0.W    UNDEFINED - ORIGINAL VALUE DESTROYED
70   70   8070              |*           (A0)    TWO-BYTE ASCII EQUIVALENT OF BINARY INPUT
71   71   8070              |*
72   72   8070 0280 0000    |TME2ASC ANDI.L   #$00FF,D0         ;CLEAR HI BYTE
72        8074 00FF         |
73   73   8076 80FC 000A    |        DIVU     #10,D0            ;DIVIDE BY 10 TO GET ONES DIGIT
74   74   807A 4840         |        SWAP     D0                ;GET THE REMAINDER
75   75   807C 0000 0030    |        ORI.B    #$30,D0           ;CONVERT TO ASCII
76   76   8080 1140 0001    |        MOVE.B   D0,1(A0)          ;WRITE THE DIGIT
77   77   8084 4240         |        CLR.W    D0                ;PREPARE FOR NEXT DIVIDE
78   78   8086 4840         |        SWAP     D0                ;GET THE PREVIOUS QUOTIENT
79   79   8088 80FC 000A    |        DIVU     #10,D0            ;DIVIDE TO GET TEN'S DIGIT
80   80   808C 4840         |        SWAP     D0                ;GET THE REMAINDER
81   81   808E 0000 0030    |        ORI.B    #$30,D0           ;CONVERT TO ASCII
82   82   8092 1080         |        MOVE.B   D0,(A0)           ;WRITE THE DIGIT
83   83   8094 4E75         |        RTS
84   84   8096              |*****************************************************************
85   85   8096              |* DATA
86   86   8096              |
87   87   8096 2020 3A20    |TIMESTR DC.B     '  :  ',$0D       ;OUTPUT STRING
87   87   809A 200D         |
88   88   809C 00           |        DC.B     0
89   89   809D              |
90   90   809D              |        END
```

Figure 12.15(b)   *(continued)*

```
abs. rel.   LC    obj. code     source line
---- ----   ----  ---------     -----------
   1    1   0000                |         TTL      'FIGURE 12.15(c)'
   2    2   0000                |         LLEN     100
   3    3   3600                |         ORG      $3600
   4    4   3600                |         OPT      P=M68000
   5    5   3600                |****************************************************
   6    6   3600                |* REGISTER ADDRESSES FOR THE MC68230 PI/T CHIP
   7    7         00FE 8021     |TCR      EQU      $FE8021 ;TIMER CONTROL REGISTER
   8    8         00FE 8023     |TIVR     EQU      $FE8023 ;TIMER INTERRUPT VECTOR REGISTER
   9    9         00FE 8027     |CPRH     EQU      $FE8027 ;COUNTER PRELOAD REG HIGH
  10   10         00FE 8029     |CPRM     EQU      $FE8029 ;COUNTER PRELOAD REG MIDDLE
  11   11         00FE 802B     |CPRL     EQU      $FE802B ;COUNTER PRELOAD REG LOW
  12   12         00FE 802F     |CNTRH    EQU      $FE802F ;COUNT REG HIGH
  13   13         00FE 8031     |CNTRM    EQU      $FE8031 ;COUNT REG MIDDLE
  14   14         00FE 8033     |CNTRL    EQU      $FE8033 ;COUNT REG LOW
  15   15         00FE 8035     |TSR      EQU      $FE8035 ;TIMER STATUS REG
  16   16   3600                |****************************************************
  17   17   3600                |*
  18   18   3600                |* THIS INTERRUPT ROUTINE EXECUTES WHEN COUNTER GOES TO ZERO.
  19   19   3600                |* THE TIMER IS SET INTERRUPT ONCE A SECOND.
  20   20   3600                |*
  21   21   3600 13FC 00A0      |PITINIT MOVE.B   #$A0,TCR        ;TOUT AND TIACK ENABLED,
  21        3604 00FE 8021      |
  22   22   3608                |                                 ; RELOAD AFTER ZERO DETECT,
  23   23   3608                |                                 ; INTERNAL CLOCK, TIMER DISABLED
  24   24   3608 13FC 0003      |         MOVE.B   #$03,CPRH       ;COUNTER SET TO 250000
  24        360C 00FE 8027      |
  25   25   3610 13FC 00D0      |         MOVE.B   #$D0,CPRM       ;   SINCE EACH TICK IS 32*125NSEC
  25        3614 00FE 8029      |
  26   26   3618 13FC 0090      |         MOVE.B   #$90,CPRL       ;   COUNTER DURATION IS 1 SEC
  26        361C 00FE 802B      |
  27   27   3620 13FC 0040      |         MOVE.B   #$40,TIVR       ;INTERRUPT VECTOR 64
  27        3624 00FE 8023      |
  28   28   3628 4239 0000      |         CLR.B    TIME_MM         ;SET INITIAL TIME TO 00:00
  28        362C 36C9           |
  29   29   362E 4239 0000      |         CLR.B    TIME_SS         ;
  29        3632 36CA           |
  30   30   3634 4E75           |         RTS
  31   31   3636                |
  32   32   3636                |****************************************************
  33   33   3636                |* TRAP #1 STARTS THE TIMER
  34   34   3636                |*
  35   35   3636 13FC 00A1      |PITSTRT MOVE.B   #$A1,TCR         ;TOUT AND TIACK ENABLED,
  35        363A 00FE 8021      |
  36   36   363E                |                                 ; RELOAD AFTER ZERO DETECT,
  37   37   363E                |                                 ; INTERNAL CLOCK, TIMER ENABLED
  38   38   363E 13FC 00FF   .  |         MOVE.B   #-1,F_TICK      ;SET NEW SECOND FLAG
  38        3642 0000 36C8      |
  39   39   3646 4E73           |         RTE
  40   40   3648                |****************************************************
  41   41   3648                |* TRAP #3 RESETS THE TIME VALUES AND RESTARTS THE TIMER
  42   42   3648                |*
```

Figure 12.15(c)   Initialization, TRAP, and interrupt routines.

```
43   43   3648 13FC 00A0  |PITRSET MOVE.B  #$A0,TCR        ;TOUT AND TIACK ENABLED,
43        364C 00FE 8021  |
44   44   3650            |                                ;  RELOAD AFTER ZERO DETECT,
45   45   3650            |                                ;  INTERNAL CLOCK, TIMER DISABLED
46   46   3650 4239 0000  |        CLR.B   TIME_MM         ;SET INITIAL TIME TO 00:00
46        3654 36C9       |
47   47   3656 4239 0000  |        CLR.B   TIME_SS         ;
47        365A 36CA       |
48   48   365C 13FC 00FF  |        MOVE.B  #-1,F_TICK      ;SET NEW SECOND FLAG
48        3660 0000 36C8  |
49   49   3664 13FC 00A1  |        MOVE.B  #$A1,TCR        ;TOUT AND TIACK ENABLED,
49        3668 00FE 8021  |
50   50   366C            |                                ;  RELOAD AFTER ZERO DETECT,
51   51   366C            |                                ;  INTERNAL CLOCK, TIMER ENABLED
52   52   366C 4E73       |        RTE
53   53   366E            |********************************************************************
54   54   366E            |* TIMER INTERRUPT ROUTINE
55   55   366E 2F00       |INTPIT  MOVE.L  D0,-(SP)        ;SAVE REGISTER
56   56   3670 13FC 0001  |        MOVE.B  #$01,TSR        ;RESET THE ZERO DETECT STATUS
56        3674 00FE 8035  |
57   57   3678            |*
58   58   3678 1039 0000  |        MOVE.B  TIME_SS,D0      ;GET THE LAST SECONDS VALUE
58        367C 36CA       |
59   59   367E 5200       |        ADDI.B  #1,D0           ;INCREMENT SECONDS
60   60   3680 0C00 003B  |        CMPI.B  #59,D0          ;HAVE WE COUNTED 60 SECONDS?
61   61   3684 6200 000C  |        BHI     PMIN            ;YES, SO INCREMENT MINUTE
62   62   3688 13C0 0000  |        MOVE.B  D0,TIME_SS      ;NO, SO SET NEW SECONDS VALUE
62        368C 36CA       |
63   63   368E 6000 002C  |        BRA     PEXIT           ;  AND GET OUT
64   64   3692            |*
65   65   3692 13FC 0000  |PMIN    MOVE.B  #0,TIME_SS      ;SET SECONDS TO ZERO
65        3696 0000 36CA  |
66   66   369A 1039 0000  |        MOVE.B  TIME_MM,D0      ;GET THE LAST MINUTES VALUE
66        369E 36C9       |
67   67   36A0 5200       |        ADDI.B  #1,D0           ;INCREMENT MINUTES
68   68   36A2 0C00 003B  |        CMPI.B  #59,D0          ;HAVE WE COUNTED 60 MINTUES?
69   69   36A6 6200 000C  |        BHI     POVER           ;YES, SO RESET TO ZERO
70   70   36AA 13C0 0000  |        MOVE.B  D0,TIME_MM      ;NO, SO SET NEW MINUTES VALUE
70        36AE 36C9       |
71   71   36B0 6000 000A  |        BRA     PEXIT           ;  AND GET OUT
72   72   36B4            |*
73   73   36B4 13FC 0000  |POVER   MOVE.B  #0,TIME_MM      ;RESET MINUTES TO ZERO
73        36B8 0000 36C9  |
74   74   36BC            |*
75   75   36BC 13FC 00FF  |PEXIT   MOVE.B  #-1,F_TICK      ;SET NEW SECOND FLAG
75        36C0 0000 36C8  |
76   76   36C4 201F       |        MOVE.L  (SP)+,D0        ;RESTORE REGISTER
77   77   36C6 4E73       |        RTE
78   78   36C8            |
79   79   36C8            |
80   80   36C8            |********************************************************************
81   81   36C8            |* DATA
82   82   36C8            |
83   83   36C8            |F_TICK  DS.B    1               ;IF SET, TIME VALUE WAS UPDATED
84   84   36C8            |
85   85   36CA            |* TIME VARIABLES
86   86   36CA            |TIME_MM DS.B    1               ;MINUTES
87   87   36CC            |TIME_SS DS.B    1               ;SECONDS
88   88   36CC            |
89   89   36CE            |        END
```

Figure 12.15(c) *(continued)*

REVIEW QUESTIONS AND PROBLEMS

Essay Questions

12.1. Example 12.2 uses queues to store the received characters and to hold the characters that are to be transmitted. Why are queues used in this application? Are there other data structures that could also be used to buffer the serial communication data?

12.2. What are the maximum and minimum resolutions of the MC68230 timer counter if the system clock is used as the timer clock source? Assume an 8-MHz system clock. Give your answer in seconds.

Programs

12.3. What changes would be required to modify the program of Example 12.2 to transmit with even parity at 9600 baud? If you have a development system available, make the modifications and test the new program.

12.4. What changes would be required to modify the program of Example 12.2 to communicate concurrently over two serial communication lines (using port A and port B of the MC68681)? If you have a development system available, make the modifications and test the new program.

12.5. The MC68230 has parallel communications capability. Refer to the Motorola documentation for the device and design a parallel communications interface between two systems, each using an MC68230. Use interrupt-driven I/O and port A for transmission and port B for receiving. Design and test the following:
(a) the initialization routine
(b) the interrupt routine
(c) the routine to initiate transmission
(d) the data structures used to buffer the I/O
(e) routines to write and read data from the I/O buffers

FURTHER READING

A brief description of each peripheral chip available from Motorola for the MC68000 family is given in the *Motorola Semiconductor Master Selection Guide.* For more details regarding a particular chip, the *User's Manual* or the data sheet for the chip should be consulted. These documents are available from the manufacturer. For example, the MC68452 Bus Arbitration Module is described in the product information brochure listed here from Motorola. The author's previous textbooks describe I/O programming techniques for several of the MC68332 modules and other peripheral chips of the MC68000 family.

Harman, Thomas L. and Barbara Lawson. *The Motorola MC68332 Microcontroller.* (Englewood Cliffs, N.J.: Prentice-Hall, 1991).

Harman, Thomas L. *The Motorola MC68020 and MC68030 Microprocessors*. (Englewood Cliffs, N.J.: Prentice-Hall, 1989).

MC68452 Bus Arbitration Module, Product Information ADI-696. Motorola, Inc.

The *User's Manual* for the Arnewsh SBC68K contains complete details about the single-board computer and its peripheral chips. (The authors used Revision 1.1 for the examples in this chapter.)

SBC68K User's Manual. (Fort Collins, Colo.: Arnewsh, Inc.).

CHAPTER 13

System Design and
Hardware Considerations

During system design, calculation of the timing requirements for programs is sometimes important to predict system performance. The exact time taken by any instruction or the time to initiate exception processing can be determined by using the instruction execution times published in the Motorola *User's Manual* for the MC68000. A number of examples in Section 13.1 show how these times may be used to determine timing in MC68000 systems.

Understanding of the signal lines of the CPU is necessary to design interfaces that connect to the system bus. Various signal lines are used for data transfer, memory protection and management, interrupt processing, bus control, and the detection of hardware errors. The MC68000 also provides special signal lines used to control peripheral chips from the 8-bit MC6800 family of products by Motorola. Section 13.2 treats the CPU signal lines from the hardware designer's point of view.

13.1 INSTRUCTION EXECUTION TIMES

The MC68000 CPU timing is controlled by a master clock. This clock is a circuit that generates a periodic sequence of pulses that synchronize all changes in the input and output signal lines. The timing of all the operations of the MC68000 is determined by the number of clock pulses required for the operation. If the number of pulses per second or pulse rate is increased (decreased), the speed of operation of the processor is increased (decreased).

Figure 13.1 shows the clock signal for MC68000 systems. The table lists the cycle time for several versions of the processor. All the processors will operate at a minimum pulse rate of 2×10^6 pulses per second except the 12.5-MHz version, which has a minimum of 4×10^6 pulses per second. The maximum pulse rate for any version is the reciprocal of the cycle time.[1] The MC68000L8 (8 MHz) version, for example,

---

[1] The frequency of operation in hertz is considered the reciprocal of the cycle time. More precisely, it refers to the frequency of the fundamental sinusoidal wave in a Fourier series analysis of the clock pulse train.

Characteristic	Symbol	4 MHz		6 MHz		8 MHz		10 MHz		12.5 MHz		Unit
		Min	Max	Min	Max	Min	Max	Min	Max	Min	Max	
Frequency of Operation	F	2.0	4.0	2.0	6.0	2.0	8.0	2.0	10.0	4.0	12.5	MHz
Cycle Time	t_{cyc}	250	500	167	500	125	500	100	500	80	250	ns
Clock Pulse Width	t_{CL}	115	250	75	250	55	250	45	250	35	125	ns
	t_{CH}	115	250	75	250	55	250	45	250	35	125	
Rise and Fall Times	t_{Cr}	−	10	−	10	−	10	−	10	−	5	ns
	t_{Cf}	−	10	−	10	−	10	−	10	−	5	

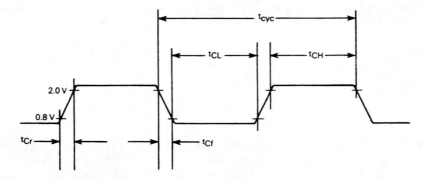

Figure 13.1   Clock waveform.

operates with a frequency between 2 and 8 MHz. The cycle time corresponding to this range has a maximum of 500 nanoseconds and a minimum of 125 nanoseconds.

The cycle time, designated t_c in the text, is used to calculate the execution time of MC68000 instructions and other operations, such as interrupt response time. The calculation is based on the time to fetch an instruction, compute the effective addresses, and fetch or store the operand. The times for the MC68000 instruction set are listed in Appendix IV. A short summary is given in Table 13.1.

From Table 13.1, the time to read or write a word from memory requires $4t_c$ because a minimum of four clock pulses is necessary to complete the operation. In

Table 13.1   Instruction Execution Time (MC68000)

Operation	Time
Write to memory	$4t_c$ word
	$8t_c$ longword
Read from memory	$4t_c$ word
	$8t_c$ longword
Calculate effective address of source and fetch operand	0 to $12t_c$ byte, word
	0 to $16t_c$ longword
Instruction execution (including instruction fetch)	$4t_c$ to $158t_c$
Interrupt response	$44t_c$
TRAP instruction	$38t_c$

practice, the time could be longer if the memory cannot respond in the minimum time. The source effective address can be calculated with no delay for register operations or up to $16t_c$ for a long operand being addressed by an absolute long address. This long word operation takes two reads ($8t_c$) to fetch the address and two more reads to fetch the operand ($8t_c$), for a total of 16 cycles.

Instruction execution, including the fetch of the op code, requires at least four cycles. The DIVS instruction is one of the longest executing instructions. It requires a maximum of 158 cycles to execute. Interrupt processing requires 44 cycles before the interrupt routine begins executing. This does not include any time required for the present instruction to complete. The TRAP instruction takes 38 cycles to push (PC) and (SR) onto the system stack and replace the contents of the PC with the TRAP vector contents.

Although the general principles presented in this section apply also to the MC68010, the exact timing of MC68010 instructions are available in the MC68010 manual from Motorola. A number of the MC68010 instructions have improved execution time as compared to those of the MC68000. The MC68010 also has a "loop mode" feature using the DBcc instruction. Under certain conditions, the mode allows a one-word length instruction to be executed in a loop without fetching the instruction repeatedly from memory.

Example 13.1

The instruction to transfer a 32-bit immediate value

MOVE.L   #⟨d$_{32}$⟩,(An)

requires 20 cycles according to the information in the MC68000 *User's Manual* (see Appendix IV). This is calculated as follows:

Instruction fetch	$4t_c$
Fetch of immediate value	$8t_c$
Store operand	$8t_c$
	$20t_c$

For the 10-MHz version of the MC68000, the time required is

$t = 20 \times 10^{-7}$ seconds

or 2 microseconds.

EXERCISES

13.1.1. Compute the minimum response time in microseconds for an interrupt with the following processors.
   (a) MC68000L6
   (b) MC68000L8
   (c) MC68000L12

13.1.2. (a) What is the worst-case timing for interrupt response if an instruction is just beginning to be fetched as the interrupt occurs?

(b) What is this time for the 8-MHz version of the MC68000?

13.1.3. Use the No Operation (NOP) instruction of the MC68000 to create a delay loop with variable delays from 100 microseconds to 1 second. The NOP requires four cycles, including fetching and execution.

13.2 HARDWARE CONSIDERATIONS

The electrical connection between elements in a MC68000 system is via the system bus. The bus contains address signal lines, data signal lines, and control signal lines. A few miscellaneous signal lines, such as a clock signal, power (+5 volts), and a ground reference, are also included. All of the peripheral devices and the processor are connected in parallel to the bus signals. In most applications, the MC68000 processor acts as the system controller and determines the use of the bus for transfer of data and control signals. The MC68000 signal lines and the dual-in-line pin (DIP) configuration is shown in Figure 13.2. The signal lines are defined by function, mnemonic, and characteristics in Figure 13.3. The signal lines that are described as "three-state" in the figure are put in a high-impedance or hi-Z state when the processor relinquishes the use of the bus to other devices. These signal lines are, effectively, disconnected electrically from the bus when in this state.

An understanding of the use of the MC68000 signal lines for fundamental operations is necessary in order to design interfaces. Data transfer operations and the use the function code lines during processor accesses to memory are described in this section. This information is particularly important to the system designer when a memory management scheme is used to protect and segment various areas of memory. The interrupt hardware sequence, the control of the bus by other devices, and interfacing to the 8-bit M6800 family devices are also discussed.

It should be emphasized that the information in this section describes the functional operation of the MC68000 signal lines. The precise timing diagrams needed by an interface designer are not presented. The source of these diagrams is Motorola's data sheets for the MC68000, containing the electrical specifications for the device. The minimum and maximum times for signal changes and those times between changes on different signal lines (to determine allowable timing margins) are given in the data sheets.

The signal lines of the MC68010 are identical to those of the MC68000. However, the use and timing of the MC68010 signal lines are not always the same as for the MC68000. In particular, the function code signal lines of the MC68010 have a more general use.

Before discussing various hardware operations, the conventions used to specify the signal lines must be presented. Further details of hardware design are presented in several references in the Further Reading section at the end of the chapter. The electrical characteristics of the MC68000 are summarized in Apendix II of this text.

Designation of Signal Lines. The electrical signals generated or received by the MC68000 can be specified in terms of their electrical characteristics and also by their functional or logical use. These electrical characteristics include the voltage level,

Figure 13.2 Input and output signals
of the MC68000. (Courtesy of Motorola,
Inc.)

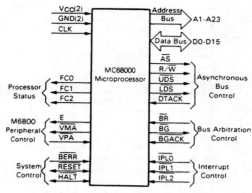

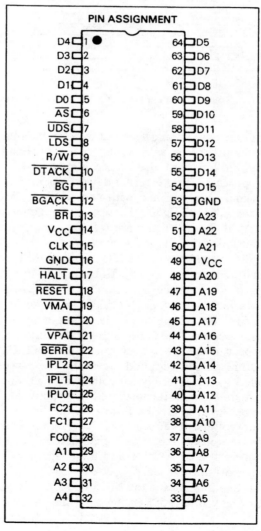

Signal Name	Mnemonic	Input/Output	Active State	Three State
Address Bus	A1-A23	Output	High	Yes
Data Bus	D0-D15	Input/Output	High	Yes
Address Strobe	$\overline{AS}$	Output	Low	Yes
Read/Write	$R/\overline{W}$	Output	Read-High Write-Low	Yes
Upper and Lower Data Strobes	$\overline{UDS}$, $\overline{LDS}$	Output	Low	Yes
Data Transfer Acknowledge	$\overline{DTACK}$	Input	Low	No
Bus Request	$\overline{BR}$	Input	Low	No
Bus Grant	$\overline{BG}$	Output	Low	No
Bus Grant Acknowledge	$\overline{BGACK}$	Input	Low	No
Interrupt Priority Level	$\overline{IPL0}$, $\overline{IPL1}$, $\overline{IPL2}$	Input	Low	No
Bus Error	$\overline{BERR}$	Input	Low	No
Reset	$\overline{RESET}$	Input/Output	Low	No*
Halt	$\overline{HALT}$	Input/Output	Low	No*
Enable	E	Output	High	No
Valid Memory Address	$\overline{VMA}$	Output	Low	Yes
Valid Peripheral Address	$\overline{VPA}$	Input	Low	No
Function Code Output	FC0, FC1, FC2	Output	High	Yes
Clock	CLK	Input	High	No
Power Input	V_{CC}	Input	–	–
Ground	GND	Input	–	–

*Open Drain

Figure 13.3   Signal lines of the MC68000. (Courtesy of Motorola, Inc.)

current requirements, and speed of switching. Other characteristics of the signal generated by the processor or by an external device and propagated along the signal lines of the system bus must also be considered.

The processor and external devices respond to the signals according to the state of the signal. A signal line with two states is designated as active/inactive, HIGH/LOW, true/false, or {1}/{0}. These four designations are considered equivalent when the electrical characteristics are transistor-transistor logic (TTL) compatible and the positive-true logic definition is employed.

All the signal lines of the MC68000 are TTL-compatible, so the processor can be electrically connected to any TTL chip as long as certain loading conditions are recognized. Because the TTL family is the most popular logic type employed today for interfacing circuitry, the MC68000 processor is easily connected to a large number of TTL circuits that are available to implement various logical functions. The TTL signal line is considered to be in the LOW state when its steady-state voltage is 0 volts with respect to the ground reference. The TTL HIGH state is indicated by a voltage level with respect to a ground of 5 volts. The power supply voltage for a TTL system, designated V_{CC}, is typically 5.0 volts. Any TTL line designated by its functional name (usually a mnemonic) is considered to be active or true or indicates a logical {1} when the voltage level is HIGH. Similarly, the line is inactive or false or indicates a {0} when the voltage level is LOW. Any signal, designated as the logical NOT of its function, represents the opposite conditions. In this case, a LOW voltage represents an active or true or {1} state and these signals will be designated with a bar (logical NOT symbol) above them.

Thus the data lines of the MC68000 are designated D0, D1, . . . , D7 with a TTL HIGH on any line representing a {1} and a TTL LOW representing a {0}. On the other hand, the signal line Address Strobe ($\overline{\text{AS}}$) in a TTL HIGH state indicates that the address strobe is not asserted (not active or not true). When the voltage is LOW on the $\overline{\text{AS}}$ line, the address strobe line is asserted (active or true). This use of an active-LOW signal is quite common with TTL logic and provides better immunity to electrical noise under certain conditions. Such a line is often referred to as an "active-low" signal line and is spoken of as "Address Strobe NOT." In print, such a distinction is not necessary because the form $\overline{\text{AS}}$ indicates the logical operation of the signal line.

13.2.1 Data Transfer Operations

The read and write operations of the MC68000 processor are presented here. A read operation requires that data from an external device or memory be placed on the bus in response to signal lines controlled by the MC68000. During a write operation the processor presents data on the bus to be stored in memory or output to an external device. The processor controls the bus to initiate data transfer operations in either direction, but waits for the selected device to acknowledge the transfer request. Such data transfer operations are termed *asynchronous* because the timing of the memory or peripheral device determines the timing of the transfer rather than the CPU timing. These devices are typically slower to transfer data than the processor.

The MC68000 operation, as bus master, for a read cycle of a byte or a word is illustrated in Figure 13.4. The read lines ($\text{R}/\overline{\text{W}}$), the address lines, and the function code lines are asserted (set to their true values) before the transfer request. Then address strobe signal line ($\overline{\text{AS}}$) is asserted (held LOW), followed by the proper data strobe signal line ($\overline{\text{UDS}}$ or $\overline{\text{LDS}}$ or both) to select the upper 8 bits, the lower 8 bits, or all 16 bits of the data signal lines. The device presents data on the data signal lines and asserts the data transfer acknowledge signal ($\overline{\text{DTACK}}$) when it is ready. The processor then reads the data value, negates its strobe signals, and waits for the $\overline{\text{DTACK}}$ signal line to be negated. The processor then initiates the next cycle. The functional timing diagram for the read operations is also shown in Figure 13.4. The read cycle is four clock cycles long if the device responding negates $\overline{\text{DTACK}}$ in the fourth clock cycle. Otherwise, wait states one clock cycle in length are generated until the device responds. The CPU will wait even if the device never responds unless external circuitry indicates that the device did not respond in the allotted time. This type of circuitry is sometimes called a "watchdog" timer.

The write operation is defined in Figure 13.5 and is similar to the read cycle just described. The read/write signal line being LOW indicates a write request for the CPU. The processor does not remove valid data from the data signal lines until the device acknowledges the transfer with $\overline{\text{DTACK}}$ negated. The write operation requires a minimum of four clock cycles to complete.

13.2.2 Function Code Lines and Memory Usage

The signal lines FC0, FC1, and FC2 present a function code to external devices or memory indicating the type of activity occurring on the address or data signal lines. The function code indicates the processor mode (supervisor or user) and the type of access (data or program) each time the processor initiates a read or write operation. The code also indicates if an interrupt is being acknowledged.

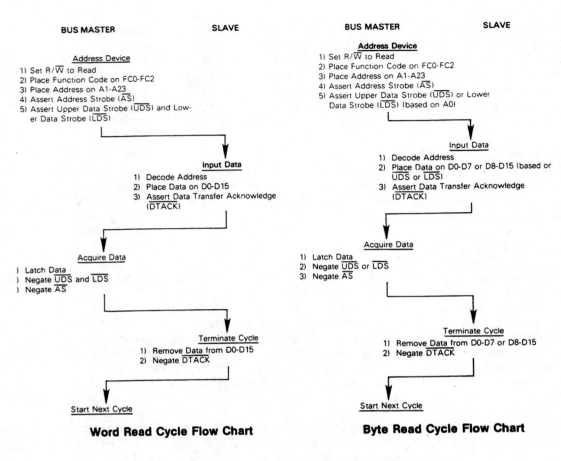

Word Read Cycle Flow Chart

Byte Read Cycle Flow Chart

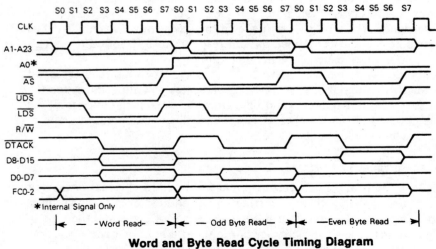

Word and Byte Read Cycle Timing Diagram

Figure 13.4 Read operations. (Courtesy of Motorola, Inc.)

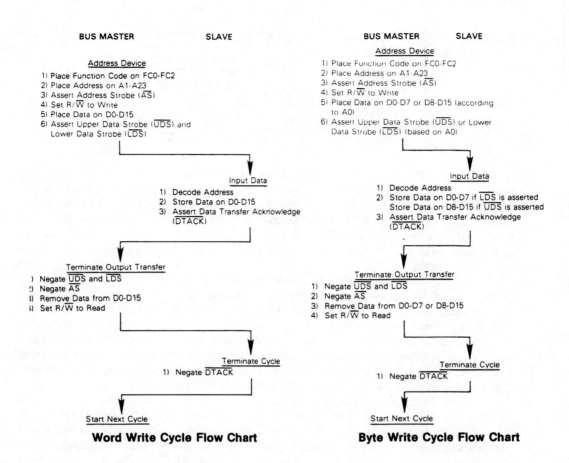

Word Write Cycle Flow Chart

Byte Write Cycle Flow Chart

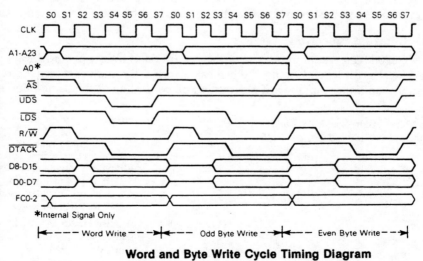

Word and Byte Write Cycle Timing Diagram

Figure 13.5   Write operations. (Courtesy of Motorola, Inc.)

Figure 13.6 Function code references. (Courtesy of Motorola, Inc.)

FC2	FC1	FC0	Cycle Type
Low	Low	Low	(Undefined, Reserved)
Low	Low	High	User Data
Low	High	Low	User Program
Low	High	High	(Undefined, Reserved)
High	Low	Low	(Undefined, Reserved)
High	Low	High	Supervisor Data
High	High	Low	Supervisor Program
High	High	High	Interrupt Acknowledge

Figure 13.6 lists the electrical state of the function code signal lines of the MC68000 for each type of reference. The states not specified are not defined. The processor mode is always determined by the setting of the supervisor status bit in the status register, (SR)[13]. The distinction between data and program references is determined by the addressing mode and the instruction being executed. Table 13.2 defines the addressing modes that cause the selection of various function code states during normal execution.

Using the function code lines to select memory areas, the system memory can be segmented into supervisor and user space. These spaces, in turn, can be segmented into program and data areas. A simplified scheme to accomplish this is shown in Figure 13.7. Each distinct function code allows a selected memory block, or group of blocks, to be accessed using a decoder and selection circuitry. More sophisticated memory protection, including write protection of certain memory areas, can be designed using a Memory Management Chip from Motorola. This chip separates supervisor and user space and also protects memory based on the address signal lines.

The function code signal lines of the MC68000 respond to internally generated conditions within the CPU and the use of these signals is not alterable. Therefore, if memory is protected as shown in Figure 13.7, the supervisor program could not access the user's data or program space unless special logic circuits were included to recognize the supervisor mode access. By way of contrast, the MC68010 allows a supervisor program to define the memory space being accessed by specifying the function code.

13.2.3 Interrupt Processing

The MC68000 interrupt circuitry allows seven levels of interrupt priorities, numbered from 1 to 7, with level 7 being the highest priority. By adding external circuitry, an essentially unlimited number of devices can be connected at the same interrupt level.

Table 13.2 Addressing Modes and Function Codes

Function Code Reference	Addressing Modes
Data	All indirect modes, unless used with JMP or JSR instructions Absolute modes, unless used with JMP or JSR instructions
Program	Relative modes Indirect and absolute modes with JMP or JSR instructions

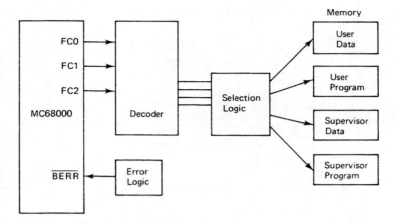

Notes:
   (1) The decoder asserts a single line corresponding to the binary
      value of the Function Code bits.
   (2) The selection logic enables selected memory blocks. It could
      be used to allow a Supervisor Program to access all of memory,
      for example, but still restrict user's access to the proper block.
   (3) The error logic is necessary to indicate an error if the wrong
      block is accessed.
   (4) Address, data, and control signal lines to memory are not shown.

Figure 13.7   Memory protection.

The subpriorities within any CPU interrupt level must be determined by special circuitry. As described in Chapter 11, the MC68000 interrupt circuitry accommodates both user vectored interrupts and autovectored interrupts.

In the user interrupt vectored mode, Motorola allows 192 vectors to be distributed among the seven priority levels as determined by the system designer. The seven autovector interrupts have vectors at fixed locations in the exception vector table (locations \$064 through \$07C). The CPU selects the proper autovector location based on the priority level requested by the external device.

An interrupt request is made to the CPU by encoding the interrupt request level, as a binary number, on the three interrupt request signal lines ($\overline{\text{IPL0}}$, $\overline{\text{IPL1}}$, $\overline{\text{IPL2}}$). An autovector interrupt is requested when the interface asserts the Valid Peripheral Address ($\overline{\text{VPA}}$) signal line in addition to the requested level. If $\overline{\text{VPA}}$ is not asserted, the number used to calculate the location of the vector in the exception vector table must be supplied by the peripheral device. If the interface does not respond to the interrupt acknowledge sequence from the CPU, a "spurious interrupt" processing routine is executed (vector location \$60).

The interrupt system of the MC68000 responds to the three signal lines, $\overline{\text{IPL0}}$ to $\overline{\text{IPL2}}$, to service an interrupt request. The external device encodes the priority level on the signal lines as

$$\text{level} = \overline{\text{IPL2}} \times 2^2 + \overline{\text{IPL1}} \times 2^1 + \overline{\text{IPL0}}$$

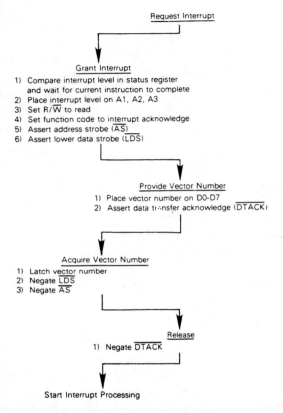

Figure 13.8   Interrupt sequence in vectored mode. (Courtesy of Motorola, Inc.)

where a LOW signal is considered a {1} and a HIGH signal a {0} in the equation. A level of zero indicates there is currently no interrupt request.

The sequence of processor and external device operations for *vectored* interrupts is shown in Figure 13.8. In the sequence, the processor places the interrupt level acknowledged on the three address signal lines (A1–A3) and indicates an interrupt acknowledge by outputting the function code {111}. For the vectored case shown, the external device should respond to the read request of the processor with the *vector number* on the lower data signal lines (D0–D7). The vector number is converted to an address in the interrupt vector table. After the transfer is complete, the processor begins interrupt processing. Figure 13.9(a) shows the timing for the interrupt request and acknowledge sequence.

An alternative mode of operation is for the processor to provide the vector address in the *autovector* mode, which requires less complicated circuitry in the interface. The

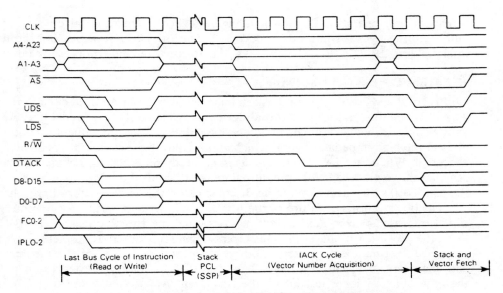

(a) Interrupt Acknowledge Sequence Timing Diagram

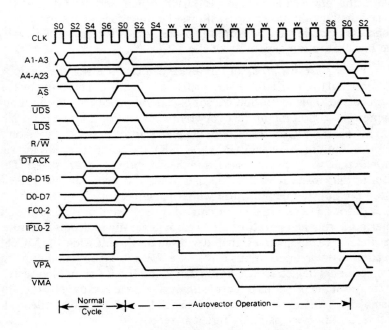

(b) Autovector Operation Timing Diagram

Figure 13.9 Interrupt timing. (Courtesy of Motorola, Inc.)

timing of that operation is shown in Figure 13.9(b). The external device requests autovectoring by asserting $\overline{\text{VPA}}$ (LOW). The priority of interrupts and the use of interrupt lines $\overline{\text{IPL0}}$ to $\overline{\text{IPL2}}$ is identical to that for vectored interrupts as just described.

13.2.4 Bus Control and Bus Error

Control of the MC68000 system bus can be taken over by an external device in several ways. One method is for the external device to assert the $\overline{\text{HALT}}$ signal line. As long as it is held LOW, the processor keeps its address, data, and function code signal lines in the high-impedance state, which allows other bus activity independent of the processor. When the $\overline{\text{HALT}}$ signal is again negated, the processor resumes execution. An alternative method is for an external device to control the bus through the use of the MC68000's bus arbitration logic. In most cases, this is the better approach.[2]

The bus arbitration feature of the MC68000 allows another device to request use of the bus, as shown in Figure 13.10(a). Asserting the Bus Request signal ($\overline{\text{BR}}$) will force the processor to assert the Bus Grant signal ($\overline{\text{BG}}$) after it completes its current bus cycle. If a number of external devices can request control of the bus and each acts as master, external circuitry must be provided to determine their priority. Once the requesting device is in control, the Bus Grant Acknowledge signal ($\overline{\text{BGACK}}$) is held LOW and the bus is used by the device as necessary. When the operations are complete, the device negates the $\overline{\text{BGACK}}$ signal and the MC68000 then resumes control of the bus. Figure 13.10(b) shows the timing sequence for a device requesting DMA access by cycle stealing from the processor. This represents a typical application for the bus arbitration capability of the MC68000.

Bus Error and Halt. If an external device detects an error, the device can assert the Bus Error signal line ($\overline{\text{BERR}}$). This signal line held LOW causes the bus error exception to be processed. This exception causes exception processing in the manner described in Chapter 11. In general, no recovery by the exception routine is possible from a bus error in MC68000 systems. However, diagnostic information is saved on the supervisor stack. In addition to the values (PC) and (SR), the contents of the instruction register, the address that was being accessed, and the function codes are saved. This information is useful to diagnose the error. The use of the bus error signal of the MC68010 was discussed previously in connection with virtual memory.

Except for an occasional error caused by excessive noise on the signal lines for a brief period during an operation, a bus error generally indicates a serious failure in the system. In fact, if a bus error is signaled without the $\overline{\text{HALT}}$ while the MC68000 is processing an address error, another bus error, or a $\overline{\text{RESET}}$ exception, the processor will halt.

If an external device asserts both $\overline{\text{BERR}}$ and $\overline{\text{HALT}}$ simultaneously, the MC68000 will try to rerun the previous bus cycle unless the cycle is caused by the TAS instruction. By design, the TAS instruction has an indivisible read–modify–write cycle. The processor will rerun the cycle as many times as requested.

---

[2] The MC68000 can be operated in a single-cycle mode, which is the hardware equivalent of the trace mode for single-instruction execution. The circuitry to accomplish this using the $\overline{\text{HALT}}$ signal line is discussed in the MC68000 *User's Manual.*

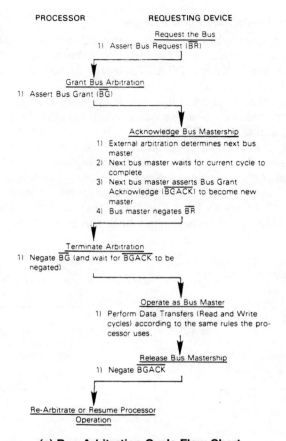

PROCESSOR   REQUESTING DEVICE

Request the Bus
1) Assert Bus Request ($\overline{BR}$)

Grant Bus Arbitration
1) Assert Bus Grant ($\overline{BG}$)

Acknowledge Bus Mastership
1) External arbitration determines next bus
   master
2) Next bus master waits for current cycle to
   complete
3) Next bus master asserts Bus Grant
   Acknowledge ($\overline{BGACK}$) to become new
   master
4) Bus master negates $\overline{BR}$

Terminate Arbitration
1) Negate $\overline{BG}$ (and wait for $\overline{BGACK}$ to be
   negated)

Operate as Bus Master
1) Perform Data Transfers (Read and Write
   cycles) according to the same rules the processor uses.

Release Bus Mastership
1) Negate $\overline{BGACK}$

Re-Arbitrate or Resume Processor
Operation

(a) Bus Arbitration Cycle Flow-Chart

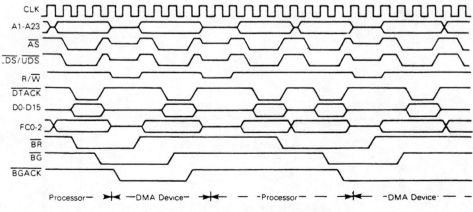

(b) Bus Arbitration Cycle Timing Diagram

Figure 13.10   Bus arbitration. (Courtesy of Motorola, Inc.)

13.2.5 Interfacing to MC6800 Family Chips

The signal activity previously described between the CPU and a device for data transfer is commonly termed *handshaking*. Using this method, every CPU activity is acknowledged (via the signal line $\overline{\text{DTACK}}$), by the device and vice versa. The transfer is asynchronous because the speed of operation is determined by the response of the device and is not based on a fixed number of cycles of the system clock. In contrast, the 8-bit M6800 peripheral chips are designed to transfer data synchronously based on cycles of an enabling signal (E), which acts as a clock for these chips. The MC68000 provides this signal as a constant-frequency clock that has one-tenth the frequency of the system clock. For example, the 10-MHz version of the MC68000 provides a 1-MHz clock on its enable signal line when the CPU is running at its maximum speed.

As shown in Figure 13.11, the 8-bit chip requests synchronous transfer by asserting Valid Peripheral Address ($\overline{\text{VPA}}$). The CPU then asserts Valid Memory Address ($\overline{\text{VMA}}$) after setting up the other signal lines for ordinary transfer. The CPU uses

Figure 13.11 MC68000 operation with M6800 peripheral chips. (Courtesy of Motorola, Inc.)

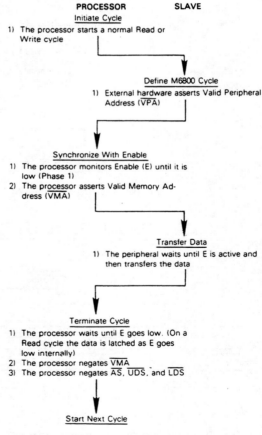

(a) M6800 Interfacing Flow Chart

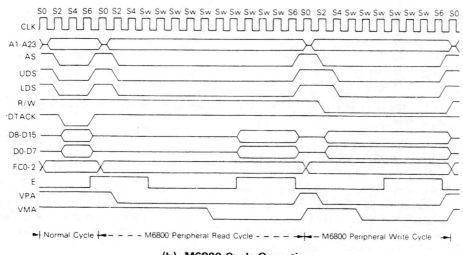

(b) M6800 Cycle Operation

Figure 13.11 *(continued)*

a number of wait states to synchronize with the enable signal, which has a period of 10 CPU cycles.

EXERCISES

13.2.1. The instruction

    MOVE.B   D1,$2001

is executed. Define the value or condition for the address lines, data strobe lines, and other signal lines as the operand is transferred to memory.

13.2.2. Determine the type of memory reference that occurs and the values of the function code lines for each of the following instructions.

    (a) MOVE.W   D1,(A1)   ;IN USER MODE
    (b) JSR   (A1)   ;IN USER MODE
    (c) MOVE.W   D1,DISP (PC)   ;IN SUPERVISOR MODE
    (d) CLR.L $2000   ;IN SUPERVISOR MODE

13.2.3. In Chapter 9 it was shown that the use of PC relative addresses for an alterable destination location was prohibited. If the memory is segmented into program space and data space, could such destination addressing be used even if the PC addressing modes were allowed to specify memory locations as destinations? How is position-independent code achieved in a system with protected memory?

13.2.4. What vector address in the exception vector table is being requested by a device that supplied vector number 64 in response to an interrupt? Could an interface supply vector numbers between 2 and 47, and if so, what problems might arise in the system?

13.2.5. Define the operation of the system if the interrupt request lines receive the following levels:

$\overline{\text{IPL2}}$ is LOW
$\overline{\text{IPL1}}$ is HIGH
$\overline{\text{IPL0}}$ is HIGH

Define the sequence of operation and the values on the address lines and the function code lines.

REVIEW QUESTIONS AND PROBLEMS

13.1. Referencing the MC68000 *User's Manual,* determine the execution time of the interrupt routine in the program of Example 12.2 when it is responding to an interrupt caused by the receipt of a character by the MC68681. Assuming that no output is performed and the input data is continuous, what is the percent CPU utilization at the following data rates for serial I/O?
(a) 38.4K baud
(b) 19.2K baud
(c) 9600 baud

13.2. Given the situation in Question 13.1, what percent of the interrupt routine is taken up by the ENQUEUE subroutine? How much time would be saved by moving the subroutine in-line, instead of performing the JSR (and associated MOVEM and RTS instructions)?

13.3. In Chapter 12, the MC68681 and MC68230 peripheral chips were discussed and examples were given showing interrupts generated by those chips. Although not required, vectored interrupts were used in both cases. On the authors' SBC68K and IDP development boards, a push-button switch is tied to circuits that trigger an autovector interrupt at level 7. Why was an autovectored interrupt used for the push-button switch?

13.4. Consider the components of a microprocessor-based product. What components would be connected directly to the data bus? What components would be connected to the data bus through a peripheral chip?

13.5. Problem 12.5 at the end of Chapter 12 asked you to design the routines to perform parallel communications between two systems using a pair of MC68230 PI/T peripheral chips. Draw the timing diagrams showing the MC68230 data and handshaking signals used to send and receive data between the two devices. You will need to refer to the Motorola documentation for the MC68230.

FURTHER READING

The textbook by Slater considers many aspects of hardware circuit design. The text discusses design with several chips in the 16-bit M68000 family.

Slater, Michael. *Microprocessor–Based Design.* (Mountain View, Calif.: Mayfield Publishing Company, 1987).

ASCII Character Set and Powers of Two and Sixteen

TABLE I.1 ASCII CHARACTER SET

Character	Comments	Hex value
NUL	Null or tape feed	00
SOH	Start of heading	01
STX	Start of Text	02
ETX	End of Text	03
EOT	End of Transmission	04
ENQ	Enquire (who are you, WRU)	05
ACK	Acknowledge	06
BEL	Bell	07
BS	Backspace	08
HT	Horizontal Tab	09
LF	Line Feed	0A
VT	Vertical Tab	0B
FF	Form Feed	0C
RETURN	Carriage Return	0D
SO	Shift Out (to red ribbon)	0E
SI	Shift In (to black ribbon)	0F
DLE	Data Link Escape	10
DC1	Device Control 1	11
DC2	Device Control 2	12
DC3	Device Control 3	13
DC4	Device Control 4	14
NAK	Negative Acknowledge	15
SYN	Synchronous Idle	16
ETB	End of Transmission Block	17
CAN	Cancel	18
EM	End of Medium	19
SUB	Substitute	1A
ESC	Escape, prefix	1B
FS	File Separator	1C
GS	Group Separator	1D
RS	Record Separator	1E
US	Unit Separator	1F
SP	Space or Blank	20

TABLE I.1   (continued)

Character	Comments	Hex value
!	Exclamation point	21
"	Quotation marks (dieresis)	22
#	Number sign	23
$	Dollar sign	24
%	Percent sign	25
&	Ampersand	26
'	Apostrophe (acute accent, closing single quote)	27
(	Opening parenthesis	28
)	Closing parenthesis	29
*	Asterisk	2A
+	Plus sign	2B
,	Comma (cedilla)	2C
-	Hyphen (minus)	2D
.	Period (decimal point)	2E
/	Slant	2F
0	Digit 0	30
1	Digit 1	31
2	Digit 2	32
3	Digit 3	33
4	Digit 4	34
5	Digit 5	35
6	Digit 6	36
7	Digit 7	37
8	Digit 8	38
9	Digit 9	39
:	Colon	3A
;	Semicolon	3B
<	Less than	3C
=	Equals	3D
>	Greater than	3E
?	Question mark	3F
@	Commercial at	40

TABLE I.1 (continued)

Character	Comments	Hex value
A	Upper-case letter A	41
B	Upper-case letter B	42
C	Upper-case letter C	43
D	Upper-case letter D	44
E	Upper-case letter E	45
F	Upper-case letter F	46
G	Upper-case letter G	47
H	Upper-case letter H	48
I	Upper-case letter I	49
J	Upper-case letter J	4A
K	Upper-case letter K	4B
L	Upper-case letter L	4C
M	Upper-case letter M	4D
N	Upper-case letter N	4E
O	Upper-case letter O	4F
P	Upper-case letter P	50
Q	Upper-case letter Q	51
R	Upper-case letter R	52
S	Upper-case letter S	53
T	Upper-case letter T	54
U	Upper-case letter U	55
V	Upper-case letter V	56
W	Upper-case letter W	57
X	Upper-case letter X	58
Y	Upper-case letter Y	59
Z	Upper-case letter Z	5A
[	Opening bracket	5B
\	Reverse slant	5C
]	Closing bracket	5D
^	Circumflex	5E
_	Underline	5F

TABLE I.1   (continued)

Character	Comments	Hex value
'	Quotation mark	60
a	Lower-case letter a	61
b	Lower-case letter b	62
c	Lower-case letter c	63
d	Lower-case letter d	64
e	Lower-case letter e	65
f	Lower-case letter f	66
g	Lower-case letter g	67
h	Lower-case letter h	68
i	Lower-case letter i	69
j	Lower-case letter j	6A
k	Lower-case letter k	6B
l	Lower-case letter l	6C
m	Lower-case letter m	6D
n	Lower-case letter n	6E
o	Lower-case letter o	6F
p	Lower-case letter p	70
q	Lower-case letter q	71
r	Lower-case letter r	72
s	Lower-case letter s	73
t	Lower-case letter t	74
u	Lower-case letter u	75
v	Lower-case letter v	76
w	Lower-case letter w	77
x	Lower-case letter x	78
y	Lower-case letter y	79
z	Lower-case letter z	7A
{	Opening brace	7B
|	Vertical line	7C
}	Closing brace	7D
~	Equivalent	7E
	Delete	7F

TABLE I.2 POWERS OF TWO AND SIXTEEN

16^k 2^n	n	k	2^{-n}
1	0	0	1.0
2	1		0.5
4	2		0.25
8	3		0.125
16	4	1	0.062 5
32	5		0.031 25
64	6		0.015 625
128	7		0.007 812 5
256	8	2	0.003 906 25
512	9		0.001 953 125
1 024	10		0.000 976 562 5
2 048	11		0.000 488 281 25
4 096	12	3	0.000 244 140 625
8 192	13		0.000 122 070 312 5
16 384	14		0.000 061 035 156 25
32 768	15		0.000 030 517 578 125
65 536	16	4	0.000 015 258 789 062 5
131 072	17		0.000 007 629 394 531 25
262 144	18		0.000 003 814 697 265 625
524 288	19		0.000 001 907 348 632 812 5
1 048 576	20	5	0.000 000 953 674 316 406 25
2 097 152	21		0.000 000 476 837 158 203 125
4 194 304	22		0.000 000 238 418 579 101 562 5
8 388 608	23		0.000 000 119 209 289 550 781 25
16 777 216	24	6	0.000 000 059 604 664 775 390 625
33 554 432	25		0.000 000 029 802 322 387 695 312 5
67 108 864	26		0.000 000 014 901 161 193 847 656 25
134 217 728	27		0.000 000 007 450 580 596 923 828 125
268 435 456	28	7	0.000 000 003 725 290 298 461 914 062 5
536 870 912	29		0.000 000 001 862 645 149 230 957 031 25
1 073 741 824	30		0.000 000 000 931 322 574 615 478 515 625
2 147 483 648	31		0.000 000 000 465 661 287 307 739 257 812 5
4 294 967 296	32	8	0.000 000 000 232 830 643 653 869 628 906 25

MC68000 Characteristics

APPENDIX IIA: ELECTRICAL CHARACTERISTICS

TABLE IIA.1 SIGNAL SUMMARY

Signal Name	Mnemonic	Input/Output	Active State	Three State
Address Bus	A1-A23	Output	High	Yes
Data Bus	D0-D15	Input/Output	High	Yes
Address Strobe	$\overline{AS}$	Output	Low	Yes
Read/Write	$R/\overline{W}$	Output	Read-High Write-Low	Yes
Upper and Lower Data Strobes	$\overline{UDS}$, $\overline{LDS}$	Output	Low	Yes
Data Transfer Acknowledge	$\overline{DTACK}$	Input	Low	No
Bus Request	$\overline{BR}$	Input	Low	No
Bus Grant	$\overline{BG}$	Output	Low	No
Bus Grant Acknowledge	$\overline{BGACK}$	Input	Low	No
Interrupt Priority Level	$\overline{IPL0}$, $\overline{IPL1}$, $\overline{IPL2}$	Input	Low	No
Bus Error	$\overline{BERR}$	Input	Low	No
Reset	$\overline{RESET}$	Input/Output	Low	No*
Halt	$\overline{HALT}$	Input/Output	Low	No*
Enable	E	Output	High	No
Valid Memory Address	$\overline{VMA}$	Output	Low	Yes
Valid Peripheral Address	$\overline{VPA}$	Input	Low	No
Function Code Output	FC0, FC1, FC2	Output	High	Yes
Clock	CLK	Input	High	No
Power Input	V_{CC}	Input	—	—
Ground	GND	Input	—	—

*Open Drain

TABLE IIA.2 DATA STROBE CONTROL OF DATA BUS

$\overline{\text{UDS}}$	$\overline{\text{LDS}}$	R/$\overline{\text{W}}$	D8-D15	D0-D7
High	High	—	No Valid Data	No Valid Data
Low	Low	High	Valid Data Bits 8-15	Valid Data Bits 0-7
High	Low	High	No Valid Data	Valid Data Bits 0-7
Low	High	High	Valid Data Bits 8-15	No Valid Data
Low	Low	Low	Valid Data Bits 8-15	Valid Data Bits 0-7
High	Low	Low	Valid Data Bits 0-7*	Valid Data Bits 0-7
Low	High	Low	Valid Data Bits 8-15	Valid Data Bits 8-15*

*These conditions are a result of current implementation and may not appear on future devices.

TABLE IIA.3 FUNCTION CODE OUTPUTS

FC2	FC1	FC0	Cycle Type
Low	Low	Low	(Undefined, Reserved)
Low	Low	High	User Data
Low	High	Low	User Program
Low	High	High	(Undefined, Reserved)
High	Low	Low	(Undefined, Reserved)
High	Low	High	Supervisor Data
High	High	Low	Supervisor Program
High	High	High	Interrupt Acknowledge

APPENDIX IIB: DATA ORGANIZATION IN MEMORY

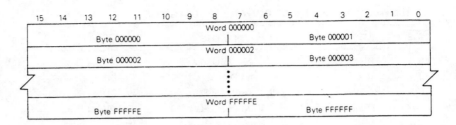

Word Organization In Memory

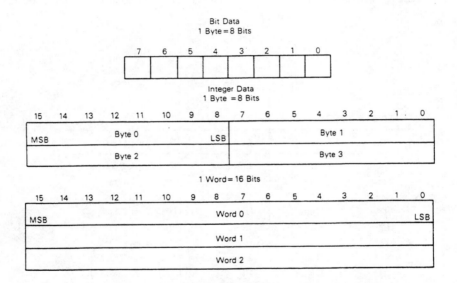

Data Organization In Memory

1 Long Word = 32 Bits

15	14	13	12	11	10	9	3	7	6	5	4	3	2	1	0

MSB

High Order

— — Long Word 0 — — — — — — — — — — — — — — —

Low Order

LSB

— — Long Word 1 — — — — — — — — — — — — — — —

— — Long Word 2 — — — — — — — — — — — — — — —

Addresses
1 Address = 32 Bits

15	14	13	12	11	10	9	8	7	6	5	4	3	2	1	0

MSB

High Order

— — Address 0 — — — — — — — — — — — — — — —

Low Order

LSB

— — Address 1 — — — — — — — — — — — — — — —

— — Address 2 — — — — — — — — — — — — — — —

MSB = Most Significant Bit
LSB = Least Significant Bit

Decimal Data
2 Binary Coded Decimal Digits = 1 Byte

15	14	13	12	11	10	9	8	7	6	5	4	3	2	1	0

MSD

| BCD 0 | BCD 1 | LSD | BCD 2 | BCD 3 |
| BCD 4 | BCD 5 | BCD 6 | BCD 7 |

MSD = Most Significant Digit
LSD = Least Significant Digit

APPENDIX IIC: EXCEPTION VECTOR ASSIGNMENTS

Vector Number(s)	Dec	Address Hex	Space	Assignment
0	0	000	SP	Reset: Initial SSP[2]
	4	004	SP	Reset: Initial PC[2]
2	8	008	SD	Bus Error
3	12	00C	SD	Address Error
4	16	010	SD	Illegal Instruction
5	20	014	SD	Zero Divide
6	24	018	SD	CHK Instruction
7	28	01C	SD	TRAPV Instruction
8	32	020	SD	Privilege Violation
9	36	024	SD	Trace
10	40	028	SD	Line 1010 Emulator
11	44	02C	SD	Line 1111 Emulator
12[1]	48	030	SD	(Unassigned, Reserved)
13[1]	52	034	SD	(Unassigned, Reserved)
14[1]	56	038	SD	(Unassigned, Reserved)
15	60	03C	SD	Uninitialized Interrupt Vector
16-23[1]	64	040	SD	(Unassigned, Reserved)
	95	05F		—
24	96	060	SD	Spurious Interrupt[3]
25	100	064	SD	Level 1 Interrupt Autovector
26	104	068	SD	Level 2 Interrupt Autovector
27	108	06C	SD	Level 3 Interrupt Autovector
28	112	070	SD	Level 4 Interrupt Autovector
29	116	074	SD	Level 5 Interrupt Autovector
30	120	078	SD	Level 6 Interrupt Autovector
31	124	07C	SD	Level 7 Interrupt Autovector
32-47	128	080	SD	TRAP Instruction Vectors[4]
	191	0BF		
48-63[1]	192	0C0	SD	(Unassigned, Reserved)
	255	0FF		—
64-255	256	100	SD	User Interrupt Vectors
	1023	3FF		—

NOTES:
1. Vector numbers 12, 13, 14, 16 through 23, and 48 through 63 are reserved for future enhancements by Motorola. No user peripheral devices should be assigned these numbers.
2. Reset vector (0) requires four words, unlike the other vectors which only require two words, and is located in the supervisor program space.
3. The spurious interrupt vector is taken when there is a bus error indication during interrupt processing. Refer to Paragraph 5.5.2.
4. TRAP #n uses vector number 32 + n.

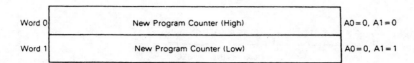

Figure IIC.1 Exception vector format (Courtesy of Motorola, Inc.)

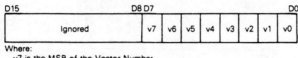

Where:
v7 is the MSB of the Vector Number
v0 is the LSB of the Vector Number

Figure IIC.2 Peripheral vector number format (Courtesy of Motorola, Inc.)

A23				A10	A9	A8	A7	A6	A5	A4	A3	A2	A1	A0
All Zeroes					v7	v6	v5	v4	v3	v2	v1	v0	0	0

Figure IIC.3 Address translated from 8-bit vector number. (Courtesy of Motorola, Inc.)

APPENDIX IID: PROGRAMMING MODEL

TABLE IID.1   REGISTER SET

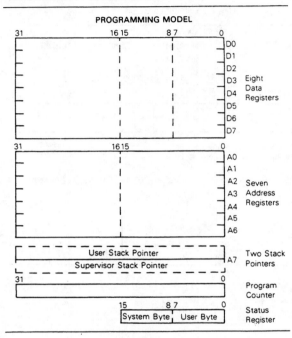

TABLE IID.2   STATUS REGISTER

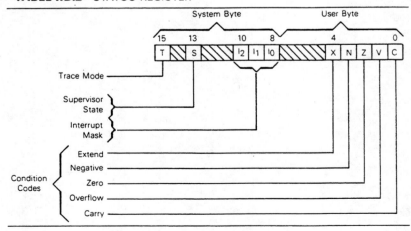

APPENDIX III

Assembly Language

TABLE III.1 CHARACTER SET

The character set recognized by the Motorola MC68000 Resident Structured Assembler is a subset of ASCII (American Standard Code for Information Interchange, 1968). The characters listed below are recognized by the assembler, and the ASCII Code.

1. The upper case letters A through Z
2. The integers 0 through 9
3. Four arithmetic operators: + − * /
4. The logical operators: >> << & !
5. Parentheses used in expressions ( )
6. Characters used as special prefixes:
   # (pound sign) specifies the immediate mode of addressing
   $ (dollar sign) specifies a hexadecimal number
   ' (apostrophe) specifies an ASCII literal character
7. The special characters used in macros: < > \ @
8. Three separating characters:
   SPACE
   , (comma)
   . (period)
9. A comment in a source statement may include any characters with ASCII hexadecimal values from 20 (SP) through 5F (＿).
10. Character used as a special suffix:
    : (colon) specifies the end of a label

TABLE III.2   ASSEMBLER INSTRUCTION SET

Mnemonic	Operation	Assembler syntax	Cond. codes				
			X	N	Z	V	C
ABCD	Add Decimal With Extend	ABCD Dy,Dx ABCD —(Ay),—(Ax)	*	U	*	U	*
ADD	Add Binary	ADD <ea>,Dn ADD Dn,<ea>	*	*	*	*	*
ADDA	Add Address	ADDA <ea>,An	—	—	—	—	—
ADDI	Add Immediate	ADDI #<data>,<ea>	*	*	*	*	*
ADDQ	Add Quick	ADDQ #<data>,<ea>	*	*	*	*	*
ADDX	Add Extended	ADDX Dy,Dx ADDX —(Ay), —(Ax)	*	*	*	*	*
AND	AND Logical	AND <ea>,Dn AND Dn,<ea>	—	*	*	0	0
ANDI	AND Immediate	ANDI #<data>,<ea>	—	*	*	0	0
ASL, ASR	Arithmetic Shift	ASd Dx,Dy ASd #<data>,Dy ASd <ea>	*	*	*	*	*
Bcc	Branch Conditionally	Bcc <label>	—	—	—	—	—
BCHG	Test a Bit and Change	BCHG Dn,<ea> BCHG #<data>,<ea>	—	—	*	—	—
BCLR	Test a Bit and Clear	BCLR Dn,<ea> BCLR #<data>,<ea>	—	—	*	—	—
BRA	Branch Always	BRA <label>	—	—	—	—	—
BSET	Test a Bit and Set	BSET Dn,<ea> BSET #<data>,<ea>	—	—	*	—	—
BSR	Branch to Subroutine	BSR <label>	—	—	—	—	—
BTST	Test a Bit	BTST Dn,<ea> BTST #<data>,<ea>	—	—	*	—	—
CHK	Check Register Against Bounds	CHK <ea>,Dn	—	*	U	U	U
CLR	Clear an Operand	CLR <ea>	—	0	1	0	0
CMP	Arithmetic Compare	CMP <ea>,Dn	—	*	*	*	*
CMPA	Arithmetic Compare Address	CMPA <ea>,An	—	*	*	*	*
CMPI	Compare Immediate	CMPI #<data>,<ea>	—	*	*	*	*
CMPM	Compare Memory	CMPM (Ay)+,(Ax)+	—	*	*	*	*
DBcc	Test Condition Decrement and Branch	DBcc Dn,<label>	—	—	—	—	—
DIVS	Signed Divide	DIVS <ea>,Dn	—	*	*	*	0
DIVU	Unsigned Divide	DIVU <ea>,Dn	—	*	*	*	0
EOR	Exclusive OR Logical	EOR Dn,<ea>	—	*	*	0	0
EORI	Exclusive OR Immediate	EORI #<data>,<ea>	—	*	*	0	0
EXG	Exchange Registers	EXG Rx,Ry	—	—	—	—	—

TABLE III.2 (continued)

Mnemonic	Operation	Assembler syntax	X	N	Z	V	C
			— Cond. codes				
EXT	Sign Extend	EXT Dn	—	*	*	0	0
JMP	Jump	JMP < ea >	—	—	—	—	—
JSR	Jump to Subroutine	JSR < ea >	—	—	—	—	—
LEA	Load Effective Address	LEA < ea >,An	—	—	—	—	—
LINK	Link and Allocate	LINK An,# < displacement >	—	—	—	—	—
LSR, LSL	Logical Shift	LSd Dx,Dy LSd # < data >,Dy LSd < ea >	*	*	*	0	*
MOVE	Move Data from Source to Destination	MOVE < ea >,< ea >	—	*	*	0	0
MOVE from SR	Move from the Status Register	MOVE SR, < ea >	—	—	—	—	—
MOVE to CC	Move to Condition Codes	MOVE < ea >,CCR	*	*	*	*	*
MOVE to SR	Move to the Status Register	MOVE < ea >,SR	*	*	*	*	*
MOVE USP	Move User Stack Pointer	MOVE USP,An MOVE An,USP	—	—	—	—	—
MOVEA	Move Address	MOVEA < ea >,An	—	—	—	—	—
MOVEM	Move Multiple Registers	MOVEM < register list >,< ea > MOVEM < ea >,< register list >	—	—	—	—	—
MOVEP	Move Peripheral Data	MOVEP Dx,d(Ay) MOVEP d(Ay),Dx	—	—	—	—	—
MOVEQ	Move Quick	MOVEQ # < data >,Dn	—	*	*	0	0
MULS	Signed Multiply	MULS < ea >,Dn	—	*	*	0	0
MULU	Unsigned Multiply	MULU < ea >,Dn	—	*	*	0	0
NBCD	Negate Decimal with Extend	NBCD < ea >	*	U	*	U	*
NEG	Two's Complement Negation	NEG < ea >	*	*	*	*	*
NEGX	Negate with Extend	NEGX < ea >	*	*	*	*	*
NOP	No Operation	NOP	—	—	—	—	—
NOT	Logical Complement	NOT < ea >	—	*	*	0	0
OR	Inclusive OR Logical	OR < ea >,Dn OR Dn, < ea >	—	*	*	0	0
ORI	Inclusive OR Immediate	ORI # < data >, < ea >	—	*	*	0	0
PEA	Push Effective Address	PEA < ea >	—	—	—	—	—
RESET	Reset External Devices	RESET	—	—	—	—	—

TABLE III.2 (continued)

Mnemonic	Operation	Assembler syntax	X	N	Z	V	C
			\multicolumn{5}{Cond. codes}				
ROL, ROR	Rotate without Extend	ROd Dx,Dy ROd # <data>,Dy ROd <ea>	—	*	*	0	*
ROXL, ROXR	Rotate with Extend	ROXd Dx,Dy ROXd # <data>,Dy ROXd <ea>	*	*	*	0	*
RTE	Return from Exception	RTE	*	*	*	*	*
RTR	Return and Restore Condition Codes	RTR	*	*	*	*	*
RTS	Return from Subroutine	RTS	—	—	—	—	—
SBCD	Subtract Decimal with Extend	SBCD Dy,Dx SBCD —(Ay),—(Ax)	*	U	*	U	*
Scc	Set according to Condition	Scc <ea>	—	—	—	—	—
STOP	Stop Program Execution	STOP #xxx	*	*	*	*	*
SUB	Subtract Binary	SUB <ea>,Dn SUB Dn,<ea>	*	*	*	*	*
SUBA	Subtract Address	SUBA <ea>,An	—	—	—	—	—
SUBI	Subtract Immediate	SUBI # <data>,<ea>	*	*	*	*	*
SUBQ	Subtract Quick	SUBQ # <data>,<ea>	*	*	*	*	*
SUBX	Subtract with Extend	SUBX Dy,Dx SUBX —(Ay),—(Ax)	*	*	*	*	*
SWAP	Swap Register Halves	SWAP Dn	—	*	*	0	0
TAS	Test and Set an Operand	TAS <ea>	—	*	*	0	0
TRAP	Trap	TRAP # <vector>	—	—	—	—	—
TRAPV	Trap on Overflow	TRAPV	—	—	—	—	—
TST	Test an Operand	TST <ea>	—	*	*	0	0
UNLK	Unlink	UNLK An	—	—	—	—	—

NOTE: For condition codes, — indicates that code is unchanged and * indicates that code is set according to the result.

Machine Language Characteristics of the MC68000, MC68008, and MC68010

Appendix IV is taken from *16-Bit Microprocessor, User's Manual*, 3rd ed. and from *MC68000 16/32-Bit Microprocessor, Programmer's Reference Manual*, 4th ed. Courtesy of Motorola, Inc.

APPENDIX IVA: CONDITION CODES COMPUTATION

A.1 INTRODUCTION

This appendix provides a discussion of how the condition codes were developed, the meanings of each bit, how they are computed, and how they are represented in the instruction set details.

Two criteria were used in developing the condition codes:
- Consistency — across instruction, uses, and instances
- Meaningful Results — no change unless it provides useful information

The consistency across instructions means that instructions which are special cases of more general instructions affect the condition codes in the same way. Consistency across instances means that if an instruction ever affects a condition code, it will always affect that condition code. Consistency across uses means that whether the condition codes were set by a compare, test, or move instruction, the conditional instructions test the same situation. The tests used for the conditional instructions and the code computations are given in paragraph A.5.

A.2 CONDITION CODE REGISTER

The condition code register portion of the status register contains five bits:
- N — Negative
- Z — Zero
- V — Overflow
- C — Carry
- X — Extend

The first four bits are true condition code bits in that they reflect the condition of the result of a processor operation. The X bit is an operand for multiprecision computations. The carry bit (C) and the multiprecision operand extend bit (X) are separate in the MC68000 to simplify the programming model.

A.3 CONDITION CODE REGISTER NOTATION

In the instruction set details given in Appendix B, the description of the effect on the condition codes is given in the following form:

X	N	Z	V	C

Condition Codes:

where:

N (negative)	Set if the most significant bit of the result is set. Cleared otherwise.
Z (zero)	Set if the result equals zero. Cleared otherwise.
V (overflow)	Set if there was an arithmetic overflow. This implies that the result is not representable in the operand size. Cleared otherwise.
C (carry)	Set if a carry is generated out of the most significant bit of the operands for an addition. Also set if a borrow is generated in a subtraction. Cleared otherwise.
X (extend)	Transparent to data movement. When affected, it is set the same as the C bit.

The notational convention that appears in the representation of the condition code register is:

* set according to the result of the operation
— not affected by the operation
0 cleared
1 set
U undefined after the operation

A.4 CONDITION CODE COMPUTATION

Most operations take a source operand and a destination operand, compute, and store the result in the destination location. Unary operations take a destination operand, compute, and store the result in the destination location. Table A-1 details how each instruction sets the condition codes.

Table A-1. Condition Code Computations

Operations	X	N	Z	V	C	Special Definition
ABCD	•	U	?	U	?	C = Decimal Carry $Z = Z \cdot \overline{R_m} \cdot \ldots \cdot \overline{R_0}$
ADD, ADDI, ADDQ	•	•	•	?	?	$V = S_m \cdot D_m \cdot \overline{R_m} + \overline{S_m} \cdot \overline{D_m} \cdot R_m$ $C = S_m \cdot D_m + \overline{R_m} \cdot D_m + S_m \cdot \overline{R_m}$
ADDX	•	•	?	?	?	$V = S_m \cdot D_m \cdot \overline{R_m} + \overline{S_m} \cdot \overline{D_m} \cdot R_m$ $C = S_m \cdot D_m + R_m \cdot \overline{D_m} + S_m \cdot \overline{R_m}$ $Z = Z \cdot \overline{R_m} \cdot \ldots \cdot \overline{R_0}$
AND, ANDI, EOR, EORI, MOVEQ, MOVE, OR, ORI, CLR, EXT, NOT, TAS, TST	−	•	•	0	0	
CHK	−	•	U	U	U	
SUB, SUBI, SUBQ	•	•	•	?	?	$V = \overline{S_m} \cdot D_m \cdot \overline{R_m} + S_m \cdot \overline{D_m} \cdot R_m$ $C = S_m \cdot \overline{D_m} + R_m \cdot \overline{D_m} + S_m \cdot R_m$
SUBX	•	•	?	?	?	$V = \overline{S_m} \cdot D_m \cdot \overline{R_m} + S_m \cdot \overline{D_m} \cdot R_m$ $C = S_m \cdot \overline{D_m} + R_m \cdot \overline{D_m} + S_m \cdot R_m$ $Z = Z \cdot \overline{R_m} \cdot \ldots \cdot \overline{R_0}$
CMP, CMPI, CMPM	−	•	•	?	?	$V = \overline{S_m} \cdot D_m \cdot \overline{R_m} + S_m \cdot \overline{D_m} \cdot R_m$ $C = S_m \cdot \overline{D_m} + R_m \cdot \overline{D_m} + S_m \cdot R_m$
DIVS, DIVU	−	•	•	?	0	V = Division Overflow
MULS, MULU	−	•	•	0	0	
SBCD, NBCD	•	U	?	U	?	C = Decimal Borrow $Z = Z \cdot \overline{R_m} \cdot \ldots \cdot \overline{R_0}$
NEG	•	•	•	?	?	$V = D_m \cdot R_m$, $C = D_m + R_m$
NEGX	•	•	?	?	?	$V = D_m \cdot R_m$, $C = D_m + R_m$ $Z = Z \cdot \overline{R_m} \cdot \ldots \cdot \overline{R_0}$
BTST, BCHG, BSET, BCLR	−	−	?	−	−	$Z = \overline{D_n}$
ASL	•	•	•	?	?	$V = D_m \cdot (\overline{D_{m-1}} + \ldots + \overline{D_{m-r}})$ $\quad + \overline{D_m} \cdot (D_{m-1} + \ldots + D_{m-r})$ $C = D_{m-r+1}$
ASL (r=0)	−	•	•	0	0	
LSL, ROXL	•	•	•	0	?	$C = D_{m-r+1}$
LSR (r=0)	−	•	•	0	0	
ROXL (r=0)	−	•	•	0	?	$C = X$
ROL	−	•	•	0	?	$C = D_{m-r+1}$
ROL (r=0)	−	•	•	0	0	
ASR, LSR, ROXR	•	•	•	0	?	$C = D_{r-1}$
ASR, LSR (r=0)	−	•	•	0	0	
ROXR (r=0)	−	•	•	0	?	$C = X$
ROR	−	•	•	0	?	$C = D_{r-1}$
ROR (r=0)	−	•	•	0	0	

− Not affected
U Undefined
? Other — see Special Definition

*General Case:
$X = C$
$N = R_m$
$Z = \overline{R_m} \cdot \ldots \cdot \overline{R_0}$

Sm Source Operand — most significant bit
Dm Destination operand — most significant bit
Rm Result operand — most significant bit
n bit number
r shift count

A.5 CONDITIONAL TESTS

Table A-2 lists the condition names, encodings, and tests for the conditional branch and set instructions. The test associated with each condition is a logical formula based on the current state of the condition codes. If this formula evaluates to 1, the condition succeeds, or is true. If the formula evaluates to 0, the condition is unsuccessful, or false. For example, the T condition always succeeds, while the EQ condition succeeds only if the Z bit is currently set in the condition codes.

Table A-2. Conditional Tests

Mnemonic	Condition	Encoding	Test
T	true	0000	1
F	false	0001	0
HI	high	0010	$\overline{C} \cdot \overline{Z}$
LS	low or same	0011	$C + Z$
CC (HS)	carry clear	0100	$\overline{C}$
CS (LO)	carry set	0101	C
NE	not equal	0110	$\overline{Z}$
EQ	equal	0111	Z
VC	overflow clear	1000	$\overline{V}$
VS	overflow set	1001	V
PL	plus	1010	$\overline{N}$
MI	minus	1011	N
GE	greater or equal	1100	$N \cdot V + \overline{N} \cdot \overline{V}$
LT	less than	1101	$N \cdot \overline{V} + \overline{N} \cdot V$
GT	greater than	1110	$N \cdot V \cdot \overline{Z} + \overline{N} \cdot \overline{V} \cdot \overline{Z}$
LE	less or equal	1111	$Z + N \cdot \overline{V} + \overline{N} \cdot V$

APPENDIX IVB: INSTRUCTION SET DETAILS

B.1 INTRODUCTION

This appendix contains detailed information about each instruction in the MC68000 instruction set. They are arranged in alphabetical order with the mnemonic heading set in large bold type for easy reference.

B.2 ADDRESSING CATEGORIES

Effective address modes may be categorized by the ways in which they may be used. The following classifications will be used in the instruction definitions.

Data   If an effective address mode may be used to refer to data operands, it is considered a data addressing effective address mode.

Memory   If an effective address mode may be used to refer to memory operands, it is considered a memory addressing effective address mode.

Alterable   If an effective address mode may be used to refer to alterable (writable) operands, it is considered an alterable addressing effective address mode.

Control   If an effective address mode may be used to refer to memory operands without an associated size, it is considered a control addressing effective address mode.

Table B-1 shows the various categories to which each of the effective address modes belong.

Table B-1. Effective Addressing Mode Categories

Addressing Mode	Mode	Register	Addressing Categories				Assembler Syntax
			Data	Memory	Control	Alterable	
Data Register Direct	000	reg. no.	X	—	—	X	Dn
Address Register Direct	001	reg. no.	—	—	—	X	An
Address Register Indirect	010	reg. no.	X	X	X	X	(An)
Address Register Indirect with Postincrement	011	reg. no.	X	X	—	X	(An) +
Address Register Indirect with Predecrement	100	reg. no.	X	X	—	X	– (An)
Address Register Indirect with Displacement	101	reg. no	X	X	X	X	d(An)
Address Register Indirect with Index	110	reg. no.	X	X	X	X	d(An, ix)
Absolute Short	111	000	X	X	X	X	xxx.W
Absolute Long	111	001	X	X	X	X	xxx.L
Program Counter with Displacement	111	010	X	X	X	—	d(PC)
Program Counter with Index	111	011	X	X	X	—	d(PC, ix)
Immediate	111	100	X	X	—	—	#xxx

NOTE:
Other notation is also used for the Assembler Syntax. See Chapter 5 for details.

These categories may be combined so that additional, more restrictive, classifications may be defined. For example, the instruction descriptions use such classifications as alterable memory or data alterable. The former refers to those addressing modes which are both alterable and memory addresses, and the latter refers to addressing modes which are both data and alterable.

B.3 INSTRUCTION DESCRIPTION

The formats of each instruction are given in the following pages. Figure B-1 illustrates what information is given.

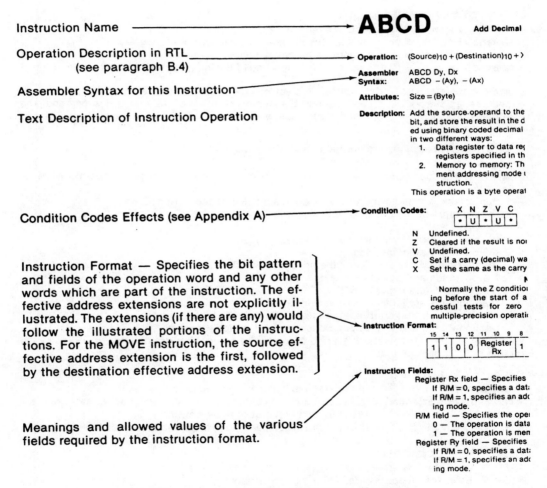

Figure B-1. Instruction Description Format

B.4 REGISTER TRANSFER LANGUAGE DEFINITIONS

The following register transfer language definitions are used for the operation description in the details of the instruction set.

OPERANDS:

An — address register
Dn — data register
Rn — any data or address register
PC — program counter
SR — status register
CCR — condition codes (low order byte of status register)

SSP — supervisor stack pointer
USP — user stack pointer
SP — active stack pointer (equivalent to A7)
X — extend operand (from condition codes)
Z — zero condition code
V — overflow condition code

Immediate Data — immediate data from the instruction
d — address displacement
Source — source effective address
Destination — destination effective address
Vector — location of exception vector

SUBFIELDS AND QUALIFIERS:

<bit>OF<operand>	selects a single bit of the operand
<operand>[<bit number>:<bit number>]	selects a subfield of an operand
(<operand>)	the contents of the referenced location
<operand>$_{10}$	the operand is binary coded decimal; operations are to be performed in decimal.
(<address register>)	the register indirect operator which indicates that the operand register points to the memory location of the instruction operand. The optional mode qualifiers are $-$, $+$, (d) and (d, ix); these are explained in Section 2.
$-$(<address register>)	
(<address register>)$+$	

OPERATIONS: Operations are grouped into binary, unary, and other.

Binary — These operations are written <operand> <op> <operand> where <op> is one of the following:

$\rightarrow$ the left operand is moved to the location specified by the right operand
$\leftrightarrow$ the contents of the two operands are exchanged
$+$ the operands are added
$-$ the right operand is subtracted from the left operand
$*$ the operands are multiplied
$/$ the first operand is divided by the second operand
Λ the operands are logically ANDed
v the operands are logically ORed
$\oplus$ the operands are logically exclusively ORed
$<$ relational test, true if left operand is less than right operand
$>$ relational test, true if left operand is not equal to right operand
shifted by the left operand is shifted or rotated by the number of positions specified by the
rotated by right operand

Unary:

$\sim$<operand>	the operand is logically complemented
<operand>sign-extended	the operand is sign extended, all bits of the upper half are made equal to high order bit of the lower half
<operand>tested	the operand is compared to 0, the results are used to set the condition codes

Other:

TRAP equivalent to PC$\rightarrow$(SSP)$-$; SR$\rightarrow$(SSP)$-$; (vector)$\rightarrow$PC
STOP enter the stopped state, waiting for interrupts

If <condition> **then** <operations> **else** <operations>   The condition is tested. If true, the operations after the "then" are performed. If the condition is false and the optional "else" clause is present, the operations after the "else" are performed. If the condition is false and the optional "else" clause is absent, the instruction performs no operation.

ABCD   Add Decimal with Extend   ABCD

Operation: $(Source)_{10} + (Destination)_{10} + X \rightarrow Destination$

Assembler   ABCD Dy, Dx
Syntax: ABCD $-(Ay), -(Ax)$

Attributes: Size = (Byte)

Description: Add the source operand to the destination operand along with the extend
bit, and store the result in the destination location. The addition is performed using binary coded decimal arithmetic. The operands may be addressed
in two different ways:
1. Data register to data register: The operands are contained in the data
    registers specified in the instruction.
2. Memory to memory: The operands are addressed with the predecre-
    ment addressing mode using the address registers specified in the in-
    struction.
This operation is a byte operation only.

Condition Codes:

X	N	Z	V	C
*	U	*	U	*

N   Undefined.
Z   Cleared if the result is non-zero. Unchanged otherwise.
V   Undefined.
C   Set if a carry (decimal) was generated. Cleared otherwise.
X   Set the same as the carry bit.

NOTE

Normally the Z condition code bit is set via programming before the start of an operation. This allows successful tests for zero results upon completion of
multiple-precision operations.

Instruction Format:

15	14	13	12	11	10	9	8	7	6	5	4	3	2	1	0
1	1	0	0	Register Rx			1	0	0	0	0	R/M	Register Ry		

Instruction Fields:

Register Rx field — Specifies the destination register:
  If R/M = 0, specifies a data register.
  If R/M = 1, specifies an address register for the predecrement address-
  ing mode.
R/M field — Specifies the operand addressing mode:
  0 — The operation is data register to data register.
  1 — The operation is memory to memory.
Register Ry field — Specifies the source register:
  If R/M = 0, specifies a data register.
  If R/M = 1, specifies an address register for the predecrement address-
  ing mode.

ADD

Add Binary

ADD

Operation: (Source) + (Destination)→ Destination

Assembler
Syntax: ADD <ea>, Dn
ADD Dn, <ea>

Attributes: Size = (Byte, Word, Long)

Description: Add the source operand to the destination operand, and store the result in the destination location. The size of the operation may be specified to be byte, word, or long. The mode of the instruction indicates which operand is the source and which is the destination as well as the operand size.

Condition Codes:

X	N	Z	V	C
*	*	*	*	*

N Set if the result is negative. Cleared otherwise.
Z Set if the result is zero. Cleared otherwise.
V Set if an overflow is generated. Cleared otherwise.
C Set if a carry is generated. Cleared otherwise.
X Set the same as the carry bit.

Instruction Format:

15	14	13	12	11 10 9	8 7 6	5 4 3	2 1 0
1	1	0	1	Register	Op-Mode	Effective Address Mode	Register

Instruction Fields:

Register field — Specifies any of the eight data registers.
Op-Mode field —

Byte	Word	Long	Operation
000	001	010	(<Dn>)+(<ea>)→ <Dn>
100	101	110	(<ea>)+(<Dn>)→ <ea>

Effective Address field — Determines addressing mode:
  a. If the location specified is a source operand, then all addressing modes are allowed as shown:

Addressing Mode	Mode	Register	Addressing Mode	Mode	Register
Dn	000	register number	d(An, Xi)	110	register number
An*	001	register number	Abs.W	111	000
(An)	010	register number	Abs.L	111	001
(An)+	011	register number	d(PC)	111	010
−(An)	100	register number	d(PC, Xi)	111	011
d(An)	101	register number	Imm	111	100

*Word and Long only.

— Continued —

ADD

Add Binary

ADD

Effective Address field (Continued)

> b. If the location specified is a destination operand, then only alterable memory addressing modes are allowed as shown:

Addressing Mode	Mode	Register	Addressing Mode	Mode	Register
Dn	—	—	d(An, Xi)	110	register number
An	—	—	Abs.W	111	000
(An)	010	register number	Abs.L	111	001
(An)+	011	register number	d(PC)	—	—
−(An)	100	register number	d(PC, Xi)	—	—
d(An)	101	register number	Imm	—	—

Notes:
1. If the destination is a data register, then it cannot be specified by using the destination <ea> mode, but must use the destination Dn mode instead.
2. ADDA is used when the destination is an address register. ADDI and ADDQ are used when the source is immediate data. Most assemblers automatically make this distinction.

ADDA

Add Address

ADDA

Operation: (Source) + (Destination) → Destination

**Assembler
Syntax:** ADDA <ea>, An

Attributes: Size = (Word, Long)

Description: Add the source operand to the destination address register, and store the result in the address register. The size of the operation may be specified to be word or long. The entire destination address register is used regardless of the operation size.

Condition Codes: Not affected.

Instruction Format:

15	14	13	12	11 10 9	8 7 6	5 4 3	2 1 0
1	1	0	1	Register	Op-Mode	Effective Address Mode	Register

Instruction Fields:

Register field — Specifies any of the eight address registers. This is always the destination.

Op-Mode field — Specifies the size of the operation:

011 — word operation. The source operand is sign-extended to a long operand and the operation is performed on the address register using all 32 bits.

111 — long operation.

Effective Address field — Specifies the source operand. All addressing modes are allowed as shown:

Addressing Mode	Mode	Register	Addressing Mode	Mode	Register
Dn	000	register number	d(An, Xi)	110	register number
An	001	register number	Abs.W	111	000
(An)	010	register number	Abs.L	111	001
(An) +	011	register number	d(PC)	111	010
− (An)	100	register number	d(PC, Xi)	111	011
d(An)	101	register number	Imm	111	100

ADDI

Add Immediate

ADDI

Operation: Immediate Data + (Destination) → Destination

**Assembler
Syntax:** ADDI #<data>,<ea>

Attributes: Size = (Byte, Word, Long)

Description: Add the immediate data to the destination operand, and store the result in the destination location. The size of the operation may be specified to be byte, word, or long. The size of the immediate data matches the operation size.

Condition Codes:

X	N	Z	V	C
*	*	*	*	*

N   Set if the result is negative. Cleared otherwise.
Z   Set if the result is zero. Cleared otherwise.
V   Set if an overflow is generated. Cleared otherwise.
C   Set if a carry is generated. Cleared otherwise.
X   Set the same as the carry bit.

Instruction Format:

15	14	13	12	11	10	9	8	7	6	5	4	3	2	1	0
0	0	0	0	0	1	1	0	Size		Effective Address Mode | Register					
Word Data (16 bits)								Byte Data (8 bits)							
Long Data (32 bits, including previous word)															

Instruction Fields:
Size field — Specifies the size of the operation:
00 — byte operation.
01 — word operation.
10 — long operation.
Effective Address field — Specifies the destination operand. Only data alterable addressing modes are allowed as shown:

Addressing Mode	Mode	Register	Addressing Mode	Mode	Register
Dn	000	register number	d(An, Xi)	110	register number
An	—	—	Abs.W	111	000
(An)	010	register number	Abs.L	111	001
(An)+	011	register number	d(PC)	—	—
−(An)	100	register number	d(PC, Xi)	—	—
d(An)	101	register number	Imm	—	—

Immediate field — (Data immediately following the instruction):
If size = 00, then the data is the low order byte of the immediate word.
If size = 01, then the data is the entire immediate word.
If size = 10, then the data is the next two immediate words.

ADDQ

Add Quick

ADDQ

Operation: Immediate Data + (Destination) → Destination

Assembler Syntax: ADDQ #<data>, <ea>

Attributes: Size = (Byte, Word, Long)

Description: Add the immediate data to the operand at the destination location. The data range is from 1 to 8. The size of the operation may be specified to be byte, word, or long. Word and long operations are also allowed on the address registers and the condition codes are not affected. The entire destination address register is used regardless of the operation size.

Condition Codes:

X	N	Z	V	C
*	*	*	*	*

N   Set if the result is negative. Cleared otherwise.
Z   Set if the result is zero. Cleared otherwise.
V   Set if an overflow is generated. Cleared otherwise.
C   Set if a carry is generated. Cleared otherwise.
X   Set the same as the carry bit.

The condition codes are not affected if an addition to an address register is made.

Instruction Format:

15	14	13	12	11	10	9	8	7	6	5	4	3	2	1	0
0	1	0	1	Data			0	Size		Effective Address Mode			Register		

Instruction Fields:

Data field — Three bits of immediate data, 0, 1-7 representing a range of 8, 1 to 7 respectively.

Size field — Specifies the size of the operation:
  00 — byte operation.
  01 — word operation.
  10 — long operation.

Effective Address field — Specifies the destination location. Only alterable addressing modes are allowed as shown:

Addressing Mode	Mode	Register	Addressing Mode	Mode	Register
Dn	000	register number	d(An, Xi)	110	register number
An*	001	register number	Abs.W	111	000
(An)	010	register number	Abs.L	111	001
(An)+	011	register number	d(PC)	—	—
−(An)	100	register number	d(PC, Xi)	—	—
d(An)	101	register number	Imm	—	—

*Word and Long only.

ADDX

Add Extended

ADDX

Operation: (Source) + (Destination) + X → Destination

Assembler   ADDX Dy, Dx
Syntax: ADDX − (Ay), − (Ax)

Attributes: Size = (Byte, Word, Long)

Description: Add the source operand to the destination operand along with the extend
bit and store the result in the destination location. The operands may be addressed in two different ways:

1. Data register to data register: the operands are contained in data
   registers specified in the instruction.
2. Memory to memory: the operands are addressed with the predecrement addressing mode using the address registers specified in the
   instruction.

The size of the operation may be specified to be byte, word, or long.

Condition Codes:

X	N	Z	V	C
*	*	*	*	*

N   Set if the result is negative. Cleared otherwise.
Z   Cleared if the result is non-zero. Unchanged otherwise.
V   Set if an overflow is generated. Cleared otherwise.
C   Set if a carry is generated. Cleared otherwise.
X   Set the same as the carry bit.

NOTE

Normally the Z condition code bit is set via programming before the start of an operation. This allows successful tests for zero results upon completion of
multiple-precision operations.

Instruction Format:

15	14	13	12	11 10 9	8	7 6	5	4	3	2 1 0
1	1	0	1	Register Rx	1	Size	0	0	R/M	Register Ry

Instruction Fields:

Register Rx field — Specifies the destination register:
 If R/M = 0, specifies a data register.
 If R/M = 1, specifies an address register for the predecrement addressing mode.
Size field — Specifies the size of the operation:
 00 — byte operation.
 01 — word operation.
 10 — long operation.

— Continued —

ADDX

Add Extended

ADDX

Instruction Fields: (Continued)

R/M field — Specifies the operand addressing mode:

0 — The operation is data register to data register.

1 — The operation is memory to memory.

Register Ry field — Specifies the source register:

If R/M = 0, specifies a data register.

If R/M = 1, specifies an address register for the predecrement addressing mode.

AND

AND Logical

AND

Operation: (Source)∧(Destination)→ Destination

Assembler AND <ea>, Dn
Syntax: AND Dn, <ea>

Attributes: Size = (Byte, Word, Long)

Description: AND the source operand to the destination operand and store the result in the destination location. The size of the operation may be specified to be byte, word, or long. The contents of an address register may not be used as an operand.

Condition Codes:

X	N	Z	V	C
—	*	*	0	0

N   Set if the most significant bit of the result is set. Cleared otherwise.
Z   Set if the result is zero. Cleared otherwise.
V   Always cleared.
C   Always cleared.
X   Not affected.

Instruction Format:

15	14	13	12	11 10 9	8 7 6	5 4 3	2 1 0
1	1	0	0	Register	Op-Mode	Effective Address Mode	Register

Instruction Fields:

Register field — Specifies any of the eight data registers.
Op-Mode field —

Byte	Word	Long	Operation
000	001	010	(<Dn>) ∧ (<ea>)→ <Dn>
100	101	110	(<ea>) ∧ (<Dn>)→ <ea>

Effective Address field — Determines addressing mode:
If the location specified is a source operand then only data addressing modes are allowed as shown:

Addressing Mode	Mode	Register	Addressing Mode	Mode	Register
Dn	000	register number	d(An, Xi)	110	register number
An	—	—	Abs.W	111	000
(An)	010	register number	Abs.L	111	001
(An)+	011	register number	d(PC)	111	010
−(An)	100	register number	d(PC, Xi)	111	011
d(An)	101	register number	Imm	111	100

— Continued —

AND   AND Logical   AND

Effective Address field (Continued)

If the location specified is a destination operand then only alterable memory addressing modes are allowed as shown:

Addressing Mode	Mode	Register	Addressing Mode	Mode	Register
Dn	—	—	d(An, Xi)	110	register number
An	—	—	Abs.W	111	000
(An)	010	register number	Abs.L	111	001
(An)+	011	register number	d(PC)	—	—
−(An)	100	register number	d(PC, Xi)	—	—
d(An)	101	register number	Imm	—	—

Notes:
1. If the destination is a data register, then it cannot be specified by using the destination <ea> mode, but must use the destination Dn mode instead.
2. ANDI is used when the source is immediate data. Most assemblers automatically make this distinction.

ANDI

AND Immediate

ANDI

Operation: Immediate Data Λ (Destination)→ Destination

**Assembler
Syntax:** ANDI #<data>, <ea>

Attributes: Size = (Byte, Word, Long)

Description: AND the immediate data to the destination operand and store the result in the destination location. The size of the operation may be specified to be byte, word, or long. The size of the immediate data matches the operation size.

Condition Codes:

X	N	Z	V	C
—	*	*	0	0

N Set if the most significant bit of the result is set. Cleared otherwise.
Z Set if the result is zero. Cleared otherwise.
V Always cleared.
C Always cleared.
X Not affected.

Instruction Format:

15	14	13	12	11	10	9	8	7	6	5	4	3	2	1	0
0	0	0	0	0	0	1	0	Size		Effective Address Mode | Register					
Word Data (16 bits)								Byte Data (8 bits)							
Long Data (32 bits, including previous word)															

Instruction Fields:

Size field — Specifies the size of the operation:
00 — byte operation.
01 — word operation.
10 — long operation.

Effective Address field — Specifies the destination operand. Only data alterable addressing modes are allowed as shown:

Addressing Mode	Mode	Register	Addressing Mode	Mode	Register
Dn	000	register number	d(An, Xi)	110	register number
An	—	—	Abs.W	111	000
(An)	010	register number	Abs.L	111	001
(An)+	011	register number	d(PC)	—	—
–(An)	100	register number	d(PC, Xi)	—	—
d(An)	101	register number	Imm	—	—

Immediate field — (Data immediately following the instruction):
If size = 00, then the data is the low order byte of the immediate word.
If size = 01, then the data is the entire immediate word.
If size = 10, then the data is the next two immediate words.

ANDI to CCR

AND Immediate to Condition Codes

ANDI to CCR

Operation: $(Source) \wedge CCR \rightarrow CCR$

Assembler Syntax: ANDI #xxx, CCR

Attributes: Size = (Byte)

Description: AND the immediate operand with the condition codes and store the result in the low-order byte of the status register.

Condition Codes:

X	N	Z	V	C
*	*	*	*	*

N   Cleared if bit 3 of immediate operand is zero. Unchanged otherwise.
Z   Cleared if bit 2 of immediate operand is zero. Unchanged otherwise.
V   Cleared if bit 1 of immediate operand is zero. Unchanged otherwise.
C   Cleared if bit 0 of immediate operand is zero. Unchanged otherwise.
X   Cleared if bit 4 of immediate operand is zero. Unchanged otherwise.

Instruction Format:

15	14	13	12	11	10	9	8	7	6	5	4	3	2	1	0
0	0	0	0	0	0	1	0	0	0	1	1	1	1	0	0
0	0	0	0	0	0	0	0	Byte Data (8 bits)							

ANDI to SR

AND Immediate to the Status Register
(Privileged Instruction)

ANDI to SR

Operation: If supervisor state
then (Source)∧SR → SR
else TRAP

Assembler
Syntax: ANDI #xxx, SR

Attributes: Size = (Word)

Description: AND the immediate operand with the contents of the status register and store the result in the status register. All bits of the status register are affected.

Condition Codes:

X	N	Z	V	C
*	*	*	*	*

N Cleared if bit 3 of immediate operand is zero. Unchanged otherwise.
Z Cleared if bit 2 of immediate operand is zero. Unchanged otherwise.
V Cleared if bit 1 of immediate operand is zero. Unchanged otherwise.
C Cleared if bit 0 of immediate operand is zero. Unchanged otherwise.
X Cleared if bit 4 of immediate operand is zero. Unchanged otherwise.

Instruction Format:

15	14	13	12	11	10	9	8	7	6	5	4	3	2	1	0
0	0	0	0	0	0	1	0	0	1	1	1	1	1	0	0
Word Data (16 bits)															

ASL, ASR   <small>Arithmetic Shift</small>   ASL, ASR

Operation: (Destination) Shifted by <count> → Destination

Assembler   ASd Dx, Dy
Syntax: ASd #<data>, Dy
            ASd <ea>

Attributes: Size = (Byte, Word, Long)

Description: Arithmetically shift the bits of the operand in the direction specified. The carry bit receives the last bit shifted out of the operand. The shift count for the shifting of a register may be specified in two different ways:

     1. Immediate: the shift count is specified in the instruction (shift range, 1-8).
     2. Register: the shift count is contained in a data register specified in the instruction.

The size of the operation may be specified to be byte, word, or long. The content of memory may be shifted one bit only and the operand size is restricted to a word.

For ASL, the operand is shifted left; the number of positions shifted is the shift count. Bits shifted out of the high order bit go to both the carry and the extend bits; zeroes are shifted into the low order bit. The overflow bit indicates if any sign changes occur during the shift.

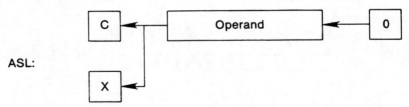

ASL:

For ASR, the operand is shifted right; the number of positions shifted is the shift count. Bits shifted out of the low order bit go to both the carry and the extend bits; the sign bit is replicated into the high order bit.

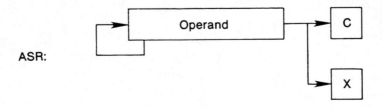

ASR:

— Continued —

ASL, ASR   _{Arithmetic Shift}   ASL, ASR

Condition Codes:

X	N	Z	V	C
*	*	*	*	*

N   Set if the most significant bit of the result is set. Cleared otherwise.
Z   Set if the result is zero. Cleared otherwise.
V   Set if the most significant bit is changed at any time during the shift operation. Cleared otherwise.
C   Set according to the last bit shifted out of the operand. Cleared for a shift count of zero.
X   Set according to the last bit shifted out of the operand. Unaffected for a shift count of zero.

Instruction Format (Register Shifts):

15	14	13	12	11	10	9	8	7	6	5	4	3	2	1	0
1	1	1	0	Count/Register			dr	Size		i/r	0	0	Register		

Instruction Fields (Register Shifts):

Count/Register field — Specifies shift count or register where count is located:

     If i/r = 0, the shift count is specified in this field. The values 0, 1-7 represent a range of 8, 1 to 7 respectively.

     If i/r = 1, the shift count (modulo 64) is contained in the data register specified in this field.

dr field — Specifies the direction of the shift:

     0 — shift right.
     1 — shift left.

Size field — Specifies the size of the operation:

     00 — byte operation.
     01 — word operation.
     10 — long operation.

i/r field —

     If i/r = 0, specifies immediate shift count.
     if i/r = 1, specifies register shift count.

Register field — Specifies a data register whose content is to be shifted.

Instruction Format (Memory Shifts):

15	14	13	12	11	10	9	8	7	6	5	4	3	2	1	0
1	1	1	0	0	0	0	dr	1	1	Effective Address Mode			Register		

— Continued —

ASL, ASR   Arithmetic Shift   ASL, ASR

Instruction Fields (Memory Shifts):

dr field — Specifies the direction of the shift:

0 — shift right.

1 — shift left.

Effective Address field — Specifies the operand to be shifted. Only memory alterable addressing modes are allowed as shown:

Addressing Mode	Mode	Register	Addressing Mode	Mode	Register
Dn	—	—	d(An, Xi)	110	register number
An	—	—	Abs.W	111	000
(An)	010	register number	Abs.L	111	001
(An) +	011	register number	d(PC)	—	—
− (An)	100	register number	d(PC, Xi)	—	—
d(An)	101	register number	Imm	—	—

Bcc

Branch Conditionally

Bcc

Operation: If (condition true) then PC + d → PC

Assembler
Syntax: Bcc < label>

Attributes: Size = (Byte, Word)

Description: If the specified condition is met, program execution continues at location (PC) + displacement. Displacement is a two's-complement integer which counts the relative distance in bytes. The value in PC is the current instruction location plus two. If the 8-bit displacement in the instruction word is zero, then the 16-bit displacement (word immediately following the instruction) is used. "cc" may specify the following conditions:

CC	carry clear	0100	$\overline{C}$	LS	low or same	0011	$C + Z$
CS	carry set	0101	C	LT	less than	1101	$N \cdot \overline{V} + \overline{N} \cdot V$
EQ	equal	0111	Z	MI	minus	1011	N
GE	greater or equal	1100	$N \cdot V + \overline{N} \cdot \overline{V}$	NE	not equal	0110	$\overline{Z}$
GT	greater than	1110	$N \cdot V \cdot \overline{Z} + \overline{N} \cdot \overline{V} \cdot \overline{Z}$	PL	plus	1010	$\overline{N}$
HI	high	0010	$\overline{C} \cdot \overline{Z}$	VC	overflow clear	1000	$\overline{V}$
LE	less or equal	1111	$Z + N \cdot \overline{V} + \overline{N} \cdot V$	VS	overflow set	1001	V

Condition Codes: Not affected.

Instruction Format:

15	14	13	12	11	10	9	8	7	6	5	4	3	2	1	0
0	1	1	0	Condition				8-bit Displacement							
16-bit Displacement if 8-bit Displacement = 0															

Instruction Fields:

Condition field — One of fourteen conditions discussed in description.

8-bit Displacement field — Two's-complement integer specifying the relative distance (in bytes) between the branch instruction and the next instruction to be executed if the condition is met.

16-bit Displacement field — Allows a larger displacement than 8 bits. Used only if the 8-bit displacement is equal to zero.

Note: A short branch to the immediately following instruction cannot be done because it would result in a zero offset which forces a word branch instruction definition.

BCHG

Test a Bit and Change

BCHG

Operation: ~(<bit number>) OF Destination → Z;
~(<bit number>) OF Destination → <bit number> OF Destination

Assembler BCHG Dn, <ea>
Syntax: BCHG #<data>, <ea>

Attributes: Size'=(Byte, Long)

Description: A bit in the destination operand is tested and the state of the specified bit
is reflected in the Z condition code. After the test, the state of the specified
bit is changed in the destination. If a data register is the destination, then
the bit numbering is modulo 32 allowing bit manipulation on all bits in a
data register. If a memory location is the destination, a byte is read from
that location, the bit operation performed using the bit number modulo 8,
and the byte written back to the location with zero referring to the least-
significant bit. The bit number for this operation may be specified in two
different ways:

1. Immediate — the bit number is specified in a second word of the instruction.
2. Register — the bit number is contained in a data register specified in
the instruction.

Condition Codes:

X	N	Z	V	C
—	—	*	—	—

N   Not affected.
Z   Set if the bit tested is zero. Cleared otherwise.
V   Not affected.
C   Not affected.
X   Not affected.

Instruction Format (Bit Number Dynamic):

15	14	13	12	11	10	9	8	7	6	5	4	3	2	1	0
0	0	0	0	Register			1	0	1	Effective Address Mode			Register		

Instruction Fields (Bit Number Dynamic):

Register field — Specifies the data register whose content is the bit
number.
Effective Address field — Specifies the destination location. Only data
alterable addressing modes are allowed as shown:

Addressing Mode	Mode	Register	Addressing Mode	Mode	Register
Dn*	000	register number	d(An, Xi)	110	register number
An	—	—	Abs.W	111	000
(An)	010	register number	Abs.L	111	001
(An)+	011	register number	d(PC)	—	—
−(An)	100	register number	d(PC, Xi)	—	—
d(An)	101	register number	Imm	—	—

*Long only; all others are byte only.

— Continued —

464

BCHG

Test a Bit and Change

BCHG

Instruction Format (Bit Number Static):

15	14	13	12	11	10	9	8	7	6	5	4	3	2	1	0
0	0	0	0	1	0	0	0	0	1	Effective Address Mode | Register					
0	0	0	0	0	0	0	0	bit number							

Instruction Fields (Bit Number Static):

Effective Address field — Specifies the destination location. Only data alterable addressing modes are allowed as shown:

Addressing Mode	Mode	Register	Addressing Mode	Mode	Register
Dn	000	register number	d(An, Xi)	110	register number
An	—	—	Abs.W	111	000
(An)	010	register number	Abs.L	111	001
(An) +	011	register number	d(PC)	—	—
– (An)	100	register number	d(PC, Xi)	—	—
d(An)	101	register number	Imm	—	—

*Long only; all others are byte only.

bit number field — Specifies the bit numbers.

BCLR

Test a Bit and Clear

Operation: ~(<bit number>) OF Destination)→ Z;
0→ <bit number> OF Destination

Assembler   BLCR Dn, <ea>
Syntax: BCLR #<data>, <ea>

Attributes: Size = (Byte, Long)

Description: A bit in the destination operand is tested and the state of the specified bit is reflected in the Z condition code. After the test, the specified bit is cleared in the destination. If a data register is the destination, then the bit numbering is modulo 32 allowing bit manipulation on all bits in a data register. If a memory location is the destination, a byte is read from that location, the bit operation performed using the bit number modulo 8, and the byte written back to the location with zero referring to the least-significant bit. The bit number for this operation may be specified in two different ways:

1. Immediate — the bit number is specified in a second word of the instruction.
2. Register — the bit number is contained in a data register specified in the instruction.

Condition Codes:

X	N	Z	V	C
—	—	*	—	—

N   Not affected.
Z   Set if the bit tested is zero. Cleared otherwise.
V   Not affected.
C   Not affected.
X   Not affected.

Instruction Format (Bit Number Dynamic):

15	14	13	12	11	10	9	8	7	6	5	4	3	2	1	0
0	0	0	0	Register			1	1	0	Effective Address Mode			Register		

Instruction Fields (Bit Number Dynamic):

Register field — Specifies the data register whose content is the bit number.

Effective Address field — Specifies the destination location. Only data alterable addressing modes are allowed as shown:

Addressing Mode	Mode	Register	Addressing Mode	Mode	Register
Dn*	000	register number	d(An, Xi)	110	register number
An	—	—	Abs.W	111	000
(An)	010	register number	Abs.L	111	001
(An) +	011	register number	d(PC)	—	—
− (An)	100	register number	d(PC, Xi)	—	—
d(An)	101	register number	Imm	—	—

*Long only; all others are byte only.

— Continued —

466

BCLR

Test a Bit and Clear

BCLR

Instruction Format (Bit Number Static):

15	14	13	12	11	10	9	8	7	6	5 4 3	2 1 0
0	0	0	0	1	0	0	0	1	0	Effective Address Mode	Register
0	0	0	0	0	0	0	0			bit number	

Instruction Fields (Bit Number Static):

Effective Address field — Specifies the destination location. Only data alterable addressing modes are allowed as shown:

Addressing Mode	Mode	Register	Addressing Mode	Mode	Register
Dn*	000	register only	d(An, Xi)	110	register number
An	—	—	Abs.W	111	000
(An)	010	register number	Abs.L	111	001
(An) +	011	register number	d(PC)	—	—
– (An)	100	register number	d(PC, Xi)	—	—
d(An)	101	register number	Imm	—	—

*Long only; all others are byte only.

bit number field — Specifies the bit number.

BRA

Branch Always

BRA

Operation: PC + d → PC

**Assembler
Syntax:**   BRA < label >

Attributes: Size = (Byte, Word)

Description: Program execution continues at location (PC) + displacement. Displacement is a two's-complement integer which counts the relative distance in bytes. The value in PC is the current instruction location plus two. If the 8-bit displacement in the instruction word is zero, then the 16-bit displacement (word immediately following the instruction) is used.

Condition Codes: Not affected.

Instruction Format:

15	14	13	12	11	10	9	8	7	6	5	4	3	2	1	0
0	1	1	0	0	0	0	0	\multicolumn{8}{c}{8-bit Displacement}
\multicolumn{16}{c}{16-bit Displacement if 8-bit Displacement = 0}

Instruction Fields:

8-bit Displacement field — Two's-complement integer specifying the relative distance (in bytes) between the branch instruction and the next instruction to be executed if the condition is met.

16-bit Displacement field — Allows a larger displacement than 8 bits. Used only if the 8-bit displacement is equal to zero.

Note: A short branch to the immediately following instruction cannot be done because it would result in a zero offset which forces a word branch instruction definition.

BSET

Test a Bit and Set

BSET

Operation: ~(<bit number>) OF Destination→Z
1→<bit number> OF Destination

Assembler BSET Dn, <ea>
Syntax: BSET #<data>, <ea>

Attributes: Size = (Byte, Long)

Description: A bit in the destination operand is tested and the state of the specified bit is reflected in the Z condition code. After the test, the specified bit ıs set in the destination. If a data register is the destination, then the bit numbering is modulo 32, allowing bit manipulation on all bits in a data register. If a memory location is the destination, a byte is read from that location, the bit operation performed using the bit number modulo 8, and the byte written back to the location with zero referring to the least-significant bit. The bit number for this operation may be specified in two different ways:
1. Immediate — the bit number is specified in a second word of the instruction.
2. Register — the bit number is contained in a data register specified in the instruction.

Condition Codes:

X	N	Z	V	C
—	—	*	—	—

N Not affected.
Z Set if the bit tested is zero. Cleared otherwise.
V Not affected.
C Not affected.
X Not affected.

Instruction Format (Bit Number Dynamic):

15	14	13	12	11	10	9	8	7	6	5	4	3	2	1	0
0	0	0	0	Register			1	1	1	Effective Address Mode			Register		

Instruction Fields (Bit Number Dynamic):

Register field — Specifies the data register whose content is the bit number.

Effective Address field — Specifies the destination location. Only data alterable addressing modes are allowed as shown:

Addressing Mode	Mode	Register	Addressing Mode	Mode	Register
Dn*	000	register number	d(An, Xi)	110	register number
An	—	—	Abs.W	111	000
(An)	010	register number	Abs.L	111	001
(An)+	011	register number	d(PC)	—	—
−(An)	100	register number	d(PC, Xi)	—	—
d(An)	101	register number	Imm	—	—

*Long only; all others are byte only

— Continued —

BSET

Test a Bit and Set

BSET

Instruction Format (Bit Number Static):

15	14	13	12	11	10	9	8	7	6	5 4 3	2 1 0
0	0	0	0	1	0	0	0	1	1	Effective Address Mode	Register
0	0	0	0	0	0	0	0			bit number	

Instruction Fields (Bit Number Static):

Effective Address field — Specifies the destination location. Only data alterable addressing modes are allowed as shown:

Addressing Mode	Mode	Register	Addressing Mode	Mode	Register
Dn*	000	register number	d(An, Xi)	110	register number
An	—	—	Abs.W	111	000
(An)	010	register number	Abs.L	111	001
(An) +	011	register number	d(PC)	—	—
− (An)	100	register number	d(PC, Xi)	—	—
d(An)	101	register number	Imm	—	—

*Long only; all others are byte only.

bit number field — Specifies the bit number.

BSR

Branch to Subroutine

BSR

Operation: PC→ − (SP); PC + d → PC

**Assembler
Syntax:** BSR < label >

Attributes: Size = (Byte, Word)

Description: The long word address of the instruction immediately following the BSR instruction is pushed onto the system stack. Program execution then continues at location (PC) + displacement. Displacement is a two's-complement integer which counts the relative distances in bytes. The value in PC is the current instruction location plus two. If the 8-bit displacement in the instruction word is zero, then the 16-bit displacement (word immediately following the instruction) is used.

Condition Codes: Not affected.

Instruction Format:

15	14	13	12	11	10	9	8	7	6	5	4	3	2	1	0
0	1	1	0	0	0	0	1				8-bit Displacement				
16-bit Displacement if 8-bit Displacement = 0															

Instruction Fields:

8-bit Displacement field — Two's-complement integer specifying the relative distance (in bytes) between the branch instruction and the next instruction to be executed if the condition is met.

16-bit Displacement field — Allows a larger displacement than 8 bits. Used only if the 8-bit displacement is equal to zero.

Note: A short subroutine branch to the immediately following instruction cannot be done because it would result in a zero offset which forces a word branch instruction definition.

BTST

Test a Bit

BTST

Operation: ~(<bit number>) OF Destination → Z

Assembler
Syntax: BTST Dn, <ea>
BTST #<data>, <ea>

Attributes: Size = (Byte, Long)

Description: A bit in the destination operand is tested and the state of the specified bit is reflected in the Z condition code. If a data register is the destination, then the bit numbering is modulo 32, allowing bit manipulation on all bits in a data register. If a memory location is the destination, a byte is read from that location, and the bit operation performed using the bit number modulo 8 with zero referring to the least-significant bit. The bit number for this operation may be specified in two different ways:

1. Immediate — the bit number is specified in a second word of the instruction.
2. Register — the bit number is contained in a data register specified in the instruction.

Condition Codes:

X	N	Z	V	C
—	—	*	—	—

N Not affected.
Z Set if the bit tested is zero. Cleared otherwise.
V Not affected.
C Not affected.
X Not affected.

Instruction Format (Bit Number Dynamic):

15	14	13	12	11	10	9	8	7	6	5	4	3	2	1	0
0	0	0	0	Register			1	0	0	Effective Address Mode | Register					

Instruction Fields (Bit Number Dynamic):

Register field — Specifies the data register whose content is the bit number.

Effective Address field — Specifies the destination location. Only data addressing modes are allowed as shown:

Addressing Mode	Mode	Register	Addressing Mode	Mode	Register
Dn*	000	register number	d(An, Xi)	110	register number
An	—	—	Abs.W	111	000
(An)	010	register number	Abs.L	111	001
(An)+	011	register number	d(PC)	111	010
-(An)	100	register number	d(PC, Xi)	111	011
d(An)	101	register number	Imm	111	100

*Long only; all others are byte only.

— Continued —

BTST

Test a Bit

BTST

Instruction Format (Bit Number Static):

15	14	13	12	11	10	9	8	7	6	5 4 3	2 1 0
0	0	0	0	1	0	0	0	0	0	Effective Address Mode	Register
0	0	0	0	0	0	0	0			bit number	

Instruction Fields (Bit Number Static):

Effective Address field — Specifies the destination location. Only data addressing modes are allowed as shown:

Addressing Mode	Mode	Register	Addressing Mode	Mode	Register
Dn*	000	register number	d(An, Xi)	110	register number
An	—	—	Abs.W	111	000
(An)	010	register number	Abs.L	111	001
(An)+	011	register number	d(PC)	111	010
−(An)	100	register number	d(PC, Xi)	111	011
d(An)	101	register number	Imm	—	—

*Long only; all others are byte only.

bit number field — Specifies the bit number.

CHK   Check Register Against Bounds   CHK

Operation: If Dn<0 or Dn> (<ea>) then TRAP

**Assembler
Syntax:** CHK <ea>, Dn

Attributes: Size = (Word)

Description: The content of the low order word in the data register specified in the instruction is examined and compared to the upper bound. The upper bound is a two's-complement integer. If the register value is less than zero or greater than the upper bound contained in the operand word, then the processor initiates exception processing. The vector number is generated to reference the CHK instruction exception vector.

Condition Codes:

X	N	Z	V	C
—	*	U	U	U

N   Set if Dn<0; cleared if Dn> (<ea>). Undefined otherwise.
Z   Undefined.
V   Undefined.
C   Undefined.
X   Not affected.

Instruction Format:

15	14	13	12	11	10	9	8	7	6	5	4	3	2	1	0
0	1	0	0	Register			1	1	0	Effective Address Mode | Register					

Instruction Fields:
Register field — Specifies the data register whose content is checked.
Effective Address field — Specifies the upper bound operand word. Only data addressing modes are allowed as shown:

Addressing Mode	Mode	Register	Addressing Mode	Mode	Register
Dn	000	register number	d(An, Xi)	110	register number
An	—	—	Abs.W	111	000
(An)	010	register number	Abs.L	111	001
(An) +	011	register number	d(PC)	111	010
− (An)	100	register number	d(PC, Xi)	111	011
d(An)	101	register number	Imm	111	100

CLR

Clear an Operand

CLR

Operation: 0 → Destination

**Assembler
Syntax:** CLR < ea >

Attributes: Size = (Byte, Word, Long)

Description: The destination is cleared to all zero bits. The size of the operation may be specified to be byte, word, or long.

Condition Codes:

X	N	Z	V	C
—	0	1	0	0

N Always cleared.
Z Always set.
V Always cleared.
C Always cleared.
X Not affected.

Instruction Format:

15	14	13	12	11	10	9	8	7	6	5	4	3	2	1	0
0	1	0	0	0	0	1	0	Size		Effective Address Mode | Register					

Instruction Fields:

Size field — Specifies the size of the operation:
00 — byte operation.
01 — word operation.
10 — long operation.

Effective Address field — Specifies the destination location. Only data alterable addressing modes are allowed as shown:

Addressing Mode	Mode	Register	Addressing Mode	Mode	Register
Dn	000	register number	d(An, Xi)	110	register number
An	—	—	Abs.W	111	000
(An)	010	register number	Abs.L	111	001
(An)+	011	register number	d(PC)	—	—
−(An)	100	register number	d(PC, Xi)	—	—
d(An)	101	register number	Imm	—	—

Note: A memory destination is read before it is written to.

CMP

Compare

CMP

Operation: (Destination) − (Source)

**Assembler
Syntax:** CMP <ea>, Dn

Attributes: Size = (Byte, Word, Long)

Description: Subtract the source operand from the destination operand and set the condition codes according to the result; the destination location is not changed. The size of the operation may be specified to be byte, word, or long.

Condition Codes:

```
X  N  Z  V  C
—  *  *  *  *
```

N   Set if the result is negative. Cleared otherwise.
Z   Set if the result is zero. Cleared otherwise.
V   Set if an overflow is generated. Cleared otherwise.
C   Set if a borrow is generated. Cleared otherwise.
X   Not affected.

Instruction Format:

15	14	13	12	11	10	9	8	7	6	5	4	3	2	1	0
1	0	1	1	Register			Op-Mode			Effective Address Mode			Register		

Instruction Fields:

Register field — Specifies the destination data register.
Op-Mode field —

Byte	Word	Long	Operation
000	001	010	(< Dn >) − (< ea >)

Effective Address field — Specifies the source operand. All addressing modes are allowed as shown:

Addressing Mode	Mode	Register	Addressing Mode	Mode	Register
Dn	000	register number	d(An, Xi)	110	register number
An*	001	register number	Abs.W	111	000
(An)	010	register number	Abs.L	111	001
(An)+	011	register number	d(PC)	111	010
−(An)	100	register number	d(PC, Xi)	111	011
d(An)	101	register number	Imm	111	100

*Word and Long only.

Note: CMPA is used when the destination is an address register. CMPI is used when the source is immediate data. CMPM is used for memory to memory compares. Most assemblers automatically make this distinction.

CMPA

Compare Address

CMPA

Operation: (Destination) – (Source)

**Assembler
Syntax:** CMPA <ea>, An

Attributes: Size = (Word, Long)

Description: Subtract the source operand from the destination address register and set the condition codes according to the result; the address register is not changed. The size of the operation may be specified to be word or long. Word length source operands are sign extended to 32 bit quantities before the operation is done.

Condition Code:

X	N	Z	V	C
—	*	*	*	*

N   Set if the result is negative. Cleared otherwise.
Z   Set if the result is zero. Cleared otherwise.
V   Set if an overflow is generated. Cleared otherwise.
C   Set if a borrow is generated. Cleared otherwise.
X   Not affected.

Instruction Format:

15	14	13	12	11 10 9	8 7 6	5 4 3	2 1 0
1	0	1	1	Register	Op-Mode	Effective Address Mode	Register

Instruction Fields:

Register field — Specifies the destination address register.
Op-Mode field — Specifies the size of the operation:

011 — word operation. The source operand is sign-extended to a long operand and the operation is performed on the address register using all 32 bits.

111 — long operation.

Effective Address field — Specifies the source operand. All addressing modes are allowed as shown:

Addressing Mode	Mode	Register	Addressing Mode	Mode	Register
Dn	000	register number	d(An, Xi)	110	register number
An	001	register number	Abs.W	111	000
(An)	010	register number	Abs.L	111	001
(An)+	011	register number	d(PC)	111	010
–(An)	100	register number	d(PC, Xi)	111	011
d(An)	101	register number	Imm	111	100

CMPI

Compare Immediate

CMPI

Operation: (Destination) – Immediate Data

**Assembler
Syntax:** CMPI #<data>, <ea>

Attributes: Size = (Byte, Word, Long)

Description: Subtract the immediate data from the destination operand and set the condition codes according to the result; the destination location is not changed. The size of the operation may be specified to be byte, word, or long. The size of the immediate data matches the operation size.

Condition Codes:

X	N	Z	V	C
—	*	*	*	*

N Set if the result is negative. Cleared otherwise.
Z Set if the result is zero. Cleared otherwise.
V Set if an overflow is generated. Cleared otherwise.
C Set if a borrow is generated. Cleared otherwise.
X Not affected.

Instruction Format:

15	14	13	12	11	10	9	8	7	6	5	4	3	2	1	0
0	0	0	0	1	1	0	0	Size		Effective Address Mode | Register					

Word Data (16 bits)	Byte Data (8 bits)
Long Data (32 bits, including previous word)	

Instruction Fields:

Size field — Specifies the size of the operation:
  00 — byte operation.
  01 — word operation.
  10 — long operation.
Effective Address field — Specifies the destination operand. Only data alterable addressing modes are allowed as shown:

Addressing Mode	Mode	Register	Addressing Mode	Mode	Register
Dn	000	register number	d(An, Xi)	110	register number
An	—	—	Abs.W	111	000
(An)	010	register number	Abs.L	111	001
(An)+	011	register number	d(PC)	—	—
–(An)	100	register number	d(PC, Xi)	—	—
d(An)	101	register number	Imm	—	—

Immediate field — (Data immediately following the instruction):
  If size = 00, then the data is the low order byte of the immediate word.
  If size = 01, then the data is the entire immediate word.
  If size = 10, then the data is the next two immediate words.

CMPM   Compare Memory   CMPM

Operation: (Destination) − (Source)

**Assembler
Syntax:**   CMPM (Ay) + , (Ax) +

Attributes: Size = (Byte, Word, Long)

Description: Subtract the source operand from the destination operand, and set the condition codes according to the results; the destination location is not changed. The operands are always addressed with the postincrement addressing mode using the address registers specified in the instruction. The size of the operation may be specified to be byte, word, or long.

Condition Codes:

X	N	Z	V	C
—	*	*	*	*

N   Set if the result is negative. Cleared otherwise.
Z   Set if the result is zero. Cleared otherwise.
V   Set if an overflow is generated. Cleared otherwise.
C   Set if a borrow is generated. Cleared otherwise.
X   Not affected.

Instruction Format:

15	14	13	12	11	10	9	8	7	6	5	4	3	2	1	0
1	0	1	1	Register Rx			1	Size		0	0	1	Register Ry		

Instruction Fields:

Register Rx field — (always the destination) Specifies an address register for the postincrement addressing mode.
Size field — Specifies the size of the operation:
00 — byte operation.
01 — word operation.
10 — long operation.
Register Ry field — (always the source) Specifies an address register for the postincrement addressing mode.

DBcc   Test Condition, Decrement, and Branch   DBcc

Operation: If (condition false)
      then $Dn - 1 \rightarrow Dn$;
   If $Dn \neq -1$
      then $PC + d \rightarrow PC$
else $PC + 2 \rightarrow PC$ (Fall through to next instruction)

**Assembler
Syntax:**   DBcc Dn, <label>

Attributes: Size = (Word)

Description: This instruction is a looping primitive of three parameters: a condition, a data register, and a displacement. The instruction first tests the condition to determine if the termination condition for the loop has been met, and if so, no operation is performed. If the termination condition is not true, the low order 16 bits of the counter data register are decremented by one. If the result is -1, the counter is exhausted and execution continues with the next instruction. If the result is not equal to -1, execution continues at the location indicated by the current value of PC plus the sign-extended 16-bit displacement. The value in PC is the current instruction location plus two "cc" may specify the following conditions:

CC	carry clear	0100	$\overline{C}$	LS	low or same	0011	$C + Z$
CS	carry set	0101	C	LT	less than	1101	$N \cdot \overline{V} + \overline{N} \cdot V$
EQ	equal	0111	Z	MI	minus	1011	N
F	false	0001	0	NE	not equal	0110	$\overline{Z}$
GE	greater or equal	1100	$N \cdot V + \overline{N} \cdot \overline{V}$	PL	plus	1010	$\overline{N}$
GT	greater than	1110	$N \cdot V \cdot \overline{Z} + \overline{N} \cdot \overline{V} \cdot \overline{Z}$	T	true	0000	1
HI	high	0010	$\overline{C} \cdot \overline{Z}$	VC	overflow clear	1000	$\overline{V}$
LE	less or equal	1111	$Z + N \cdot \overline{V} + \overline{N} \cdot V$	VS	overflow set	1001	V

Condition Codes: Not affected.

Instruction Format:

15	14	13	12	11	10	9	8	7	6	5	4	3	2	1	0
0	1	0	1	Condition				1	1	0	0	1	Register		
Displacement															

Instruction Fields:
Condition field — One of the sixteen conditions discussed in description.
Register field — Specifies the data register which is the counter.
Displacement field — Specifies the distance of the branch (in bytes).

Notes: 1. The terminating condition is like that defined by the UNTIL loop constructs of high-level languages. For example: DBMI can be stated as "decrement and branch until minus."

— Continued —

DBcc   Test Condition, Decrement and Branch   DBcc

Notes: (Continued)

2. Most assemblers accept DBRA for DBF for use when no condition is required for termination of a loop.

3. There are two basic ways of entering a loop; at the beginning or by branching to the trailing DBcc instruction. If a loop structure terminated with DBcc is entered at the beginning, the control index count must be one less than the number of loop executions desired. This count is useful for indexed addressing modes and dynamically specified bit operations. However, when entering a loop by branching directly to the trailing DBcc instruction, the control index should equal the loop execution count. In this case, if a zero count occurs, the DBcc instruction will not branch causing complete bypass of the main loop.

DIVS

Signed Divide

DIVS

Operation: (Destination)/(Source) → Destination

Assembler Syntax: DIVS <ea>, Dn

Attributes: Size = (Word)

Description: Divide the destination operand by the source operand and store the result in the destination. The destination operand is a long operand (32 bits) and the source operand is a word operand (16 bits). The operation is performed using signed arithmetic. The result is a 32-bit result such that:
1. The quotient is in the lower word (least significant 16-bits).
2. The remainder is in the upper word (most significant 16-bits).

The sign of the remainder is always the same as the dividend unless the remainder is equal to zero. Two special conditions may arise:
1. Division by zero causes a trap.
2. Overflow may be detected and set before completion of the instruction. If overflow is detected, the condition is flagged but the operands are unaffected.

Condition Codes:

X	N	Z	V	C
—	*	*	*	0

N   Set if the quotient is negative. Cleared otherwise. Undefined if overflow.
Z   Set if the quotient is zero. Cleared otherwise. Undefined if overflow.
V   Set if division overflow is detected. Cleared otherwise.
C   Always cleared.
X   Not affected.

Instruction Format:

15	14	13	12	11 10 9	8	7	6	5 4 3 2 1 0
1	0	0	0	Register	1	1	1	Effective Address — Mode | Register

Instruction Fields:

Register field — Specifies any of the eight data registers. This field always specifies the destination operand.

Effective Address field — Specifies the source operand. Only data addressing modes are allowed as shown:

Addressing Mode	Mode	Register	Addressing Mode	Mode	Register
Dn	000	register number	d(An, Xi)	110	register number
An	—	—	Abs.W	111	000
(An)	010	register number	Abs.L	111	001
(An)+	011	register number	d(PC)	111	010
−(An)	100	register number	d(PC, Xi)	111	011
d(An)	101	register number	Imm	111	100

Note: Overflow occurs if the quotient is larger than a 16-bit signed integer.

DIVU

Unsigned Divide

DIVU

Operation: (Destination)/(Source) → Destination

**Assembler
Syntax:** DIVU <ea>, Dn

Attributes: Size = (Word)

Description: Divide the destination operand by the source operand and store the result in the destination. The destination operand is a long operand (32 bits) and the source operand is a word (16 bit) operand. The operation is performed using unsigned arithmetic. The result is a 32-bit result such that:
1. The quotient is in the lower word (least significant 16 bits).
2. The remainder is in the upper word (most significant 16 bits).
Two special conditions may arise:
1. Division by zero causes a trap.
2. Overflow may be detected and set before completion of the instruction. If overflow is detected, the condition is flagged but the operands are unaffected.

Condition Codes:

X	N	Z	V	C
—	*	*	*	0

N Set if the most significant bit of the quotient is set. Cleared otherwise. Undefined if overflow.
Z Set if the quotient is zero. Cleared otherwise. Undefined if overflow.
V Set if division overflow is detected. Cleared otherwise.
C Always cleared.
X Not affected.

Instruction Format:

15	14	13	12	11	10	9	8	7	6	5	4	3	2	1	0
1	0	0	0	Register			0	1	1	Effective Address Mode			Register		

Instruction Fields:

Register field — specifies any of the eight data registers. This field always specifies the destination operand.

Effective Address field — Specifies the source operand. Only data addressing modes are allowed as shown:

Addressing Mode	Mode	Register	Addressing Mode	Mode	Register
Dn	000	register number	d(An, Xi)	110	register number
An	—	—	Abs.W	111	000
(An)	010	register number	Abs.L	111	001
(An) +	011	register number	d(PC)	111	010
− (An)	100	register number	d(PC, Xi)	111	011
d(An)	101	register number	Imm	111	100

Note: Overflow occurs if the quotient is larger than a 16-bit unsigned integer.

EOR

Exclusive OR Logical

EOR

Operation: (Source) ⊕ (Destination) → Destination

**Assembler
Syntax:** EOR Dn, <ea>

Attributes: Size = (Byte, Word, Long)

Description: Exclusive OR the source operand to the destination operand and store the result in the destination location. The size of the operation may be specified to be byte, word, or long. This operation is restricted to data registers as the source operand. The destination operand is specified in the effective address field.

Condition Codes:

X	N	Z	V	C
—	*	*	0	0

N   Set if the most significant bit of the result is set. Cleared otherwise.
Z   Set if the result is zero. Cleared otherwise.
V   Always cleared.
C   Always cleared.
X   Not affected.

Instruction Format:

15	14	13	12	11 10 9	8 7 6	5 4 3	2 1 0
1	0	1	1	Register	Op-Mode	Effective Address Mode	Register

Instruction Fields:

Register field — Specifies any of the eight data registers.
Op-Mode field —

Byte	Word	Long	Operation
100	101	110	(<ea>) ⊕ (<Dx>) → <ea>

Effective Address field — Specifies the destination operand. Only data alterable addressing modes are allowed as shown:

Addressing Mode	Mode	Register	Addressing Mode	Mode	Register
Dn	000	register number	d(An, Xi)	110	register number
An	—	—	Abs.W	111	000
(An)	010	register number	Abs.L	111	001
(An) +	011	register number	d(PC)	—	—
− (An)	100	register number	d(PC, Xi)	—	—
d(An)	101	register number	Imm	—	—

Note: Memory to data register operations are not allowed. EORI is used when the source is immediate data. Most assemblers automatically make this distinction.

EORI

Exclusive OR Immediate

EORI

Operation: Immediate Data ⊕ (Destination) → Destination

Assembler Syntax: EORI #<data>, <ea>

Attributes: Size = (Byte, Word, Long)

Description: Exclusive OR the immediate data to the destination operand and store the result in the destination location. The size of the operation may be specified to be byte, word, or long. The immediate data matches the operation size.

Condition Codes:

X	N	Z	V	C
—	*	*	0	0

N Set if the most significant bit of the result is set. Cleared otherwise.
Z Set if the result is zero. Cleared otherwise.
V Always cleared.
C Always cleared.
X Not affected.

Instruction Format:

15	14	13	12	11	10	9	8	7	6	5	4	3	2	1	0
0	0	0	0	1	0	1	0	Size		Effective Address Mode | Register					
Word Data (16 bits)								Byte Data (8 bits)							
Long Data (32 bits, including previous word)															

Instruction Fields:

Size field — Specifies the size of the operation:
00 — byte operation.
01 — word operation.
10 — long operation.

Effective Address field — Specifies the destination operand. Only data alterable addressing modes are allowed as shown:

Addressing Mode	Mode	Register	Addressing Mode	Mode	Register
Dn	000	register number	d(An, Xi)	110	register number
An	—	—	Abs.W	111	000
(An)	010	register number	Abs.L	111	001
(An)+	011	register number	d(PC)	—	—
−(An)	100	register number	d(PC, Xi)	—	—
d(An)	101	register number	Imm	—	—

Immediate field — (Data immediately following the instruction):
If size = 00, then the data is the low order byte of the immediate word.
If size = 01, then the data is the entire immediate word.
If size = 10, then the data is the next two immediate words.

EORI to CCR

Exclusive OR Immediate to Condition Codes

EORI to CCR

Operation: (Source) ⊕ CCR → CCR

Assembler Syntax: EORI #xxx, CCR

Attributes: Size = (Byte)

Description: Exclusive OR the immediate operand with the condition codes and store the result in the low-order byte of the status register.

Condition Codes:

X	N	Z	V	C
*	*	*	*	*

N Changed if bit 3 of immediate operand is one. Unchanged otherwise.
Z Changed if bit 2 of immediate operand is one. Unchanged otherwise.
V Changed if bit 1 of immediate operand is one. Unchanged otherwise.
C Changed if bit 0 of immediate operand is one. Unchanged otherwise.
X Changed if bit 4 of immediate operand is one. Unchanged otherwise.

Instruction Format:

15	14	13	12	11	10	9	8	7	6	5	4	3	2	1	0
0	0	0	0	1	0	1	0	0	0	1	1	1	1	0	0
0	0	0	0	0	0	0	0	Byte Data (8 bits)							

EORI
to SR

Exclusive OR Immediate to the Status Register
(Privileged Instruction)

EORI
to SR

Operation: If supervisor state
        then (Source) $\oplus$ SR $\rightarrow$ SR
        else TRAP

Assembler
Syntax: EORI #xxx, SR

Attributes: Size = (Word)

Description: Exclusive OR the immediate operand with the contents of the status register and store the result in the status register. All bits of the status register are affected.

Condition Codes:

X	N	Z	V	C
*	*	*	*	*

N   Changed if bit 3 of immediate operand is one. Unchanged otherwise.
Z   Changed if bit 2 of immediate operand is one. Unchanged otherwise.
V   Changed if bit 1 of immediate operand is one. Unchanged otherwise.
C   Changed if bit 0 of immediate operand is one. Unchanged otherwise.
X   Changed if bit 4 of immediate operand is one. Unchanged otherwise.

Instruction Format:

15	14	13	12	11	10	9	8	7	6	5	4	3	2	1	0
0	0	0	0	1	0	1	0	0	1	1	1	1	1	0	0
Word Data (16 bits)															

EXG

Exchange Registers

EXG

Operation: Rx ↔ Ry

**Assembler
Syntax:** EXG Rx, Ry

Attributes: Size = (Long)

Description: Exchange the contents of two registers. This exchange is always a long (32 bit) operation. Exchange works in three modes:
1. Exchange data registers.
2. Exchange address registers.
3. Exchange a data register and an address register.

Condition Codes: Not affected.

Instruction Format:

15	14	13	12	11	10	9	8	7	6	5	4	3	2	1	0
1	1	0	0	Register Rx			1	Op-Mode					Register Ry		

Instruction Fields:

Register Rx field — Specifies either a data register or an address register depending on the mode. If the exchange is between data and address registers, this field always specifies the data register.

Op-Mode field — Specifies whether exchanging:
01000 — data registers.
01001 — address registers.
10001 — data register and address register.

Register Ry field — Specifies either a data register or an address register depending on the mode. If the exchange is between data and address registers, this field always specifies the address register.

EXT

Sign Extend

EXT

Operation: (Destination) Sign-extended → Destination

**Assembler
Syntax:**   EXT Dn

Attributes: Size = (Word, Long)

Description: Extend the sign bit of a data register from a byte to a word or from a word to a long operand depending on the size selected. If the operation is word sized, bit [7] of the designated data register is copied to bits [15:8] of that data register. If the operation is long sized, bit [15] of the designated data register is copied to bits [31:16] of that data register.

Condition Codes:

X	N	Z	V	C
—	*	*	0	0

N   Set if the result is negative. Cleared otherwise.
Z   Set if the result is zero. Cleared otherwise.
V   Always cleared.
C   Always cleared.
X   Not affected.

Instruction Format:

15	14	13	12	11	10	9	8	7	6	5	4	3	2	1	0
0	1	0	0	1	0	0	Op-Mode			0	0	0	Register		

Instruction Fields:

Op-Mode Field — Specifies the size of the sign-extension operation:
010 — Sign-extend low order byte of data register to word.
011 — Sign-extend low order word of data register to long.
Register field — Specifies the data register whose content is to be sign-extended.

ILLEGAL   Illegal Instruction   ILLEGAL

Operation: PC→ − (SSP); SR→ − (SSP)
(Illegal Instruction Vector)→ PC

Attributes: None

Description: This bit pattern causes an illegal instruction exception. All other illegal instruction bit patterns are reserved for future extension of the instruction set.

Condition Codes: Not affected.

Instruction Format:

15	14	13	12	11	10	9	8	7	6	5	4	3	2	1	0
0	1	0	0	1	0	1	0	1	1	1	1	1	1	0	0

JMP

Jump

JMP

Operation: Destination → PC

**Assembler
Syntax:** JMP <ea>

Attributes: Unsized

Description: Program execution continues at the effective address specified by the instruction. The address is specified by the control addressing modes.

Condition Codes: Not affected.

Instruction Format:

15	14	13	12	11	10	9	8	7	6	5 4 3	2 1 0
0	1	0	0	1	1	1	0	1	1	Effective Address Mode	Register

Instruction Fields:

Effective Address field — Specifies the address of the next instruction. Only control addressing modes are allowed as shown:

Addressing Mode	Mode	Register	Addressing Mode	Mode	Register
Dn	—	—	d(An, Xi)	110	register number
An	—	—	Abs.W	111	000
(An)	010	register number	Abs.L	111	001
(An) +	—	—	d(PC)	111	010
− (An)	—	—	d(PC, Xi)	111	011
d(An)	101	register number	Imm	—	—

JSR

Jump to Subroutine

JSR

Operation: PC $\rightarrow$ $-$(SP); Destination $\rightarrow$ PC

**Assembler
Syntax:**   JSR <ea>

Attributes: Unsized

Description: The long word address of the instruction immediately following the JSR instruction is pushed onto the system stack. Program execution then continues at the address specifed in the instruction.

Condition Codes: Not affected.

Instruction Format:

15	14	13	12	11	10	9	8	7	6	5 4 3	2 1 0
0	1	0	0	1	1	1	0	1	0	Effective Address Mode	Register

Instruction Fields:

Effective Address field — Specifies the address of the next instruction. Only control addressing modes are allowed as shown:

Addressing Mode	Mode	Register	Addressing Mode	Mode	Register
Dn	—	—	d(An, Xi)	110	register number
An	—	—	Abs.W	111	000
(An)	010	register number	Abs.L	111	001
(An)+	—	—	d(PC)	111	010
-(An)	—	—	d(PC, Xi)	111	011
d(An)	101	register number	Imm	—	—

LEA

Load Effective Address

LEA

Operation: Destination → An

**Assembler
Syntax:** LEA <ea>, An

Attributes: Size = (Long)

Description: The effective address is loaded into the specified address register. All 32
bits of the address register are affected by this instruction.

Condition Codes: Not affected.

Instruction Format:

15	14	13	12	11	10	9	8	7	6	5	4	3	2	1	0
0	1	0	0	Register			1	1	1	Effective Address Mode			Register		

Instruction Fields:

Register field — Specifies the address register which is to be loaded with
the effective address.

Effective Address field — Specifies the address to be loaded into the address register. Only control addressing modes are allowed as shown:

Addressing Mode	Mode	Register	Addressing Mode	Mode	Register
Dn	—	—	d(An, Xi)	110	register number
An	—	—	Abs.W	111	000
(An)	010	register number	Abs.L	111	001
(An)+	—	—	d(PC)	111	010
−(An)	—	—	d(PC, Xi)	111	011
d(An)	101	register number	Imm	—	—

LINK

Link and Allocate

LINK

Operation: An → − (SP); SP → An; SP + d → SP

**Assembler
Syntax:** LINK An, #<displacement>

Attributes: Unsized

Description: The current content of the specified address register is pushed onto the stack. After the push, the address register is loaded from the updated stack pointer. Finally, the 16-bit sign-extended displacement is added to the stack pointer. The content of the address register occupies two words on the stack. A negative displacement is specified to allocate stack area.

Condition Codes: Not affected.

Instruction Format:

15	14	13	12	11	10	9	8	7	6	5	4	3	2	1	0
0	1	0	0	1	1	1	0	0	1	0	1	0	Register		
Displacement															

Instruction Fields:

Register field — Specifies the address register through which the link is to be constructed.

Displacement field — Specifies the two's-complement integer which is to be added to the stack pointer.

Note: LINK and UNLK can be used to maintain a linked list of local data and parameter areas on the stack for nested subroutine calls.

LSL, LSR   Logical Shift   LSL, LSR

Operation: (Destination) Shifted by <count> → Destination

**Assembler
Syntax:** LSd Dx, Dy
LSd #<data>, Dy
LSd <ea>

Attributes: Size = (Byte, Word, Long)

Description: Shift the bits of the operand in the direction specified. The carry bit receives the last bit shifted out of the operand. The shift count for the shifting of a register may be specified in two different ways:

1. Immediate — the shift count is specified in the instruction (shift range 1-8).
2. Register — the shift count is contained in a data register specified in the instruction.

The size of the operation may be specified to be byte, word, or long. The content of memory may be shifted one bit only and the operand size is restricted to a word.

For LSL, the operand is shifted left; the number of positions shifted is the shift count. Bits shifted out of the high order bit go to both the carry and the extend bits; zeroes are shifted into the low order bit.

LSL:

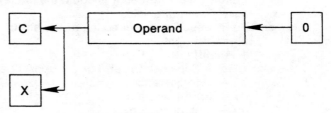

For LSR, the operand is shifted right; the number of positions shifted is the shift count. Bits shifted out of the low order bit go to both the carry and the extend bits; zeroes are shifted into the high order bit.

LSR:

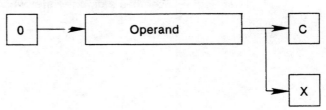

— Continued —

LSL, LSR   Logical Shift   LSL, LSR

Condition Codes:

X	N	Z	V	C
*	*	*	0	*

N   Set if the result is negative. Cleared otherwise.
Z   Set if the result is zero. Cleared otherwise.
V   Always cleared.
C   Set according to the last bit shifted out of the operand. Cleared for a shift count of zero.
X   Set according to the last bit shifted out of the operand. Unaffected for a shift count of zero.

Instruction Format (Register Shifts):

15	14	13	12	11 10 9	8	7 6	5	4	3	2 1 0
1	1	1	0	Count/ Register	dr	Size	i/r	0	1	Register

Instruction Fields (Register Shifts):

Count/Register field —
   If i/r = 0, the shift count is specified in this field. The values 0, 1-7 represent a range of 8, 1 to 7 respectively.
   If i/r = 1, the shift count (modulo 64) is contained in the data register specified in this field.

dr field — Specifies the direction of the shift:
   0 — shift right.
   1 — shift left.

Size field — Specifies the size of the operation:
   00 — byte operation.
   01 — word operation.
   10 — long operation.

i/r field —
   If i/r = 0, specifies immediate shift count.
   If i/r = 1, specifies register shift count.

Register field — Specifies a data register whose content is to be shifted.

— Continued —

LSL, LSR   Logical Shift   LSL,LSR

Instruction Format (Memory Shifts):

15	14	13	12	11	10	9	8	7	6	5	4	3	2	1	0
1	1	1	0	0	0	1	dr	1	1	Effective Address Mode | Register					

Instruction Fields (Memory Shifts):

dr field — Specifies the direction of the shift:

0 — shift right.

1 — shift left.

Effective Address field — Specifies the operand to be shifted. Only memory alterable addressing modes are allowed as shown:

Addressing Mode	Mode	Register	Addressing Mode	Mode	Register
Dn	—	—	d(An, Xi)	110	register number
An	—	—	Abs.W	111	000
(An)	010	register number	Abs.L	111	001
(An) +	011	register number	d(PC)	—	—
− (An)	100	register number	d(PC, Xi)	—	—
d(An)	101	register number	Imm	—	—

MOVE

MOVE Move Data from Source to Destination

Operation: (Source) → Destination

**Assembler
Syntax:** MOVE < ea >, < ea >

Attributes: Size = (Byte, Word, Long)

Description: Move the content of the source to the destination location. The data is examined as it is moved, and the condition codes set accordingly. The size of the operation may be specified to be byte, word, or long.

Condition Codes:

X	N	Z	V	C
—	*	*	0	0

N Set if the result is negative. Cleared otherwise.
Z Set if the result is zero. Cleared otherwise.
V Always cleared.
C Always cleared.
X Not affected.

Instruction Format:

15	14	13	12	11	10	9	8	7	6	5	4	3	2	1	0
0	0	Size		Destination Register | Mode						Source Mode | Register					

Instruction Fields:

Size field — Specifies the size of the operand to be moved:

01 — byte operation.
11 — word operation.
10 — long operation.

Destination Effective Address field — Specifies the destination location. Only data alterable addressing modes are allowed as shown:

Addressing Mode	Mode	Register	Addressing Mode	Mode	Register
Dn	000	register number	d(An, Xi)	110	register number
An	—	—	Abs.W	111	000
(An)	010	register number	Abs.L	111	001
(An) +	011	register number	d(PC)	—	—
− (An)	100	register number	d(PC, Xi)	—	—
d(An)	101	register number	Imm	—	—

— Continued —

MOVE

MOVE Move Data from Source to Destination **MOVE**

Instruction Fields: (Continued)

Source Effective Address field — Specifies the source operand. All addressing modes are allowed as shown:

Addressing Mode	Mode	Register	Addressing Mode	Mode	Register
Dn	000	register number	d(An, Xi)	110	register number
An*	001	register number	Abs.W	111	000
(An)	010	register number	Abs.L	111	001
(An)+	011	register number	d(PC)	111	010
−(An)	100	register number	d(PC, Xi)	111	011
d(An)	101	register number	Imm	111	100

*For byte size operation, address register direct is not allowed.

Notes: 1.   MOVEA is used when the destination is an address register. Most assemblers automatically make this distinction.
2.   MOVEQ can also be used for certain operations on data registers.

MOVE from CCR

**Move from the
Condition Code Register**

MOVE from CCR

Operation: CCR→Destination

**Assembler
Syntax:** MOVE CCR, <ea>

Attributes: Size = (Word)

Description: The content of the status register is moved to the destination location. The source operand is a word, but only the low order byte contains the condition codes. The upper byte is all zeros.

Condition Codes: Not affected.

Instruction Format:

15	14	13	12	11	10	9	8	7	6	5	4	3	2	1	0
0	1	0	0	0	0	1	0	1	1	Effective Mode			Address Register		

Instruction Fields:

Effective Address field — Specifies the destination location.
Only data alterable addressing modes are allowed as shown:

Addressing Mode	Mode	Register	Addressing Mode	Mode	Register
Dn	000	register number	d(An, Xi)	110	register number
An	—	—	Abs.W	111	000
(An)	010	register number	Abs.L	111	001
(An)+	011	register number	d(PC)	—	—
−(An)	100	register number	d(PC, Xi)	—	—
d(An)	101	register number	Imm	—	—

Note: MOVE to CCR is a word operation. AND, OR, and EOR to CCR are byte operations.

MC68010

MOVE to CCR

Move to Condition Codes

MOVE to CCR

Operation: (Source) → CCR

Assembler Syntax: MOVE <ea>, CCR

Attributes: Size = (Word)

Description: The content of the source operand is moved to the condition codes. The source operand is a word, but only the low order byte is used to update the condition codes. The upper byte is ignored.

Condition Codes:

X	N	Z	V	C
*	*	*	*	*

N Set the same as bit 3 of the source operand.
Z Set the same as bit 2 of the source operand.
V Set the same as bit 1 of the source operand.
C Set the same as bit 0 of the source operand.
X Set the same as bit 4 of the source operand.

Instruction Format:

15	14	13	12	11	10	9	8	7	6	5 4 3	2 1 0
0	1	0	0	0	1	0	0	1	1	Effective Address Mode	Register

Instruction Fields:

Effective Address field — Specifies the location of the source operand. Only data addressing modes are allowed as shown:

Addressing Mode	Mode	Register	Addressing Mode	Mode	Register
Dn	000	register number	d(An, Xi)	110	register number
An	—	—	Abs.W	111	000
(An)	010	register number	Abs.L	111	001
(An)+	011	register number	d(PC)	111	010
−(An)	100	register number	d(PC, Xi)	111	011
d(An)	101	register number	Imm	111	100

Note: MOVE to CCR is a word operation. AND, OR, and EOR to CCR are byte operations.

MOVE to SR

**Move to the Status Register
(Privileged Instruction)**

MOVE to SR

Operation: If supervisor state
then (Source) → SR
else TRAP

**Assembler
Syntax:** MOVE <ea>, SR

Attributes: Size = (Word)

Description: The content of the source operand is moved to the status register. The source operand is a word and all bits of the status register are affected.

Condition Codes: Set according to the source operand.

Instruction Format:

15	14	13	12	11	10	9	8	7	6	5 4 3	2 1 0
0	1	0	0	0	1	1	0	1	1	Effective Address Mode	Register

Instruction Fields:

Effective Address field — Specifies the location of the source operand. Only data addressing modes are allowed as shown:

Addressing Mode	Mode	Register	Addressing Mode	Mode	Register
Dn	000	register number	d(An, Xi)	110	register number
An	—	—	Abs.W	111	000
(An)	010	register number	Abs.L	111	001
(An) +	011	register number	d(PC)	111	010
− (An)	100	register number	d(PC, Xi)	111	011
d(An)	101	register number	Imm	111	100

MOVE from SR

Move from the Status Register

MOVE from SR

Operation: SR → Destination

Assembler Syntax: MOVE SR, <ea>

Attributes: Size = (Word)

Description: The content of the status register is moved to the destination location. The operand size is a word.

Condition Codes: Not affected.

Instruction Format:

15	14	13	12	11	10	9	8	7	6	5 4 3	2 1 0
0	1	0	0	0	0	0	0	1	1	Effective Address Mode	Register

Instruction Fields:

Effective Address field — Specifies the destination location. Only data alterable addressing modes are allowed as shown:

Addressing Mode	Mode	Register	Addressing Mode	Mode	Register
Dn	000	register number	d(An, Xi)	110	register number
An	—	—	Abs.W	111	000
(An)	010	register number	Abs.L	111	001
(An) +	011	register number	d(PC)	—	—
− (An)	100	register number	d(PC, Xi)	—	—
d(An)	101	register number	Imm	—	—

Note: A memory destination is read before it is written to.

MOVE from SR

**Move from the Status Register
(Privileged Instruction)**

MOVE from SR

Operation: If supervisor state
then SR → Destination
else TRAP

**Assembler
Syntax:** MOVE SR, <ea>

Attributes: Size = (Word)

Description: The content of the status register is moved to the destination location. The operand size is a word.

Condition Codes: Not affected.

Instruction Format:

15	14	13	12	11	10	9	8	7	6	5	4	3	2	1	0
0	1	0	0	0	0	0	0	1	1	Effective Address Mode			Register		

Instruction Fields:
Effective Address field — Specifies the destination location. Only data alterable addressing modes are allowed as shown:

Addressing Mode	Mode	Register	Addressing Mode	Mode	Register
Dn	000	register number	d(An, Xi)	110	register number
An	—	—	Abs.W	111	000
(An)	010	register number	Abs.L	111	001
(An) +	011	register number	d(PC)	—	—
– (An)	100	register number	d(PC, Xi)	—	—
d(An)	101	register number	Imm	—	—

NOTE: Use the MOVE from CCR instruction to access the condition codes.

MC68010

MOVE USP

**Move User Stack Pointer
(Privileged Instruction)**

MOVE USP

Operation: If supervisor state
             then USP → An;
                 An → USP
             else TRAP

Assembler   MOVE USP, An
Syntax: MOVE An, USP

Attributes: Size = (Long)

Description: The contents of the user stack pointer are transferred to or from the specified address register.

Condition Codes: Not affected.

Instruction Format:

15	14	13	12	11	10	9	8	7	6	5	4	3	2	1	0
0	1	0	0	1	1	1	0	0	1	1	0	dr	Register		

Instruction Fields:

dr field — Specifies the direction of transfer:
  0 — transfer the address register to the USP.
  1 — transfer the USP to the address register.
Register field — Specifies the address register to or from which the user stack pointer is to be transferred.

MOVEA   Move Address   MOVEA

Operation: (Source) → Destination

**Assembler
Syntax:** MOVEA <ea>, An

Attributes: Size = (Word, Long)

Description: Move the content of the source to the destination address register. The size of the operation may be specified to be word or long. Word size source operands are sign extended to 32 bit quantities before the operation is done.

Condition Codes: Not affected.

Instruction Format:

15	14	13	12	11	10	9	8	7	6	5	4	3	2	1	0
0	0	Size		Destination Register			0	0	1	Source Mode | Register					

Instruction Fields:

Size field — Specifies the size of the operand to be moved:
   11 — Word operation. The source operand is sign-extended to a long operand and all 32 bits are loaded into the address register.
   10 — Long operation.
Destination Register field — Specifies the destination address register.
Source Effective Address field — Specifies the location of the source operand. All addressing modes are allowed as shown:

Addressing Mode	Mode	Register	Addressing Mode	Mode	Register
Dn	000	register number	d(An, Xi)	110	register number
An	001	register number	Abs.W	111	000
(An)	010	register number	Abs.L	111	001
(An)+	011	register number	d(PC)	111	010
−(An)	100	register number	d(PC, Xi)	111	011
d(An)	101	register number	Imm	111	100

MOVEC Move to/from Control Register
(Privileged Instruction) # MOVEC

Operation: If supervisor state
then Rc → Rn, Rn → Rc
else TRAP

Assembler MOVEC Rc, Rn
Syntax: MOVEC Rn, Rc

Attributes: Size = (Long)

Description: Copy the contents of the specified control register to the specified general register or copy the contents of the specified general register to the specified control register. This is always a 32-bit transfer even though the control register may be implemented with fewer bits. Unimplemented bits are read as zeros.

Condition Codes: Not affected.

Instruction Format:

15	14	13	12	11	10	9	8	7	6	5	4	3	2	1	0
0	1	0	0	1	1	1	0	0	1	1	1	1	0	1	dr
A/D	Register			Control Register											

Instruction Fields:

dr field — Specifies the direction of the transfer:
0—control register to general register.
1—general register to control register.
A/D field — Specifies the type of general register:
0—data register.
1—address register.
Register field — Specifies the register number.
Control Register field — Specifies the control register.
Currently defined control registers are:

Binary	Hex	Name/Function
0000 0000 0000	000	Source Function Code (SFC) register.
0000 0000 0001	001	Destination Function Code (DFC) register.
1000 0000 0000	800	User Stack Pointer.
1000 0000 0001	801	Vector Base Register for exception vector table.

All other codes cause an illegal instruction exception.

MC68010

507

MOVEM Move Multiple Registers MOVEM

Operation: Registers → Destination
(Source) → Registers

Assembler MOVEM < register list > , < ea >
Syntax: MOVEM < ea > , < register list >

Attributes: Size = (Word, Long)

Description: Selected registers are transferred to or from consecutive memory location starting at the location specified by the effective address. A register is transferred if the bit corresponding to that register is set in the mask field. The instruction selects how much of each register is transferred; either the entire long word can be moved or just the low order word. In the case of a word transfer to the registers, each word is sign-extended to 32 bits (also data registers) and the resulting long word loaded into the associated register.

MOVEM allows three forms of address modes: the control modes, the predecrement mode, or the postincrement mode. If the effective address is in one of the control modes, the registers are transferred starting at the specified address and up through higher addresses. The order of transfer is from data register 0 to data register 7, then from address register 0 to address register 7.

If the effective address is in the predecrement mode, only a register to memory operation is allowed. The registers are stored starting at the specified address minus two and down through lower addresses. The order of storing is from address register 7 to address register 0, then from data register 7 to data register 0. The decremented address register is updated to contain the address of the last word stored.

If the effective address is in the postincrement mode, only a memory to register operation is allowed. The registers are loaded starting at the specified address and up through higher addresses. The order of loading is the same as for the control mode addressing. The incremented address register is updated to contain the address of the last word loaded plus two.

Condition Codes: Not affected.

Instruction Format:

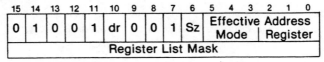

15	14	13	12	11	10	9	8	7	6	5 4 3	2 1 0
0	1	0	0	1	dr	0	0	1	Sz	Effective Address Mode	Register
Register List Mask												

— Continued —

MOVEM   Move Multiple Registers   MOVEM

Instruction Fields:

dr field:

Specifies the direction of the transfer:

0 — register to memory

1 — memory to register.

Sz field — Specifies the size of the registers being transferred:

0 — word transfer.

1 — long transfer.

Effective Address field — Specifies the memory address to or from which the registers are to be moved.

For register to memory transfer, only control alterable addressing modes or the predecrement addressing mode are allowed as shown:

Addressing Mode	Mode	Register	Addressing Mode	Mode	Register
Dn	—	—	d(An, Xi)	110	register number
An	—	—	Abs.W	111	000
(An)	010	register number	Abs.L	111	001
(An)+	—	—	d(PC)	—	—
−(An)	100	register number	d(PC, Xi)	—	—
d(An)	101	register number	Imm	—	—

For memory to register transfer, only control addressing modes or the postincrement addressing mode are allowed as shown:

Addressing Mode	Mode	Register	Addressing Mode	Mode	Register
Dn	—	—	d(An, Xi)	110	register number
An	—	—	Abs.W	111	000
(An)	010	register number	Abs.L	111	001
(An)+	011	register number	d(PC)	111	010
−(An)	—	—	d(PC, Xi)	111	011
d(An)	101	register number	Imm	—	—

Register List Mask field — Specifies which registers are to be transferred. The low order bit corresponds to the first register to be transferred; the high bit corresponds to the last register to be transferred. Thus, both for control modes and for the postincrement mode addresses, the mask correspondence is

15	14	13	12	11	10	9	8	7	6	5	4	3	2	1	0
A7	A6	A5	A4	A3	A2	A1	A0	D7	D6	D5	D4	D3	D2	D1	D0

while for the predecrement mode addresses, the mask correspondence is

15	14	13	12	11	10	9	8	7	6	5	4	3	2	1	0
D0	D1	D2	D3	D4	D5	D6	D7	A0	A1	A2	A3	A4	A5	A6	A7

Note: An extra read bus cycle occurs for memory operands. This amounts to a memory word at one address higher than expected being addressed during operation.

MOVEP   Move Peripheral Data   MOVEP

Operation: (Source) → Destination

Assembler   MOVEP Dx, d(Ay)
Syntax: MOVEP d(Ay), Dx

Attributes: Size = (Word, Long)

Description: Data is transferred between a data register and alternate bytes of memory, starting at the location specified and incrementing by two. The high order byte of the data register is transferred first and the low order byte is transferred last. The memory address is specified using the address register indirect plus displacement addressing mode. If the address is even, all the transfers are made on the high order half of the data bus; if the address is odd, all the transfers are made on the low order half of the data bus.

Example: Long transfer to/from an even address.

Byte organization in register

31	24	23	16	15	8	7	0
hi-order		mid-upper		mid-lower		low-order	

Byte organization in memory (low address at top)

15	14	13	12	11	10	9	8	7	6	5	4	3	2	1	0
hi-order															
mid-upper															
mid-lower															
low-order															

Example: Word transfer to/from an odd address.

Byte organization in register

31	24	23	16	15	8	7	0
				hi-order		low-order	

Byte organization in memory (low address at top)

15	14	13	12	11	10	9	8	7	6	5	4	3	2	1	0
								hi-order							
								low-order							

Condition Codes: Not affected.

— Continued —

MOVEP   Move Peripheral Data   MOVEP

Instruction Format:

15	14	13	12	11 10 9	8 7 6	5	4	3	2 1 0
0	0	0	0	Data Register	Op-Mode	0	0	1	Address Register
Displacement									

Instruction Fields:

Data Register field — Specifies the data register to or from which the data is to be transferred.

Op-Mode field — Specifies the direction and size of the operation:

100 — transfer word from memory to register.

101 — transfer long from memory to register.

110 — transfer word from register to memory.

111 — transfer long from register to memory.

Address Register field — Specifies the address register which is used in the address register indirect plus displacement addressing mode.

Displacement field — Specifies the displacement which is used in calculating the operand address.

MOVEQ

Move Quick

MOVEQ

Operation: Immediate Data → Destination

**Assembler
Syntax:** MOVEQ #<data>, Dn

Attributes: Size = (Long)

Description: Move immediate data to a data register. The data is contained in an 8-bit field within the operation word. The data is sign-extended to a long operand and all 32 bits are transferred to the data register.

Condition Codes:

X	N	Z	V	C
—	*	*	0	0

N Set if the result is negative. Cleared otherwise.
Z Set if the result is zero. Cleared otherwise.
V Always cleared.
C Always cleared.
X Not affected.

Instruction Format:

15	14	13	12	11	10	9	8	7	6	5	4	3	2	1	0
0	1	1	1	Register			0	Data							

Instruction Fields:

Register field — Specifies the data register to be loaded.
Data field — 8 bits of data which are sign extended to a long operand.

MOVES

Move to/from Address Space
(Privileged Instruction)

MOVES

Operation: If supervisor state
then Rn → Destination <DFC>
Source <SFC> → Rn
else TRAP

Assembler
Syntax:
MOVES Rn, <ea>
MOVES <ea>, Rn

Attributes: Size = (Byte, Word, Long)

Description: Move the byte, word, or long operand from the specified general register to a location within the address space specified by the destination function code (DFC) register. Or, move the byte, word, or long operand from a location within the address space specified by the source function code (SFC) register to the specified general register.

If the destination is a data register, the source operand replaces the corresponding low-order bits of the that data register. If the destination is an address register, the source operand is sign-extended to 32 bits and then loaded into that address register.

Condition Codes: Not affected.

Instruction Format:

15	14	13	12	11	10	9	8	7	6	5	4	3	2	1	0
0	0	0	0	1	1	1	0	Size		Effective Address					
A/D	Register			dr	0	0	0	0	0	0	0	0	0	0	0

Instruction Fields:

Size field — Specifies the size of the operation:
00—byte operation.
01—word operation.
10—long operation.

A/D field — Specifies the type of general register:
0—data register.
1—address register.
Register field — Specifies the register number.
dr field — Specifies the direction of the transfer:
0—from <ea> to general register.
1—from general register to <ea>.

MC68010

MOVES

**Move to/from Address Space
(Privileged Instruction)**

MOVES

Instruction Fields: (continued)

Effective Address field — Specifies the source or destination location within the alternate address space. Only alterable memory addressing modes are allowed as shown:

Addressing Mode	Mode	Register
Dn	—	—
An	—	—
(An)	010	register number
(An) +	011	register number
− (An)	100	register number
d(An)	101	register number

Addressing Mode	Mode	Register
d(An, Xi)	110	register number
Abs.W	111	000
Abs.L	111	001
d(PC)	—	—
d(PC, Xi)	—	—
Imm	—	—

MULS

Signed Multiply

MULS

Operation: (Source)*(Destination)→ Destination

**Assembler
Syntax:** MULS <ea>, Dn

Attributes: Size = (Word)

Description: Multiply two signed 16-bit operands yielding a 32-bit signed result. The operation is performed using signed arithmetic. A register operand is taken from the low order word; the upper word is unused. All 32 bits of the product are saved in the destination data register.

Condition Codes:

X	N	Z	V	C
—	*	*	0	0

N Set if the result is negative. Cleared otherwise.
Z Set if the result is zero. Cleared otherwise.
V Always cleared.
C Always cleared.
X Not affected.

Instruction Format:

15	14	13	12	11 10 9	8	7	6	5 4 3	2 1 0
1	1	0	0	Register	1	1	1	Effective Address Mode	Register

Instruction Fields:

Register field — Specifies one of the data registers. This field always specifies the destination.

Effective Address field — Specifies the source operand. Only data addressing modes are allowed as shown:

Addressing Mode	Mode	Register	Addressing Mode	Mode	Register
Dn	000	register number	d(An, Xi)	110	register number
An	—	—	Abs.W	111	000
(An)	010	register number	Abs.L	111	001
(An) +	011	register number	d(PC)	111	010
− (An)	100	register number	d(PC, Xi)	111	011
d(An)	101	register number	Imm	111	100

MULU

Unsigned Mulitply

MULU

Operation: (Source)*(Destination) → Destination

**Assembler
Syntax:** MULU <ea>, Dn

Attributes: Size = (Word)

Description: Multiply two unsigned 16-bit operands yielding a 32-bit unsigned result. The operation is performed using unsigned arithmetic. A register operand is taken from the low order word; the upper word is unused. All 32 bits of the product are saved in the destination data register.

Condition Codes:

X	N	Z	V	C
—	*	*	0	0

N   Set if the most significant bit of the result is set. Cleared otherwise.
Z   Set if the result is zero. Cleared otherwise.
V   Always cleared.
C   Always cleared.
X   Not affected.

Instruction Format:

15	14	13	12	11	10	9	8	7	6	5 4 3	2 1 0
1	1	0	0	Register			0	1	1	Effective Address Mode	Register

Instruction Fields:

Register field — Specifies one of the data registers. This field always specifies the destination.

Effective Address field — Specifies the source operand. Only data addressing modes are allowed as shown:

Addressing Mode	Mode	Register	Addressing Mode	Mode	Register
Dn	000	register number	d(An, Xi)	110	register number
An	—	—	Abs.W	111	000
(An)	010	register number	Abs.L	111	001
(An)+	011	register number	d(PC)	111	010
–(An)	100	register number	d(PC, Xi)	111	011
d(An)	101	register number	Imm	111	100

NBCD

Negate Decimal with Extend

NBCD

Operation: $0 - (Destination)_{10} - X \rightarrow$ Destination

**Assembler
Syntax:** NBCD <ea>

Attributes: Size = (Byte)

Description: The operand addressed as the destination and the extend bit are subtracted from zero. The operation is performed using decimal arithmetic. The result is saved in the destination location. This instruction produces the ten's complement of the destination if the extend bit is clear, the nine's complement if the extend bit is set. This is a byte operation only.

Condition Codes:

X	N	Z	V	C
*	U	*	U	*

N   Undefined.
Z   Cleared if the result is non-zero. Unchanged otherwise.
V   Undefined.
C   Set if a borrow (decimal) was generated. Cleared otherwise.
X   Set the same as the carry bit.

NOTE

Normally the Z condition code bit is set via programming before the start of an operation. This allows successful tests for zero results upon completion of multiple-precision operations.

Instruction Format:

15	14	13	12	11	10	9	8	7	6	5	4	3	2	1	0
0	1	0	0	1	0	0	0	0	0	Effective Address Mode | Register				

Instruction Fields:

Effective Address field — Specifies the destination operand. Only data alterable addressing modes are allowed as shown:

Addressing Mode	Mode	Register	Addressing Mode	Mode	Register
Dn	000	register number	d(An, Xi)	110	register number
An	—	—	Abs.W	111	000
(An)	010	register number	Abs.L	111	001
(An)+	011	register number	d(PC)	—	—
−(An)	100	register number	d(PC, Xi)	—	—
d(An)	101	register number	Imm	—	—

NEG

Negate

NEG

Operation: 0 – (Destination) → Destination

Assembler Syntax: NEG <ea>

Attributes: Size = (Byte, Word, Long)

Description: The operand addressed as the destination is subtracted from zero. The result is stored in the destination location. The size of the operation may be specified to be byte, word, or long.

Condition Codes:

X	N	Z	V	C
*	*	*	*	*

N   Set if the result is negative. Cleared otherwise.
Z   Set if the result is zero. Cleared otherwise.
V   Set if an overflow is generated. Cleared otherwise.
C   Cleared if the result is zero. Set otherwise.
X   Set the same as the carry bit.

Instruction Format:

15	14	13	12	11	10	9	8	7	6	5	4	3	2	1	0
0	1	0	0	0	1	0	0	Size		Effective Address Mode			Register		

Instruction Fields:

Size field — Specifies the size of the operation:
00 — byte operation.
01 — word operation.
10 — long operation.

Effective Address field — Specifies the destination operand. Only data alterable addressing modes are allowed as shown:

Addressing Mode	Mode	Register	Addressing Mode	Mode	Register
Dn	000	register number	d(An, Xi)	110	register number
An	—	—	Abs.W	111	000
(An)	010	register number	Abs.L	111	001
(An) +	011	register number	d(PC)	—	—
– (An)	100	register number	d(PC, Xi)	—	—
d(An)	101	register number	Imm	—	—

NEGX   Negate with Extend   NEGX

Operation: 0 − (Destination) − X → Destination

Assembler Syntax: NEGX <ea>

Attributes: Size = (Byte, Word, Long)

Description: The operand addressed as the destination and the extend bit are subtracted from zero. The result is stored in the destination location. The size of the operation may be specified to be byte, word, or long.

Condition Codes:

X	N	Z	V	C
*	*	*	*	*

N   Set if the result is negative. Cleared otherwise.
Z   Cleared if the result is non-zero. Unchanged otherwise.
V   Set if an overflow is generated. Cleared otherwise.
C   Set if a borrow is generated. Cleared otherwise.
X   Set the same as the carry bit.

NOTE

Normally the Z condition code bit is set via programming before the start of an operation. This allows successful tests for zero results upon completion of multiple-precision operations.

Instruction Format:

15	14	13	12	11	10	9	8	7	6	5	4	3	2	1	0
0	1	0	0	0	0	0	0	Size		Effective Address Mode | Register					

Instruction Fields:

Size field — Specifies the size of the operation:
00 — byte operation.
01 — word operation.
10 — long operation.

Effective Address field — Specifies the destination operand. Only data alterable addressing modes are allowed as shown:

Addressing Mode	Mode	Register	Addressing Mode	Mode	Register
Dn	000	register number	d(An, Xi)	110	register number
An	—	—	Abs.W	111	000
(An)	010	register number	Abs.L	111	001
(An) +	011	register number	d(PC)	—	—
− (An)	100	register number	d(PC, Xi)	—	—
d(An)	101	register number	Imm	—	—

NOP

No Operation

NOP

Operation: None

**Assembler
Syntax:** NOP

Attributes: Unsized

Description: No operation occurs. The processor state, other than the program counter, is unaffected. Execution continues with the instruction following the NOP instruction.

Condition Codes: Not affected.

Instruction Format:

15	14	13	12	11	10	9	8	7	6	5	4	3	2	1	0
0	1	0	0	1	1	1	0	0	1	1	1	0	0	0	1

NOT

Logical Complement

NOT

Operation: ~(Destination) → Destination

**Assembler
Syntax:** NOT <ea>

Attributes: Size = (Byte, Word, Long)

Description: The ones complement of the destination operand is taken and the result stored in the destination location. The size of the operation may be specified to be byte, word, or long.

Condition Codes:

X	N	Z	V	C
—	*	*	0	0

N   Set if the result is negative. Cleared otherwise.
Z   Set if the result is zero. Cleared otherwise.
V   Always cleared.
C   Always cleared.
X   Not affected.

Instruction Format:

15	14	13	12	11	10	9	8	7	6	5	4	3	2	1	0
0	1	0	0	0	1	1	0	Size		Effective Address Mode | Register					

Instruction Fields:

Size field — Specifies the size of the operation:
00 — byte operation.
01 — word operation.
10 — long operation.

Effective Address field — Specifies the destination operand. Only data alterable addressing modes are allowed as shown:

Addressing Mode	Mode	Register	Addressing Mode	Mode	Register
Dn	000	register number	d(An, Xi)	110	register number
An	—	—	Abs.W	111	000
(An)	010	register number	Abs.L	111	001
(An)+	011	register number	d(PC)	—	—
−(An)	100	register number	d(PC, Xi)	—	—
d(An)	101	register number	Imm	—	—

OR

Inclusive OR Logical

Operation: (Source) v (Destination) → Destination

Assembler OR <ea>, Dn
Syntax: OR Dn, <ea>

Attributes: Size = (Byte, Word, Long)

Description: Inclusive OR the source operand to the destination operand and store the result in the destination location. The size of the operation may be specified to be byte, word, or long. The contents of an address register may not be used as an operand.

Condition Codes:

X	N	Z	V	C
—	*	*	0	0

N Set if the most significant bit of the result is set. Cleared otherwise.
Z Set if the result is zero. Cleared otherwise.
V Always cleared.
C Always cleared.
X Not affected.

Instruction Format:

15	14	13	12	11 10 9	8 7 6	5 4 3	2 1 0
1	0	0	0	Register	Op-Mode	Effective Address Mode	Register

Instruction Fields:

Register field — Specifies any of the eight data registers.
Op-Mode field —

Byte	Word	Long	Operation
000	001	010	(<Dn>) v (<ea>) → <Dn>
100	101	110	(<ea>) v (<Dn>) → <ea>

Effective Address field —
If the location specified is a source operand then only data addressing modes are allowed as shown:

Addressing Mode	Mode	Register	Addressing Mode	Mode	Register
Dn	000	register number	d(An, Xi)	110	register number
An	—	—	Abs.W	111	000
(An)	010	register number	Abs.L	111	001
(An) +	011	register number	d(PC)	111	010
− (An)	100	register number	d(PC, Xi)	111	011
d(An)	101	register number	Imm	111	100

— Continued —

OR

Inclusive OR Logical

OR

Effective Address field (Continued)

If the location specified is a destination operand then only memory alterable addressing modes are allowed as shown:

Addressing Mode	Mode	Register	Addressing Mode	Mode	Register
Dn	—	—	d(An, Xi)	110	register number
An	—	—	Abs.W	111	000
(An)	010	register number	Abs.L	111	001
(An) +	011	register number	d(PC)	—	—
− (An)	100	register number	d(PC, Xi)	—	—
d(An)	101	register number	Imm	—	—

Notes:

1. If the destination is a data register, then it cannot be specified by using the destination <ea> mode, but must use the destination Dn mode instead.

2. ORI is used when the source is immediate data. Most assemblers automatically make this distinction.

ORI

Inclusive OR Immediate

ORI

Operation: Immediate Data v (Destination) → Destination

**Assembler
Syntax:** ORI #<data>, <ea>

Attributes: Size = (Byte, Word, Long)

Description: Inclusive OR the immediate data to the destination operand and store the result in the destination location. The size of the operation may be specified to be byte, word, or long. The size of the immediate data matches the operation size.

Condition Codes:

X	N	Z	V	C
—	*	*	0	0

N   Set if the most significant bit of the result is set. Cleared otherwise.
Z   Set if the result is zero. Cleared otherwise.
V   Always cleared.
C   Always cleared.
X   Not affected.

Instruction Format:

15	14	13	12	11	10	9	8	7	6	5 4 3	2 1 0
0	0	0	0	0	0	0	0	Size		Effective Address Mode | Register	
Word Data (16 bits)								Byte Data (8 bits)			
Long Data (32 bits, including previous word)											

Instruction Fields:

Size field — Specifies the size of the operation:
00 — byte operation.
01 — word operation.
10 — long operation.

Effective Address field — Specifies the destination operand. Only data alterable addressing modes are allowed as shown:

Addressing Mode	Mode	Register	Addressing Mode	Mode	Register
Dn	000	register number	d(An, Xi)	110	register number
An	—	—	Abs.W	111	000
(An)	010	register number	Abs.L	111	001
(An)+	011	register number	d(PC)	—	—
−(An)	100	register number	d(PC, Xi)	—	—
d(An)	101	register number	Imm	—	—

Immediate field — (Data immediately following the instruction):
If size = 00, then the data is the low order byte of the immediate word.
If size = 01, then the data is the entire immediate word.
If size = 10, then the data is the next two immediate words.

ORI to CCR

Inclusive OR Immediate to Condition Codes

ORI to CCR

Operation: (Source) v CCR → CCR

**Assembler
Syntax:** ORI #xxx, CCR

Attributes: Size = (Byte)

Description: Inclusive OR the immediate operand with the condition codes and store the result in the low-order byte of the status register.

Condition Codes:

X	N	Z	V	C
*	*	*	*	*

N   Set if bit 3 of immediate operand is one. Unchanged otherwise.
Z   Set if bit 2 of immediate operand is one. Unchanged otherwise.
V   Set if bit 1 of immediate operand is one. Unchanged otherwise.
C   Set if bit 0 of immediate operand is one. Unchanged otherwise.
X   Set if bit 4 of immediate operand is one. Unchanged otherwise.

Instruction Format:

15	14	13	12	11	10	9	8	7	6	5	4	3	2	1	0
0	0	0	0	0	0	0	0	0	0	1	1	1	1	0	0
0	0	0	0	0	0	0	0	Byte Data (8 bits)							

ORI
to SR

**Inclusive OR Immediate to the Status Register
(Privileged Instruction)**

ORI
to SR

Operation: If supervisor state
              then (Source) v SR → SR
              else TRAP

**Assembler
Syntax:**   ORI #xxx, SR

Attributes: Size = (Word)

Description: Inclusive OR the immediate operand with the contents of the status
register and store the result in the status register. All bits of the status
register are affected.

Condition Codes:

X	N	Z	V	C
*	*	*	*	*

N   Set if bit 3 of immediate operand is one. Unchanged otherwise.
Z   Set if bit 2 of immediate operand is one. Unchanged otherwise.
V   Set if bit 1 of immediate operand is one. Unchanged otherwise.
C   Set if bit 0 of immediate operand is one. Unchanged otherwise.
X   Set if bit 4 of immediate operand is one. Unchanged otherwise.

Instruction Format:

15	14	13	12	11	10	9	8	7	6	5	4	3	2	1	0
0	0	0	0	0	0	0	0	0	1	1	1	1	1	0	0
Word Data (16 bits)																

PEA

Push Effective Address

PEA

Operation: Destination → − (SP)

**Assembler
Syntax:** PEA < ea >

Attributes: Size = (Long)

Description: The effective address is computed and pushed onto the stack. A long word address is pushed onto the stack.

Condition Codes: Not affected.

Instruction Format:

15	14	13	12	11	10	9	8	7	6	5 4 3	2 1 0
0	1	0	0	1	0	0	0	0	1	Effective Address Mode	Register

Instruction Fields:

Effective Address field — Specifies the address to be pushed onto the stack. Only control addressing modes are allowed as shown:

Addressing Mode	Mode	Register	Addressing Mode	Mode	Register
Dn	—	—	d(An, Xi)	110	register number
An	—	—	Abs.W	111	000
(An)	010	register number	Abs.L	111	001
(An) +	—	—	d(PC)	111	010
− (An)	—	—	d(PC, Xi)	111	011
d(An)	101	register number	Imm	—	—

RESET

**Reset External Devices
(Privileged Instruction)**

RESET

Operation: If supervisor state
      then Assert RESET Line
      else TRAP

**Assembler
Syntax:** RESET

Attributes: Unsized

Description: The reset line is asserted causing all external devices to be reset. The processor state, other than the program counter, is unaffected and execution continues with the next instruction.

Condition Codes: Not affected.

Instruction Format:

15	14	13	12	11	10	9	8	7	6	5	4	3	2	1	0
0	1	0	0	1	1	1	0	0	1	1	1	0	0	0	0

Operation: (Destination) Rotated by <count> → Destination

Assembler
Syntax:
ROd Dx, Dy
ROd #<data>, Dy
ROd <ea>

Attributes: Size = (Byte, Word, Long)

Description: Rotate the bits of the operand in the direction specified. The extend bit is not included in the rotation. The shift count for the rotation of a register may be specified in two different ways:

    1. Immediate — the shift count is specified in the instruction (shift range, 1-8).
    2. Register — the shift count is contained in a data register specified in the instruction.

The size of the operation may be specified to be byte, word, or long. The content of memory may be rotated one bit only and the operand size is restricted to a word.

For ROL, the operand is rotated left; the number of positions shifted is the shift count. Bits shifted out of the high order bit go to both the carry bit and back into the low order bit. The extend bit is not modified or used.

ROL:

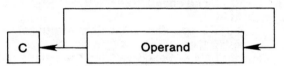

For ROR, the operand is rotated right; the number of position shifted is the shift count. Bits shifted out of the low order bit go to both the carry bit and back into the high order bit. The extend bit is not modified or used.

ROR:

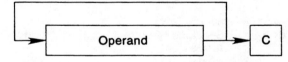

Condition Codes:

X	N	Z	V	C
—	*	*	0	*

N   Set if the most significant bit of the result is set. Cleared otherwise.
Z   Set if the result is zero. Cleared otherwise.
V   Always cleared.
C   Set according to the last bit shfited out of the operand. Cleared for a shift count of zero.
X   Not affected.

— Continued —

ROL
ROR

Rotate (Without Extend)

ROL
ROR

Instruction Format (Register Rotate):

15	14	13	12	11	10	9	8	7	6	5	4	3	2	1	0
1	1	1	0	Count/ Register			dr	Size		i/r	1	1	Register		

Instruction Fields (Register Rotate):

Count/Register field —
  if i/r = 0, the rotate count is specified in this field. The values 0, 1-7 represent a range of 8, 1 to 7 respectively.
  If i/r = 1, the rotate count (modulo 64) is contained in the data register specified in this field.
dr field — Specifies the direction of the rotate:
  0 — rotate right.
  1 — rotate left.
Size field — Specifies the size of the operation:
  00 — byte operation.
  01 — word operation.
  10 — long operation.
i/r field —
  If i/r = 0, specifies immediate rotate count.
  If i/r = 1, specifies register rotate count.
Register field — Specifies a data register whose content is to be rotated.

Instruction Format (Memory Rotate):

15	14	13	12	11	10	9	8	7	6	5	4	3	2	1	0
1	1	1	0	0	1	1	dr	1	1	Effective Address Mode			Register		

Instruction Fields (Memory Rotate):

dr field — Specifies the direction of the rotate:
  0 — rotate right
  1 — rotate left.
Effective Address field — Specifies the operand to be rotated. Only memory alterable addressing modes are allowed as shown:

Addressing Mode	Mode	Register	Addressing Mode	Mode	Register
Dn	—	—	d(An, Xi)	110	register number
An	—	—	Abs.W	111	000
(An)	010	register number	Abs.L	111	001
(An)+	011	register number	d(PC)	—	—
−(An)	100	register number	d(PC, Xi)	—	—
d(An)	101	register number	Imm	—	—

ROXL
ROXR

Rotate with Extend

ROXL
ROXR

Operation: (Destination) Rotated by <count> → Destination

Assembler ROXd Dx, Dy
Syntax: ROXd #<data>, Dy
ROXd <ea>

Attributes: Size = (Byte, Word, Long)

Description: Rotate the bits of the destination operand in the direction specified. The extend bit is included in the rotation. The shift count for the rotation of a register may be specified in two different ways:
1. Immediate — the shift count is specified in the instruction (shift range, 1-8).
2. Register — the shift count is contained in a data register specified in the instruction.

The size of the operation may be specified to be byte, word, or long. The content of memory may be rotated one bit only and the operand size is restricted to a word.

For ROXL, the operand is rotated left; the number of positions shifted is the shift count. Bits shifted out of the high order bit go to both the carry and extend bits; the previous value of the extend bit is shifted into the low order bit.

ROXL:

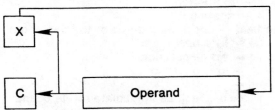

For ROXR, the operand is rotated right; the number of positions shifted is the shift count. Bits shifted out of the low order bit go to both the carry and extend bits; the previous value of the extend bit is shifted into the high order bit.

ROXR:

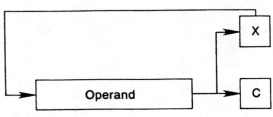

— Continued —

ROXL
ROXR

Rotate with Extend

ROXL
ROXR

Condition Codes:

X	N	Z	V	C
*	*	*	0	*

N   Set if the most significant bit of the result is set. Cleared otherwise.
Z   Set if the result is zero. Cleared otherwise.
V   Always cleared.
C   Set according to the last bit shifted out of the operand. Set to the value of the extend bit for a shift count of zero.
X   Set according to the last bit shifted out of the operand. Unaffected for a shift count of zero.

Instruction Format (Register Rotate):

15	14	13	12	11 10 9	8	7 6	5	4	3	2 1 0
1	1	1	0	Count/ Register	dr	Size	i/r	1	0	Register

Instruction Fields (Register Rotate):

Count/Register field:

If i/r = 0, the rotate count is specified in this field. The values 0, 1-7 represent range of 8, 1 to 7 respectively.

If i/r = 1, the rotate count (modulo 64) is contained in the data register specified in this field.

dr field — Specifies the direction of the rotate:

0 — rotate right.
1 — rotate left.

Size field — Specifies the size of the operation:

00 — byte operation.
01 — word operation.
10 — long operation.

i/r field —

If i/r = 0, specifies immediate rotate count.
If i/r = 1, specifies register rotate count.

Register field — Specifies a data register whose content is to be rotated.

— Continued —

ROXL
ROXR

Rotate with Extend

ROXL
ROXR

Instruction Format (Memory Rotate):

15	14	13	12	11	10	9	8	7	6	5 4 3	2 1 0
1	1	1	0	0	1	0	dr	1	1	Effective Address Mode	Register

Instruction Fields (Memory Rotate):

dr field — Specifies the direction of the rotate:

0 — rotate right.

1 — rotate left.

Effective Address field — Specifies the operand to be rotated. Only memory alterable addressing modes are allowed as shown:

Addressing Mode	Mode	Register	Addressing Mode	Mode	Register
Dn	—	—	d(An, Xi)	110	register number
An	—	—	Abs.W	111	000
(An)	010	register number	Abs.L	111	001
(An)+	011	register number	d(PC)	—	—
–(An)	100	register number	d(PC, Xi)	—	—
d(An)	101	register number	Imm	—	—

RTD

Return and Deallocate Parameters

RTD

Operation: (SP) + → PC; SP + d → SP

**Assembler
Syntax:** RTD #<displacement>

Attributes: Unsized

Description: The program counter is pulled from the stack. The previous program counter value is lost. After the program counter is read from the stack, the displacement value is sign-extended to 32 bits and added to the stack pointer.

Condition Codes: Not affected.

Instruction Format:

15	14	13	12	11	10	9	8	7	6	5	4	3	2	1	0
0	1	0	0	1	1	1	0	0	1	1	1	0	1	0	0
Displacement															

Instruction Field:
Displacement field — Specifies the two's-complement integer which is to be sign-extended and added to the stack pointer.

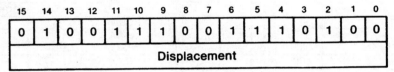

MC68010

RTE

**Return from Exception
(Privileged Instruction)**

RTE

Operation: If supervisor state
                     then (SP) + → SR; (SP) + → PC
                     else TRAP

**Assembler
Syntax:**   RTE

Attributes: Unsized

Description: The status register and program counter are pulled from the system stack. The previous status register and program counter are lost. All bits in the status register are affected.

Condition Codes: Set according to the content of the word on the stack.

Instruction Format:

15	14	13	12	11	10	9	8	7	6	5	4	3	2	1	0
0	1	0	0	1	1	1	0	0	1	1	1	0	0	1	1

RTE

**Return from Exception
(Privileged Instruction)**

RTE

Operation: If supervisor state
  then (SP) + → SR; (SP) + → PC
  If (SP) + = long format
    then full restore
  else TRAP

**Assembler
Syntax:** RTE

Attributes: Unsized

Description: The status register and program counter are pulled from the system stack. The previous status register and program counter are lost. The vector offset word is also pulled from the stack and the format field is examined to determine the amount of information to be restored.

Condition Codes: Set according to the content of the word on the stack.

Instruction Format:

15	14	13	12	11	10	9	8	7	6	5	4	3	2	1	0
0	1	0	0	1	1	1	0	0	1	1	1	0	0	1	1

Vector Offset Word Format:

15		12	11	10	9		0
Format			0	0		Vector Offset	

Vector Offset Word Format Fields:
  Format Field: — Specifies the amount of information to be restored.
    0000 — Short. Four words are to be removed from the top of the stack.
    1000 — Long. Twenty-nine words are to be removed from the top of the stack.
    Any Other
      Pattern — Error. The processor takes the format error exception.

MC68010

RTR

Return and Restore Condition Codes

RTR

Operation: (SP)+ → CC; (SP)+ → PC

Assembler
Syntax: RTR

Attributes: Unsized

Description: The condition codes and program counter are pulled from the stack. The previous condition codes and program counter are lost. The supervisor portion of the status register is unaffected.

Condition Codes: Set according to the content of the word on the stack.

Instruction Format:

15	14	13	12	11	10	9	8	7	6	5	4	3	2	1	0
0	1	0	0	1	1	1	0	0	1	1	1	0	1	1	1

RTS

Return from Subroutine

RTS

Operation: (SP) + → PC

**Assembler
Syntax:**   RTS

Attributes: Unsized

Description: The program counter is pulled from the stack. The previous program counter is lost.

Condition Codes: Not affected.

Instruction Format:

15	14	13	12	11	10	9	8	7	6	5	4	3	2	1	0
0	1	0	0	1	1	1	0	0	1	1	1	0	1	0	1

SBCD Subtract Decimal with Extend SBCD

Operation: $(\text{Destination})_{10} - (\text{Source})_{10} - X \rightarrow \text{Destination}$

Assembler
Syntax:
SBCD Dy, Dx
SBCD −(Ay), −(Ax)

Attributes: Size = (Byte)

Description: Subtract the source operand from the destination operand along with the extend bit and store the result in the destination location. The subtraction is performed using binary coded decimal arithmetic. The operands may be addressed in two different ways:
1. Data register to data register: The operands are contained in the data registers specified in the instruction.
2. Memory to memory: The operands are addressed with the predecrement addressing mode using the address registers specified in the instruction.

This operation is a byte operation only.

Condition Codes:

X	N	Z	V	C
*	U	*	U	*

N Undefined.
Z Cleared if the result is non-zero. Unchanged otherwise.
V Undefined.
C Set if a borrow (decimal) is generated. Cleared otherwise.
X Set the same as the carry bit.

NOTE

Normally the Z condition code bit is set via programming before the start of an operation. This allows successful tests for zero results upon completion of multiple-precision operations.

Instruction Format:

15	14	13	12	11	10	9	8	7	6	5	4	3	2	1	0
1	0	0	0	Register Rx			1	0	0	0	0	R/M	Register Ry		

Instruction Fields:

Register Rx field — Specifies the destination register:
If R/M = 0, specifies a data register.
If R/M = 1, specifies an address register for the predecrement addressing mode.
R/M field — Specifies the operand addressing mode:
0 — The operation is data register to data register.
1 — The operation is memory to memory.
Register Ry field — Specifies the source register:
If R/M = 0, specifies a data register.
If R/M = 1, specifies an address register for the predecrement addressing mode.

Scc

Set According to Condition

Scc

Operation: If (Condition True)
then 1s → Destination
else 0s → Destination

**Assembler
Syntax:** Scc < ea >

Attributes: Size = (Byte)

Description: The specified condition code is tested; if the condition is true, the byte specified by the effective address is set to TRUE (all ones), otherwise that byte is set to FALSE (all zeroes). "cc" may specify the following conditions:

CC	carry clear	0100	$\overline{C}$		LS	low or same	0011	$C+Z$
CS	carry set	0101	C		LT	less than	1101	$N \cdot \overline{V} + \overline{N} \cdot V$
EQ	equal	0111	Z		MI	minus	1011	N
F	false	0001	0		NE	not equal	0110	$\overline{Z}$
GE	greater or equal	1100	$N \cdot V + \overline{N} \cdot \overline{V}$		PI	plus	1010	$\overline{N}$
GT	greater than	1110	$N \cdot V \cdot \overline{Z} + \overline{N} \cdot \overline{V} \cdot \overline{Z}$		T	true	0000	1
HI	high	0010	$\overline{C} \cdot \overline{Z}$		VC	overflow clear	1000	$\overline{V}$
LE	less or equal	1111	$Z + N \cdot \overline{V} + \overline{N} \cdot V$		VS	overflow set	1001	V

Condition Codes: Not affected.

Instruction Format:

15	14	13	12	11 10 9 8	7	6	5 4 3	2 1 0
0	1	0	1	Condition	1	1	Effective Address Mode	Register

Instruction Fields:

Condition field — One of sixteen conditions discussed in description.
Effective Address field — Specifies the location in which the true/false byte is to be stored. Only data alterable addressing modes are allowed as shown:

Addressing Mode	Mode	Register	Addressing Mode	Mode	Register
Dn	000	register number	d(An, Xi)	110	register number
An	—	—	Abs.W	111	000
(An)	010	register number	Abs.L	111	001
(An) +	011	register number	d(PC)	—	—
– (An)	100	register number	d(PC, Xi)	—	—
d(An)	101	register number	Imm	—	—

Notes:
1. A memory destination is read before being written to.
2. An arithmetic one and zero result may be generated by following the Scc instruction with a NEG instruction.

STOP

Load Status Register and Stop
(Privileged Instruction)

STOP

Operation: If supervisor state
                   then Immediate Data→SR; STOP
                   else TRAP

**Assembler
Syntax:**   STOP #xxx

Attributes: Unsized

Description: The immediate operand is moved into the entire status register; the program counter is advanced to point to the next instruction and the processor stops fetching and executing instructions. Execution of instructions resumes when a trace, interrupt, or reset exception occurs. A trace exception will occur if the trace state is on when the STOP instruction is executed. If an interrupt request arrives whose priority is higher than the current processor priority, an interrupt exception occurs, otherwise the interrupt request has no effect. If the bit of the immediate data corresponding to the S-bit is off, execution of the instruction will cause a privilege violation. External reset will always initiate reset exception processing.

Condition Codes: Set according to the immediate operand.

Instruction Format:

15	14	13	12	11	10	9	8	7	6	5	4	3	2	1	0
0	1	0	0	1	1	1	0	0	1	1	1	0	0	1	0
Immediate Data															

Instruction Fields:

Immediate field — Specifies the data to be loaded into the status register.

SUB

Subtract Binary

SUB

Operation: (Destination) − (Source) → Destination

Assembler SUB <ea>, Dn
Syntax: SUB Dn, <ea>

Attributes: Size = (Byte, Word, Long)

Description: Subtract the source operand from the destination operand and store the result in the destination. The size of the operation may be specified to be byte, word, or long. The mode of the instruction indicates which operand is the source and which is the destination as well as the operand size.

Condition Codes:

X	N	Z	V	C
*	*	*	*	*

N Set if the result is negative. Cleared otherwise.
Z Set if the result is zero. Cleared otherwise.
V Set if an overflow is generated. Cleared otherwise.
C Set if a borrow is generated. Cleared otherwise.
X Set the same as the carry bit.

Instruction Format:

15	14	13	12	11	10	9	8	7	6	5	4	3	2	1	0
1	0	0	1	Register			Op-Mode			Effective Address Mode			Register		

Instruction Fields:

Register field — Specifies any of the eight data registers.

Op-Mode field —

Byte	Word	Long	Operation
000	001	010	(<Dn>) − (<ea>) → <Dn>
100	101	110	(<ea>) − (<Dn>) → <ea>

Effective Address field — Determines addressing mode:

If the location specified is a source operand, then all addressing modes are allowed as shown:

Addressing Mode	Mode	Register	Addressing Mode	Mode	Register
Dn	000	register number	d(An, Xi)	110	register number
An*	001	register number	Abs.W	111	000
(An)	010	register number	Abs.L	111	001
(An) +	011	register number	d(PC)	111	010
− (An)	100	register number	d(PC, Xi)	111	011
d(An)	101	register number	Imm	111	100

*For byte size operation, address register direct is not allowed.

— Continued —

SUB

Subtract Binary

SUB

Effective Address field (Continued)

If the location specified is a destination operand, then only alterable memory addressing modes are allowed as shown:

Addressing Mode	Mode	Register	Addressing Mode	Mode	Register
Dn	—	—	d(An, Xi)	110	register number
An	—	—	Abs.W	111	000
(An)	010	register number	Abs.L	111	001
(An)+	011	register number	d(PC)	—	—
−(An)	100	register number	d(PC, Xi)	—	—
d(An)	101	register number	Imm	—	—

Notes:
1. If the destination is a data register, then it cannot be specified by using the destination <ea> mode, but must use the destination Dn mode instead.
2. SUBA is used when the destination is an address register. SUBI and SUBQ are used when the source is immediate data. Most assemblers automatically make this distinction.

SUBA

Subtract Address

SUBA

Operation: (Destination) – (Source) → Destination

**Assembler
Syntax:** SUBA <ea>, An

Attributes: Size = (Word, Long)

Description: Subtract the source operand from the destination address register and store the result in the address register. The size of the operation may be specified to be word or long. Word size source operands are sign extended to 32 bit quantities before the operation is done.

Condition Codes: Not affected.

Instruction Format:

15	14	13	12	11 10 9	8 7 6	5 4 3	2 1 0
1	0	0	1	Register	Op-Mode	Effective Address Mode	Register

Instruction Fields:

Register field — Specifies any of the eight address registers. This is always the destination.

Op-Mode field — Specifies the size of the operation:

011 — Word operation. The source operand is sign-extended to a long operand and the operation is performed on the address register using all 32 bits.

111 — Long operations.

Effective Address field — Specifies the source operand. All addressing modes are allowed as shown:

Addressing Mode	Mode	Register	Addressing Mode	Mode	Register
Dn	000	register number	d(An, Xi)	110	register number
An	001	register number	Abs.W	111	000
(An)	010	register number	Abs.L	111	001
(An)+	011	register number	d(PC)	111	010
–(An)	100	register number	d(PC, Xi)	111	011
d(An)	101	register number	Imm	111	100

SUBI   Subtract Immediate   SUBI

Operation: (Destination) – Immediate Data → Destination

**Assembler
Syntax:** SUBI #<data>, <ea>

Attributes: Size = (Byte, Word, Long)

Description: Subtract the immediate data from the destination operand and store the result in the destination location. The size of the operation may be specified to be byte, word, or long. The size of the immediate data matches the operation size.

Condition Codes:

X	N	Z	V	C
*	*	*	*	*

N   Set if the result is negative. Cleared otherwise.
Z   Set if the result is zero. Cleared otherwise.
V   Set if an overflow is generated. Cleared otherwise.
C   Set if a borrow is generated. Cleared otherwise.
X   Set the same as the carry bit.

Instruction Format:

15	14	13	12	11	10	9	8	7	6	5	4	3	2	1	0
0	0	0	0	0	1	0	0	Size		Effective Address Mode | Register					
Word Data (16 bits)								Byte Data (8 bits)							
Long Data (32 bits, including previous word)															

Instruction Fields:

Size field — Specifies the size of the operation.
   00 — byte operation.
   01 — word operation.
   10 — long operation.
Effective Address field — Specifies the destination operand. Only data alterable addressing modes are allowed as shown:

Addressing Mode	Mode	Register	Addressing Mode	Mode	Register
Dn	000	register number	d(An, Xi)	110	register number
An	—	—	Abs.W	111	000
(An)	010	register number	Abs.L	111	001
(An)+	011	register number	d(PC)	—	—
–(An)	100	register number	d(PC, Xi)	—	—
d(An)	101	register number	Imm	—	—

Immediate field — (Data immediately following the instruction)
   If size = 00, then the data is the low order byte of the immediate word.
   If size = 01, then the data is the entire immediate word.
   If size = 10, then the data is the next two immediate words.

SUBQ

Subtract Quick

SUBQ

Operation: (Destination) − Immediate Data → Destination

**Assembler
Syntax:** SUBQ #<data>, <ea>

Attributes: Size = (Byte, Word, Long)

Description: Subtract the immediate data from the destination operand. The data range is from 1-8. The size of the operation may be specified to be byte, word, or long. Word and long operations are also allowed on the address registers and the condition codes are not affected. Word size source operands are sign extended to 32-bit quantities before the operation is done.

Condition Codes:

X	N	Z	V	C
*	*	*	*	*

N   Set if the result is negative. Cleared otherwise.
Z   Set if the result is zero. Cleared otherwise.
V   Set if an overflow is generated. Cleared otherwise.
C   Set if a borrow is generated. Cleared otherwise.
X   Set the same as the carry bit.

The condition codes are not affected if a subtraction from an address register is made.

Instruction Format:

15	14	13	12	11	10	9	8	7	6	5	4	3	2	1	0
0	1	0	1	Data			1	Size		Effective Address Mode / Register					

Instruction Fields:

Data field — Three bits of immediate data, 0, 1-7 representing a range of 8, 1 to 7 respectively.
Size field — Specifies the size of the operation:
00 — byte operation.
01 — word operation.
10 — long operation.
Effective Address field — Specifies the destination location. Only alterable addressing modes are allowed as shown:

Addressing Mode	Mode	Register	Addressing Mode	Mode	Register
Dn	000	register number	d(An, Xi)	110	register number
An*	001	register number	Abs.W	111	000
(An)	010	register number	Abs.L	111	001
(An)+	011	register number	d(PC)	—	—
−(An)	100	register number	d(PC, Xi)	—	—
d(An)	101	register number	Imm	—	—

*Word and Long only.

SUBX

Subtract with Extend

SUBX

Operation: (Destination) − (Source) − X → Destination

Assembler SUBX Dy, Dx
Syntax: SUBX − (Ay), − (Ax)

Attributes: Size = (Byte, Word, Long)

Description: Subtract the source operand from the destination operand along with the extend bit and store the result in the destination location. The operands may be addressed in two different ways:
1. Data register to data register: The operands are contained in data registers specified in the instruction.
2. Memory to memory. The operands are contained in memory and addressed with the predecrement addressing mode using the address registers specified in the instruction.
The size of the operation may be specified to be byte, word, or long.

Condition Codes:

X	N	Z	V	C
*	*	*	*	*

N   Set if the result is negative. Cleared otherwise.
Z   Cleared if the result is non-zero. Unchanged otherwise.
V   Set if an overflow is generated. Cleared otherwise.
C   Set if a carry is generated. Cleared otherwise.
X   Set the same as the carry bit.

NOTE
Normally the Z condition code bit is set via programming before the start of an operation. This allows successful tests for zero results upon completion of multiple-precision operations.

Instruction Format:

15	14	13	12	11	10	9	8	7	6	5	4	3	2	1	0
1	0	0	1	Register Rx			1	Size		0	0	R/M	Register Ry		

— Continued —

SUBX   Subtract with Extend   SUBX

Instruction Fields:

Register Rx field — Specifies the destination register:

If R/M = 0, specifies a data register.

If R/M = 1, specifies an address register for the predecrement addressing mode.

Size field — Specifies the size of the operation:

00 — byte operation.

01 — word operation.

10 — long operation.

R/M field — Specifies the operand addressing mode:

0 — The operation is data register to data register.

1 — The operation is memory to memory.

Register Ry field — Specifies the source register:

If R/M = 0, specifies a data register.

If R/M = 1, specifies an address register for the predecrement addressing mode.

SWAP

Swap Register Halves

SWAP

Operation: Register [31:16] ↔ Register [15:0]

**Assembler
Syntax:** SWAP Dn

Attributes: Size = (Word)

Description: Exchange the 16-bit halves of a data register.

Condition Codes:

X	N	Z	V	C
—	*	*	0	0

N   Set if the most significant bit of the 32-bit result is set. Cleared otherwise.
Z   Set if the 32-bit result is zero. Cleared otherwise.
V   Always cleared.
C   Always cleared.
X   Not affected.

Instruction Format:

15	14	13	12	11	10	9	8	7	6	5	4	3	2	1	0
0	1	0	0	1	0	0	0	0	1	0	0	0	Register		

Instruction Fields:

Register field — Specifies the data register to swap.

TAS

Test and Set an Operand

TAS

Operation: (Destination) Tested → CC; 1 → bit 7 OF Destination

**Assembler
Syntax:** TAS <ea>

Attributes: Size = (Byte)

Description: Test and set the byte operand addressed by the effective address field. The current value of the operand is tested and N and Z are set accordingly. The high order bit of the operand is set. The operation is indivisible (using a read-modify-write memory cycle) to allow synchronization of several processors.

Condition Codes:

X	N	Z	V	C
—	*	*	0	0

N Set if the most significant bit of the operand was set. Cleared otherwise.
Z Set if the operand was zero. Cleared otherwise.
V Always cleared.
C Always cleared.
X Not affected.

Instruction Format:

15	14	13	12	11	10	9	8	7	6	5 4 3	2 1 0
0	1	0	0	1	0	1	0	1	1	Effective Address Mode	Register

Instruction Fields:

Effective Address field — Specifies the location of the tested operand. Only data alterable addressing modes are allowed as shown:

Addressing Mode	Mode	Register	Addressing Mode	Mode	Register
Dn	000	register number	d(An, Xi)	110	register number
An	—	—	Abs.W	111	000
(An)	010	register number	Abs.L	111	001
(An)+	011	register number	d(PC)	—	—
−(An)	100	register number	d(PC, Xi)	—	—
d(An)	101	register number	Imm	—	—

Note: Bus error retry is inhibited on the read portion of the TAS read-modify-write bus cycle to ensure system integrity. The bus error exception is always taken.

TRAP

Trap

TRAP

Operation: PC→ – (SSP); SR→ – (SSP); (Vector)→ PC

**Assembler
Syntax:** TRAP #<vector>

Attributes: Unsized

Description: The processor initiates exception processing. The vector number is generated to reference the TRAP instruction exception vector specified by the low order four bits of the instruction. Sixteen TRAP instruction vectors are available.

Condition Codes: Not affected.

Instruction Format:

15	14	13	12	11	10	9	8	7	6	5	4	3	2	1	0
0	1	0	0	1	1	1	0	0	1	0	0		Vector		

Instruction Fields:

Vector field — Specifies which trap vector contains the new program counter to be loaded.

TRAPV

Trap on Overflow

TRAPV

Operation: If V then TRAP

**Assembler
Syntax:** TRAPV

Attributes: Unsized

Description: If the overflow condition is on, the processor initiates exception processing. The vector number is generated to reference the TRAPV exception vector. If the overflow condition is off, no operation is performed and execution continues with the next instruction in sequence.

Condition Codes: Not affected.

Instruction Format:

15	14	13	12	11	10	9	8	7	6	5	4	3	2	1	0
0	1	0	0	1	1	1	0	0	1	1	1	0	1	1	0

TST

Test an Operand

TST

Operation: (Destination) Tested → CC

**Assembler
Syntax:** TST <ea>

Attributes: Size = (Byte, Word, Long)

Description: Compare the operand with zero. No results are saved; however, the condition codes are set according to results of the test. The size of the operation may be specified to be byte, word, or long.

Condition Codes:

X	N	Z	V	C
—	*	*	0	0

N   Set if the operand is negative. Cleared otherwise.
Z   Set if the operand is zero. Cleared otherwise.
V   Always cleared.
C   Always cleared.
X   Not affected.

Instruction Format:

15	14	13	12	11	10	9	8	7	6	5	4	3	2	1	0
0	1	0	0	1	0	1	0	Size		Effective Address Mode			Register		

Instruction Fields:

Size field — Specifies the size of the operation:
00 — byte operation.
01 — word operation.
10 — long operation.

Effective Address field — Specifies the destination operand. Only data alterable addressing modes are allowed as shown:

Addressing Mode	Mode	Register	Addressing Mode	Mode	Register
Dn	000	register number	d(An, Xi)	110	register number
An	—	—	Abs.W	111	000
(An)	010	register number	Abs.L	111	001
(An)+	011	register number	d(PC)	—	—
−(An)	100	register number	d(PC, Xi)	—	—
d(An)	101	register number	Imm	—	—

UNLK

Unlink

UNLK

Operation: An → SP; (SP) + → An

**Assembler
Syntax:** UNLK An

Attributes: Unsized

Description: The stack pointer is loaded from the specified address register. The address register is then loaded with the long word pulled from the top of the stack.

Condition Codes: Not affected.

Instruction Format:

15	14	13	12	11	10	9	8	7	6	5	4	3	2	1	0
0	1	0	0	1	1	1	0	0	1	0	1	1	Register		

Instruction Fields:

Register field — specifies the address register through which the unlinking is to be done.

APPENDIX IVC: MC68000 INSTRUCTION FORMAT SUMMARY

C.1 INTRODUCTION

This appendix provides a summary of the first word in each instruction of the instruction set. Table C-1 is an operation code (op-code) map which illustrates how bits 15 through 12 are used to specify the operations. The remaining paragraph groups the instructions according to the op-code map.

Table C-1. Operation Code Map

Bits 15 through 12	Operation	Bits 15 through 12	Operation
0000	Bit Manipulation/MOVEP/Immediate	1000	OR/DIV/SBCD
0001	Move Byte	1001	SUB/SUBX
0010	Move Long	1010	(Unassigned)
0011	Move Word	1011	CMP/EOR
0100	Miscellaneous	1100	AND/MUL/ABCD/EXG
0101	ADDQ/SUBQ/Scc/DBcc	1101	ADD/ADDX
0110	Bcc/BSR	1110	Shift/Rotate
0111	MOVEQ	1111	(Unassigned)

Table C-2. Effective Address Encoding Summary

Addressing Mode	Mode	Register
Data Register Direct	000	register number
Address Register Direct	001	register number
Address Register Indirect	010	register number
Address Register Indirect with Postincrement	011	register number
Address Register Indirect with Predecrement	100	register number
Address Register Indirect with Displacement	101	register number
Address Register Indirect with Index	110	register number
Absolute Short	111	000
Absolute Long	111	001
Program Counter with Displacement	111	010
Program Counter with Index	111	011
Immediate or Status Register	111	100

Table C-3. Conditional Tests

Mnemonic	Condition	Encoding	Test
T	true	0000	1
F	false	0001	0
HI	high	0010	$\overline{C} \cdot \overline{Z}$
LS	low or same	0011	$C + Z$
CC(HS)	carry clear	0100	$\overline{C}$
CS(LO)	carry set	0101	C
NE	not equal	0110	$\overline{Z}$
EQ	equal	0111	Z
VC	overflow clear	1000	$\overline{V}$
VS	overflow set	1001	V
PL	plus	1010	$\overline{N}$
MI	minus	1011	N
GE	greater or equal	1100	$N \cdot V + \overline{N} \cdot \overline{V}$
LT	less than	1101	$N \cdot \overline{V} + \overline{N} \cdot V$
GT	greater than	1110	$N \cdot V \cdot \overline{Z} + \overline{N} \cdot \overline{V} \cdot \overline{Z}$
LE	less or equal	1111	$Z + N \cdot \overline{V} + \overline{N} \cdot V$

OR Immediate

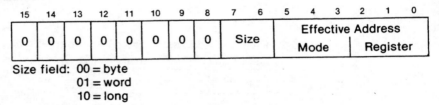

15	14	13	12	11	10	9	8	7	6	5	4	3	2	1	0
0	0	0	0	0	0	0	0	Size		Effective Address Mode			Register		

Size field: 00 = byte
01 = word
10 = long

OR Immediate to CCR

15	14	13	12	11	10	9	8	7	6	5	4	3	2	1	0
0	0	0	0	0	0	0	0	0	0	1	1	1	1	0	0

OR Immediate to SR

15	14	13	12	11	10	9	8	7	6	5	4	3	2	1	0
0	0	0	0	0	0	0	0	0	1	1	1	1	1	0	0

Dynamic Bit

15	14	13	12	11	10	9	8	7	6	5	4	3	2	1	0
0	0	0	0	Data Register			1	Type		Effective Address					
										Mode			Register		

Type field: 00 = TST
01 = CHG
10 = CLR
11 = SET

MOVEP

15	14	13	12	11	10	9	8	7	6	5	4	3	2	1	0
0	0	0	0	Data Register			Op-Mode			0	0	1	Address Register		

Op-Mode field: 100 = transfer word from memory to register
101 = transfer long from memory to register
110 = transfer word from register to memory
111 = transfer long from register to memory

AND Immediate

15	14	13	12	11	10	9	8	7	6	5	4	3	2	1	0
0	0	0	0	0	0	1	0	Size		Effective Address					
										Mode			Register		

Size field: 00 = byte
01 = word
10 = long

AND Immediate to CCR

15	14	13	12	11	10	9	8	7	6	5	4	3	2	1	0
0	0	0	0	0	0	1	0	0	0	1	1	1	1	0	0

AND Immediate to SR

15	14	13	12	11	10	9	8	7	6	5	4	3	2	1	0
0	0	0	0	0	0	1	0	0	1	1	1	1	1	0	0

SUB Immediate

15	14	13	12	11	10	9	8	7	6	5	4	3	2	1	0
										Effective Address					
0	0	0	0	0	1	0	0	Size		Mode			Register		

Size field: 00 = byte
01 = word
10 = long

ADD Immediate

15	14	13	12	11	10	9	8	7	6	5	4	3	2	1	0
										Effective Address					
0	0	0	0	0	1	1	0	Size		Mode			Register		

Size field: 00 = byte
01 = word
10 = long

Static Bit

15	14	13	12	11	10	9	8	7	6	5	4	3	2	1	0
										Effective Address					
0	0	0	0	1	0	0	0	Type		Mode			Register		

Type field: 00 = TST
01 = CHG
10 = CLR
11 = SET

EOR Immediate

15	14	13	12	11	10	9	8	7	6	5	4	3	2	1	0
										Effective Address					
0	0	0	0	1	0	1	0	Size		Mode			Register		

Size field: 00 = byte
01 = word
10 = long

EOR Immediate to CCR

15	14	13	12	11	10	9	8	7	6	5	4	3	2	1	0
0	0	0	0	1	0	1	0	0	0	1	1	1	1	0	0

EOR Immediate to SR

15	14	13	12	11	10	9	8	7	6	5	4	3	2	1	0
0	0	0	0	1	0	1	0	0	1	1	1	1	1	0	0

CMP Immediate

15	14	13	12	11	10	9	8	7	6	5	4	3	2	1	0
0	0	0	0	1	1	0	0	Size		Effective Address					
										Mode			Register		

Size field: 00 = byte
01 = word
10 = word

MOVES   MC68010

15	14	13	12	11	10	9	8	7	6	5	4	3	2	1	0
0	0	0	1	1	1	1	0	Size		Effective Address					
										Mode			Register		

Size field: 00 = byte
01 = word
10 = long

MOVE Byte

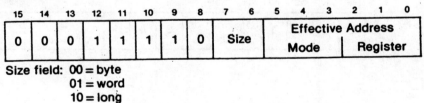

15	14	13	12	11	10	9	8	7	6	5	4	3	2	1	0
0	0	0	1	Destination						Source					
				Register			Mode			Mode			Register		

Note register and mode locations

MOVEA Long

14	14	13	12	11	10	9	8	7	6	5	4	3	2	1	0
0	0	1	0	Destination Register			0	0	1	Source					
										Mode			Register		

MOVE Long

15	14	13	12	11	10	9	8	7	6	5	4	3	2	1	0
0	0	1	0	Destination						Source					
				Register			Mode			Mode			Register		

Note register and mode locations

MOVEA Word

15	14	13	12	11	10	9	8	7	6	5	4	3	2	1	0
0	0	1	1	Destination Register			0	0	1	Source					
										Mode			Register		

MOVE Word

15	14	13	12	11	10	9	8	7	6	5	4	3	2	1	0
0	0	1	1	Destination						Source					
				Register			Mode			Mode			Register		

Note register and mode locations

NEGX

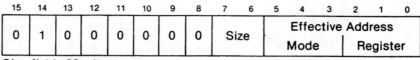

15	14	13	12	11	10	9	8	7	6	5	4	3	2	1	0
0	1	0	0	0	0	0	0	Size		Effective Address					
										Mode			Register		

Size field: 00 = byte
01 = word
10 = long

MOVE from SR

15	14	13	12	11	10	9	8	7	6	5	4	3	2	1	0
0	1	0	0	0	0	0	0	1	1	Effective Address					
										Mode			Register		

CHK

15	14	13	12	11	10	9	8	7	6	5	4	3	2	1	0
0	1	0	0	Data Register			1	1	0	Effective Address					
										Mode			Register		

LEA

15	14	13	12	11	10	9	8	7	6	5	4	3	2	1	0
0	1	0	0	Address Register			1	1	1	Effective Address					
										Mode			Register		

CLR

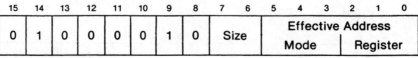

15	14	13	12	11	10	9	8	7	6	5	4	3	2	1	0
0	1	0	0	0	0	1	0	Size		Effective Address					
										Mode			Register		

Size field: 00 = byte
01 = word
10 = long

MOVE from CCR   MC68010

15	14	13	12	11	10	9	8	7	6	5	4	3	2	1	0
0	1	0	0	0	0	1	0	1	1	Effective Address Mode			Register		

NEG

15	14	13	12	11	10	9	8	7	6	5	4	3	2	1	0
0	1	0	0	0	1	0	0	Size		Effective Address Mode			Register		

Size field: 00 = byte
01 = word
10 = long

MOVE to CCR

15	14	13	12	11	10	9	8	7	6	5	4	3	2	1	0
0	1	0	0	0	1	0	0	1	1	Effective Address Mode			Register		

NOT

15	14	13	12	11	10	9	8	7	6	5	4	3	2	1	0
0	1	0	0	0	1	1	0	Size		Effective Address Mode			Register		

Size field: 00 = byte
01 = word
10 = long

MOVE to SR

15	14	13	12	11	10	9	8	7	6	5	4	3	2	1	0
0	1	0	0	0	1	1	0	1	1	Effective Address Mode			Register		

NBCD

15	14	13	12	11	10	9	8	7	6	5	4	3	2	1	0
0	1	0	0	1	0	0	0	0	0	Effective Address					
										Mode			Register		

SWAP

15	14	13	12	11	10	9	8	7	6	5	4	3	2	1	0
0	1	0	0	1	0	0	0	0	1	0	0	0	Data Register		

PEA

15	14	13	12	11	10	9	8	7	6	5	4	3	2	1	0
0	1	0	0	1	0	0	0	0	1	Effective Address					
										Mode			Register		

EXT Word

15	14	13	12	11	10	9	8	7	6	5	4	3	2	1	0
0	1	0	0	1	0	0	0	1	0	0	0	0	Data Register		

MOVEM Registers to EA

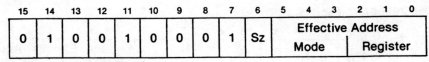

15	14	13	12	11	10	9	8	7	6	5	4	3	2	1	0
0	1	0	0	1	0	0	0	1	Sz	Effective Address					
										Mode			Register		

Sz field: 0 = word transfer
        1 = long transfer

EXT Long

15	14	13	12	11	10	9	8	7	6	5	4	3	2	1	0
0	1	0	0	1	0	0	0	1	1	0	0	0	Data Register		

TST

15	14	13	12	11	10	9	8	7	6	5	4	3	2	1	0
0	1	0	0	1	0	1	0	Size		Effective Address Mode			Register		

Size field: 00 = byte
            01 = word
            10 = long

TAS

15	14	13	12	11	10	9	8	7	6	5	4	3	2	1	0
0	1	0	0	1	0	1	0	1	1	Effective Address Mode			Register		

ILLEGAL

15	14	13	12	11	10	9	8	7	6	5	4	3	2	1	0
0	1	0	0	1	0	1	0	1	1	1	1	1	1	0	0

MOVEM EA to Registers

15	14	13	12	11	10	9	8	7	6	5	4	3	2	1	0
0	1	0	0	1	1	0	0	1	Sz	Effective Address Mode			Register		

Sz field: 0 = word transfer
          1 = long transfer

TRAP

15	14	13	12	11	10	9	8	7	6	5	4	3	2	1	0
0	1	0	0	1	1	1	0	0	1	0	0	Vector			

LINK

15	14	13	12	11	10	9	8	7	6	5	4	3	2	1	0
0	1	0	0	1	1	1	0	0	1	0	1	0	Address Register		

UNLK

15	14	13	12	11	10	9	8	7	6	5	4	3	2	1	0
0	1	0	0	1	1	1	0	0	1	0	1	1	Address Register		

MOVE to USP

15	14	13	12	11	10	9	8	7	6	5	4	3	2	1	0
0	1	0	0	1	1	1	0	0	1	1	0	0	Address Register		

MOVE from USP

15	14	13	12	11	10	9	8	7	6	5	4	3	2	1	0
0	1	0	0	1	1	1	0	0	1	1	0	1	Address Register		

RESET

15	14	13	12	11	10	9	8	7	6	5	4	3	2	1	0
0	1	0	0	1	1	1	0	0	1	1	1	0	0	0	0

NOP

15	14	13	12	11	10	9	8	7	6	5	4	3	2	1	0
0	1	0	0	1	1	1	0	0	1	1	1	0	0	0	1

STOP

15	14	13	12	11	10	9	8	7	6	5	4	3	2	1	0
0	1	0	0	1	1	1	0	0	1	1	1	0	0	1	0

RTE

15	14	13	12	11	10	9	8	7	6	5	4	3	2	1	0
0	1	0	0	1	1	1	0	0	1	1	1	0	0	1	1

RTD   MC68010

15	14	13	12	11	10	9	8	7	6	5	4	3	2	1	0
0	1	0	0	1	1	1	0	0	1	1	1	0	1	0	0

RTS

15	14	13	12	11	10	9	8	7	6	5	4	3	2	1	0
0	1	0	0	1	1	1	0	0	1	1	1	0	1	0	1

TRAPV

15	14	13	12	11	10	9	8	7	6	5	4	3	2	1	0
0	1	0	0	1	1	1	0	0	1	1	1	0	1	1	0

RTR

15	14	13	12	11	10	9	8	7	6	5	4	3	2	1	0
0	1	0	0	1	1	1	0	0	1	1	1	0	1	1	1

MOVEC   MC68010

15	14	13	12	11	10	9	8	7	6	5	4	3	2	1	0
0	1	0	0	1	1	1	0	0	1	1	1	1	0	1	dr

dr field: 0 = control register to general register
           1 = general register to control register

JSR

15	14	13	12	11	10	9	8	7	6	5	4	3	2	1	0
0	1	0	0	1	1	1	0	1	0	Effective Address					
										Mode			Register		

JMP

15	14	13	12	11	10	9	8	7	6	5	4	3	2	1	0
0	1	0	0	1	1	1	0	1	1	Effective Mode					
										Mode			Register		

ADDQ

15	14	13	12	11	10	9	8	7	6	5	4	3	2	1	0
0	1	0	1	Data			0	Size		Effective Address					
										Mode			Register		

Data field: Three bits of immediate data, 0, 1-7 representing a range of 8, 1 to 7 respectively.

Size field: 00 = byte
01 = word
10 = long

Scc

15	14	13	12	11	10	9	8	7	6	5	4	3	2	1	0
0	1	0	1	Condition				1	1	Effective Address					
										Mode			Register		

Condition field:
0000 = true   1000 = overflow clear
0001 = false   1001 = overflow set
0010 = high   1010 = plus
0011 = low or same   1011 = minus
0100 = carry clear   1100 = greater or equal
0101 = carry set   1101 = less than
0110 = not equal   1110 = greater than
0111 = equal   1111 = less or equal

DBcc

15	14	13	12	11	10	9	8	7	6	5	4	3	2	1	0
0	1	0	1	Condition				1	1	0	0	1	Data Register		

Condition field:
0000 = true   1000 = overflow clear
0001 = false   1001 = overflow set
0010 = high   1010 = plus
0011 = low or same   1011 = minus
0100 = carry clear   1100 = greater or equal
0101 = carry set   1101 = less than
0110 = not equal   1110 = greater than
0111 = equal   1111 = less or equal

SUBQ

15	14	13	12	11	10	9	8	7	6	5	4	3	2	1	0
0	1	0	1		Data		1		Size		Effective Address Mode			Register	

Data field: Three bits of immediate data, 0, 1-7 representing a range of 8, 1 to 7 respectively.

Size field: 00 = byte
01 = word
10 = long

Bcc

15	14	13	12	11	10	9	8	7	6	5	4	3	2	1	0
0	1	1	0		Condition					8-Bit Displacement					

Condition field: 0010 = high   1001 = overflow set
0011 = low or same   1010 = plus
0100 = carry clear   1011 = minus
0101 = carry set   1100 = greater or equal
0110 = not equal   1101 = less than
0111 = equal   1110 = greater than
1000 = overflow clear 1111 = less or equal

BRA

15	14	13	12	11	10	9	8	7	6	5	4	3	2	1	0
0	1	1	0	0	0	0	0			8-Bit Displacement					

BSR

15	14	13	12	11	10	9	8	7	6	5	4	3	2	1	0
0	1	1	0	0	0	0	1			8-Bit Displacement					

MOVEQ

15	14	13	12	11	10	9	8	7	6	5	4	3	2	1	0
0	1	1	1	Data Register			0	Data							

Data field: Data is sign extended to a long operand and all 32 bits are transferred to the data register.

OR

15	14	13	12	11	10	9	8	7	6	5	4	3	2	1	0
1	0	0	0	Data Register			Op-Mode			Effective Address					
										Mode			Register		

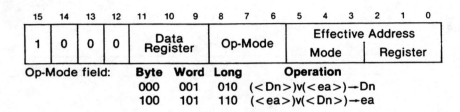

Op-Mode field:

Byte	Word	Long	Operation
000	001	010	(<Dn>)v(<ea>)→Dn
100	101	110	(<ea>)v(<Dn>)→ea

DIVU

15	14	13	12	11	10	9	8	7	6	5	4	3	2	1	0
1	0	0	0	Data Register			0	1	1	Effective Address					
										Mode			Register		

SBCD

15	14	13	12	11	10	9	8	7	6	5	4	3	2	1	0
1	0	0	0	Destination Register*			1	0	0	0	0	R/M	Source Register*		

R/M field: 0 = data register to data register
1 = memory to memory
*If R/M = 0, specifies a data register.
If R/M = 1, specifies an address register for the predecrement addressing mode.

DIVS

15	14	13	12	11 10 9	8	7	6	5 4 3	2 1 0
								Effective Address	
1	0	0	0	Data Register	1	1	1	Mode	Register

SUB

15	14	13	12	11 10 9	8 7 6	5 4 3	2 1 0
						Effective Address	
1	0	0	1	Data Register	Op-Mode	Mode	Register

Op-Mode field:

Byte	Word	Long	Operation
000	001	010	(<Dn>)−(<ea>) →Dn
100	101	110	(<ea>)−(<Dn>) →ea

SUBA

15	14	13	12	11 10 9	8 7 6	5 4 3	2 1 0
						Effective Address	
1	0	0	1	Data Register	Op-Mode	Mode	Register

Op-Mode field:

Word	Long	Operation
011	111	(<ea>)−(<An>) →An

SUBX

15	14	13	12	11 10 9	8	7 6	5	4	3	2 1 0
1	0	0	1	Destination Register*	1	Size	0	0	R/M	Source Register*

Size field: 00 = byte
             01 = word
             10 = long
R/M field: 0 = data register to data register
            1 = memory to memory
*If R/M = 0, specifies a data register.
  If R/M = 1, specifies an address register for the predecrement addressing
          mode.

CMP

15	14	13	12	11	10	9	8	7	6	5	4	3	2	1	0
1	0	1	1	Data Register			Op-Mode			Effective Address					
										Mode			Register		

Op-Mode field:

	Byte	Word	Long	Operation
	000	001	010	$(<Dn>) - (<ea>)$

CMPA

15	14	13	12	11	10	9	8	7	6	5	4	3	2	1	0
1	0	1	1	Data Register			Op-Mode			Effective Address					
										Mode			Register		

Op-Mode field:

	Word	Long	Operation
	011	111	$(<ea>) - (<An>)$

EOR

15	14	13	12	11	10	9	8	7	6	5	4	3	2	1	0
1	0	1	1	Data Register			Op-Mode			Effective Address					
										Mode			Register		

Op-Mode field:

	Byte	Word	Long	Operation
	100	101	110	$(<ea>) \oplus (<Dn>) \to ea$

CMPM

15	14	13	12	11	10	9	8	7	6	5	4	3	2	1	0
1	0	1	1	Destination Register			1	Size		0	0	1	Source Register		

Size field: 00 = byte
01 = word
10 = long

AND

15	14	13	12	11	10	9	8	7	6	5	4	3	2	1	0
1	1	0	0	Data Register			Op-Mode			Effective Address					
										Mode			Register		

Op-Mode field:	Byte	Word	Long	Operation
	000	001	010	$(<Dn>)\wedge(<ea>) \rightarrow Dn$
	100	101	110	$(<ea>)\wedge(<Dn>) \rightarrow ea$

MULU

15	14	13	12	11	10	9	8	7	6	5	4	3	2	1	0
1	1	0	0	Data Register			0	1	1	Effective Address					
										Mode			Register		

ABCD

15	14	13	12	11	10	9	8	7	6	5	4	3	2	1	0
1	1	0	0	Destination Register*			1	0	0	0	0	R/M	Source Register*		

R/M field: 0 = data register to data register
            1 = memory to memory
*If R/M = 0, specifies a data register.
If R/M = 1, specifies an address register for the predecrement addressing
            mode.

EXG Data Registers

15	14	13	12	11	10	9	8	7	6	5	4	3	2	1	0
1	1	0	0	Data Register			1	0	1	0	0	0	Data Register		

EXG Address Registers

15	14	13	12	11	10	9	8	7	6	5	4	3	2	1	0
1	1	0	0	Address Register			1	0	1	0	0	1	Address Register		

EXG Data Register and Address Register

15	14	13	12	11	10	9	8	7	6	5	4	3	2	1	0
1	1	0	0	Data Register			1	1	0	0	0	1	Address Register		

MULS

15	14	13	12	11	10	9	8	7	6	5	4	3	2	1	0
1	1	0	0	Data Register			1	1	1	Effective Address					
										Mode			Register		

ADD

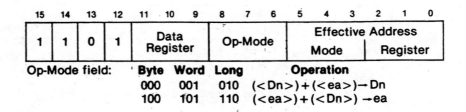

15	14	13	12	11	10	9	8	7	6	5	4	3	2	1	0
1	1	0	1	Data Register			Op-Mode			Effective Address					
										Mode			Register		

Op-Mode field:

Byte	Word	Long	Operation
000	001	010	(\<Dn\>)+(\<ea\>)→Dn
100	101	110	(\<ea\>)+(\<Dn\>)→ea

ADDA

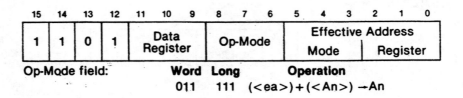

15	14	13	12	11	10	9	8	7	6	5	4	3	2	1	0
1	1	0	1	Data Register			Op-Mode			Effective Address					
										Mode			Register		

Op-Mode field:

Word	Long	Operation
011	111	(\<ea\>)+(\<An\>)→An

ADDX

15	14	13	12	11	10	9	8	7	6	5	4	3	2	1	0
1	1	0	1	Destination Register*			1	Size		0	0	R/M	Source Register*		

Size field: 00 = byte
           01 = word
           10 = long
R/M field: 0 = data register to data register
          1 = memory to memory
*If R/M = 0, specifies a data register.
 If R/M = 1, specifies an address register for the predecrement addressing
     mode.

SHIFT/ROTATE — Register

15	14	13	12	11	10	9	8	7	6	5	4	3	2	1	0
1	1	1	0	Count/Register			dr	Size		i/r	Type		Data Register		

Count/Register field: If i/r field = 0, specifies shift count
                         If i/r field = 1, specifies a data register that contains the
                                   shift count

dr field: 0 = right
        1 = left
Size field: 00 = byte
           01 = word
           10 = long
i/r field: 0 = immediate shift count
        1 = register shift count
Type field: 00 = arithmetic shift
           01 = logical shift
           10 = rotate with extend
           11 = rotate

SHIFT/ROTATE — Memory

15	14	13	12	11	10	9	8	7	6	5	4	3	2	1	0
1	1	1	0	0	Type		dr	1	1	Effective Address Mode			Register		

Type field: 00 = arithmetic shift
           01 = logical shift
           10 = rotate with extend
           11 = rotate
dr field: 0 = right
        1 = left

APPENDIX IVD: INSTRUCTION EXECUTION TIMES

D.1 INTRODUCTION

This Appendix contains listings of the instruction execution times in terms of external clock (CLK) periods. In this data, it is assumed that both memory read and write cycle times are four clock periods. A longer memory cycle will cause the generation of wait states which must be added to the total instruction time.

The number of bus read and write cycles for each instruction is also included with the timing data. This data is enclosed in parenthesis following the number of clock periods and is shown as: (r/w) where r is the number of read cycles and w is the number of write cycles included in the clock period number. Recalling that either a read or write cycle requires four clock periods, a timing number given as 18(3/1) relates to 12 clock periods for the three read cycles, plus 4 clock periods for the one write cycle, plus 2 cycles required for some internal function of the processor.

NOTE
The number of periods includes instruction fetch and all applicable operand fetches and stores.

D.2 OPERAND EFFECTIVE ADDRESS CALCULATION TIMING

Table D-1 lists the number of clock periods required to compute an instruction's effective address. It includes fetching of any extension words, the address computation, and fetching of the memory operand. The number of bus read and write cycles is shown in parenthesis as (r/w). Note there are no write cycles involved in processing the effective address.

Table D-1. Effective Address Calculation Times

Addressing Mode		Byte, Word	Long
Register			
Dn	Data Register Direct	0(0/0)	0(0/0)
An	Address Register Direct	0(0/0)	0(0/0)
Memory			
(An)	Address Register Indirect	4(1/0)	8(2/0)
(An) +	Address Register Indirect with Postincrement	4(1/0)	8(2/0)
− (An)	Address Register Indirect with Predecrement	6(1/0)	10(2/0)
d(An)	Address Register Indirect with Displacement	8(2/0)	12(3/0)
d(An, ix)*	Address Register Indirect with Index	10(2/0)	14(3/0)
xxx.W	Absolute Short	8(2/0)	12(3/0)
xxx.L	Absolute Long	12(3/0)	16(4/0)
d(PC)	Program Counter with Displacement	8(2/0)	12(3/0)
d(PC, ix)*	Program Counter with Index	10(2/0)	14(3/0)
#xxx	Immediate	4(1/0)	8(2/0)

*The size of the index register (ix) does not affect execution time.

576

D.3 MOVE INSTRUCTION EXECUTION TIMES

Tables D-2 and D-3 indicate the number of clock periods for the move instruction. This data includes instruction fetch, operand reads, and operand writes. The number of bus read and write cycles is shown in parenthesis as (r/w).

Table D-2. Move Byte and Word Instruction Execution Times

Source	Destination								
	Dn	An	(An)	(An)+	−(An)	d(An)	d(An, ix)*	xxx.W	xxx.L
Dn	4(1/0)	4(1/0)	8(1/1)	8(1/1)	8(1/1)	12(2/1)	14(2/1)	12(2/1)	16(3/1)
An	4(1/0)	4(1/0)	8(1/1)	8(1/1)	8(1/1)	12(2/1)	14(2/1)	12(2/1)	16(3/1)
(An)	8(2/0)	8(2/0)	12(2/1)	12(2/1)	12(2/1)	16(3/1)	18(3/1)	16(3/1)	20(4/1)
(An)+	8(2/0)	8(2/0)	12(2/1)	12(2/1)	12(2/1)	16(3/1)	18(3/1)	16(3/1)	20(4/1)
−(An)	10(2/0)	10(2/0)	14(2/1)	14(2/1)	14(2/1)	18(3/1)	20(3/1)	18(3/1)	22(4/1)
d(An)	12(3/0)	12(3/0)	16(3/1)	16(3/1)	16(3/1)	20(4/1)	22(4/1)	20(4/1)	24(5/1)
d(An, ix)*	14(3/0)	14(3/0)	18(3/1)	18(3/1)	18(3/1)	22(4/1)	24(4/1)	22(4/1)	26(5/1)
xxx.W	12(3/0)	12(3/0)	16(3/1)	16(3/1)	16(3/1)	20(4/1)	22(4/1)	20(4/1)	24(5/1)
xxx.L	16(4/0)	16(4/0)	20(4/1)	20(4/1)	20(4/1)	24(5/1)	26(5/1)	24(5/1)	28(6/1)
d(PC)	12(3/0)	12(3/0)	16(3/1)	16(3/1)	16(3/1)	20(4/1)	22(4/1)	20(4/1)	24(5/1)
d(PC, ix)*	14(3/0)	14(3/0)	18(3/1)	18(3/1)	18(3/1)	22(4/1)	24(4/1)	22(4/1)	26(5/1)
#xxx	8(2/0)	8(2/0)	12(2/1)	12(2/1)	12(2/1)	16(3/1)	18(3/1)	16(3/1)	20(4/1)

*The size of the index register (ix) does not affect execution time.

Table D-3. Move Long Instruction Execution Times

Source	Destination								
	Dn	An	(An)	(An)+	−(An)	d(An)	d(An, ix)*	xxx.W	xxx.L
Dn	4(1/0)	4(1/0)	12(1/2)	12(1/2)	12(1/2)	16(2/2)	18(2/2)	16(2/2)	20(3/2)
An	4(1/0)	4(1/0)	12(1/2)	12(1/2)	12(1/2)	16(2/2)	18(2/2)	16(2/2)	20(3/2)
(An)	12(3/0)	12(3/0)	20(3/2)	20(3/2)	20(3/2)	24(4/2)	26(4/2)	24(4/2)	28(5/2)
(An)+	12(3/0)	12(3/0)	20(3/2)	20(3/2)	20(3/2)	24(4/2)	26(4/2)	24(4/2)	28(5/2)
−(An)	14(3/0)	14(3/0)	22(3/2)	22(3/2)	22(3/2)	26(4/2)	28(4/2)	26(4/2)	30(5/2)
d(An)	16(4/0)	16(4/0)	24(4/2)	24(4/2)	24(4/2)	28(5/2)	30(5/2)	28(5/2)	32(6/2)
d(An, ix)*	18(4/0)	18(4/0)	26(4/2)	26(4/2)	26(4/2)	30(5/2)	32(5/2)	30(5/2)	34(6/2)
xxx.W	16(4/0)	16(4/0)	24(4/2)	24(4/2)	24(4/2)	28(5/2)	30(5/2)	28(5/2)	32(6/2)
xxx.L	20(5/0)	20(5/0)	28(5/2)	28(5/2)	28(5/2)	32(6/2)	34(6/2)	32(6/2)	36(7/2)
d(PC)	16(4/0)	16(4/0)	24(4/2)	24(4/2)	24(4/2)	28(5/2)	30(5/2)	28(5/2)	32(5/2)
d(PC, ix)*	18(4/0)	18(4/0)	26(4/2)	26(4/2)	26(4/2)	30(5/2)	32(5/2)	30(5/2)	34(6/2)
#xxx	12(3/0)	12(3/0)	20(3/2)	20(3/2)	20(3/2)	24(4/2)	26(4/2)	24(4/2)	28(5/2)

*The size of the index register (ix) does not affect execution time.

D.4 STANDARD INSTRUCTION EXECUTION TIMES

The number of clock periods shown in Table D-4 indicates the time required to perform the operations, store the results, and read the next instruction. The number of bus read and write cycles is shown in parenthesis as (r/w). The number of clock periods and the number of read and write cycles must be added respectively to those of the effective address calculation where indicated.

In Table D-4 the headings have the following meanings: An = address register operand, Dn = data register operand, ea = an operand specified by an effective address, and M = memory effective address operand.

Table D-4. Standard Instruction Execution Times

Instruction	Size	op<ea>, Ant	op<ea>, Dn	op Dn, <M>
ADD	Byte, Word	8(1/0) +	4(1/0) +	8(1/1) +
	Long	6(1/0) + **	6(1/0) + **	12(1/2) +
AND	Byte, Word	−	4(1/0) +	8(1/1) +
	Long	−	6(1/0) + **	12(1/2) +
CMP	Byte, Word	6(1/0) +	4(1/0) +	−
	Long	6(1/0) +	6(1/0) +	−
DIVS	−	−	158(1/0) + *	−
DIVU	−	−	140(1/0) + *	−
EOR	Byte, Word	−	4(1/0) ***	8(1/1) +
	Long	−	8(1/0) ***	12(1/2) +
MULS	−	−	70(1/0) + *	−
MULU	−	−	70(1/0) + *	−
OR	Byte, Word	−	4(1/0) +	8(1/1) +
	Long	−	6(1/0) + **	12(1/2) +
SUB	Byte, Word	8(1/0) +	4(1/0) +	8(1/1) +
	Long	6(1/0) + **	6(1/0) + **	12(1/2) +

NOTES:
+ add effective address calculation time
† word or long only
* indicates maximum value
** The base time of six clock periods is increased to eight if the effective address mode is register direct or immediate (effective address time should also be added).
*** Only available effective address mode is data register direct.
DIVS, DIVU — The divide algorithm used by the MC68000 provides less than 10% difference between the best and worst case timings.
MULS, MULU — The multiply algorithm requires 38 + 2n clocks where n is defined as:
  MULU: n = the number of ones in the <ea>
  MULS: n = concatenate the <ea> with a zero as the LSB; n is the resultant number of 10 or 01 patterns in the 17-bit source; i.e., worst case happens when the source is $5555.

D.5 IMMEDIATE INSTRUCTION EXECUTION TIMES

The number of clock periods shown in Table D-5 includes the time to fetch immediate operands, perform the operations, store the results, and read the next operation. The number of bus read and write cycles is shown in parenthesis as (r/w). The number of clock periods and the number of read and write cycles must be added respectively to those of the effective address calculation where indicated.

In Table D-5, the headings have the following meanings: # = immediate operand, Dn = data register operand, An = address register operand, and M = memory operand. SR = status register.

Table D-5. Immediate Instruction Execution Times

Instruction	Size	op #, Dn	op #, An	op #, M
ADDI	Byte, Word	8(2/0)	–	12(2/1) +
	Long	16(3/0)	–	20(3/2) +
ADDQ	Byte, Word	4(1/0)	8(1/0) *	8(1/1) +
	Long	8(1/0)	8(1/0)	12(1/2) +
ANDI	Byte, Word	8(2/0)	–	12(2/1) +
	Long	16(3/0)	–	20(3/1) +
CMPI	Byte, Word	8(2/0)	–	8(2/0) +
	Long	14(3/0)	–	12(3/0) +
EORI	Byte, Word	8(2/0)	–	12(2/1) +
	Long	16(3/0)	–	20(3/2) +
MOVEQ	Long	4(1/0)	–	–
ORI	Byte, Word	8(2/0)	–	12(2/1) +
	Long	16(3/0)	–	20(3/2) +
SUBI	Byte, Word	8(2/0)	–	12(2/1) +
	Long	16(3/0)	–	20(3/2) +
SUBQ	Byte, Word	4(1/0)	8(1/0) *	8(1/1) +
	Long	8(1/0)	8(1/0)	12(1/2) +

+ add effective address calculation time
* word only

D.6 SINGLE OPERAND INSTRUCTION EXECUTION TIMES

Table D-6 indicates the number of clock periods for the single operand instructions. The number of bus read and write cycles is shown in parenthesis as (r/w). The number of clock periods and the number of read and write cycles must be added respectively to those of the effective address calculation where indicated.

Table D-6. Single Operand Instruction Execution Times

Instruction	Size	Register	Memory
CLR	Byte, Word	4(1/0)	8(1/1) +
	Long	6(1/0)	12(1/2) +
NBCD	Byte	6(1/0)	8(1/1) +
NEG	Byte, Word	4(1/0)	8(1/1) +
	Long	6(1/0)	12(1/2) +
NEGX	Byte, Word	4(1/0)	8(1/1) +
	Long	6(1/0)	12(1/2) +
NOT	Byte, Word	4(1/0)	8(1/1) +
	Long	6(1/0)	12(1/2) +
S_{CC}	Byte, False	4(1/0)	8(1/1) +
	Byte, True	6(1/0)	8(1/1) +
TAS	Byte	4(1/0)	10(1/1) +
TST	Byte, Word	4(1/0)	4(1/0) +
	Long	4(1/0)	4(1/0) +

+ add effective address calculation time

D.7 SHIFT/ROTATE INSTRUCTION EXECUTION TIMES

Table D-7 indicates the number of clock periods for the shift and rotate instructions. The number of bus read and write cycles is shown in parenthesis as (r/w). The number of clock periods and the number of read and write cycles must be added respectively to those of the effective address calculation where indicated.

Table D-7. Shift/Rotate Instruction Execution Times

Instruction	Size	Register	Memory
ASR, ASL	Byte, Word	6 + 2n(1/0)	8(1/1) +
	Long	8 + 2n(1/0)	—
LSR, LSL	Byte, Word	6 + 2n(1/0)	8(1/1) +
	Long	8 + 2n(1/0)	—
ROR, ROL	Byte, Word	6 + 2n(1/0)	8(1/1) +
	Long	8 + 2n(1/0)	—
ROXR, ROXL	Byte, Word	6 + 2n(1/0)	8(1/1) +
	Long	8 + 2n(1/0)	—

+ add effective address calculation time
n is the shift count

D.8 BIT MANIPULATION INSTRUCTION EXECUTION TIMES

Table D-8 indicates the number of clock periods required for the bit manipulation instructions. The number of bus read and write cycles is shown in parenthesis as (r/w). The number of clock periods and the number of read and write cycles must be added respectively to those of the effective address calculation where indicated.

Table D-8. Bit Manipulation Instruction Execution Times

Instruction	Size	Dynamic		Static	
		Register	Memory	Register	Memory
BCHG	Byte	—	8(1/1) +	—	12(2/1) +
	Long	8(1/0) *	—	12(2/0) *	—
BCLR	Byte	—	8(1/1) +	—	12(2/1) +
	Long	10(1/0) *	—	14(2/0) *	—
BSET	Byte	—	8(1/1) +	—	12(2/1) +
	Long	8(1/0) *	—	12(2/0) *	—
BTST	Byte	—	4(1/0) +	—	8(2/0) +
	Long	6(1/0)	—	10(2/0)	—

+ add effective address calculation time
* indicates maximum value

D.9 CONDITIONAL INSTRUCTION EXECUTION TIMES

Table D-9 indicates the number of clock periods required for the conditional instructions. The number of bus read and write cycles is indicated in parenthesis as (r/w). The number of clock periods and the number of read and write cycles must be added respectively to those of the effective address calculation where indicated.

Table D-9. Conditional Instruction Execution Times

Instruction	Displacement	Branch Taken	Branch Not Taken
BCC	Byte	10(2/0)	8(1/0)
	Word	10(2/0)	12(2/0)
BRA	Byte	10(2/0)	—
	Word	10(2/0)	—
BSR	Byte	18(2/2)	—
	Word	18(2/2)	—
DBCC	CC true	—	12(2/0)
	CC false	10(2/0)	14(3/0)

+ add effective address calculation time
* indicates maximum value

D.10 JMP, JSR, LEA, PEA, AND MOVEM INSTRUCTION EXECUTION TIMES

Table D-10 indicates the number of clock periods required for the jump, jump-to-subroutine, load effective address, push effective address, and move multiple registers instructions. The number of bus read and write cycles is shown in parenthesis as (r/w).

Table D-10. JMP, JSR, LEA, PEA, and MOVEM Instruction Execution Times

Instr	Size	(An)	(An) +	– (An)	d(An)	d(An, ix) +	xxx.W	xxx.L	d(PC)	d(PC, ix)*
JMP	–	8(2/0)	–	–	10(2/0)	14(3/0)	10(2/0)	12(3/0)	10(2/0)	14(3/0)
JSR	–	16(2/2)	–	–	18(2/2)	22(2/2)	18(2/2)	20(3/2)	18(2/2)	22(2/2)
LEA	–	4(1/0)	–	–	8(2/0)	12(2/0)	8(2/0)	12(3/0)	8(2/0)	12(2/0)
PEA	–	12(1/2)	–	–	16(2/2)	20(2/2)	16(2/2)	20(3/2)	16(2/2)	20(2/2)
MOVEM M → R	Word	12 + 4n (3 + n/0)	12 + 4n (3 + n/0)	–	16 + 4n (4 + n/0)	18 + 4n (4 + n/0)	16 + 4n (4 + n/0)	20 + 4n (5 + n/0)	16 + 4n (4 + n/0)	18 + 4n (4 + n/0)
	Long	12 + 8n (3 + 2n/0)	12 + 8n (3 + 2n/0)	–	16 + 8n (4 + 2n/0)	18 + 8n (4 + 2n/0)	16 + 8n (4 + 2n/0)	20 + 8n (5 + 2n/0)	16 + 8n (4 + 2n/0)	18 + 8n (4 + 2n/0)
MOVEM R → M	Word	8 + 4n (2/n)	–	8 + 4n (2/n)	12 + 4n (3/n)	14 + 4n (3/n)	12 + 4n (3/n)	16 + 4n (4/n)	–	–
	Long	8 + 8n (2/2n)	–	8 + 8n (2/2n)	12 + 8n (3/2n)	14 + 8n (3/2n)	12 + 8n (3/2n)	16 + 8n (4/2n)	–	–

n is the number of registers to move
* is the size of the index register (ix) does not affect the instruction's execution time

D 11 MULTI-PRECISION INSTRUCTION EXECUTION TIMES

Table D-11 indicates the number of clock periods for the multi-precision instructions. The number of clock periods includes the time to fetch both operands, peform the operations, store the results, and read the next instructions. The number of read and write cycles is shown in parenthesis as (r/w).

In Table D-11, the headings have the following meanings: Dn = data register operand and M = memory operand.

Table D-11. Multi-Precision Instruction Execution Times

Instruction	Size	op Dn, Dn	op M, M
ADDX	Byte, Word	4(1/0)	18(3/1)
	Long	8(1/0)	30(5/2)
CMPM	Byte, Word	–	12(3/0)
	Long	–	20(5/0)
SUBX	Byte, Word	4(1/0)	18(3/1)
	Long	8(1/0)	30(5/2)
ABCD	Byte	6(1/0)	18(3/1)
SBCD	Byte	6(1/0)	18(3/1)

D.12 MISCELLANEOUS INSTRUCTION EXECUTION TIMES

Tables D-12 and D-13 indicate the number of clock periods for the following miscellaneous instructions. The number of bus read and write cycles is shown in parenthesis as (r/w). The number of clock periods plus the number of read and write cycles must be added to those of the effective address calculation where indicated.

Table D-12. Miscellaneous Instruction Execution Times

Instruction	Size	Register	Memory
ANDI to CCR	Byte	20(3/0)	—
ANDI to SR	Word	20(3/0)	—
CHK	—	10(1/0) +	—
EORI to CCR	Byte	20(3/0)	—
EORI to SR	Word	20(3/0)	—
ORI to CCR	Byte	20(3/0)	—
ORI to SR	Word	20(3/0)	—
MOVE from SR	—	6(1/0)	8(1/1) +
MOVE to CCR	—	12(2/0)	12(2/0) +
MOVE to SR	—	12(2/0)	12(2/0) +
EXG	—	6(1/0)	—
EXT	Word	4(1/0)	—
	Long	4(1/0)	—
LINK	—	16(2/2)	—
MOVE from USP	—	4(1/0)	—
MOVE to USP	—	4(1/0)	—
NOP	—	4(1/0)	—
RESET	—	132(1/0)	—
RTE	—	20(5/0)	—
RTR	—	20(5/0)	—
RTS	—	16(4/0)	—
STOP	—	4(0/0)	—
SWAP	—	4(1/0)	—
TRAPV	—	4(1/0)	—
UNLK	—	12(3/0)	—

+ add effective address calculation time

Table D-13. Move Peripheral Instruction Execution Times

Instruction	Size	Register → Memory	Memory → Register
MOVEP	Word	16(2/2)	16(4/0)
	Long	24(2/4)	24(6/0)

D.13 EXCEPTION PROCESSING EXECUTION TIMES

Table D-14 indicates the number of clock periods for exception processing. The number of clock periods includes the time for all stacking, the vector fetch, and the fetch of the first two instruction words of the handler routine. The number of bus read and write cycles is shown in parenthesis as (r/w).

Table D-14. Exception Processing Execution Times

Exception	Periods
Address Error	50(4/7)
Bus Error	50(4/7)
CHK Instruction	44(5/4) +
Divide by Zero	42(5/4)
Illegal Instruction	34(4/3)
Interrupt	44(5/3) *
Privilege Violation	34(4/3)
RESET * *	40(6/0)
Trace	34(4/3)
TRAP Instruction	38(4/4)
TRAPV Instruction	34(4/3)

+ add effective address calculation time

* The interrupt acknowledge cycle is assumed to take four clock periods.

* * Indicates the time from when $\overline{RESET}$ and $\overline{HALT}$ are first sampled as negated to when instruction execution starts.

APPENDIX IVE: MC68008 INSTRUCTION EXECUTION TIMES

E.1 INTRODUCTION

This Appendix contains listings of the instruction execution times in terms of external clock (CLK) periods. In this data, it is assumed that both memory read and write cycle times are four clock periods. A longer memory cycle will cause the generation of wait states which must be added to the total instruction time.

The number of bus read and write cycles for each instruction is also included with the timing data. This data is enclosed in parenthesis following the number of clock periods and is shown as: (r/w) where r is the number of read cycles and w is the number of write cycles included in the clock period number. Recalling that either a read or write cycle requires four clock periods, a timing number given as 18(3/1) relates to 12 clock periods for the three read cycles, plus 4 clock periods for the one write cycle, plus 2 cycles required for some internal function of the processor.

NOTE
The number of periods includes instruction fetch and all applicable operand fetches and stores.

E.2 OPERAND EFFECTIVE ADDRESS CALCULATION TIMES

Table E-1 lists the number of clock periods required to compute an instruction's effective address. It includes fetching of any extension words, the address computation, and fetching of the memory operand. The number of bus read and write cycles is shown in parenthesis as (r/w). Note there are no write cycles involved in processing the effective address.

Table E-1. Effective Address Calculation Times

Addressing Mode		Byte	Word	Long
	Register			
Dn	Data Register Direct	0(0/0)	0(0/0)	0(0/0)
An	Address Register Direct	0(0/0)	0(0/0)	0(0/0)
	Memory			
(An)	Address Register Indirect	4(1/0)	8(2/0)	16(4/0)
(An)+	Address Register Indirect with Postincrement	4(1/0)	8(2/0)	16(4/0)
−(An)	Address Register Indirect with Predecrement	6(1/0)	10(2/0)	18(4/0)
d(An)	Address Register Indirect with Displacement	12(3/0)	16(4/0)	24(6/0)
d(An, ix)*	Address Register Indirect with Index	14(3/0)	18(4/0)	26(6/0)
xxx.W	Absolute Short	12(3/0)	16(4/0)	24(6/0)
xxx.L	Absolute Long	20(5/0)	24(6/0)	32(8/0)
d(PC)	Program Counter with Displacement	12(3/0)	16(4/0)	24(6/0)
d(PC, ix)	Program Counter with Index	14(3/0)	18(4/0)	26(6/0)
#xxx	Immediate	8(2/0)	8(2/0)	16(4/0)

*The size of the index register (ix) does not affect execution time.

E.3 MOVE INSTRUCTION EXECUTION TIMES

Tables E-2, E-3, and E-4 indicate the number of clock periods for the move instruction. This data includes instruction fetch, operand reads, and operand writes. The number of bus read and write cycles is shown in parenthesis as: (r/w).

Table E-2. Move Byte Instruction Execution Times

Source	Destination								
	Dn	An	(An)	(An)+	–(An)	d(An)	d(An, x)*	xxx.W	xxx.L
Dn	8(2/0)	8(2/0)	12(2/1)	12(2/1)	12(2/1)	20(4/1)	22(4/1)	20(4/1)	28(6/1)
An	8(2/0)	8(2/0)	12(2/1)	12(2/1)	12(2/1)	20(4/1)	22(4/1)	20(4/1)	28(6/1)
(An)	12(3/0)	12(3/0)	16(3/1)	16(3/1)	16(3/1)	24(5/1)	26(5/1)	24(5/1)	32(7/1)
(An)+	12(3/0)	12(3/0)	16(3/1)	16(3/1)	16(3/1)	24(5/1)	26(5/1)	24(5/1)	32(7/1)
–(An)	14(3/0)	14(3/0)	18(3/1)	18(3/1)	18(3/1)	26(5/1)	28(5/1)	26(5/1)	34(7/1)
d(An)	20(5/0)	20(5/0)	24(5/1)	24(5/1)	24(5/1)	32(7/1)	34(7/1)	32(7/1)	40(9/1)
d(An, ix)*	22(5/0)	22(5/0)	26(5/1)	26(5/1)	26(5/1)	34(7/1)	36(7/1)	34(7/1)	42(9/1)
xxx.W	20(5/0)	20(5/0)	24(5/1)	24(5/1)	24(5/1)	32(7/1)	34(7/1)	32(7/1)	40(9/1)
xxx.L	28(7/0)	28(7/0)	32(7/1)	32(7/1)	32(7/1)	40(9/1)	42(9/1)	40(9/1)	48(11/1)
d(PC)	20(5/0)	20(5/0)	24(5/1)	24(5/1)	24(5/1)	32(7/1)	34(7/1)	32(7/1)	40(9/1)
d(PC, ix)*	22(5/0)	22(5/0)	26(5/1)	26(5/1)	26(5/1)	34(7/1)	36(7/1)	34(7/1)	42(9/1)
#xxx	16(4/0)	16(4/0)	20(4/1)	20(4/1)	20(4/1)	28(6/1)	30(6/1)	28(6/1)	36(8/1)

*The size of the index register (ix) does not affect execution time.

Table E-3. Move Word Instruction Execution Times

Source	Destination								
	Dn	An	(An)	(An)+	–(An)	d(An)	d(An, ix)*	xxx.W	xxx.L
Dn	8(2/0)	8(2/0)	16(2/2)	16(2/2)	16(2/2)	24(4/2)	26(4/2)	20(4/2)	32(6/2)
An	8(2/0)	8(2/0)	16(2/2)	16(2/2)	16(2/2)	24(4/2)	26(4/2)	20(4/2)	32(6/2)
(An)	16(4/0)	16(4/0)	24(4/2)	24(4/2)	24(4/2)	32(6/2)	34(6/2)	32(6/2)	40(8/2)
(An)+	16(4/0)	16(4/0)	24(4/2)	24(4/2)	24(4/2)	32(6/2)	34(6/2)	32(6/2)	40(8/2)
–(An)	18(4/0)	18(4/0)	26(4/2)	26(4/2)	26(4/2)	34(6/2)	36(6/2)	34(6/2)	42(8/2)
d(An)	24(6/0)	24(6/0)	32(6/2)	32(6/2)	32(6/2)	40(8/2)	42(8/2)	40(8/2)	48(10/2)
d(An, ix)*	26(6/0)	26(6/0)	34(6/2)	34(6/2)	34(6/2)	42(8/2)	44(8/2)	42(8/2)	50(10/2)
xxx.W	24(6/0)	24(6/0)	32(6/2)	32(6/2)	32(6/2)	40(8/2)	42(8/2)	40(8/2)	48(10/2)
xxx.L	32(8/0)	32(8/0)	40(8/2)	40(8/2)	40(8/2)	48(10/2)	50(10/2)	48(10/2)	56(12/2)
d(PC)	24(6/0)	24(6/0)	32(6/2)	32(6/2)	32(6/2)	40(8/2)	42(8/2)	40(8/2)	48(10/2)
d(PC, ix)*	26(6/0)	26(6/0)	34(6/2)	34(6/2)	34(6/2)	42(8/2)	44(8/2)	42(8/2)	50(10/2)
#xxx	16(4/0)	16(4/0)	24(4/2)	24(4/2)	24(4/2)	32(6/2)	34(6/2)	32(6/2)	40(8/2)

*The size of the index register (ix) does not affect execution time.

Table E-4. Move Long Instruction Execution Times

| Source | Destination | | | | | | | | |
	Dn	An	(An)	(An)+	−(An)	d(An)	d(An, ix)*	xxx.W	xxx.L
Dn	8(2/0)	8(2/0)	24(2/4)	24(2/4)	24(2/4)	32(4/4)	34(4/4)	32(4/4)	40(6/4)
An	8(2/0)	8(2/0)	24(2/4)	24(2/4)	24(2/4)	32(4/4)	34(4/4)	32(4/4)	40(6/4)
(An)	24(6/0)	24(6/0)	40(6/4)	40(6/4)	40(6/4)	48(8/4)	50(8/4)	48(8/4)	56(10/4)
(An)+	24(6/0)	24(6/0)	40(6/4)	40(6/4)	40(6/4)	48(8/4)	50(8/4)	48(8/4)	56(10/4)
−(An)	26(6/0)	26(6/0)	42(6/4)	42(6/4)	42(6/4)	50(8/4)	52(8/4)	50(8/4)	58(10/4)
d(An)	32(8/0)	32(8/0)	48(8/4)	48(8/4)	48(8/4)	56(10/4)	58(10/4)	56(10/4)	64(12/4)
d(An, ix)*	34(8/0)	34(8/0)	50(8/4)	50(8/4)	50(8/4)	58(10/4)	60(10/4)	58(10/4)	66(12/4)
xxx.W	32(8/0)	32(8/0)	48(8/4)	48(8/4)	48(8/4)	56(10/4)	58(10/4)	56(10/4)	64(12/4)
xxx.L	40(10/0)	40(10/0)	56(10/4)	56(10/4)	56(10/4)	64(12/4)	66(12/4)	64(12/4)	72(14/4)
d(PC)	32(8/0)	32(8/0)	48(8/4)	48(8/4)	48(8/4)	56(10/4)	58(10/4)	56(10/4)	64(12/4)
d(PC, ix)*	34(8/0)	34(8/0)	50(8/4)	50(8/4)	50(8/4)	58(10/4)	60(10/4)	58(10/4)	66(12/4)
#xxx	24(6/0)	24(6/0)	40(6/4)	40(6/4)	40(6/4)	48(8/4)	50(8/4)	48(8/4)	56(10/4)

*The size of the index register (ix) does not affect execution time.

E.4 STANDARD INSTRUCTION EXECUTION TIMES

The number of clock periods shown in Table E-5 indicates the time required to perform the operations, store the results, and read the next instruction. The number of bus read and write cycles is shown in parenthesis as: (r/w). The number of clock periods and the number of read and write cycles must be added respectively to those of the effective address calculation where indicated. In Table E-5 the headings have the following meanings: An = address register operand, Dn = data register operand, ea = an operand specified by an effective address, and M = memory effective address operand.

Table E-5. Standard Instruction Execution Times

Instruction	Size	op <ea>, An	op <ea>, Dn	op Dn, <M>
ADD	Byte	−	8(2/0) +	12(2/1) +
	Word	12(2/0) +	8(2/0) +	16(2/2) +
	Long	10(2/0) + **	10(2/0) + **	24(2/4) +
AND	Byte	−	8(2/0) +	12(2/1) +
	Word	−	8(2/0) +	16(2/2) +
	Long	−	10(2/0) + **	24(2/4) +
CMP	Byte	−	8(2/0) +	−
	Word	10(2/0) +	8(2/0) +	−
	Long	10(2/0) +	10(2/0) +	−
DIVS		−	162(2/0) + *	−
DIVU		−	144(2/0) + *	−
EOR	Byte	−	8(2/0) + ***	12(2/1) +
	Word	−	8(2/0) + ***	16(2/2) +
	Long	−	12(2/0) + ***	24(2/4) +
MULS		−	74(2/0) + *	−
MULU		−	74(2/0) + *	−
OR	Byte	−	8(2/0) +	12(2/1) +
	Word	−	8(2/0) +	16(2/2) +
	Long	−	10(2/0) + **	24(2/4) +
SUB	Byte	−	8(2/0) +	12(2/1) +
	Word	12(2/0) +	8(2/0) +	16(2/2) +
	Long	10(2/0) + **	10(2/0) + **	24(2/4) +

NOTES:
- + Add effective address calculation time
- * Indicates maximum value
- ** The base time of 10 clock periods is increased to 12 if the effective address mode is register direct or immediate (effective address time should also be added).
- *** Only available effective address mode is data register direct

DIVS, DIVU — The divide algorithm used by the MC68008 provides less than 10% difference between the best and worst case timings.

MULS, MULU — The multiply algorithm requires 42 + 2n clocks where n is defined as:
MULS: n = tag the <ea> with a zero as the MSB; n is the resultant number of 10 or 01 patterns in the 17-bit source, i.e., worst case happens when the source is $5555.
MULU: n = the number of ones in the <ea>

E.5 IMMEDIATE INSTRUCTION EXECUTION TIMES

The number of clock periods shown in Table E-6 includes the time to fetch immediate operands, perform the operations, store the results, and read the next operation. The number of bus read and write cycles is shown in parenthesis as: (r/w). The number of clock periods and the number of read and write cycles must be added respectively to those of the effective address calculation where indicated. In Table E-6, the headings have the following meanings: # = immediate operand, Dn = data register operand, An = address register operand, and M = memory operand.

Table E-6. Immediate Instruction Clock Periods

Instruction	Size	op#, Dn	op#, An	op#, M
ADDI	Byte	16(4/0)	—	20(4/1) +
	Word	16(4/0)	—	24(4/2) +
	Long	28(6/0)	—	40(6/4) +
ADDQ	Byte	8(2/0)	—	12(2/1) +
	Word	8(2/0)	12(2/0)	16(2/2) +
	Long	12(2/0)	12(2/0)	24(2/4) +
ANDI	Byte	16(4/0)	—	20(4/1) +
	Word	16(4/0)	—	24(4/2) +
	Long	28(6/0)	—	40(6/4) +
CMPI	Byte	16(4/0)	—	16(4/0) +
	Word	16(4/0)	—	16(4/0) +
	Long	26(6/0)	—	24(6/0) +
EORI	Byte	16(4/0)	—	20(4/1) +
	Word	16(4/0)	—	24(4/2) +
	Long	28(6/0)	—	40(6/4) +
MOVEQ	Long	8(2/0)	—	—
ORI	Byte	16(4/0)	—	20(4/1) +
	Word	16(4/0)	—	24(4/2) +
	Long	28(6/0)	—	40(6/4) +
SUBI	Byte	16(4/0)	—	12(2/1) +
	Word	16(4/0)	—	16(2/2) +
	Long	28(6/0)	—	24(2/4) +
SUBQ	Byte	8(2/0)	—	20(4/1) +
	Word	8(2/0)	12(2/0)	24(4/2) +
	Long	12(2/0)	12(2/0)	40(6/4) +

+ add effective address calculation time

E.6 SINGLE OPERAND INSTRUCTION EXECUTION TIMES

Table E-7 indicates the number of clock periods for the single operand instructions. The number of bus read and write cycles is shown in parenthesis as (r/w). The number of clock periods and the number of read and write cycles must be added respectively to those of the effective address calculation where indicated.

Table E-7. Single Operand Instruction Execution Times

Instruction	Size	Register	Memory
CLR	Byte	8(2/0)	12(2/1) +
	Word	8(2/0)	16(2/2) +
	Long	10(2/0)	24(2/4) +
NBCD	Byte	10(2/0)	12(2/1) +
NEG	Byte	8(2/0)	12(2/1) +
	Word	8(2/0)	16(2/2) +
	Long	10(2/0)	24(2/4) +
NEGX	Byte	8(2/0)	12(2/1) +
	Word	8(2/0)	16(2/2) +
	Long	10(2/0)	24(2/4) +
NOT	Byte	8(2/0)	12(2/1) +
	Word	8(2/0)	16(2/2) +
	Long	10(2/0)	24(2/4) +
S$_{CC}$	Byte, False	8(2/0)	12(2/1) +
	Byte, True	10(2/0)	12(2/1) +
TAS	Byte	8(2/0)	14(2/1) +
TST	Byte	8(2/0)	8(2/0) +
	Word	8(2/0)	8(2/0) +
	Long	8(2/0)	8(2/0) +

+ add effective address calculation time.

E.7 SHIFT/ROTATE INSTRUCTION EXECUTION TIMES

Table E-8 indicates the number of clock periods for the shift and rotate instructions. The number of bus read and write cycles is shown in parenthesis as: (r/w). The number of clock periods and the number of read and write cycles must be added respectively to those of the effective address calculation where indicated.

Table E-8. Shift/Rotate Instruction Clock Periods

Instruction	Size	Register	Memory
ASR, ASL	Byte	10 + 2n(2/0)	—
	Word	10 + 2n(2/0)	16(2/2) +
	Long	12 + 2n(2/0)	—
LSR, LSL	Byte	10 + 2n(2/0)	—
	Word	10 + 2n(2/0)	16(2/2) +
	Long	12 + 2n(2/0)	—
ROR, ROL	Byte	10 + 2n(2/0)	—
	Word	10 + 2n(2/0)	16(2/2) +
	Long	12 + 2n(2/0)	—
ROXR, ROXL	Byte	10 + 2n(2/0)	—
	Word	10 + 2n(2/0)	16(2/2) +
	Long	12 + 2n(2/0)	—

+ add effective address calculation time
n is the shift count

E.8 BIT MANIPULATION INSTRUCTION EXECUTION TIMES

Table E-9 indicates the number of clock periods required for the bit manipulation instructions. The number of bus read and write cycles is shown in parenthesis as: (r/w). The number of clock periods and the number of read and write cycles must be added respectively to those of the effective address calculation where indicated.

Table E-9. Bit Manipulation Instruction Execution Times

Instruction	Size	Dynamic		Static	
		Register	Memory	Register	Memory
BCHG	Byte	–	12(2/1) +	–	20(4/1) +
	Long	12(2/0) *	–	20(4/0) *	–
BCLR	Byte	–	12(2/1) +	–	20(4/1) +
	Long	14(2/0) *	–	22(4/0) *	–
BSET	Byte	–	12(2/1) +	–	20(4/1) +
	Long	12(2/0) *	–	20(4/0) *	–
BTST	Byte	–	8(2/0) +	–	16(4/0) +
	Long	10(2/0)	–	18(4/0)	–

\+ add effective address calculation time
* indicates maximum value

E.9 CONDITIONAL INSTRUCTION EXECUTION TIMES

Table E-10 indicates the number of clock periods required for the conditional instructions. The number of bus read and write cycles is indicated in parenthesis as: (r/w). The number of clock periods and the number of read and write cycles must be added respectively to those of the effective address calculation where indicated.

Table E-10. Conditional Instruction Execution Times

Instruction	Displacement	Trap or Branch Taken	Trap or Branch Not Taken
B$_{CC}$	Byte	18(4/0)	12(2/0)
	Word	18(4/0)	20(4/0)
BRA	Byte	18(4/0)	–
	Word	18(4/0)	–
BSR	Byte	34(4/4)	–
	Word	34(4/4)	–
DBCC	CC True	–	20(4/0)
	CC False	18(4/0)	26(6/0)
CHK	–	68(8/6) + *	14(2/0) +
TRAP	–	62(8/6)	–
TRAPV	–	66(10/6)	8(2/0)

\+ add effective address calculation time
* indicates maximum value

E.10 JMP, JSR, LEA, PEA, AND MOVEM INSTRUCTION EXECUTION TIMES

Table E-11 indicates the number of clock periods required for the jump, jump-to-subroutine, load effective address, push effective address, and move multiple registers instructions. The number of bus read and write cycles is shown in parenthesis as: (r/w).

Table E-11. JMP, JSR, LEA, PEA, and MOVEM Instruction Execution Times

Instruction	Size	(An)	(An) +	– (An)	d(An)	d(An, ix)*	xxx.W	xxx.L
JMP	–	16(4/0)	–	–	18(4/0)	22(4/0)	18(4/0)	24(6/0)
JSR	–	32(4/4)	–	–	34(4/4)	38(4/4)	34(4/4)	40(6/4)
LEA	–	8(2/0)	–	–	16(4/0)	20(4/0)	16(4/0)	24(6/0)
PEA	–	24(2/4)	–	–	32(4/4)	36(4/4)	32(4/4)	40(6/4)
MOVEM M → R	Word	24 + 8n (6 + 2n/0)	24 + 8n (6 + 2n/0)	– –	32 + 8n (8 + 2n/0)	34 + 8n (8 + 2n/0)	32 + 8n (10 + n/0)	40 + 8n (10 + 2n/0)
	Long	24 + 16n (6 + 4n/0)	24 + 16n (6 + 4n/0)	– –	32 + 16n (8 + 4n/0)	32 + 16n (8 + 4n/0)	32 + 16n (8 + 4n/0)	40 + 16n (8 + 4n/0)
MOVEM R → M	Word	16 + 8n (4/2n)	– –	16 + 8n (4/2n)	24 + 8n (6/2n)	26 + 8n (6/2n)	24 + 8n (6/2n)	32 + 8n (8/2n)
	Long	16 + 16n (4/4n)	– –	16 + 16n (4/4n)	24 + 16n (6/4n)	26 + 16n	24 + 16n (8/4n)	32 + 16n (6/4n)

n is the number of registers to move
* is the size of the index register (ix) does not affect the instruction's execution time

E.11 MULTI-PRECISION INSTRUCTION EXECUTION TIMES

Table E-12 indicates the number of clock periods for the multi-precision instructions. The number of clock periods includes the time to fetch both operands, perform the operations, store the results, and read the next instructions. The number of read and write cycles is shown in parenthesis as: (r/w).

In Table E-12, the headings have the following meanings: Dn = data register operand and M = memory operand.

Table E-12. Multi-Precision Instruction Execution Times

Instruction	Size	op Dn, Dn	op M, M
ADDX	Byte	8(2/0)	22(4/1)
	Word	8(2/0)	50(6/2)
	Long	12(2/0)	58(10/4)
CMPM	Byte	–	16(4/0)
	Word	–	24(6/0)
	Long	–	40(10/0)
SUBX	Byte	8(2/0)	22(4/1)
	Word	8(2/0)	50(6/2)
	Long	12(2/0)	58(10/4)
ABCD	Byte	10(2/0)	20(4/1)
SBCD	Byte	10(2/0)	20(4/1)

E.12 MISCELLANEOUS INSTRUCTION EXECUTION TIMES

Tables E-13 and E-14 indicate the number of clock periods for the following miscellaneous instructions. The number of bus read and write cycles is shown in parenthesis as: (r/w). The number of clock periods plus the number of read and write cycles must be added to those of the effective address calculation where indicated.

Table E-13. Miscellaneous Instruction Execution Times

Instruction	Register	Memory
ANDI to CCR	32(6/0)	—
ANDI to SR	32(6/0)	—
EORI to CCR	32(6/0)	—
EORI to SR	32(6/0)	—
EXG	10(2/0)	—
EXT	8(2/0)	—
LINK	32(4/4)	—
MOVE to CCR	18(4/0)	18(4/0) +
MOVE to SR	18(4/0)	18(4/0) +
MOVE from SR	10(2/0)	16(2/2) +
MOVE to USP	8(2/0)	—
MOVE from USP	8(2/0)	—
NOP	8(2/0)	—
ORI to CCR	32(6/0)	—
ORI to SR	32(6/0)	—
RESET	136(2/0)	—
RTE	40(10/0)	—
RTR	40(10/0)	—
RTS	32(8/0)	—
STOP	4(0/0)	—
SWAP	8(2/0)	—
UNLK	24(6/0)	—

+ add effective address calculation time

Table E-14. Move Peripheral Instruction Execution Times

Instruction	Size	Register → Memory	Memory → Register
MOVEP	Word	24(4/2)	24(6/0)
	Long	32(4/4)	32(8/0)

+ add effective address calculation time

E.13 EXCEPTION PROCESSING EXECUTION TIMES

Table E-15 indicates the number of clock periods for exception processing. The number of clock periods includes the time for all stacking, the vector fetch, and the fetch of the first instruction of the handler routine. The number of bus read and write cycles is shown in parenthesis as: (r/w).

Table E-15. Exception Processing Execution Times

Exception	Periods
Address Error	94(8/14)
Bus Error	94(8/14)
Interrupt	72(9/6) *
Illegal Instruction	62(8/6)
Privileged Instruction	62(8/6)
Trace	62(8/6)

* The interrupt acknowledge bus cycle is assumed to take four external clock periods.

APPENDIX IVF: MC68010 INSTRUCTION EXECUTION TIMES

F.1 INTRODUCTION

This Appendix contains listings of the instruction execution times in terms of external clock (CLK) periods. In this data, it is assumed that both memory read and write cycle times are four clock periods. A longer memory cycle will cause the generation of wait states which must be added to the total instruction time.

The number of bus read and write cycles for each instruction is also included with the timing data. This data is enclosed in parenthesis following the number of clock periods and is shown as: (r/w) where r is the number of read cycles and w is the number of write cycles included in the clock period number. Recalling that either a read or write cycle requires four clock periods, a timing number given as 18(3/1) relates to 12 clock periods for the three read cycles, plus 4 clock periods for the one write cycle, plus 2 cycles required for some internal function of the processor.

NOTE

The number of periods includes instruction fetch and all applicable operand fetches and stores.

F.2 OPERAND EFFECTIVE ADDRESS CALCULATION TIMES

Table F-1 lists the number of clock periods required to compute an instruction's effective address. It includes fetching of any extension words, the address computation, and fetching of the memory operand if necessary. Several instructions do not need the operand at an effective address to be fetched and thus require fewer clock periods to calculate a given effective address than the instructions that do fetch the effective address operand. The number of bus read and write cycles is shown in parenthesis as (r/w). Note there are no write cycles involved in processing the effective address.

Table F-1. Effective Address Calculation Times

Addressing Mode		Byte, Word		Long	
		Fetch	No Fetch	Fetch	No Fetch
	Register				
Dn	Data Register Direct	0(0/0)	—	0(0/0)	—
An	Address Register Direct	0(0/0)	—	0(0/0)	—
	Memory				
(An)	Address Register Indirect	4(1/0)	2(0/0)	8(2/0)	2(0/0)
(An)+	Address Register Indirect with Postincrement	4(1/0)	4(0/0)	8(2/0)	4(0/0)
−(An)	Address Register Indirect with Predecrement	6(1/0)	4(0/0)	10(2/0)	4(0/0)
d(An)	Address Register Indirect with Displacement	8(2/0)	4(0/0)	12(3/0)	4(1/0)
d(An, ix)*	Address Register Indirect with Index	10(2/0)	8(1/0)	14(3/0)	8(1/0)
xxx.W	Absolute Short	8(2/0)	4(1/0)	12(3/0)	4(1/0)
xxx.L	Absolute Long	12(3/0)	8(2/0)	16(4/0)	8(2/0)
d(PC)	Program Counter with Displacement	8(2/0)	—	12(3/0)	—
d(PC, ix)	Program Counter with Index	10(2/0)	—	14(3/0)	—
#xxx	Immediate	4(1/0)	—	8(2/0)	—

*The size of the index register (ix) does not affect execution time.

F.3 MOVE INSTRUCTION EXECUTION TIMES

Tables F-2, F-3, F-4, and F-5 indicate the number of clock periods for the move instruction. This data includes instruction fetch, operand reads, and operand writes. The number of bus read and write cycles is shown in parenthesis as (r/w).

Table F-2. Move Byte and Word Instruction Execution Times

Source	Destination								
	Dn	An	(An)	(An)+	−(An)	d(An)	d(An, ix)*	xxx.W	xxx.L
Dn	4(1/0)	4(1/0)	8(1/1)	8(1/1)	8(1/1)	12(2/1)	14(2/1)	12(2/1)	16(3/1)
An	4(1/0)	4(1/0)	8(1/1)	8(1/1)	8(1/1)	12(2/1)	14(2/1)	12(2/1)	16(3/1)
(An)	8(2/0)	8(2/0)	12(2/1)	12(2/1)	12(2/1)	16(3/1)	18(3/1)	16(3/1)	20(4/1)
(An)+	8(2/0)	8(2/0)	12(2/1)	12(2/1)	12(2/1)	16(3/1)	18(3/1)	16(3/1)	20(4/1)
−(An)	10(2/0)	10(2/0)	14(2/1)	14(2/1)	14(2/1)	18(3/1)	20(3/1)	18(3/1)	22(4/1)
d(An)	12(3/0)	12(3/0)	16(3/1)	16(3/1)	16(3/1)	20(4/1)	22(4/1)	20(4/1)	24(5/1)
d(An, ix)*	14(3/0)	14(3/0)	18(3/1)	18(3/1)	18(3/1)	22(4/1)	24(4/1)	22(4/1)	26(5/1)
xxx.W	12(3/0)	12(3/0)	16(3/1)	16(3/1)	16(3/1)	20(4/1)	22(4/1)	20(4/1)	24(5/1)
xxx.L	16(4/0)	16(4/0)	20(4/1)	20(4/1)	20(4/1)	24(5/1)	26(5/1)	24(5/1)	28(6/1)
d(PC)	12(3/0)	12(3/0)	16(3/1)	16(3/1)	16(3/1)	20(4/1)	22(4/1)	20(4/1)	24(5/1)
d(PC, ix)*	14(3/0)	14(3/0)	18(3/1)	18(3/1)	18(3/1)	22(4/1)	24(4/1)	22(4/1)	26(5/1)
#xxx	8(2/0)	8(2/0)	12(2/1)	12(2/1)	12(2/1)	16(3/1)	18(3/1)	16(3/1)	20(4/1)

*The size of the index register (ix) does not affect execution time.

Table F-3. Move Byte and Word Instruction Loop Mode Execution Times

	Loop Continued			Loop Terminated					
	Valid Count, cc False			Valid Count, cc True			Expired Count		
				Destination					
Source	(An)	(An)+	−(An)	(An)	(An)+	−(An)	(An)	(An)+	−(An)
Dn	10(0/1)	10(0/1)	—	18(2/1)	18(2/1)	—	16(2/1)	16(2/1)	—
An*	10(0/1)	10(0/1)	—	18(2/1)	18(2/1)	—	16(2/1)	16(2/1)	—
(An)	14(1/1)	14(1/1)	16(1/1)	20(3/1)	20(3/1)	22(3/1)	18(3/1)	18(3/1)	20(3/1)
(An)+	14(1/1)	14(1/1)	16(1/1)	20(3/1)	20(3/1)	22(3/1)	18(3/1)	18(3/1)	20(3/1)
−(An)	16(1/1)	16(1/1)	18(1/1)	22(3/1)	22(3/1)	24(3/1)	20(3/1)	20(3/1)	22(3/1)

*Word only.

Table F-4. Move Long Instruction Execution Times

Source	Destination								
	Dn	An	(An)	(An)+	−(An)	d(An)	d(An, ix)*	xxx.W	xxx.L
Dn	4(1/0)	4(1/0)	12(1/2)	12(1/2)	14(1/2)	16(2/2)	18(2/2)	16(2/2)	20(3/2)
An	4(1/0)	4(1/0)	12(1/2)	12(1/2)	14(1/2)	16(2/2)	18(2/2)	16(2/2)	20(3/2)
(An)	12(3/0)	12(3/0)	20(3/2)	20(3/2)	20(3/2)	24(4/2)	26(4/2)	24(4/2)	28(5/2)
(An)+	12(3/0)	12(3/0)	20(3/2)	20(3/2)	20(3/2)	24(4/2)	26(4/2)	24(4/2)	28(5/2)
−(An)	14(3/0)	14(3/0)	22(3/2)	22(3/2)	22(3/2)	26(4/2)	28(4/2)	26(4/2)	30(5/2)
d(An)	16(4/0)	16(4/0)	24(4/2)	24(4/2)	24(4/2)	28(5/2)	30(5/2)	28(5/2)	32(6/2)
d(An, ix)*	18(4/0)	18(4/0)	26(4/2)	26(4/2)	26(4/2)	30(5/2)	32(5/2)	30(5/2)	34(6/2)
xxx.W	16(4/0)	16(4/0)	24(4/2)	24(4/2)	24(4/2)	28(5/2)	30(5/2)	28(5/2)	32(6/2)
xxx.L	20(5/0)	20(5/0)	28(5/2)	28(5/2)	28(5/2)	32(6/2)	34(6/2)	32(6/2)	36(7/2)
d(PC)	16(4/0)	16(4/0)	24(4/2)	24(4/2)	24(4/2)	28(5/2)	30(5/2)	28(5/2)	32(5/2)
d(PC, ix)*	18(4/0)	18(4/0)	26(4/2)	26(4/2)	26(4/2)	30(5/2)	32(5/2)	30(5/2)	34(6/2)
#xxx	12(3/0)	12(3/0)	20(3/2)	20(3/2)	20(3/2)	24(4/2)	26(4/2)	24(4/2)	28(5/2)

*The size of the index register (ix) does not affect execution time.

Table F-5. Move Long Instruction Loop Mode Execution Times

	Loop Continued			Loop Terminated					
	Valid Count, cc False			Valid Count, cc True			Expired Count		
				Destination					
Source	(An)	(An)+	− (An)	(An)	(An)+	− (An)	(An)	(An)+	− (An)
Dn	14(0/2)	14(0/2)	−	20(2/2)	20(2/2)	−	18(2/2)	18(2/2)	−
An	14(0/2)	14(0/2)	−	20(2/2)	20(2/2)	−	18(2/2)	18(2/2)	−
(An)	22(2/2)	22(2/2)	24(2/2)	28(4/2)	28(4/2)	30(4/2)	24(4/2)	24(4/2)	26(4/2)
(An)+	22(2/2)	22(2/2)	24(2/2)	28(4/2)	28(4/2)	30(4/2)	24(4/2)	24(4/2)	26(4/2)
− (An)	24(2/2)	24(2/2)	26(2/2)	30(4/2)	30(4/2)	32(4/2)	26(4/2)	26(4/2)	28(4/2)

F.4 STANDARD INSTRUCTION EXECUTION TIMES

The number of clock periods shown in Tables F-6 and F-7 indicate the time required to perform the operations, store the results, and read the next instruction. The number of bus read and write cycles is shown in parenthesis as (r/w). The number of clock periods and the number of read and write cycles must be added respectively to those of the effective address calculation where indicated.

In Tables F-6 and F-7 the headings have the following meanings: An = address register operand, Dn = data register operand, ea = an operand specified by an effective address, and M = memory effective address operand.

Table F-6. Standard Instruction Execution Times

Instruction	Size	op<ea>, An***	op<ea>, Dn	op Dn, <M>
ADD	Byte, Word	8(1/0) +	4(1/0) +	8(1/1) +
	Long	6(1/0) +	6(1/0) +	12(1/2) +
AND	Byte, Word	−	4(1/0) +	8(1/1) +
	Long	−	6(1/0) +	12(1/2) +
CMP	Byte, Word	6(1/0) +	4(1/0) +	−
	Long	6(1/0) +	6(1/0) +	−
DIVS	−	−	122(1/0) +	−
DIVU	−	−	108(1/0) +	−
EOR	Byte, Word	−	4(1/0) **	8(1/1) +
	Long	−	6(1/0) **	12(1/2) +
MULS	−	−	42(1/0) + *	−
MULU	−	−	40(1/0) +	−
OR	Byte, Word	−	4(1/0) +	8(1/1) +
	Long	−	6(1/0) +	12(1/2) +
SUB	Byte, Word	8(1/0) +	4(1/0) +	8(1/1) +
	Long	6(1/0) +	6(1/0) +	12(1/2) +

NOTES:
+ add effective address calculation time
* indicates maximum value
** only available addressing mode is data register direct
*** word or long only

Table F-7. Standard Instruction Loop Mode Execution Times

Instruction	Size	Loop Continued			Loop Terminated					
		Valid Count cc False			Valid Count cc True			Expired Count		
		op <ea>, An*	op <ea>, Dn	op Dn, <ea>	op <ea>, An*	op <ea>, Dn	op Dn, <ea>	op <ea>, An*	op <ea>, Dn	op Dn, <ea>
ADD	Byte, Word	18(1/0)	16(1/0)	16(1/1)	24(3/0)	22(3/0)	22(3/1)	22(3/0)	20(3/0)	20(3/1)
	Long	22(2/0)	22(2/0)	24(2/2)	28(4/0)	28(4/0)	30(4/2)	26(4/0)	26(4/0)	28(4/2)
AND	Byte, Word	—	16(1/0)	16(1/1)	—	22(3/0)	22(3/1)	—	20(3/0)	20(3/1)
	Long	—	22(2/0)	24(2/2)	—	28(4/0)	30(4/2)	—	26(4/0)	28(4/2)
CMP	Byte, Word	12(1/0)	12(1/0)	—	18(3/0)	18(3/0)	—	16(3/0)	16(4/0)	—
	Long	18(2/0)	18(2/0)	—	24(4/0)	24(4/0)	—	20(4/0)	20(4/0)	—
EOR	Byte, Word	—	—	16(1/0)	—	—	22(3/1)	—	—	20(3/1)
	Long	—	—	24(2/2)	—	—	30(4/2)	—	—	28(4/2)
OR	Byte, Word	—	16(1/0)	16(1/0)	—	22(3/0)	22(3/1)	—	20(3/0)	20(3/1)
	Long	—	22(2/0)	24(2/2)	—	28(4/0)	30(4/2)	—	26(4/0)	28(4/2)
SUB	Byte, Word	18(1/0)	16(1/0)	16(1/1)	24(3/0)	22(3/0)	22(3/1)	22(3/0)	20(3/0)	20(3/1)
	Long	22(2/0)	20(2/0)	24(2/2)	28(4/0)	26(4/0)	30(4/2)	26(4/0)	24(4/0)	28(4/2)

*Word or long only.
<ea> may be (An), + (An), or − (An) only. Add two clock periods to the table value if <ea> is − (An).

F.5 IMMEDIATE INSTRUCTION EXECUTION TIMES

The number of clock periods shown in Table F-8 includes the time to fetch immediate operands, perform the operations, store the results, and read the next operation. The number of bus read and write cycles is shown in parenthesis as (r/w). The number of clock periods and the number of read and write cycles must be added respectively to those of the effective address calculation where indicated.

In Table F-8, the headings have the following meanings: # = immediate operand, Dn = data register operand, An = address register operand, and M = memory operand.

Table F-8 Immediate Instruction Execution Times

Instruction	Size	op #, Dn	op #, An	op #, M
ADDI	Byte, Word	8(2/0)	—	12(2/1) +
	Long	14(3/0)	—	20(3/2) +
ADDQ	Byte, Word	4(1/0)	4(1/0)*	8(1/1) +
	Long	8(1/0)	8(1/0)	12(1/2) +
ANDI	Byte, Word	8(2/0)	—	12(2/1) +
	Long	14(3/0)	—	20(3/1) +
CMPI	Byte, Word	8(2/0)	—	8(2/0) +
	Long	12(3/0)	—	12(3/0) +
EORI	Byte, Word	8(2/0)	—	12(2/1) +
	Long	14(3/0)	—	20(3/2) +
MOVEQ	Long	4(1/0)	—	—
ORI	Byte, Word	8(2/0)	—	12(2/1) +
	Long	14(3/0)	—	20(3/2) +
SUBI	Byte, Word	8(2/0)	—	12(2/1) +
	Long	14(3/0)	—	20(3/2) +
SUBQ	Byte, Word	4(1/0)	4(1/0)*	8(1/1) +
	Long	8(1/0)	8(1/0)	12(1/2) +

+ add effective address calculation time.
* word only

F.6 SINGLE OPERAND INSTRUCTION EXECUTION TIMES

Tables F-9, F-10, and F-11 indicate the number of clock periods for the single operand instructions. The number of bus read and write cycles is shown in parenthesis as (r/w). The number of clock periods and the number of read and write cycles must be added respectively to those of the effective address calculation where indicated.

Table F-9. Single Operand Instruction Execution Times

Instruction	Size	Register	Memory
NBCD	Byte	6(1/0)	8(1/1) +
NEG	Byte, Word	4(1/0)	8(1/1) +
	Long	6(1/0)	12(1/2) +
NEGX	Byte, Word	4(1/0)	8(1/1) +
	Long	6(1/0)	12(1/2) +
NOT	Byte, Word	4(1/0)	8(1/1) +
	Long	6(1/0)	12(1/2) +
Scc	Byte, False	4(1/0)	8(1/1) + *
	Byte, True	4(1/0)	8(1/1) + *
TAS	Byte	4(1/0)	14(2/1) + *
TST	Byte, Word	4(1/0)	4(1/0)
	Long	4(1/0)	4(1/0) +

+ add effective address calculation time
* Use non-fetching effective address calculation time.

Table F-10. Clear Instruction Execution Times

	Size	Dn	An	(An)	(An) +	– (An)	d(An)	d(An, ix)*	xxx.W	xxx.L
CLR	Byte, Word	4(1/0)	–	8(1/1)	8(1/1)	10(1/1)	12(2/1)	16(2/1)	12(2/1)	16(3/1)
	Long	6(1/0)	–	12(1/2)	12(1/2)	14(1/2)	16(2/2)	20(2/2)	16(2/2)	20(3/2)

* The size of the index register (ix) does not affect execution time.

Table F-11. Single Operand Instruction Loop Mode Execution Times

Instruction	Size	Loop Continued			Loop Terminated					
		Valid Count, cc False			Valid Count, cc True			Expired Count		
		(An)	(An) +	– (An)	(An)	(An) +	– (An)	(An)	(An) +	– (An)
CLR	Byte, Word	10(0/1)	10(0/1)	12(0/1)	18(2/1)	18(2/1)	20(2/0)	16(2/1)	16(2/1)	18(2/1)
	Long	14(0/2)	14(0/2)	16(0/2)	22(2/2)	22(2/2)	24(2/2)	20(2/2)	20(2/2)	22(2/2)
NBCD	Byte	18(1/1)	18(1/1)	20(1/1)	24(3/1)	24(3/1)	26(3/1)	22(3/1)	22(3/1)	24(3/1)
NEG	Byte, Word	16(1/1)	16(1/1)	18(2/2)	22(3/1)	22(3/1)	24(3/1)	20(3/1)	20(3/1)	22(3/1)
	Long	24(2/2)	24(2/2)	26(2/2)	30(4/2)	30(4/2)	32(4/2)	28(4/2)	28(4/2)	30(4/2)
NEGX	Byte, Word	16(1/1)	16(1/1)	18(2/2)	22(3/1)	22(3/1)	24(3/1)	20(3/1)	20(3/1)	22(3/1)
	Long	24(2/2)	24(2/2)	26(2/2)	30(4/2)	30(4/2)	32(4/2)	28(4/2)	28(4/2)	30(4/2)
NOT	Byte, Word	16(1/1)	16(1/1)	18(2/2)	22(3/1)	22(3/1)	24(3/1)	20(3/1)	20(3/1)	22(3/1)
	Long	24(2/2)	24(2/2)	26(2/2)	30(4/2)	30(4/2)	32(4/2)	28(4/2)	28(4/2)	30(4/2)
TST	Byte, Word	12(1/0)	12(1/0)	14(1/0)	18(3/0)	18(3/0)	20(3/0)	16(3/0)	16(3/0)	18(3/0)
	Long	18(2/0)	18(2/0)	20(2/0)	24(4/0)	24(4/0)	26(4/0)	20(4/0)	20(4/0)	22(4/0)

F.7 SHIFT/ROTATE INSTRUCTION EXECUTION TIMES

Tables F-12 and F-13 indicate the number of clock periods for the shift and rotate instructions. The number of bus read and write cycles is shown in parenthesis as (r/w). The number of clock periods and the number of read and write cycles must be added respectively to those of the effective address calculation where indicated.

Table F-12. Shift/Rotate Instruction Execution Times

Instruction	Size	Register	Memory*
ASR, ASL	Byte, Word	6 + 2n(1/0)	8(1/1) +
	Long	8 + 2n(1/0)	—
LSR, LSL	Byte, Word	6 + 2n(1/0)	8(1/1) +
	Long	8 + 2n(1/0)	—
ROR, ROL	Byte, Word	6 + 2n(1/0)	8(1/1) +
	Long	8 + 2n(1/0)	—
ROXR, ROXL	Byte, Word	6 + 2n(1/0)	8(1/1) +
	Long	8 + 2n(1/0)	—

+ add effective address calculation time
n is the shift or rotate count
* word only

Table F-13. Shift/Rotate Instruction Loop Mode Execution Times

Instruction	Size	Loop Continued			Loop Terminated					
		Valid Count, cc False			Valid Count, cc True			Expired Count		
		(An)	(An) +	− (An)	(An)	(An) +	− (An)	(An)	(An) +	− (An)
ASR, ASL	Word	18(1/1)	18(1/1)	20(1/1)	24(3/1)	24(3/1)	26(3/1)	22(3/1)	22(3/1)	24(3/1)
LSR, LSL	Word	18(1/1)	18(1/1)	20(1/1)	24(3/1)	24(3/1)	26(3/1)	22(3/1)	22(3/1)	24(3/1)
ROR, ROL	Word	18(1/1)	18(1/1)	20(1/1)	24(3/1)	24(3/1)	26(3/1)	22(3/1)	22(3/1)	24(3/1)
ROXR, ROXL	Word	18(1/1)	18(1/1)	20(1/1)	24(3/1)	24(3/1)	26(3/1)	22(3/1)	22(3/1)	24(3/1)

F.8 BIT MANIPULATION INSTRUCTION EXECUTION TIMES

Table F-14 indicates the number of clock periods required for the bit manipulation instructions. The number of bus read and write cycles is shown in parenthesis as (r/w). The number of clock periods and the number of read and write cycles must be added respectively to those of the effective address calculation where indicated.

Table F-14. Bit Manipulation Instruction Execution Times

Instruction	Size	Dynamic		Static	
		Register	Memory	Register	Memory
BCHG	Byte	−	8(1/1) +	−	12(2/1) +
	Long	8(1/0) *	−	12(2/0) *	−
BCLR	Byte	−	10(1/1) +	−	14(2/1) +
	Long	10(1/0) *	−	14(2/0) *	−
BSET	Byte	−	8(1/1) +	−	12(2/1) +
	Long	8(1/0) *	−	12(2/0) *	−
BTST	Byte	−	4(1/0) +	−	8(2/0) +
	Long	6(1/0) *	−	10(2/0)	−

+ add effective address calculation time
* indicates maximum value

F.9 CONDITIONAL INSTRUCTION EXECUTION TIMES

Table F-15 indicates the number of clock periods required for the conditional instructions. The number of bus read and write cycles is indicated in parenthesis as (r/w). The number of clock periods and the number of read and write cycles must be added respectively to those of the effective address calculation where indicated.

Table F-15. Conditional Instruction Execution Times

Instruction	Displacement	Branch Taken	Branch Not Taken
BCC	Byte	10(2/0)	6(1/0)
	Word	10(2/0)	10(2/0)
BRA	Byte	10(2/0)	−
	Word	10(2/0)	−
BSR	Byte	18(2/2)	−
	Word	18(2/2)	−
DBCC	CC true	−	10(2/0)
	CC false	10(2/0)	16(3/0)

+ add effective address calculation time
* indicates maximum value

F.10 JMP, JSR, LEA, PEA, AND MOVEM INSTRUCTION EXECUTION TIMES

Table F-16 indicates the number of clock periods required for the jump, jump-to-subroutine, load effective address, push effective address, and move multiple registers instructions. The number of bus read and write cycles is shown in parenthesis as (r/w).

Table F-16. JMP, JSR, LEA, PEA, and MOVEM Instruction Execution Times

Instr	Size	(An)	(An) +	− (An)	d(An)	d(An, ix) +	xxx.W	xxx.L	d(PC)	d(PC, ix)*
JMP	−	8(2/0)	−	−	10(2/0)	14(3/0)	10(2/0)	12(3/0)	10(2/0)	14(3/0)
JSR	−	16(2/2)	−	−	18(2/2)	22(2/2)	18(2/2)	20(3/2)	18(2/2)	22(2/2)
LEA	−	4(1/0)	−	−	8(2/0)	12(2/0)	8(2/0)	12(3/0)	8(2/0)	12(2/0)
PEA	−	12(1/2)	−	−	16(2/2)	20(2/2)	16(2/2)	20(3/2)	16(2/2)	20(2/2)
MOVEM M → R	Word	12 + 4n (3 + n/0)	12 + 4n (3 + n/0)	−	16 + 4n (4 + n/0)	18 + 4n (4 + n/0)	16 + 4n (4 + n/0)	20 + 4n (5 + n/0)	16 + 4n (4 + n/0)	18 + 4n (4 + n/0)
	Long	12 + 8n (3 + 2n/0)	12 + 8n (3 + 2n/0)	−	16 + 8n (4 + 2n/0)	18 + 8n (4 + 2n/0)	16 + 8n (4 + 2n/0)	20 + 8n (5 + 2n/0)	16 + 8n (4 + 2n/0)	18 + 8n (4 + 2n/0)
MOVEM R → M	Word	8 + 4n (2/n)	−	8 + 4n (2/n)	12 + 4n (3/n)	14 + 4n (3/n)	12 + 4n (3/n)	16 + 4n (4/n)	−	−
	Long	8 + 8n (2/2n)	−	8 + 8n (2/2n)	12 + 8n (3/2n)	14 + 8n (3/2n)	12 + 8n (3/2n)	16 + 8n (4/2n)	−	−

n is the number of registers to move
* is the size of the index register (ix) does not affect the instruction's execution time

F.11 MULTI-PRECISION INSTRUCTION EXECUTION TIMES

Table F-17 indicates the number of clock periods for the multi-precision instructions. The number of clock periods includes the time to fetch both operands, perform the operations, store the results, and read the next instructions. The number of read and write cycles is shown in parenthesis as (r/w).

In Table F-17, the headings have the following meanings: Dn = data register operand and M = memory operand.

Table F-17. Multi-Precision Instruction Execution Times

Instruction	Size	Non-Looped		Loop Mode		
				Continued	Terminated	
				Valid Count, cc False	Valid Count, cc True	Expired Count
		op Dn, Dn	op M, M*			
ADDX	Byte, Word	4(1/0)	18(3/10)	22(2/1)	28(4/1)	26(4/1)
	Long	6(1/0)	30(5/2)	32(4/2)	38(6/2)	36(6/2)
CMPM	Byte, Word	−	12(3/0)	14(2/0)	20(4/0)	18(4/0)
	Long	−	20(5/0)	24(4/0)	30(6/0)	26(6/0)
SUBX	Byte, Word	4(1/0)	18(3/1)	22(2/1)	28(4/1)	26(4/1)
	Long	6(1/0)	30(5/2)	32(4/2)	38(6/2)	36(6/2)
ABCD	Byte	6(1/0)	18(3/1)	24(2/1)	30(4/1)	28(4/1)
SBCD	Byte	6(1/0)	18(3/1)	24(2/1)	30(4/1)	28(4/1)

* Source and destination ea is (An) + for CMPM and − (An) for all others.

F.12 MISCELLANEOUS INSTRUCTION EXECUTION TIMES

Table F-18 indicates the number of clock periods for the following miscellaneous instructions. The number of bus read and write cycles is shown in parenthesis as (r/w). The number of clock periods plus the number of read and write cycles must be added to those of the effective address calculation where indicated.

Table F-18. Miscellaneous Instruction Execution Times

Instruction	Size	Register	Memory	Register → Destination**	Source** → Register
ANDI to CCR	−	16(2/0)	−	−	−
ANDI to SR	−	16(2/0)	−	−	−
CHK	−	8(1/0) +	−	−	−
EORI to CCR	−	16(2/0)	−	−	−
EORI to SR	−	16(2/0)	−	−	−
EXG	−	6(1/0)	−	−	−
EXT	Word	4(1/0)	−	−	−
	Long	4(1/0)	−	−	−
LINK	−	16(2/2)	−	−	−
MOVE from CCR	−	4(1/0)	8(1/1) + *	−	−
MOVE to CCR	−	12(2/0)	12(2/0) +	−	−
MOVE from SR	−	4(1/0)	8(1/1) + *	−	−
MOVE to SR	−	12(2/0)	12(2/0) +	−	−
MOVE from USP	−	6(1/0)	−	−	−
MOVE to USP	−	6(1/0)	−	−	−
MOVEC	−	−	−	10(2/0)	12(2/0)
MOVEP	Word	−	−	16(2/2)	16(4/0)
	Long	−	−	24(2/4)	24(6/0)
NOP	−	4(1/0)	−	−	−
ORI to CCR	−	16(2/0)	−	−	−
ORI to SR	−	16(2/0)	−	−	−
RESET	−	130(1/0)	−	−	−
RTD	−	16(4/0)	−	−	−
RTE	Short	24(6/0)	−	−	−
	Long, Retry Read	112(27/10)	−	−	−
	Long, Retry Write	112(26/1)	−	−	−
	Long, No Retry	110(26/0)	−	−	−
RTR	−	20(5/0)	−	−	−
RTS	−	16(4/0)	−	−	−
STOP	−	4(0/0)	−	−	−
SWAP	−	4(1/0)	−	−	−
TRAPV	−	4(1/0)	−	−	−
UNLK	−	12(3/0)	−	−	−

+ add effective address calculation time.
* use non-fetching effective address calculation time.
** Source or destination is a memory location for the MOVEP instruction and a control register for the MOVEC instruction.

F.13 EXCEPTION PROCESSING EXECUTION TIMES

Table F-19 indicates the number of clock periods for exception processing. The number of clock periods includes the time for all stacking, the vector fetch, and the fetch of the first two instruction words of the handler routine. The number of bus read and write cycles is shown in parenthesis as (r/w).

Table F-19. Exception Processing Execution Times

Exception	
Address Error	126(4/26)
Breakpoint Instruction*	42(5/4)
Bus Error	126(4/26)
CHK Instruction**	44(5/4) +
Divide By Zero	42(5/4)
Illegal Instruction	38(4/4)
Interrupt*	46(5/4)
MOVEC, Illegal Cr**	46(5/4)
Privilege Violation	38(4/4)
Reset***	40(6/0)
RTE, Illegal Format	50(7/4)
RTE, Illegal Revision	70(12/4)
Trace	38(4/4)
TRAP Instruction	38(4/4)
TRAPV Instruction	40(5/4)

+ add effective address calculation time.
  *The interrupt acknowledge and breakpoint cycles are assumed to take four clock periods.
  **Indicates maximum value.
  ***Indicates the time from when $\overline{\text{RESET}}$ and $\overline{\text{HALT}}$ are first sampled as negated to when instruction execution starts.

APPENDIX IVG: MC68010 LOOP MODE OPERATION

The MC68010 has several features that provide efficient execution of program loops. One of these features is the DBcc looping primitive instruction. The DBcc instruction operates on three operands, a loop counter, a branch condition, and a branch displacement. When the DBcc is executed in loop mode, the contents of the low order word of the register specified as the loop counter is decremented by one and compared to minus one. If equal to minus one, the result of the decrement is placed back into the count register and the next sequential instruction is executed, otherwise the condition code register is checked against the specified branch condition. If the condition is true, the result of the decrement is discarded and the next sequential instruction is executed. Finally, if the count register is not equal to minus one and the branch condition is false, the branch displacement is added to the program counter and instruction execution continues at that new address. Note that this is slightly different than non-looped execution; however, the results are the same.

An example of using the DBcc instruction in a simple loop for moving a block of data is shown in Figure G-1. In this program, the block of data 'LENGTH' words long at address 'SOURCE' is to be moved to address 'DEST' provided that none of the words moved are equal to zero. When the effect of instruction prefetch on this loop is examined it can be seen that the bus activity during the loop execution would be:

1. Fetch the MOVE.W instruction,
2. Fetch the DBEQ instruction,
3. Read the operand where A0 points,
4. Write the operand where A1 points,
5. Fetch the DBEQ branch displacement, and
6. If loop conditions are met, return to step 1.

```
         LEA      SOURCE, A0      Load A Pointer To Source Data
         LEA      DEST, A1        Load A Pointer To Destination
         MOVE.W   #LENGTH, D0     Load The Counter Register
LOOP     MOVE.W   (A0) + , (A1) + Loop To Move The Block Of Data
         DBEQ     D0, LOOP        Stop If Data Word Is Zero
```

Figure G-1. DBcc Loop Program Example

During this loop, five bus cycles are executed; however, only two bus cycles perform the data movement. Since the MC68010 has a two word prefetch queue in addition to a one word instruction decode register, it is evident that the three instruction fetches in this loop could be eliminated by placing the MOVE.W word in the instruction decode register and holding the DBEQ instruction and its branch displacement in the prefetch queue. The MC68010 has the ability to do this by entering the loop mode of operation. During loop mode operation, all opcode fetches are suppressed and only operand reads and writes are performed until an exit loop condition is met.

Loop mode operation is transparent to the programmer, with only two conditions required for the MC68010 to enter the loop mode. First, a DBcc instruction must be executed with both branch conditions met and a branch displacement of minus four; which indicates that the branch is to a one word instruction preceding the DBcc instruction. Second, when the processor fetches the instruction at the branch address, it is checked to determine whether it is one of the allowed looping instructions. If it is, the loop mode is entered. Thus, the single word looped instruction and the first word of the DBcc instruction will each be fetched twice when the loop is entered; but no instruction fetches will occur again until the DBcc loop conditions fail.

In addition to the normal termination conditions for a loop, there are several conditions that will cause the MC68010 to exit loop mode operation. These conditions are interrupts, trace exceptions, reset errors, and bus errors. Interrupts are honored after each execution of the DBcc instruction, but not after the execution of the looped instruction. If an interrupt exception occurs, loop mode operation is terminated and can be restarted on return from the interrupt handler. If the T bit is set, trace exceptions will occur at the end of both the loop instruction and the DBcc instruction and thus loop mode operation is not available. Reset will abort all processing, including the loop mode. Bus errors during the loop mode will be treated the same as in normal processing; however, when the RTE instruction is used to continue the execution of the looped instruction, the three word loop will not be re-fetched.

The loopable instructions available on the MC68010 are listed in Table G-1. These instructions may use the three address register indirect modes to form one word looping instructions; (An), (An)+, and −(An).

Table G-1. MC68010 Loopable Instructions

Opcodes	Applicable Addressing Modes	
MOVE [BWL]	(Ay) to (Ax) (Ay) to (Ax)+ (Ay) to −(Ax) (Ay)+ to (Ax) (Ay)+ to (Ax)+ (Ay)+ to −(Ax)	−(Ay) to (Ax) −(Ay) to (Ax)+ −(Ay) to −(Ax) Ry to (Ax) Ry to (Ax)+
ADD [BWL] AND [BWL] CMP [BWL] OR [BWL] SUB [BWL]	(Ay) to Dx (Ay)+ to Dx −(Ay) to Dx	
ADDA [WL] CMPA [WL] SUBA [WL]	(Ay) to Ax −(Ay) to Ax (Ay)+ to Ax	
ADD [BWL] AND [BWL] EOR [BWL] OR [BWL] SUB [BWL]	Dx to (Ay) Dx to (Ay)+ Dx to −(Ay)	

Opcodes	Applicable Addressing Modes
ABCD [B] ADDX [BWL] SBCD [B] SUBX [BWL]	−(Ay) to −(Ax)
CMP [BWL]	(Ay)+ to (Ax)+
CLR [BWL] NEG [BWL] NEGX [BWL] NOT [BWL] TST [BWL] NBCD [B]	(Ay) (Ay)+ −(Ay)
ASL [W] ASR [W] LSL [W] LSR [W] ROL [W] ROR [W] ROXL [W] ROXR [W]	(Ay) by #1 (Ay)+ by #1 −(Ay) by #1

NOTE
[B, W, or L] indicate an operand size of byte, word, or long word.

APPENDIX IVH: MC68000 PREFETCH

H.1 INTRODUCTION

The MC68000 uses a two-word tightly coupled instruction prefetch mechanism to enhance performance. This mechanism is described in terms of the microcode operations involved. If the execution is defined to begin when the microroutine for that instruction is entered, some features of the prefetch mechanism can be described.

1. When execution of an instruction begins, the operation word and the word following have already been fetched. The operation word is in the instruction decoder.

2. In the case of multiword instructions, as each additional word of the instruction is used internally, a fetch is made to the instruction stream to replace it.

3. The last fetch from the instruction stream is made when the operation word is discarded and decoding is started on the next instruction.

4. If the instruction is a single-word instruction causing a branch, the second word is not used. But because this word is fetched by the previous instruction, it is impossible to avoid this superfluous fetch. In the case of an interrupt or trace exception, both words are not used.

5. The program counter usually points to the last word fetched from the instruction stream.

H.2 INSTRUCTION PREFETCH

The following example illustrates many of the features of instruction prefetch. The contents of memory are assumed to be as illustrated in Figure H-1.

```
                ORG       0                       DEFINE RESTART VECTOR

                DC.L      INISSP                  INITIAL SYSTEM STACK POINTER
                DC.L      RESTART                 RESTART SYSTEM ENTRY POINT

                ORG       INTVECTOR               DEFINE AN INTERRUPT VECTOR
                DC.L      INTHANDLER              HANDLER ADDRESS FOR THIS VECTOR

                ORG                               SYSTEM RESTART CODE
    RESTART:
                NOP                               NO OPERATION EXAMPLE
                BRA.S     LABEL                   SHORT BRANCH
                ADD.W     D0,D1                   ADD REGISTER TO REGISTER
    LABEL:
                SUB.W     DISP(A0),A1             SUBTRACT REGISTER INDIRECT WITH OFFSET
                CMP.W     D2,D3                   COMPARE REGISTER TO REGISTER
                SGE.B     D7                      Scc TO REGISTER
                ...
                ...
    INTHANDLER:
                MOVE.W    LONGADR1,LONGADR2       MOVE WORD FROM AND TO LONG ADDRESS
                NOP                               NO OPERATION
                SWAP.W                            REGISTER SWAP
```

Figure H-1. Instruction Prefetch Example, Memory Contents

The sequence we shall illustrate consists of the power-up reset, the execution of NOP, BRA, SUB, the taking of an interrupt, and the execution of the MOVE.W xxx.L to yyy.L. The order of operations described within each microroutine is not exact, but is intended for illustrative purpose only.

Microroutine	Operation	Location	Operand
Reset	Read	0	SSP High
	Read	2	SSP Low
	Read	4	PC High
	Read	6	PC Low
	Read	(PC)	NOP
	Read	+ (PC)	BRA
	< begin NOP >		
NOP	Read	+ (PC)	ADD
	< begin BRA >		
BRA	PC = PC + d		
	Read	(PC)	SUB
	Read	+ (PC)	DISP
	< begin SUB >		
SUB	Read	+ (PC)	CMP
	Read	DISP(A0)	< src >
	Read	+ (PC)	SGE
	< begin CMP >	< take INT >	
INTERRUPT	Write	– (SSP)	PC Low
	Write	– (SSP)	PC High
	Read	< INT ACK >	Vector #
	Write	– (SSP)	SR
	Read	(VR)	PC High
	Read	+ (VR)	PC Low
	Read	(PC)	MOVE
	Read	+ (PC)	xxx High
	< begin MOVE >		
MOVE	Read	+ (PC)	xxx Low
	Read	+ (PC)	yyy High
	Read	xxx	< src >
	Read	+ (PC)	yyy Low
	Read	+ (PC)	NOP
	Write	yyy	< dest >
	Read	+ (PC)	SWAP
	< begin NOP >		

Figure H-2. Instruction Prefetch Example

H.3 DATA PREFETCH

Normally the MC68000 prefetches only instructions and not data. However, when the MOVEM instruction is used to move data from memory to registers, the data stream is prefetched in order to optimize performance. As a result, the processor reads one extra word beyond the higher end of the source area. For example, the instruction sequence in Figure H-3 will operate as shown in Figure H-4.

```
          . . .
        MOVEM.L   A,D0/D1
          . . .
A       DC.W      1
B       DC.W      2
C       DC.W      3
D       DC.W      4
E       DC.W      5
F       DC.W      6
```

MOVE TWO
LONGWORDS
INTO REGISTERS

WORD 1
WORD 2
WORD 3
WORD 4
WORD 5
WORD 6

Assume Effective Address Evaluation is Already Done

Microroutine	Operation	Location	Other Operations
	. . .		
MOVEM	Read	A	
			Prepare to Fill D0
	Read	B	A → D0H
	Read	C	B → D0L
			Prepare to Fill D1
	Read	D	C → D1H
	Read	E	D → D1L
			Detect Register List Complete

Figure H-3.
MOVEM Example, Memory Contents

Figure H-4.
MOVEM Example, Operation Sequence

Comparison of M68000 Family Members

The summary of Appendix V represents a comparison of the MC68000, MC68008, MC68010, MC68020 and MC68030 processors. Important features of the processors are listed to indicate the differences between the central processor units of the Motorola M68000 family. These processors were introduced in Chapter 1 of this textbook.

Courtesy of Motorola publications.

Appendix V summarizes the characteristics of the microprocessors in the M68000 Family. M68000 UM/AD, *M68000 User's Manual* Sixth Edition includes more detailed information about the MC68000 and MC68010 differences.

	MC68000	**MC68010**	**CPU32**	**MC68020**
Data Bus Size (Bits)	16	16	8, 16	8, 16, 32
Address Bus Size (Bits)	24	24	24	32
Instruction Cache (In Words)		3*	3*	128

* Three-word cache for the loop mode.

Virtual Memory/Machine
MC68000   None
MC68010   Bus Error Detection, Instruction Continuation
CPU32   Bus Error Detection, Instruction Restart
MC68020   Bus Error Detection, Instruction Continuation

Coprocessor Interface
MC68000   Emulated in Software
MC68010   Emulated in Software
CPU32   Emulated in Software
MC68020   In Microcode

Word/Longword Data Alignment
MC68000   Word/Longword Data, Instructions, and Stack Must Be Word
          Aligned
MC68010   Word/Longword Data, Instructions, and Stack Must Be Word
          Aligned
CPU32   Word/Longword Data, Instructions, and Stack Must Be Word
          Aligned
MC68020   Only Instructions Must Be Word Aligned (Data Alignment Improves
          Performance)

Control Registers
MC68000   None
MC68010   SFC, DFC, VBR
CPU32   SFC, DFC, VBR
MC68020   SFC, DFC, VBR, CACR, CAAR

Stack Pointers
MC68000   USP, SSP
MC68010   USP, SSP

CPU32   USP, SSP
MC68020   USP, SSP (MSP, ISP)

Status Register Bits
MC68000   T, S, I0/I1/I2, X/N/Z/V/C
MC68010   T, S, I0/I1/I2, X/N/Z/V/C
CPU32   T1/T0, S, I0/I1/I2, X/N/Z/V/C
MC68020   T1/T0, S, M, I0/I1/I2, X/N/Z/V/C

Function Code/Address Space
MC68000   FC0–FC2 = 7 is Interrupt Acknowledge Only
MC68010   FC0–FC2 = 7 is CPU Space
CPU32   FC0–FC2 = 7 is CPU Space
MC68020   FC0–FC2 = 7 is CPU Space

Indivisible Bus Cycles
MC68000   Use $\overline{\text{AS}}$ Signal
MC68010   Use $\overline{\text{AS}}$ Signal
CPU32   Use $\overline{\text{RMC}}$ Signal
MC68020   Use $\overline{\text{RMC}}$ Signal

Stack Frames
MC68000   Supports Original Set
MC68010   Supports Formats $0, $8
CPU32   Supports Formats $0, $2, $C
MC68020   Supports Formats $0, $1, $2, $9, $A, $B

M68000 Instruction Set Extensions

Mnemonic	Description	CPU32	M68020
Bcc	Supports 32-Bit Displacements	✔	✔
BFxxxx	Bit Field Instructions (BFCHG, BFCLR, BFEXTS, BFEXTU, BFFFO, BFINS, BFSET, BFTST)		✔
BGND	Background Operation	✔	
BKPT	New Instruction Functionality	✔	✔
BRA	Supports 32-Bit Displacements	✔	✔
BSR	Supports 32-Bit Displacements	✔	✔
CALLM	New Instruction		✔
CAS, CAS2	New Instructions		✔
CHK	Supports 32-Bit Operands	✔	✔
CHK2	New Instruction	✔	✔
CMP1	Supports Program Counter Relative Addressing	✔	✔
CMP2	New Instruction	✔	✔
cp	Coprocessor Instructions		✔
DIVS/DIVU	Supports 32-Bit and 64-Bit Operations	✔	✔
EXTB	Supports 8-Bit Extend to 32 Bits	✔	✔
LINK	Supports 32-Bit Displacements	✔	✔
LPSTOP	New Instruction	✔	
MOVEC	Supports New Control Registers	✔	✔
MULS/MULU	Supports 32-Bit Operands, 64-Bit Results	✔	✔
PACK	New Instruction		✔
RTM	New Instruction		✔
TABLE	New Instruction	✔	
TST	Supports Program Counter Relative, Immediate, and an Addressing	✔	✔
TRAPcc	New Instruction	✔	✔
UNPK	New Instruction		✔

M68000 Addressing Modes

Mode	Mnemonic	MC68010/ MC68000	CPU32	MC68020
Register Direct	Rn	✔	✔	✔
Address Register Indirect	(An)	✔	✔	✔
Address Register Indirect with Postincrement	(An)+	✔	✔	✔
Address Register Indirect with Predecrement	−(An)	✔	✔	✔
Address Register Indirect with Displacement	(d_{16},An)	✔	✔	✔
Address Register Indirect with Index (8-Bit Displacement)	(d_8,An,Xn)	✔	✔	✔
Address Register Indirect with Index (Base Displacement)	(bd,An,Xn*SCALE)		✔	✔
Memory Indirect with Postincrement	([bd,An],Xn,od)			✔
Memory Indirect with Predecrement	([bd,An,Xn],od)			✔
Absolute Short	(xxx).W	✔	✔	✔
Absolute Long	(xxx).L	✔	✔	✔
Program Counter Indirect with Displacement	(d_{16},PC)	✔	✔	✔
Program Counter Indirect with Index (8-Bit) Displacement	(d_8,PC,Xn)	✔	✔	✔
Program Counter Indirect with Index (Base Displacement)	(bd,PC,Xn*SCALE)		✔	✔
Immediate	#(data)	✔	✔	✔
Program Counter Memory Indirect with Postincrement	([bd,PC],Xn,od)			✔
Program Counter Memory Indirect with Predecrement	([bd,PC,Xn],od)			✔

Answers to Selected Questions

CHAPTER 1

1.2 *Advantages:* Less distinct parts generally mean cheaper production and greater reliability. Integration of modules means that development is easier because intermodule interface is already in place.

Disadvantages: Restricted to set of integrated modules. You could choose others, but that defeats the purpose of using microcontroller.

1.4 MC indicates a particular chip within a family of processors.

1.6 M6801, M6804, M6805, M68HC11.

1.8 M6800

1.10 Instruction set and electrical characteristics.

1.12 Maximum speed of the system clock; 8, 10, and 12 MHz, respectively.

1.14 *Advantages:* You can take advantage of power, speed, and storage of a general-purpose computer. You can test some aspects of the system independently of the targeted hardware.

Disadvantages: You cannot debug timing problems or test interfaces that connect to peripherals and I/O lines.

1.16 2 million instructions.

1.18 Processor, memory, monitor program in ROM, some peripherals such as serial I/O, parallel I/O, maybe ADCs. To make a usable computer system requires, at a minimum, a mass storage device, terminal and keyboard.

1.20 To reduce the cost of the product and to allow the user to customize the product for specific applications. To make the product both independent of and compatible with changes and advances in peripheral devices.

CHAPTER 2

2.1.2 Address bus size, data bus size, speed of operation.

2.1.6 No, it's an odd address.

2.2.2 32 bits.

2.2.6 23 bits.

2.3.2 Jumps and branches to subroutines, indexed and indirect addressing modes, traps, exceptions, and interrupts.

2.3.4 The address space of the operating system can be such that it can only be accessed by supervisor-mode programs. Then the address decoding hardware can process the Function Code lines to only allow access to operating system addresses by supervisor-mode programs. Note that the exception vector table (addresses 000–3FF) should be protected from user-mode access as well.

2.3.6 Trap, exception, and interrupt.

2.2 Unpredictable results, which is why the 68000 family generates an Illegal Instruction exception.

2.4 1 word, 2 words, 3 words, 4 words, 5 words.

2.6 Interrupt signal lines indicate that devices need attention.

Function Code lines are used for memory protection.

Bus Grant and Request lines are used for bus sharing.

2.8 **(a)** User mode—no privileged operations required, no I/O.
(b) Supervisor mode—needs to be managed by the operating system.
(c) Supervisor mode—needs to be managed by the operating system.
(d) Supervisor mode—interrupt driven, I/O.
(e) User mode—not an OS function, no privileged services required.
(f) User mode—no I/O, no privileged services required.
(g) Supervisor mode—I/O, terminal driver.
(h) Supervisor mode—I/O, printer driver.
(i) User mode—no I/O, no privileged services.

CHAPTER 3

3.1.1.2 The binary number $(1111\ 1111.1111\ 1111\ 1111\ 1111)_2$ has the decimal value $(2^8 - 1) + (1 - 2^{-16}) = 255.999984$.

3.1.1.4 If $111_x = 31$, then $x^2 + x + 1 = 31$; $x = 5$.

3.1.2.2 The complement of the most negative number is the same number. This is an out-of-range condition in the two's-complement system.

3.1.2.4 **(a)** To sign-extend a positive number, repeat the leading zero to the left for m additional bits.

(b) To sign-extend a negative number, repeat the leading 1 to the left for m additional bits.

3.1.2.6 The positive numbers have the range 0000 to 0999. The negative numbers have the range -1 to -1000 written as 9999, 9998, . . . , 9001, 9000.

3.1.3.2 The octal fraction 0.333 . . . has the value:

$3(1/8) + 3(1/64) + . . .$ or

$3/8 (1 + 1/8 + 1/64 + . . .)$ or

$3/8 (1/(1 - 1/8)) = 0.42857143$

3.1.3.4 The binary value 1.111 1111$_2$ is the fraction:

$-1 + 2^{-1} + . . . + 2^{-7}$

or

$-1 + 1 - 2^{-7} = -0.0078125$

3.2.1.2 **(a)** $1(1000) + 9(100) + 7(10) = 1970$

(b) This is an invalid BCD value.

3.2.1.4 Test the range of the sum or difference to be less than or equal to 10^L.

3.2.2.2 The sum in binary is

$(0101)(0110\ 0100) + 0 + (1001)(1) = 1\ 1111\ 1101_2$

3.2.3.2 The range of BCD values is -10^{L-1} to $10^{L-1} - 1$ where L is the number of decimal digits; $L = m/4$ for m binary digits.

(a) 8-bit binary is 2-digit BCD; range is -10 to 9.

(b) 16-bit binary is 4-digit BCD; range is -1000 to 999.

(c) 32-bit binary is 8-digit BCD; range is $-10,000,000$ to $9,999,999$.

3.3.1.2 To convert 1/32 to floating-point:

$1/32 = (.0001)_2 = (1.0)_2 \times 2^{-4}$

The exponent with bias is

$124 = 0111\ 1100_2$

The result is

$0\ 0111\ 1100\ 0000 . . ._2$

with the leading bit of the fraction implied.

3.3.2.2 The largest value is $(1.1111 . . . 1111)_2 \times 2^{127}$ which is approximately 2×10^{38}.

3.4.2 The text "THE MOTOROLA MC68000" has the ASCII equivalent

54 48 45 20	THE
4D 4F 54 4F 52 4F 4C 41 20	MOTOROLA
4D 43 36 38 30 30 30	MC68000

3.4.4 **(a)** The binary value of 45 is given by converting $4(10) + 5$ to binary; thus, the result is $0010\ 1101_2$.

(b) Add 30_{16} to each digit to obtain 34 35.

(c) -045 is 9955 in ten's-complement notation. In ASCII, the result is found by converting 9955 to the positive value 045 and preceding it with a minus sign; thus, 2D 30 34 35 is the internal representation in ASCII.

CHAPTER 4

4.2.2 $2400

4.2.4 After level 1 interrupt:

   ($7FFE) = $101C
   ($7FFC) = $0000
   ($7FFA) = $0000

   After level 2 interrupt:

   ($7FFE) = $101C
   ($7FFC) = $0000
   ($7FFA) = $0000
   ($7FF8) = $200C
   ($7FF6) = $0000
   ($7FF4) = $2100

4.3.2 **(a)** (D2) = $00000801
   (b) ($1001) = $01
   (c) ($03E8) = $0000
   (d) (D2) = $00000806
   (e) (D1) = $00002514

4.4.2 **(a)** Move the value $03E8 to D1.W.
   (b) Move the value $FFE0 to D1.W.
   (c) Move the value $FFE0 to D1.W.
   (d) Move the value $03E8 to D1.W.

4.4.4   MOVE.W   D1,1002
          MOVE.B   1002,1001
          MOVE.B   D1,1002

4.5.2 **(a)** ($1000) = $0000
   (b) ($1000) = $0000
   (c) ($1000) = $1000
   (d) D1.W = $0010
   (e) D1.W = $1000
   (f) D1.B = $00

4.5.4 **(a)** $4240
   (b) $2008
   (c) $DA00

4.6.2 **(a)** ($1000) = $10
       ($1001) = $20
       ($1002) = $30
       ($1003) = $40
   (b) ($1000) = $02
       ($1001) = $00

($1002) = $00
($1003) = $FC
(c) ($1000) = $41
($1001) = $42
($1002) = $43
($1003) = $44

4.2 ii

4.4 i

4.6 iii

4.8 iii

4.10 ii

4.12 ii

4.14 iii

4.16 **(a)** In user mode, A7 holds the user-stack pointer. Subroutine calls place return addresses on the stack; operations using −(A7) place data on the stack. Return from subroutine and (A7)+ operations remove data from the stack.

In supervisor mode, A7 holds the supervisor stack pointer. Subroutine calls, interrupts, and traps all place data on the supervisor stack; returns from subroutines, interrupts, and traps take data off the supervisor stack.

(b) The system stack grows down in memory (toward lower addresses).

(c) −(SP) and (SP)+

4.18 The operation is limited to one word, so there is a limited number of bit patterns that can be used to represent an operation. In order to increase the number of operations that could be offered, the number of addressing modes for each operation must be limited.

4.20 Source-code compatibility is desirable when implementing an application on multiple processors. The source code can be compiled and linked for each computer, letting the compiler and system libraries provide the customization required for each environment.

Object-code compatibility is desirable when implementing an application in a ROM chip that might be used on different systems using processors in the same family (such as the 68000 family or the 80 × 86 family).

CHAPTER 5

5.1.2 **(a)** Useful for small changes to a program being debugged on a development board. Not useful for a significant program change or a large number of program changes.

(b) Useful for writing and changing applications, and for testing portions of the application. Not useful for developing or debugging hardware interfaces or any timing-related problems.

(c) Good for most application development and testing. Resident assembler is not as useful as host-based assembler because it may be harder to upgrade or replace and it makes the development board more complicated.

(d) Linking loader is better for project work because less coordination is required between members of a project; absolute loader requires more coordination between team members to avoid overwriting each other's memory spaces and to make sure the addresses of shared data remain correct throughout the development cycle.

5.2.2 Line 2 should use the operand #$2000 rather than $2000. Line 4 should use D1 as the accumulator; therefore the program should have cleared D1 in first line of the program. Line 4 should have the label LOOP. The DC directive should be placed just before the END directive—after the return to the monitor.

5.2.4 **(a)** $0FFE

(b) $1000

(c) $0800

(d) $0555

(e) $4000

5.2.6 Conditional assembly can be used to assign different values to the constants that specify the configuration of memory and the disk units. Such constants might include size of memory, size of disk, sectors per track, tracks per cylinder, number of cylinders, disk controller protocol, etc.

5.2.8 Every time a routine exceeded its assigned memory space, other routine's addresses would need to change to accommodate the larger module. Furthermore, the changing addresses of the modules and the shared memory locations (a.k.a. external variables) would need to be communicated correctly to each developer on the team.

A relocating assembler resolves addresses at random and can dynamically concatenate multiple routines without requiring that the user specify an address for each routine. This relieves the team of the burden of managing and validating the memory addresses of routines and external variables.

5.3.2 This is not a problem, because we can use a MOVE.L #0,Ax instead. To address location zero, one can MOVE zero to an address register, or hardcode an absolute address of zero in the instruction. You can also use a negative displacement— say A1 = $0002, then −2(A1) will address location zero.

5.3.4 *Advantages:* Absolute short addressing requires less memory. Because it sign-extends, it can be used to split memory into two physical sections ($000000–007FFFF and $FF8000–FFFFFF) without requiring a long address.

Disadvantage: Cannot address memory locations between the range $008000 and $FF7FFF.

5.3.6 **(a)** Moves the longword that starts 4 bytes beyond (A0) to the address (A0).

(b) Moves the word at $9000 to D1.

(c) Moves to D1.B the byte at the effective address obtained by adding D1.W and A0.L.

5.2 iii

5.4 ii

5.6 ii

5.8 ii

5.10 iii

5.12 i

5.14 iii

CHAPTER 6

6.1.4 It takes up less memory, and for three or more registers uses fewer CPU clock cycles.

6.2.2 (a) Jump to the address stored in A2.

(b) Move the longword stored at the address in A2 to A1.

6.2.4
```
          MOVE.W   CNT,D1
LOOP      . . .
          (body of loop)
          DBEQ   D1,LOOP
```

6.2.8
```
$606C            ⟶ BRA.B    LOOP2
$6000 0F90       ⟶ BRA.W    LOOP3
$60FF 000005FE   ⟶ BRA.L    LOOP4
$60FF FFFFE9FE   ⟶ BRA.L    LOOP1
```

6.3.2
($7FFC) = $8006
($7FFA) = $0000
($7FF8) = CCR contents
($7FF6) = A6 lo word
($7FF4) = A6 hi word
($7FF2) = A5 lo word
($7FE0) = A5 hi word
($7FEE) = A4 lo word
($7FEC) = A4 hi word
($7FEA) = A3 lo word
($7FF8) = A3 hi word
($7FE6) = A2 lo word
($7FE4) = A2 hi word
($7FE2) = A1 lo word
($7FE0) = A1 hi word
($7FDE) = A0 lo word
($7FDC) = A0 hi word
($7FDA) = D7 lo word
($7FD8) = D7 hi word

($7FD6) = D6 lo word
($7FD4) = D6 hi word
($7FD2) = D5 lo word
($7FD0) = D5 hi word
($7FCE) = D4 lo word
($7FCC) = D4 hi word
($7FCA) = D3 lo word
($7FC8) = D3 hi word
($7FC6) = D2 lo word
($7FC4) = D2 hi word
($7FC2) = D1 lo word
($7FC0) = D1 hi word
($7FBE) = D0 lo word
($7FBC) = D0 hi word

6.3.4 Change line 15 to

    JSR   (A1)

CHAPTER 7

7.1.2 When $V = \{1\}$, the logical tests of the BGE and BPL instructions are as follows:

    BGE: N AND V = N   ;true if N = {1}
    BPL: N   ;true if N = {0}

Thus, the instructions test the opposite results if the overflow bit is set.

7.1.4 In the subtraction $N = N1 - N2$ one can write the sum of $N1$ and the two's-complement of $N2$ as

$$N1 + (r^m - N2) = r^m + (N1 - N2)$$

which is $(N1 - N2)$ in modulus r^m arithmetic when the result is valid.

7.2.2 The maximum value resulting from the addition of two unsigned integers is

$$2^{m-1} + 2^{m-1} = 2^{m+1} - 2.$$

This is less than $2^{m+1} - 1$, which is the maximum value using m bits plus the carry bit.

7.3.2 Assume that $N2$ is negative in the multiplication $N1 \cdot N2$. Then the unsigned product is

$$(N1)(r^m - |N2|) = (N1)(r^m) - |N1||N2|$$

which is incorrect. The correct result would be obtained by subtracting $(N1)r^m$ from the result. If both values are negative, unsigned multiplication results in the product

$$r^{2m} - r^m(|N1|+|N2|) + |N1||N2|$$

which is corrected by adding $r^m(|N1|+|N2|)$ or subtracting $r^m(N1 + N2)$ because both $N1$ and $N2$ are negative.

7.3.6 **(a)** Q = $00000000
   (b) Q = $00000004
   (c) Q = $FFFFFFFF
   (d) Q = $FFFD   R = $0001
   (e) Q = $FFFF   R = $0001

7.5.2 **(a)** 0435
   (b) 2845, X = {1}. If only unsigned integers are allowed, the result is an overflow as indicated by X = {1}.

7.2 iii

7.4 i

7.6 iii

7.8 iii

7.10 ii

7.12 iii

7.14 i

CHAPTER 8

8.2.4 The value of the one's-complement integer can be written as

$$(1 - 2^{m-1})d_{m-1} = d_{m-2}(2^{m-1}) + \ldots + d_0$$

according to the formula given in Chapter 3. For a positive value to yield a positive result, it is required that:

$$d_{m-1} = d_{m-2} = \{0\}.$$

The left cyclic shift results in the value

$$d_{m-3}(2_{m-2}) + \ldots d_0(2) + 0$$

which is twice the original value, as proven by subtracting the original value from the shifted value. Similar results are obtained with negative integers which require that

$$d_{m-1} = d_{m-2} = \{1\}$$

for a valid result after shifting.

8.3.2 **(a)** BCLR   #15,D2
   (b) BSET   #15,D2
   (c) BCHG   #15,D2
   (d) The TST instruction tests for zero and negative. BTST only tests one bit for zero or nonzero.

8.2 i

8.4 i

8.6 ii

8.8 iii

8.10 ii

CHAPTER 9

9.1.2 The CMP instruction computes

$$(A1) - (A0) = \$10F00.$$

Therefore, the following instructions will cause a branch:

BCC, BGE, BGT, BHI, BNE, BPL, BVC.

9.1.4 The LEA and PEA instructions permit the calculation of any address used to access memory except those addresses obtained by the predecrement and/or postincrement modes. For program debugging, LEA and PEA can be used to determine the address from which an operand was fetched to verify that the proper values were used in the address calculation.

9.2.2 The subroutine is statically position independent because the pointer (A0) is reloaded with the address of DATA each time the routine is loaded into memory and executed.

9.2.4 Indirect addressing is typically used to allow an instruction to operate on a different location each time the instruction executes. Base addressing allows address modification in order to change the location of a program and the data to which its instructions refer.

9.4.2 An in-line parameter must be a fixed value because it is defined during assembly. This method cannot be used if the value must be changed during execution of a program. The advantage is that the address of the parameter is held on the system stack after a subroutine call so that the address need not be passed to the subroutine before the subroutine call.

9.2 i

9.4 iii

9.6 iii

9.8 ii

9.10 iii

9.12 i

9.14 i

CHAPTER 10

10.2.2 **(a)** Set N, Z to {1}; clear X, V, and C.

(b) C unchanged; clear X, N, Z, and V.

(c) Privilege violation (trap).

10.3.2 The sequence shown puts the CPU in user mode before the JMP is executed. The JMP instruction is in the supervisor memory space.

10.2 ii

10.4 ii

10.6 i

10.8 iii

CHAPTER 11

11.3.2 An illegal instruction might occur if a program terminated improperly or otherwise caused a transfer of control to an area of memory containing data values. A memory failure might cause illegal instructions to be fetched.

11.4.2 192 user interrupt vectors are available, with the others reserved for TRAP instructions and other exceptions.

Nothing prevents the designer from using one of the reserved vectors as an interrupt vector. As long as the normal exception processing for a reserved vector is desired when a specific interrupt occurs, it may be a legitimate technique. However, in general this is an unusual practice.

11.2 i

11.4 i

11.6 ii

11.8 iii

11.10 ii

11.12 ii

11.14 i

CHAPTER 12

12.2.2 The transfer of 10 characters takes approximately 1 second. For each character, the CPU is idle for 0.1 second out of $(0.1 + 10^{-5})$ seconds or about 99.99% of the time.

12.2.4 **(a)** $2(1200/10) = 240$ characters per second (duplex).

(b) The CPU must respond before the next character is received in $10/1200 = 8.33$ ms.

(c) For 240 cps, the CPU services interrupts a total of $240(10 \times 10^{-6})$ seconds or about 0.24% of the time.

CHAPTER 13

13.1.2 Assume the instruction being fetched was the DIVS instruction. Because it has a maximum duration for 158 clock cycles and interrupt processing requires 44 clock cycles, the worst-case delay is 202 cycles. On the MC68000L8 running at 8 MHz, this would be 125(202) nanoseconds or 25.25 microseconds.

13.2.2 **(a)** user data
    (b) user data
    (c) supervisor program
    (d) supervisor data

13.2.4 The vector address is $0100. Although devices could supply vectors 2 through 47, this is not a good idea, because these vectors are assigned to specific CPU internal operations.

Index

For individual chips, see entries such as M6800 or MC68000. Software-related issues for the MC68000 are listed under CPU 68000.